GROUND WATER
QUALITY PROTECTION

Larry W. Canter, Robert C. Knox,
and Deborah M. Fairchild

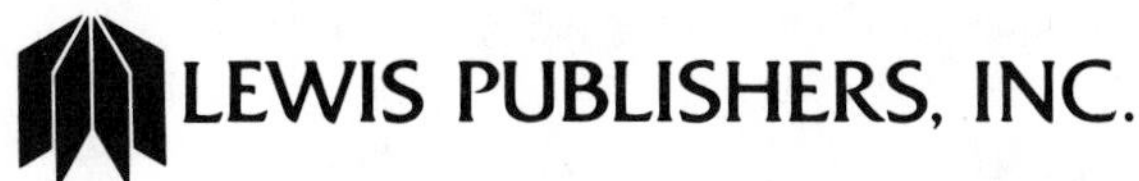

Library of Congress Cataloging-in-Publication Data

Canter, Larry W.
Ground water quality protection

Includes bibliographies and index.
1. Water, Underground—Quality. I. Knox, Robert C.
II. Fairchild, D. M. (Deborah M.) III. Title.
TD426.C37 1987 363.7′394 87-4090
ISBN 0-87371-018-5

Second Printing 1988

LEWIS PUBLISHERS, INC.
121 South Main Street, Chelsea, Michigan 48118

PRINTED IN THE UNITED STATES OF AMERICA

Preface

Protection and appropriate management of the quality of the nation's ground water resources have become a significant public issue within the last decade. The importance of ground water quality protection is expected to increase as a combined result of several factors, such as increased demands for ground water resource development and usage, the recognition of deteriorated ground water quality resulting from point and diffuse sources of pollution, and the continued development and passage of laws and regulations directed toward pollution source control and ground water cleanup. One of the major national needs in ground water management is for dissemination of information to practicing ground water professionals and to university students planning to enter this field of endeavor. This book has been prepared to summarize the rapidly expanding body of knowledge on ground water pollution sources and their evaluation and control.

Many ground water quality management programs are being developed within federal agencies, state and local governments, and the private sector. The planning and implementation of these programs should be based on sound technical information and knowledge, and this book can be considered as a resource document to aid in the process.

This book is organized into 11 chapters, with the first three containing basic information, Chapters 4–9 addressing ground water pollution sources and their evaluation, and Chapters 10 and 11 focusing on pollution control and cleanup and ground water quality management. Chapter 1 illustrates the importance of ground water as a resource by including information on the locations of major aquifer systems and the growing usage of ground water for domestic, industrial, and agricultural purposes. An overview of some basic principles and concepts in ground water hydrology is contained in Chapter 2. Attention is given to major types of aquifer systems and their characteristics and testing. Chapter 3 addresses information sources for ground water studies: key reports, periodicals, and books; computerized systems for data and bibliographic information retrieval; and pertinent associations and organizations interested in ground water.

Selected point and diffuse sources of ground water pollution are summarized in Chapter 4. The point sources included are septic tank systems, underground storage tanks, hazardous waste sites, landfills, surface impoundments, abandoned wells, and drilling mud disposal. Agricultural and acid precipitation diffuse sources are also discussed.

Basic information on pollutant transport and fate is the subject of Chapter 5. Hydrodynamic, abiotic, and biotic processes are described; in addition, information is also summarized on the subsurface behavior of bacteria, viruses, nitrogen, phosphorus, metals, organics, pesticides, and radionuclides.

Ground water pollution sources can be evaluated via mathematical modeling and/or source prioritization methodologies. Chapter 6 is related to flow and solute transport modeling. Physical and numerical models are described from a categorical perspective; in addition, examples of model usage for a variety of purposes are presented. Models can be used for resource characterization, water usage evaluation, and assessment of the water quality impacts of pollution sources.

Prioritization methodologies for pollution sources in relation to aquifer vulnerability are the focus of Chapter 7. Nine methodologies are described, as well as the generic steps involved in the development of a source-oriented methodology.

Monitoring can also be used for ground water pollution source evaluation, and is frequently required for this purpose. Chapter 8 includes information on planning basinwide and source monitoring networks. Detailed information on monitoring well design, ground water sampling equipment, and quality assurance/quality control is contained in Chapter 9. It should be noted that ground water quality data are frequently needed in the calibration and usage of solute transport models; these data can also be used to confirm pollution source prioritization methodologies. Many decisions are necessary in the planning and implementation of a ground water monitoring program; Chapters 8 and 9 include information useful in identifying the advantages and disadvantages of various choices relative to the number and design of monitoring wells, the type of sampling equipment, and the necessary analytical testing.

Technologies for ground water pollution control are described in Chapter 10. This chapter summarizes the more extensive information on this subject in *Ground Water Pollution Control* (Lewis Publishers, Inc., 1985). Physical control measures, in situ technologies, and treatment processes for extracted ground water are highlighted in Chapter 10. A ground water cleanup program is one aspect of an institutional ground water quality management program; Chapter 11 describes institutional programs at the federal, state, and local government levels. Chapter 11 also contains information on developing the institutional capacity for managing ground water, and on the foundation and action elements of a comprehensive ground water quality management program.

This book has been written in a format conducive to its usage as a textbook in an upper division or graduate course addressing ground water quality protection. The course could be offered through the auspices of an engineering, geology, or environmental science department, or it could be offered as an interdisciplinary effort of several departments. The book does not assume that the user has an extensive base of knowledge of ground water science; therefore, it can be used in introductory or singular courses dealing with ground water quality. The book could also be used by practicing hydrogeologists, chemical and/or environmental engineers, environmental scientists, and chemists and microbiologists working on ground water quality issues.

The authors wish to express their appreciation to several individuals who indirectly contributed to the development of this book. First, Dr. Gary Miller, assistant director, Illinois Hazardous Waste Research and Information Center, Champaign, Illinois, was a faculty colleague who participated in the initial planning for the book. Several former graduate students conducted their research on topical areas addressed herein, including Raj Mahadevaiah, Dr. George Deeley, Dr. Sam Atkinson, and Barbara Wilson. The authors are extremely grateful to Mrs. Wilma Clark, technical typist, Environmental and Ground Water Institute, for her typing skills and commitment to the preparation of the manuscript. The typing contributions of Mrs. Mittie Durham, Mrs. Linda Pierce, and Mrs. Darlene Scallon are also acknowledged. The authors also want to express their appreciation to Lewis Publishers for their understanding in conjunction with this book.

The authors gratefully acknowledge the support and encouragement of the College of Engineering at the University of Oklahoma relative to faculty writing endeavors. Most important, the authors thank their families for their patience and understanding.

Larry W. Canter

Robert C. Knox

Deborah M. Fairchild

To Donna, Doug, Steve, and Greg

To Ruth, Cheryl, Pamela, and Kevin

To my daughter, Jean Ruth

Larry W. Canter

LARRY W. CANTER, P.E., is the Sun Company Professor of Ground Water Hydrology, and director, Environmental and Ground Water Institute, at the University of Oklahoma at Norman. Dr. Canter received a PhD in environmental health engineering from the University of Texas in 1967, an MS in sanitary engineering from the University of Illinois in 1962, and a BE in civil engineering from Vanderbilt University in 1961. Before joining the faculty of the University of Oklahoma in 1969, he was on the faculty at Tulane University and was a sanitary engineer in the U.S. Public Health Service. He served as director of the School of Civil Engineering and Environmental Science at the University of Oklahoma from 1971 to 1979.

Dr. Canter has published several books and has written chapters in other books; he is also the author or co-author of numerous papers and research reports. His research interests include environmental impact assessment and ground water pollution control; he has received two research awards at the University of Oklahoma.

Dr. Canter currently serves on the U.S. Army Corps of Engineers Environmental Advisory Board. He has conducted research, presented short courses, or served as advisor to institutions in Algeria, Colombia, France, Germany, Greece, Italy, Kuwait, Panama, Peru, Mexico, Morocco, Scotland, Thailand, The Netherlands, the People's Republic of China, Tunisia, Turkey, and Venezuela.

Robert C. Knox

ROBERT C. KNOX, P.E., is an assistant professor of civil engineering and environmental science at the University of Oklahoma in Norman, Oklahoma. Dr. Knox came to Oklahoma from McNeese State University in Lake Charles, Louisiana, where he was awarded the Dow Chemical Outstanding Young Faculty Award for the Gulf-Southwest Section of the American Society for Engineering Education. Before joining the faculty at McNeese, Dr. Knox was a research engineer at the Environmental and Ground Water Institute at the University of Oklahoma. Dr. Knox received his BS, MS, and PhD degrees in civil engineering from the University of Oklahoma.

Dr. Knox's research interests include ground water contamination and pollution control, environmental impact assessment, and wastewater treatment. Dr. Knox's dissertation research involved one of the first assessments of a ground water pollution control technology focusing on subsurface impermeable barriers. Dr. Knox has published several technical reports and articles concerning ground water pollution control.

Deborah M. Fairchild

DEBORAH M. FAIRCHILD is a senior environmental scientist at the Environmental and Ground Water Institute at the University of Oklahoma and has been with the institute for six years. She received a BS in biology and chemistry from the State University of New York at Oswego in 1975. She worked as a chemist for the U.S. Department of Agriculture while obtaining an MS in soil microbiology from West Virginia University in 1979. Since coming to the University of Oklahoma, she has worked on many ground water availability and quality studies focused on petroleum production and agriculture.

In addition, she has worked on several environmental impact statements and has served as conference organizer, information and data management specialist, and quality assurance officer for laboratory studies. Her professional activities have resulted in over 30 publications and presentations.

Contents

List of Figures

List of Tables

CHAPTER 1

Ground Water—An Important Resource

Ground water quality protection and management has recently emerged as a national public concern within the United States. Just a few short years ago, hydrogeologists and water resource managers were unaware of many potential pollution sources and the extent of ground water contamination. Through a combination of federal laws, court cases, and increased attention from the media, the institutional, scientific, and public perceptions of ground water have been enhanced. Today's awareness of ground water is evidenced by increasing activities at both local and state levels of government, in the Congress, in universities, and in a variety of federal agencies that have responsibilities in this area. The general public is becoming more knowledgeable about the nature of ground water, the extent of its usage, and that its future usefulness is affected not only by our waste disposal activities but also by many consumer choices. These choices encourage the production of new synthetic organic chemicals each year, the disposal of which may ultimately lead to ground water pollution. There are also growing realizations of the importance of protecting ground water for its beneficial uses—drinking, irrigation, and industrial supply.

GROUND WATER AND AQUIFER SYSTEMS

Ground water may be defined as subsurface water that occurs beneath the water table in soils and geologic formations that are fully saturated. In order for ground water to be used as a water supply, the formations must have adequate permeability (porosity and fractures) to transmit and yield water. A geological formation having these characteristics is called an aquifer. Aquifers may be mixes of sand and gravel layers known as unconsolidated or alluvial aquifers; or they can be consolidated aquifers consisting of permeable

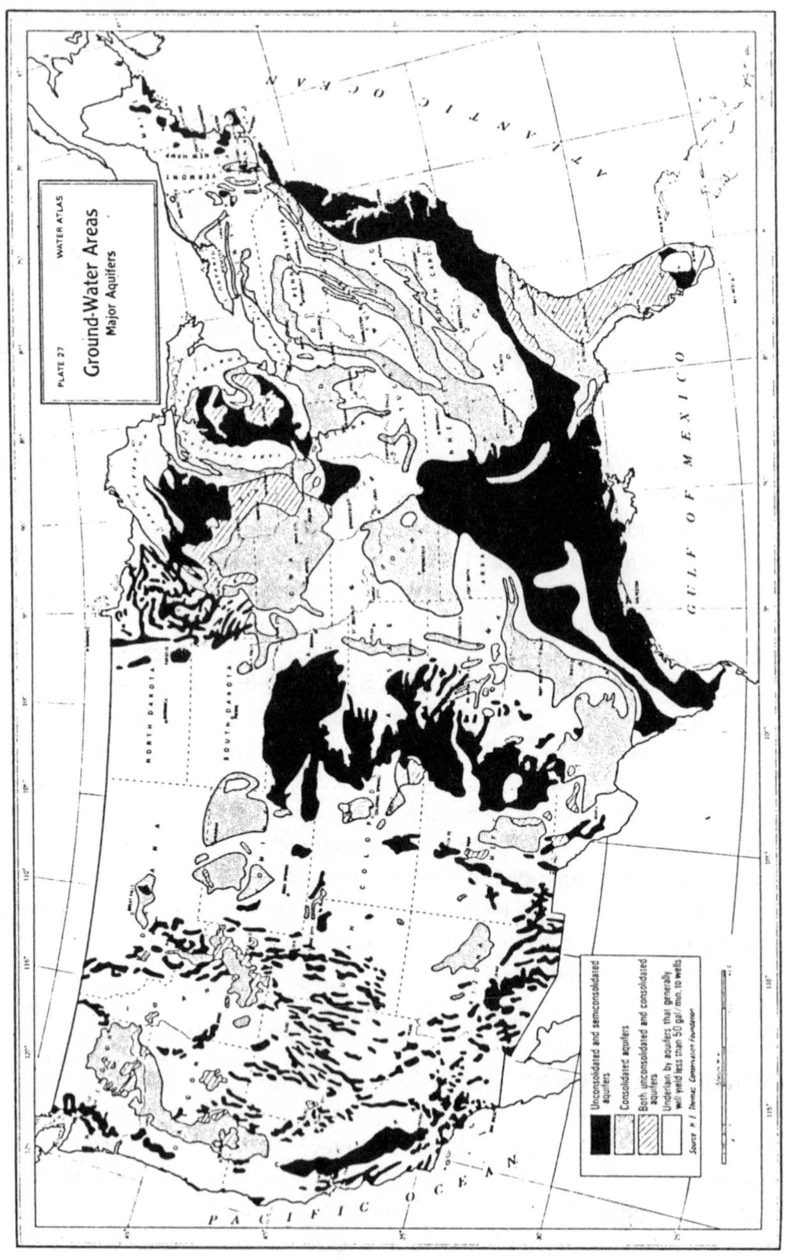

Figure 1.1. Ground water areas, major aquifers.[4]

sandstone, shale, granite, basalt, or limestone. Aquifers vary greatly in their size and extent, thickness ranges from less than 1m to greater than 100 m. They may be long and narrow (such as alluvial valleys along river beds) or cover large areas of up to several million km^2. The depth from the land surface to ground water ranges from less than 1 m to more than 800 m. The depth to the aquifer is one of the major factors determining the cost of obtaining ground water for subsequent usage.

The quantity of ground water underlying the continental United States is immense. The quantity which can be retrieved with current technologies is at least six times greater than all the water stored in surface lakes and reservoirs. The amount within 800 m of the surface is equal to about 35 years of all surface water runoff, or about 400 times the annual consumptive use. Estimates of water in storage within 800 m of the land surface in the conterminous United States range from 33 quadrillion gal[2] to 100 quadrillion gal.[3] More than one-third of the nation is underlain by aquifers capable of yielding at least 100,000 gpd to an individual well, and at almost any location ground water can provide a supply sufficient for single-family domestic usage. Figure 1.1 depicts the major ground water areas in the conterminous United States divided into unconsolidated, semiconsolidated, and consolidated aquifers.[4] Even the unshaded areas shown may yield ground water up to 50 gpm in wells. Figure 1.2 shows alluvial aquifers within the conterminous United States—those narrow aquifers related to river valleys.[4]

There are two main types of aquifers—unconfined and confined. An unconfined or water table aquifer contains water under atmospheric pressure. The upper surface of the water may rise and fall according to the volume of water stored, which is dependent upon natural recharge (replenishment). The confined or artesian aquifer is bounded on the top and bottom by layers of relatively impermeable materials called aquitards or confining layers. The water is under greater than atmospheric pressure and recharge is not uniform but takes place over only a portion of the aquifer.

Ground water is an integral part of the hydrologic cycle because the global cycle accounts for water movement between the oceans, surface water, atmosphere, and ground water. The process is illustrated in Figure 1.3.[5] Rainfall or melting snow infiltrates the ground and percolates down through the unsaturated soil to the zone of saturation or the water table level. About two-thirds of the precipitation evaporates or is transpired by vegetation. The remaining 1,450 billion gallons per day (bgd) accumulates in ground and surface waters, flows to the sea (or across national boundaries), is consumed, or evaporates from reservoirs.[2] Of the 1,450 bgd, only 675 bgd are usually available for intensive beneficial uses. Table 1.1 contains an estimate of the water balance of the world.[1] About 3% of the total water is fresh water, and of this ground water comprises 95%, surface water 3.5%, and soil moisture

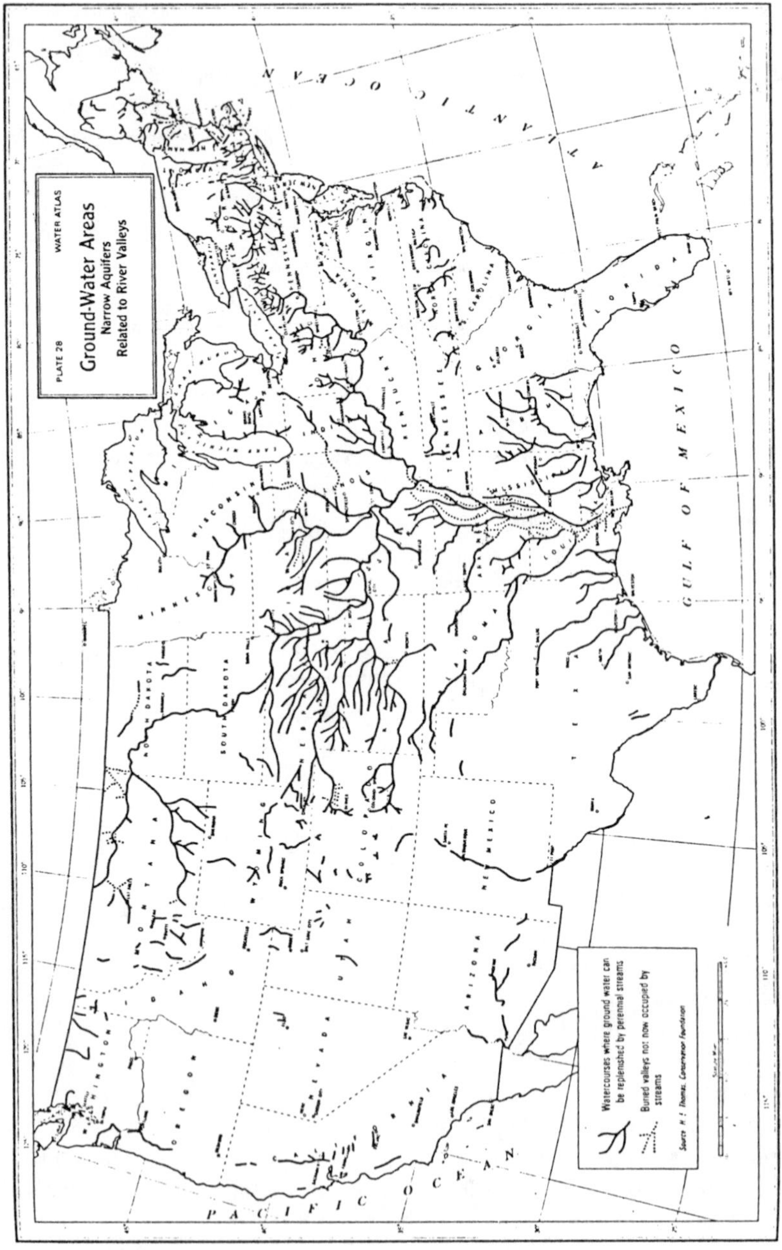

Figure 1.2 Ground water areas—narrow aquifers related to river valleys.[4]

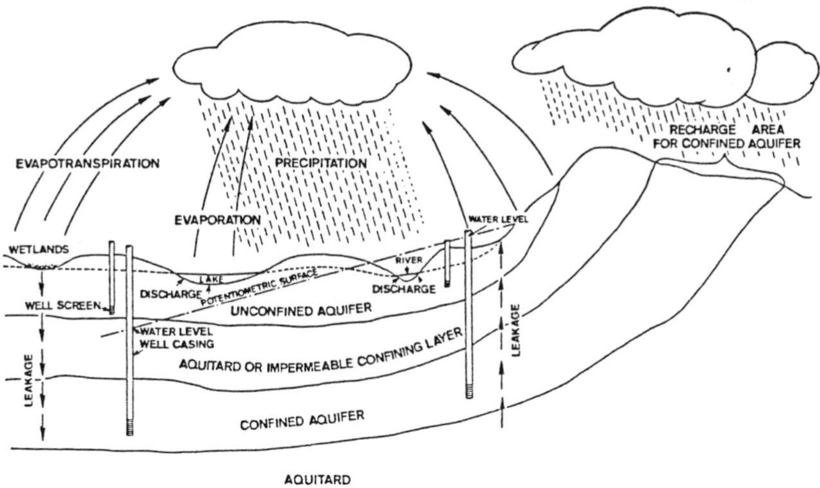

Figure 1.3. Movement of water between the oceans and the surface water, the atmosphere and the land.[5]

1.5%. Out of all the fresh water on earth, only 0.36% is easily available to users.[6]

IMPORTANCE AND USE OF GROUND WATER

Ground water is important as a source of drinking water as well as for irrigation and industrial use. Ground water is a major natural resource in the United States and is often more readily available than surface water. It is estimated that 50% of the population uses ground water as its primary source of drinking water.[7] Approximately 36% of all municipal public drinking water supplies come from ground water. Ninety-five percent of the rural population is dependent upon ground water for drinking purposes. The drinking water supply of about three-fourths of the major cities in the United States is totally or partially dependent on ground water. The states vary in their dependence upon ground water as a source of drinking water as shown in Table 1.2.[5] New Mexico has the highest level of dependence, with 92% of its population using ground water as its source of drinking water, as compared with only 30% of Maryland's population.

Table 1.1. Water Balance of the World.

Parameter	Surface area $(km^2) \times 10^6$	Volume $(km^3) \times 10^6$	Volume (%)	Equivalent depth (m)[a]	Residence time
Oceans and seas	361	1370	94	2500	~4000 years
Lakes and reservoirs	1.55	0.13	<0.01	0.25	~10 years
Swamps	<0.1	<0.01	<0.01	0.007	1–10 years
River channels	<0.1	<0.01	<0.01	0.003	~2 weeks
Soil moisture	130	0.07	<0.01	0.13	2 weeks–1 year
Ground water	130	60	4	120	2 weeks–10,000 years
Icecaps and glaciers	17.8	30	2	60	10–1000 years
Atmospheric water	504	0.01	<0.01	0.025	~10 days
Biospheric water	<0.1	<0.01	<0.01	0.001	~1 week

[a]Computed as though storage were uniformly distributed over the entire surface of the earth.

Table 1.2. Dependence of the United States Population on Ground Water as a Source of Drinking Water for 1970.[5]

State	Total Population (Thousands)	% Of Total Population Relying On Ground Water	Population Served By Ground Water From Public Supplies (Thousands)	Population Served By Ground Water From Rural Supplies (Thousands)	% Of Public Supply Population Relying On Ground Water	% Of Rural Population Relying On Ground Water
Alabama	3,444	59	884	1,139	38	100
Alaska	302	63	62	130	49	74
Arizona	1,772	71	989	274	66	100
Arkansas	1,923	67	605	681	49	100
California	19,953	46	8,000	1,164	43	93
Colorado	2,207	23	306	197	16	87
Connecticut	3,032	37	590	531	24	98
Delaware	548	65	217	138	53	100
Florida	6,789	91	4,819	1,379	89	100
Georgia	4,590	70	825	2,376	38	98
Hawaii	770	87	662	11	95	14
Idaho	713	88	407	218	87	90
Illinois	11,114	38	3,880	358	36	82
Indiana	5,194	58	1,497	1,494	43	87
Iowa	2,825	82	1,524	794	75	100
Kansas	2,249	62	817	565	50	93
Kentucky	3,219	39	304	936	14	88
Louisiana	3,643	62	1,291	927	48	100

Table 1.2, continued

State	Total Population (Thousands)	% Of Total Population Relying On Ground Water	Population Served By Ground Water From Public Supplies (Thousands)	Population Served By Ground Water From Rural Supplies (Thousands)	% Of Public Supply Population Relying On Ground Water	% Of Rural Population Relying On Ground Water
Maine	994	37	153	219	20	91
Maryland	3,922	30	368	803	12	100
Massachusetts	5,689	31	1,464	286	27	100
Michigan	8,875	38	1,365	2,006	20	100
Minnesota	3,805	67	1,498	1,057	55	100
Mississippi	2,217	90	1,177	827	85	100
Missouri	4,677	31	942	529	24	74
Montana	694	47	152	174	30	93
Nebraska	1,484	86	908	365	81	100
Nevada	489	64	265	46	60	97
New Hampshire	738	61	261	191	48	98
New Jersey	7,168	53	3,032	746	47	100
New Mexico	1,016	92	645	293	91	96
New York	18,191	32	4,152	1,734	25	100
North Carolina	5,082	60	660	2,374	25	99
North Dakota	618	66	189	222	48	99
Ohio	10,652	40	2,475	1,754	29	80
Oklahoma	2,559	40	551	475	28	85

Table 1.2, continued

Oregon	2,091	56	355	821	30	92
Pennsylvania	11,794	30	1,351	2,144	14	100
Rhode Island	950	33	213	103	25	100
South Carolina	2,591	61	221	1,359	18	100
South Dakota	666	79	284	240	69	94
Tennessee	3,924	51	1,194	805	38	100
Texas	11,197	58	4,584	1,949	50	100
Utah	1,059	58	502	113	53	99
Vermont	445	56	100	150	35	96
Virginia	4,648	34	477	1,109	14	98
Washington	3,409	44	1,072	433	38	78
West Virginia	1,744	53	380	551	32	97
Wisconsin	4,418	64	1,536	1,311	49	100
Wyoming	332	61	130	73	52	89
District of Columbia	757	0	0	0	0	0
Puerto Rico	2,712	26	396	302	17	80
United States Total	205,897	48	60,600	38,568	37	94

Figure 1.4 shows ground water as a percentage of total water use for all purposes in 1975.[8] States west of the Mississippi, where irrigated agriculture is prevalent, depend heavily on ground water. The five states using the most ground water per day as of 1985, in decreasing order, include California (14.6 bgd), Texas (9.7 bgd), Nebraska (7.1 bgd), Idaho (6.3 bgd), and Kansas (5.6 bgd).[9] Figure 1.5 shows the actual usage figures for the various states in mgd for public water systems (1970) and total ground water usage for all purposes (1975).[5,8] The western states are again shown as the main users of ground water.

Three times more ground water (90 bgd) is used now than in 1950, but there is still three times more surface water used than ground water. Withdrawals of surface water (exclusive of thermoelectric power generation) have increased 30% since 1950. This rapid increase in the use of ground water relative to surface water underscores the growing importance of ground water to the nation's future.[10] In 1980, total water usage in the United States was $1{,}700 \times 10^6$ m^3/day, and it is expected to double by the year 2000. Within the home each person uses approximately 200 gpd (70% for toilet flushing and bathing).[11] When taking into consideration the total amount of water each person needs for products and services used, the quantity increases to 2000 gpd/person.

Figure 1.6 displays historical trends in the use of both ground and surface water.[12] In 1984, ground water provided 34% of public water supply, 79% of rural water supply, 40% of the irrigation, and about 27% of industrial freshwater withdrawals—excluding thermoelectric power uses. The total irrigated land in the United States is about 50 million acres. Eighty percent of the irrigation occurs in the western states, while 84% of the water used for industrial purposes occurs in the east.

Ground water is also an important resource because of discharge to rivers and streams; at any one time this accounts for approximately 30% of the stream flow. Ground water nourishes and maintains many ecosystems valued for fish production, wildlife habitat, recreational opportunities, and other attributes. In periods of drought especially, ground water serves important ecological functions, providing fresh water for many lakes, rivers, inland wetlands, bays, and estuaries. In times of drought, all the stream flow in the low-flow months may be due to ground water.[2]

Excessive depletion of ground water can cause intrusion of water having undesirable quality, interfere with water rights, increase pumping charges, decrease stream flow, and intensify land subsidence.[1] Any withdrawal in excess of a safe yield is called an overdraft and is considered to be serious when it continues indefinitely. Such problems have developed in California; the High Plains regions of Texas, New Mexico, Oklahoma, Kansas, Colorado, and Nebraska; Florida; and Arizona.[13]

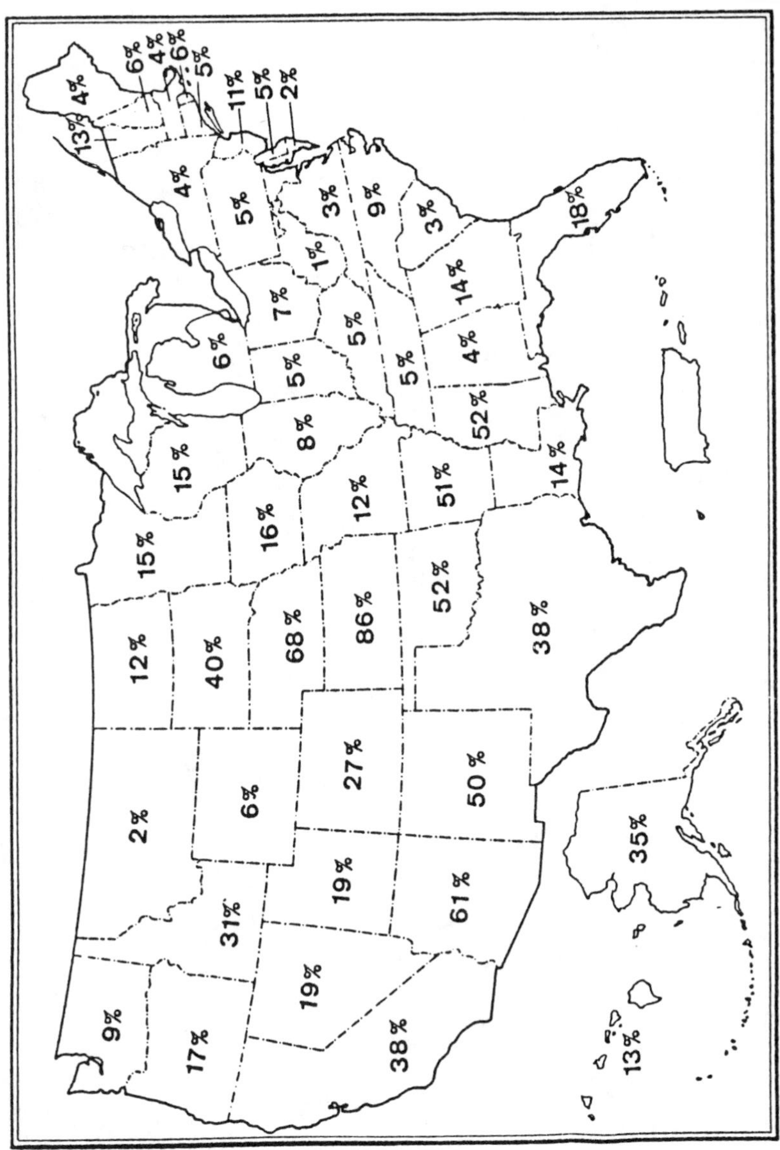

Figure 1.4. Ground water as a percentage of total water use in 1975.[8]

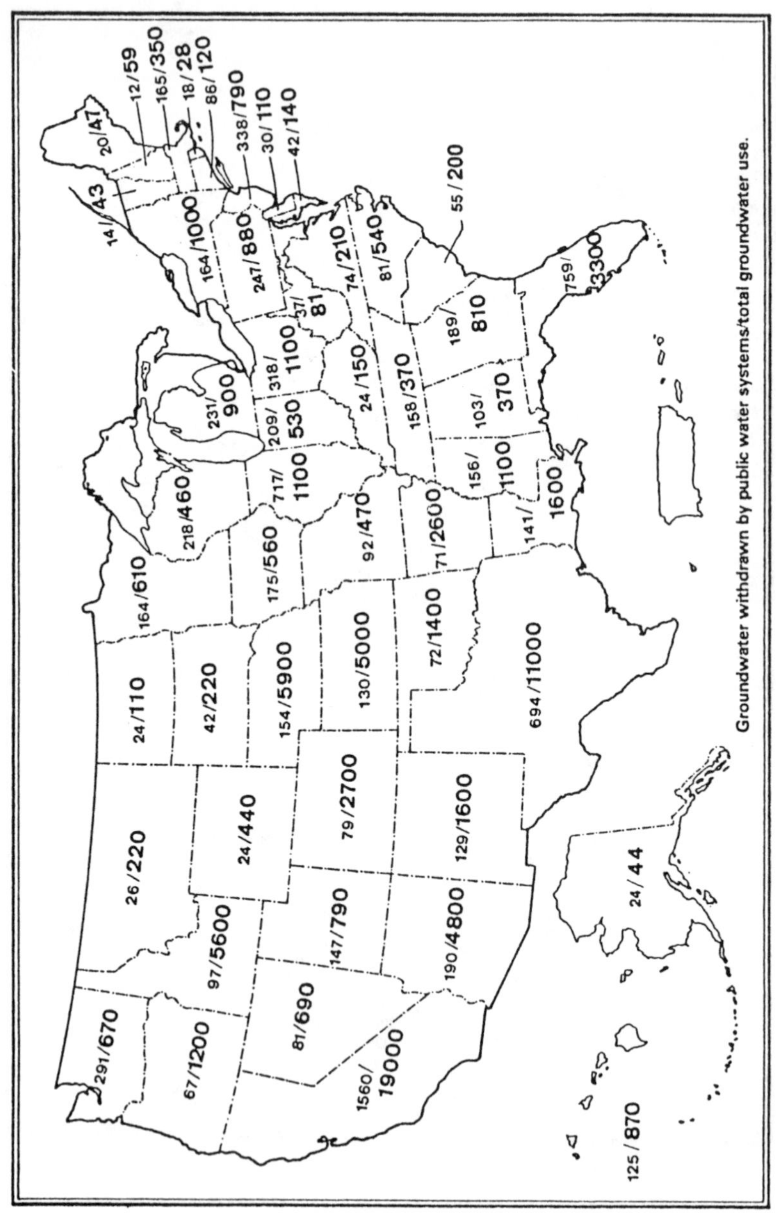

Figure 1.5 Ground water usage in the United States, mgd.[5,8]

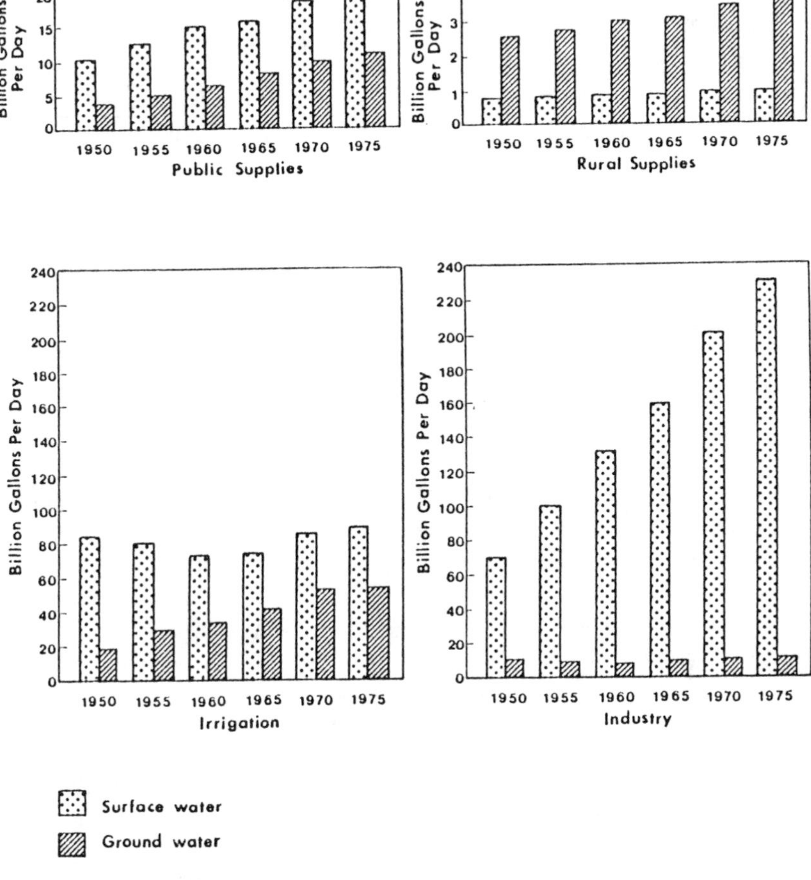

Figure 1.6. Graphs showing trends in use of water for public supplies, rural supplies, irrigation, and self-supplied industry, 1950–75.[12]

NATURAL QUALITY AND MAN-MADE CONTAMINATION

Water quality is usually defined in terms of the concentration of its chemical constituents relative to a variety of potential uses. Table 1.3 shows four simple ground water classifications based on total dissolved solids (TDS),[1] while Table 1.4 disaggregates water components into major, minor, and trace constituents.[14] As water moves through the hydrologic cycle it interacts with the atmosphere, soils, and subsurface geologic formations. This affects its chemical composition and constituent concentrations. Freeze and

Table 1.3. Simple Ground Water Classification Based on Total Dissolved Solids.[1]

Category	Total Dissolved Solids (mg/l or g/m^3)
Fresh water	0– 1,000
Brackish water	1,000– 10,000
Saline water	10,000–100,000
Brine water	More than 100,000

Cherry[1] discussed how precipitation that is naturally slightly acidic (pH = 5.6) causes chemical reactions with geologic materials. Through this process the natural recharge of an aquifer may add from 10 to 50 mg/l TDS to the ground water. Acid rain with a pH as low as 3.5 could cause even greater increases in the TDS of ground water.

Rainwater percolating through soil may increase in acidity due to plant decomposition and microbial respiration producing carbon dioxide. The chemical quality of ground water depends upon its age and the geologic formations encountered in its flow history. The leaching of soils, organic matter, and rocks is influenced by pH. Carbonate formations can increase the magnesium, calcium, and bicarbonate content of ground water due to the dissolution of calcite ($CaCO_3$) and dolomite [$Mg{\cdot}Ca(CO_3)_2$]. Aluminosilicate minerals can increase the concentrations of sodium, potassium, magnesium, calcium, and silicon hydroxide in the percolating water. Ground water quality is often described in terms of hardness and salinity. Naturally occurring hardness values are shown in Figure 1.7.[4] Hardness reflects the calcium and magnesium content and is usually expressed as the equivalent amount of calcium carbonate. It can be used to assess the appropriateness of ground water for domestic and industrial use.

There are tens of thousands of community water supply wells and millions of domestic wells in the United States. The U.S. Public Health Service Standard for TDS in drinking water is 500 mg/l. Until the Safe Drinking Water Act (SDWA) was passed in 1974, there was no formalized national program for analyzing potential toxic elements in public water supplies. The SDWA considers waters containing up to 10,000 mg/l TDS as potential sources of drinking water.[5] The SDWA requires regular testing of public water supplies and wells and has established drinking water standards to protect public health. However, SDWA regulations do not require analysis of most of the synthetic organics, and there is no testing program for individual and small system domestic wells that serve some 40 million people. The four types of naturally occurring ground water that often exceed 10,000 mg/l TDS are connate water, intruded seawater, magmatic and geothermal water, and

Table 1.4. Classification of Dissolved Inorganic Constituents in Ground Water.[14]

Major constituents (greater than 5 mg/l)	
Bicarbonate	Silicon
Calcium	Sodium
Chloride	Sulfate
Magnesium	Carbonic acid
Minor constituents (0.01–10.0 mg/l)	
Boron	Nitrate
Carbonate	Potassium
Fluoride	Strontium
Iron	
Trace constituents (less than 0.1 mg/l)	
Aluminum	Molybdenum
Antimony	Nickel
Arsenic	Niobium
Barium	Phosphate
Beryllium	Platinum
Bismuth	Radium
Bromide	Rubidium
Cadmium	Ruthenium
Cerium	Scandium
Cesium	Selenium
Chromium	Silver
Cobalt	Thallium
Copper	Thorium
Gallium	Tin
Germanium	Titanium
Gold	Tungsten
Indium	Uranium
Iodide	Vanadium
Lanthanum	Ytterbium
Lead	Yttrium
Lithium	Zinc
Manganese	Zirconium

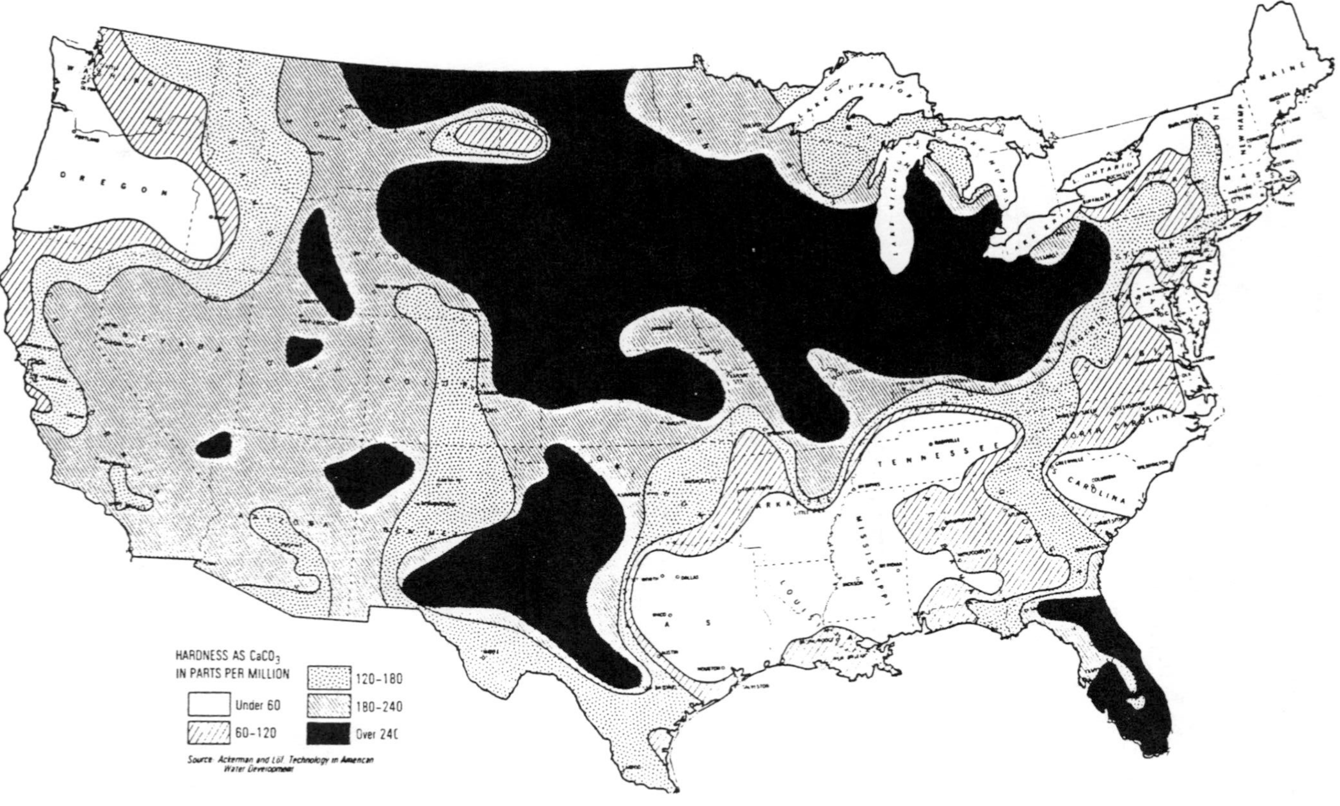

Figure 1.7. Hardness of ground water.[4]

water affected by salt leaching and evapotranspiration. In many areas of the United States ground water can be used without any pretreatment. In other areas it may be necessary to remove color or hardness. Chlorination is typically used to provide protection against pathogenic microorganisms. Water from some aquifers is unusable due to its salinity or the presence of naturally occurring substances such as arsenic or radionuclides.

Ground water can become contaminated in a multitude of ways, including some of the most ordinary human activities. Prevalent sources of contamination include, but are not limited to: (1) waste disposal; (2) storage, transportation, and handling of commercial materials; (3) mining operations (some of which involve the disposal of mining and processing wastes); (4) agricultural operations; and (5) other nonpoint sources. Contaminants may range in type from simple inorganic ions to complex synthetic organic chemicals.

Federal and state laws and state and local ordinances designed to protect ground water have focused on chemical landfills, surface impoundments at industrial facilities, and septic tank systems. Nonindustrial sources of ground water pollutants also include road-runoff, municipal landfills, junkyards, and domestic wastewater. Household products, containing many soluble organic chemicals are discharged into septic tanks, cesspools, and leaky sewer lines, and from there migrate to the water table. Common commercial operations such as automotive service and repair, dry cleaning, and printing are often major contributors to local and regional ground water contamination.[15]

Attention to date on hazardous waste sites has been primarily focused on those sites that were created accidentally or with planning by the private sector. Ground water clean-up efforts through Superfund have focused on these sites. Another significant source of hazardous waste sites, from a nationwide perspective, is associated with various military activities. Numerous military facilities have disposed of potentially hazardous or toxic waste materials for several decades, and in many cases, the records of disposal activities are sparse if existent at all. Installation restoration programs being conducted by the Department of Defense are focusing greater attention in this area of concern. These problems may be heightened in areas where the military base is dependent upon ground water as a source of supply; or where lands adjacent to the military bases, as well as associated communities, depend upon ground water for a variety of uses.

An emerging ground water pollution source of potentially large national importance is nonpoint source pollution. Traditional attention has been given to the surface water implications of nonpoint source pollution from both urban and rural areas. The ground water pollution potential of the land application of fertilizers and pesticides in both urban and rural areas is potentially significant. There are many locations in agricultural states where nitrate levels and pesticide levels in ground water have been noted as being excessive. As additional monitoring programs are implemented to focus on these potential sources, it is anticipated that still more information will be assembled to demonstrate the widespread occurrence of ground water pollution from agricultural chemicals. This potential source is expected to be a major future concern due to the importance of the agricultural industry in the United States as well as the importance of the chemical industry relative to fertilizer and pesticide manufacturing.

Ground water contamination is typically a local phenomenon affecting only the uppermost aquifers. Contamination generally occurs in an area less than a mile long and a half-mile wide with pollutants moving at an average rate of less than 0.3 m/day.[15] Ground water contamination incidents have been reported in all parts of the United States. Contamination problems vary from region to region and are influenced by climate, population density, intensity of industrial and agricultural activities, the hydrogeology of the region, and the status and enforcement of federal and state regulations that can be used to protect ground water.[16] Estimates of the percentage of usable ground water near the surface which has been contaminated range from less than 1% to 2%, and take into account only contamination from certain point sources.[3] However, nationwide estimates or averages are not very meaningful on a regional or local basis.

Several federal statutes addressing ground water quality and/or sources of ground water pollution have been enacted in recent years. The SDWA includes programs to: (1) regulate the quality of drinking water at the tap from ground or surface sources; (2) regulate the underground injection of wastes; and (3) protect particularly essential and vulnerable aquifers that are sole sources of community drinking water. The Resource Conservation and Recovery Act (RCRA) controls the generation, treatment, storage, and disposal of hazardous wastes. It also regulates underground storage tanks to prevent leaks. The Superfund law focuses on the clean-up of hazardous waste sites. Other federal laws dealing with surface water quality, strip mining of coal, and pesticide regulation can also control some sources of ground water pollution. In 1984 the U.S. Environmental Protection Agency issued a ground water protection strategy that attempts to coordinate the laws for which it is responsible. Some states have adopted more comprehensive ground water protection programs, imposing additional source controls and using a variety of planning and regulatory techniques.

SELECTED REFERENCES

1. Freeze, R. A. and Cherry, J. A. *Groundwater,* 1979, Prentice-Hall Book Company, Englewood Cliffs, New Jersey.
2. U.S. Water Resources Council, "The Nations Water Resources, 1975–2000, Vol. 1: Summary," Second National Water Assessment, 1978, Washington, DC.
3. Lehr, J. H. "How Much Ground Water Have We Really Polluted?" in *Ground Water Monitoring Review,* Winter 1982, pp. 4–5.
4. Geraghty, J. J., et al. *Water Atlas of the United States,* 3rd. ed., 1973, Water Information Center, Inc., Syosset, New York.
5. U.S. Environmental Protection Agency, "The Report to Congress—Waste Disposal Practices and Their Effects on Ground Water," 1977, Washington, DC.
6. Leopold, L. B. *Water: A Primer,* 1974, W. H. Freeman and Co., San Francisco, California.
7. Conservation Foundation, "Groundwater—Saving the Unseen Resource, Proposed Conclusions and Recommendations," Nov. 1985, Washington, DC.
8. Lehr, J. H. "Groundwater in the Eighties," in *Water and Engineering Management,* Vol. 123, No. 3, 1981, pp. 30–33.
9. U.S. Water News, "California Uses Most Groundwater," Vol. 2, No. 5, Nov. 1985, p. 1.
10. U.S. Geological Survey, "National Water Resource Summary—1984, Hydrologic Events, Selected Water Quality Trends and the Ground Water Resource," Water Supply Paper-2275, 1984, Reston, Virginia.
11. Last, J. M. *Public Health and Preventive Medicine,* 11th ed., 1980, Appleton-Century-Crofts, Norwalk, Connecticut.
12. Murray, C. R. and Reeves, E. B. "Estimated Use of Water in the United States in 1975," U.S. Geological Survey Circular No. 765, 1977, Washington, DC.
13. U.S. General Accounting Office, "Ground Water Overdrafting Must be Controlled, A Report to Congress of the U.S. by the Comptroller General," CED-80-96, Sept. 1980, Washington, DC.
14. Davis, S. N. and DeWiest, R. J. *Hydrogeology,* 1966, John Wiley and Sons, Inc., New York, New York.
15. Miller, D. W. "Groundwater Contamination—A Special Report," 1984, Geraghty and Miller, Inc., Syosset, New York.
16. Pye, V. I., Patrick, R. and Quarles, J. *Groundwater Contamination in the United States,* 1983, University of Pennsylvania Press, Philadelphia, Pennsylvania.

CHAPTER 2

Overview of Ground Water Hydrology

Definitions of ground water generally tend to be subjective. However, most definitions of ground water include some statement as to "the portion of underground water that flows freely into open holes." More elaborate definitions state that ground water "is that portion of subsurface water that is at greater than atmospheric pressure." A comprehensive definition of ground water could be as follows: that portion of water beneath the surface of the earth that is under pressure greater than atmospheric such that it will flow into open holes dug into the earth or will naturally move to the earth's surface in the form of seepage or springs.

It is obvious from the above discussion that not all underground moisture can be classified as ground water. Significant quantities of subsurface moisture are "held" within the soil matrix, at pressures below atmospheric, by physical or chemical forces. Since this moisture is not readily collectible, it is usually considered not to be ground water.

The relationships between subsurface moisture and other entities of the hydrologic cycle are shown in Figure 2.1. Sources of fresh ground water include infiltration from direct precipitation, seepage from surface impoundments, and artificial recharge. Ground water can be removed from the subsurface environment by seepage into streams, evapotranspiration, and man-made removal systems such as water supply wells.

As shown in Figure 2.2, the soil beneath the earth's surface consists of two distinct hydrogeologic zones; the vadose zone and the saturated zone. The vadose zone consists of all materials up from and including the capillary fringe to the earth's surface. The capillary fringe consists of that water above the water table, that has risen up in the soil matrix by capillary forces caused by the surface tension properties of water. Capillary water is usually considered to be nonrecoverable. However, there has been research in recent years as to the feasibility of recovering capillary water in areas of falling water tables by pressure air injection.[1]

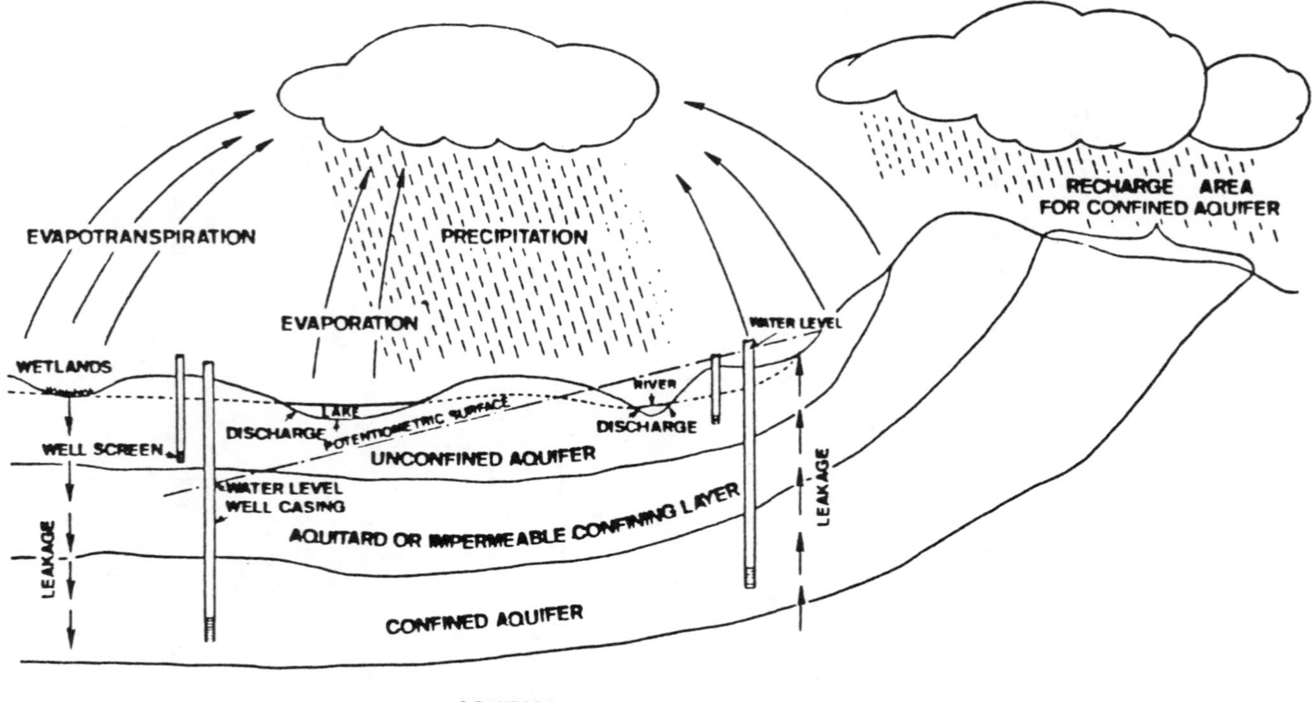

Figure 2.1. Illustration of relationships within the hydrologic system.

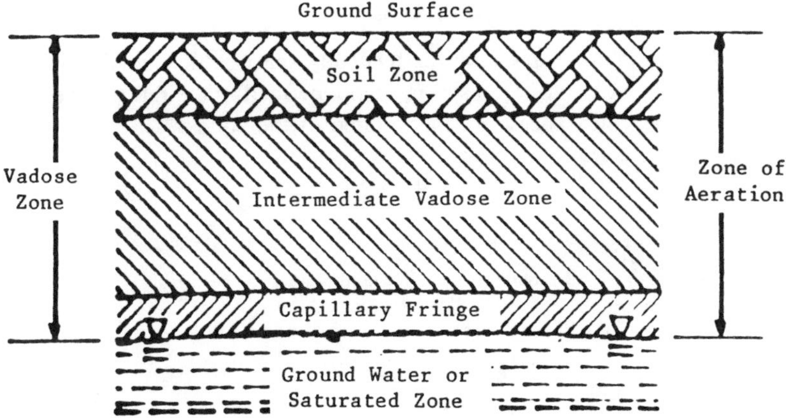

Figure 2.2. Idealized cross-section of vadose and ground water zones.[2]

The saturated zone consists of all materials beneath the water table. The water table is defined as the surface on which the fluid pressure in the pores of the porous medium (soil) is exactly atmospheric. The water table also represents the level to which water will rise in a shallow open well.

SOIL CHARACTERISTICS

The presence of water (and air) in subsurface soils is attributable to the fact that the sursurface is a "porous medium." In other words, soils consist of a matrix of solid granular material and interconnecting pore (open) spaces. The most appropriate analogy is that of a sponge, however, ground water formations are not nearly so deformable.

The two soil characteristics of most interest from the hydrogeological standpoint are porosity and void ratio. The porosity of a soil is simply the volume of the soil that is available for storage of water, i.e., the amount of pore spaces that could be occupied by water. In equation form, porosity can be defined as:

$$N = \frac{V_t - V_s}{V_t} = \frac{V_v}{V_t} = \frac{V_v}{V_v + V_s} \tag{2.1}$$

where:

N = porosity (dimensionless decimal fraction)
V_t = total volume of the soil (L^3)
V_s = volume of the solids within the soil (L^3)
V_v = volume of void spaces within the soil (L^3)

Another means of expressing the available space within a soil is by the void ratio. The void ratio of a soil is simply defined as the ratio between the volume of the void spaces and the volume of the solids. Thus:

$$e = \frac{V_v}{V_s} \tag{2.2}$$

where:

e = void ratio (dimensionless decimal fraction)

The relationships between porosity and void ratio are:

$$e = \frac{N}{1-N} \text{ and } N = \frac{e}{e+1} \tag{2.3}$$

AQUIFER SYSTEMS

Although many subsurface formations may be saturated, i.e., the available pore spaces are occupied by water, not all saturated formations are considered to be aquifers. An aquifer can be defined as a ground water bearing formation sufficiently permeable so as to transmit and yield water in useable quantities. The point of distinction is whether or not the formation yields water in useable quantities. Hence, an impermeable formation, such as a clay layer, may indeed be saturated and capable of yielding water, but at a very low rate. Highly impermeable or flow-restricting formations are referred to as aquicludes or aquitards.

There are two major types of aquifers; unconfined and confined. An unconfined aquifer, as depicted in Figure 2.3, is characterized by the absence of flow restricting material above it and the presence of a free water surface (water table) that can rise and fall.

A confined aquifer, like that in Figure 2.4, is defined as a water/bearing formation between two confining layers such that no free surface exits. If penetrated by a well, borehole, or fracture, a confined aquifer will cause water to rise in the intrusion above the top of the formation. The locus of points connecting the free surfaces of water in wells penetrating a confined aquifer is called the piezometric surface.

A special case of the unconfined aquifer is the perched aquifer. As shown in Figure 2.5, a perched aquifer is a temporary unconfined aquifer formed by the "ponding" of infiltration on a flow-restricting layer. Perched aquifers

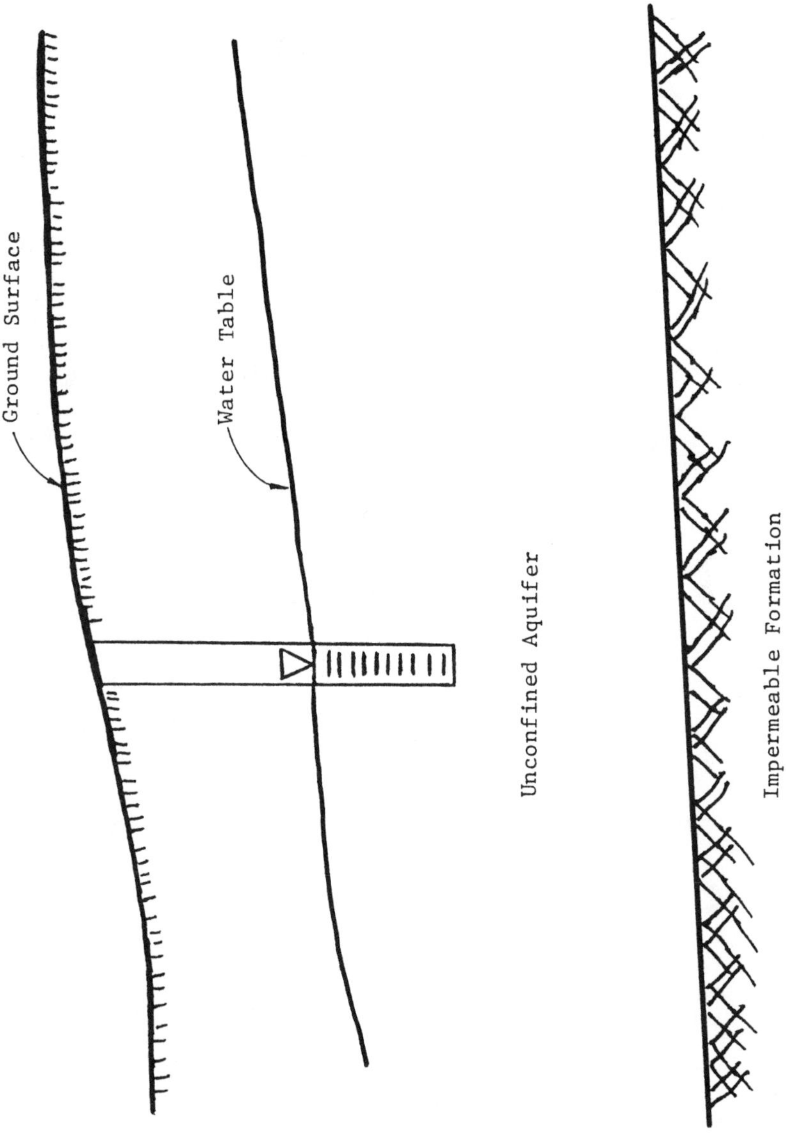

Figure 2.3. Well penetrating an unconfined aquifer.

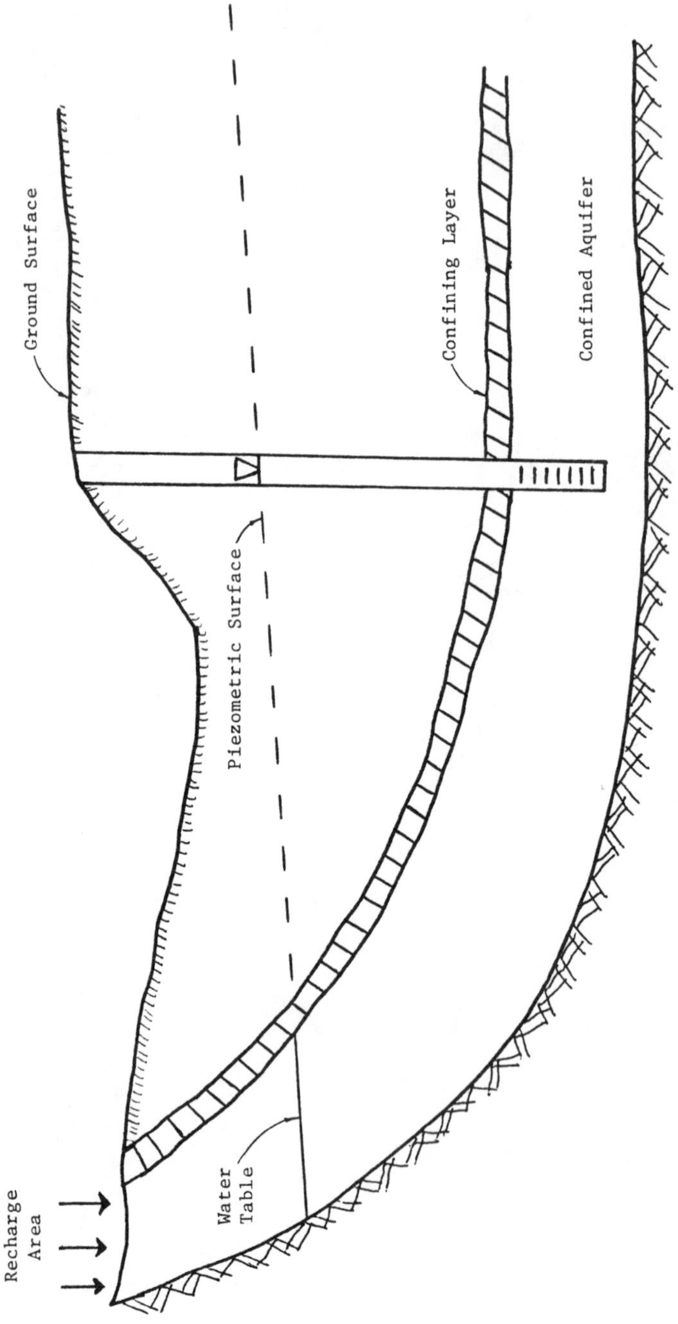

Figure 2.4. Well penetrating a confined aquifer.

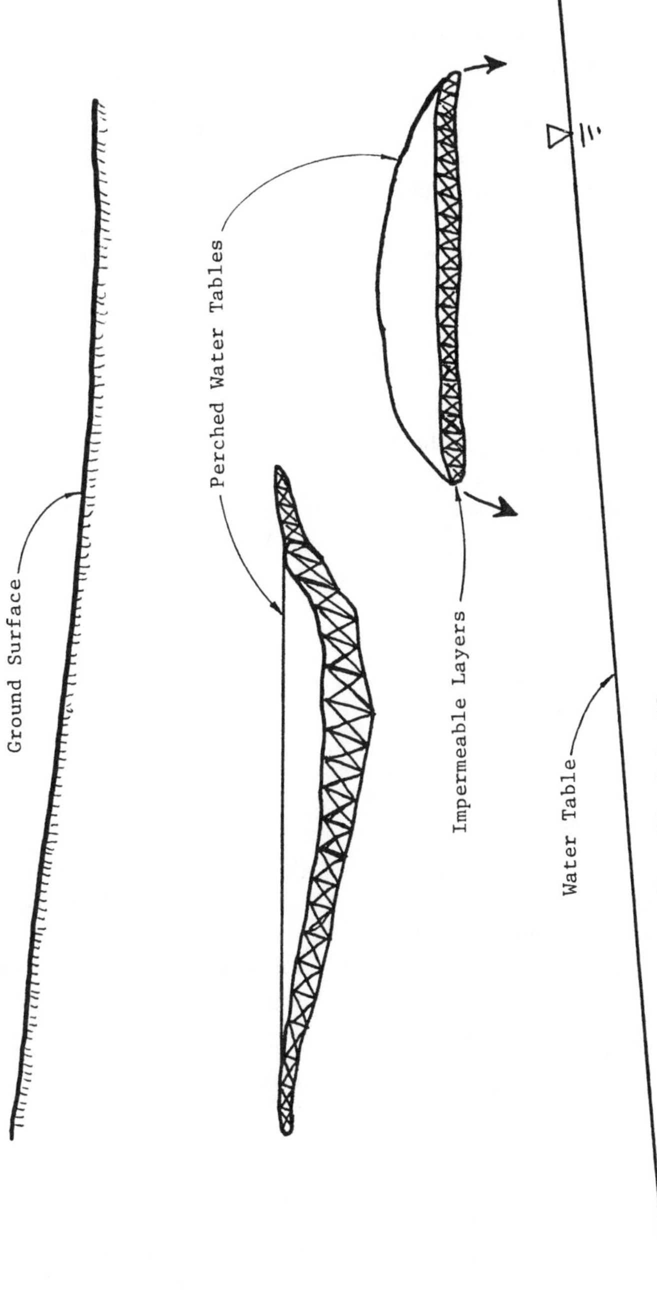

Figure 2.5. Perched aquifer.

can be localized or of significant areal extent. However, perched aquifers are temporary in that after infiltration ceases, the perched aquifer will eventually disappear as the water flows over or through the perching layer.

Leaky aquifers are basically those aquifers that lose water due to downward vertical seepage through underlying formations that are not truly impermeable. Figure 2.6 shows a leaky unconfined aquifer overlying a confined (actually semiconfined) aquifer. Note that vertical leakage will only occur if a driving force exists. In this instance, a driving force exists only if the water table level of the unconfined aquifer and piezometric surface level of the confined aquifer do not coincide.

DARCY'S LAW

The flow of fluids through porous media is governed by Darcy's Law. Darcy's Law, referring to Figure 2.7, states that the velocity of flow of fluid through a porous medium is proportional to the hydraulic gradient. Thus:

$$V = K\frac{(h_1 + z_1) - (h_2 + z_2)}{L} \tag{2.4}$$

where:

V = Darcy velocity of fluid (L/t)
h_1 = pressure head at point 1 (L)
h_2 = pressure head at point 2 (L)
z_1 = elevation head at point 1 (L)
z_2 = elevation head at point 2 (L)
L = distance of flow between points 1 and 2 along a streamline (L)
K = hydraulic conductivity of medium (L/t)

There are three key points to be made about Darcy's Law. First, the Darcy velocity calculated by Equation 2.4 is not the actual velocity of the water particles. The Darcy velocity is the overall apparent velocity of the ground water. In essence, it is the average velocity of all the water particles in going from point 1 to point 2. The second point to be made is that the driving force for flow is the difference in total head at the two points. Total head comprises both the pressure head due to the vertical column of water above the point and the elevation head of the point relative to some datum. The third point is that the distance over which the head difference is taken is along the streamline. In fact, the streamline between points 1 and 2 is not the actual path taken by the water molecules, but rather the net average of the tortuous paths taken by the molecules.

Knowledge of the relationship between hydraulic head and velocity is fundamental to a thorough understanding of ground water behavior. Figure 2.8 is a graphical depiction of the dashed lines of constant head (equipotentials)

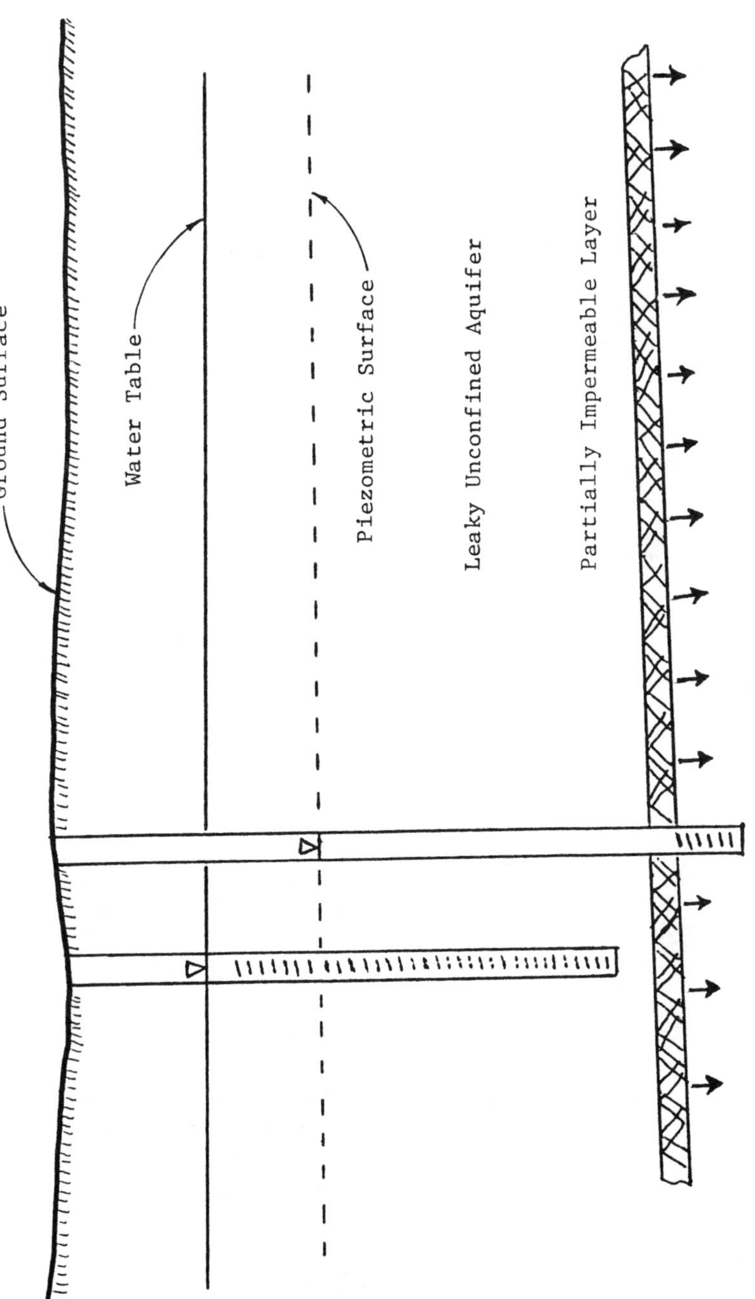

Figure 2.6. Leaky aquifer.

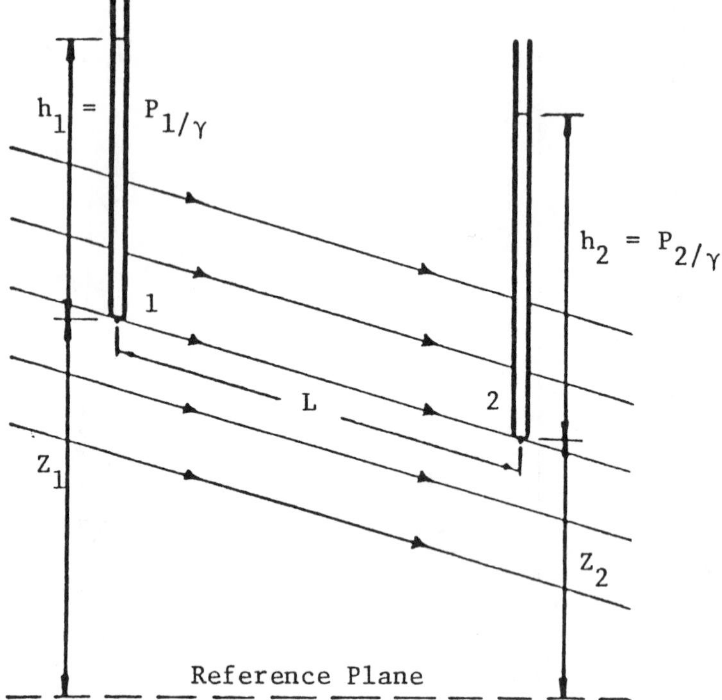

Figure 2.7. Vertical cross-section of ground water flow.

and the solid streamlines or flow lines in a two-dimensional effluent stream model. In this model, the effluent stream in the middle of the figure is receiving recharge from both sides. Hence, the edges of the model are constant head boundaries.

Several points relative to Figure 2.8 can further an understanding of Darcy's Law and ground water behavior. The first point to be made is that streamlines or flowpaths are always normal to equipotential lines. This is true in all ground water flow systems.

A second point relative to Figure 2.8 involves the bottommost streamline on either side of the figure. It appears that the streamline shows ground water going uphill. In fact, the ground water is going uphill, but only in a topographic sense; it is going downhill in a hydraulic sense. The total head at the boundary is greater than the total head in the stream. Hence, flow is from high to low hydraulic head.

A third point relative to the figure involves the uppermost and bottommost streamlines on either side of the model. In the actual modeling process,[3] it

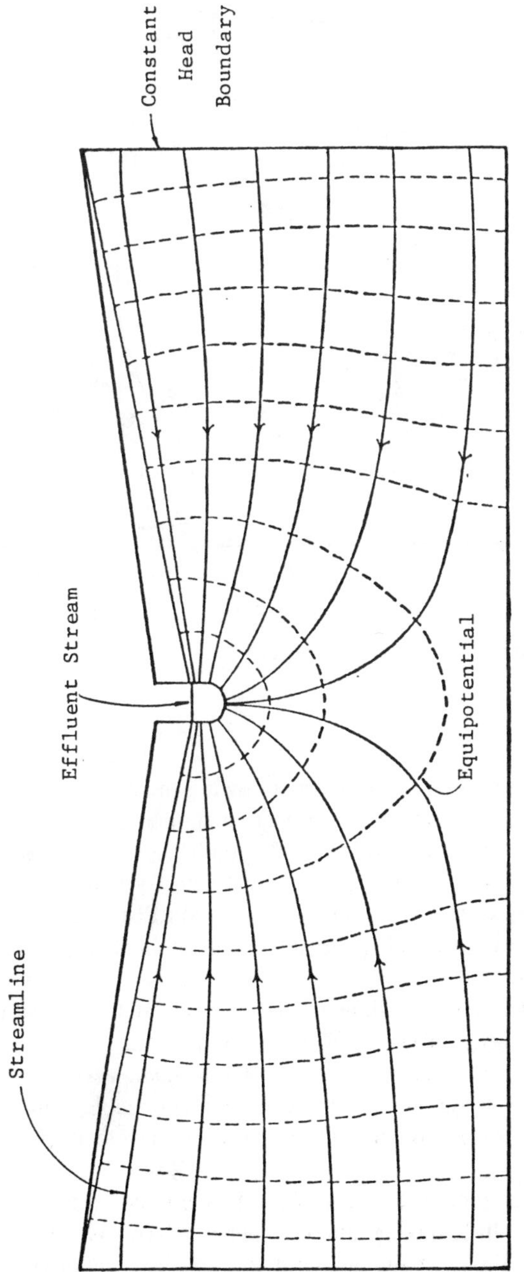

Figure 2.8. Equipotentials and streamlines for an effluent stream.[3]

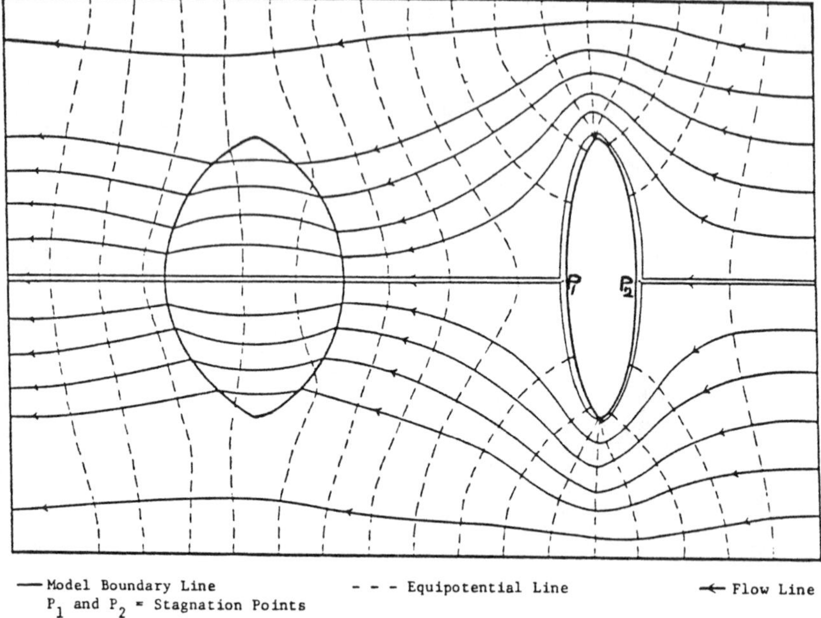

Figure 2.9. Flow diagram of horizontal lens model.[4]

is shown that the velocity of ground water following the uppermost streamline is significantly higher than the lowermost streamline even though both streamlines experience the same change in hydraulic head (headloss). The explanation lies in Darcy's Law which states that velocity is proportional to the hydraulic gradient, which is a change in head over distance. The actual length of the lowermost flow path is greater, hence a smaller gradient and lower velocity of flow.

Figure 2.9 is a depiction of two-dimensional ground water flow that points out some hydraulic features that can affect contaminant transport. In the figure, flow is from right to left through a permeable formation which contains two distinct lenses. The lense on the right is more impermeable than the surrounding formation; the lens on the left is more permeable than the surrounding formation.

The overall point of Figure 2.9 is to show that ground water flow is a laminar flow with very little mixing and/or dispersion.[4] This is evidenced by the fact that the streamlines remain parallel before, during, and after passage through materials of varying permeabilities. In the modeling process,[4] the streamlines are shown by injecting dyed water, and the model shows no mixing of the different colors.

A second point to note about Figure 2.9 is the refraction phenomenon encountered in the left-hand lens. In going from one formation to another with differing permeabilities, ground water molecules are deflected. The angle of defection obeys a tangent law.

AQUIFER CHARACTERISTICS

The factor of proportionality (K) in Darcy's Law (Equation 2.4) is called the hydraulic conductivity and is a property of the soil or rock material through which the water is flowing. The value of K depends on the number and size of the pore openings in the aquifer. Hydraulic conductivities have the dimensions of velocity, hence they represent the Darcy velocity in the aquifer under a unit hydraulic gradient. Table 2.1 shows ranges of K for various soil and aquifer materials.

Not all water occupying pore spaces in a saturated aquifer is readily removed. One measure of the water yielding ability of an aquifer is the storage coefficient. The storage coefficient of an aquifer is defined as the volume of water yielded per unit horizontal area by a unit drop in the water table level (unconfined aquifers) or the piezometric surface (confined aquifers). Specific yield is the term sometimes applied to unconfined aquifers.

Table 2.1. Hydraulic Conductivities of Selected Materials.[5]

Granular Material		Consolidated Material	
Clay soils (surface)	0.01–0.2 m/day	Sandstone	0.001–1 m/day
Deep clay beds	10^{-8}–10^{-2} m/day	Carbonate rock with secondary porosity	0.01–1 m/day
Loam soils (surface)	0.1–1 m/day	Shale	10^{-7} m/day
Fine sand	1–5 m/day	Dense, solid rock	10^{-5} m/day
Medium sand	5–20 m/day	Fractured or weathered rock (aquifers)	0.001–10 m/day
Coarse sand	20–100 m/day	Fractured or weathered rock (core samples)	almost 0–300 m/day
Gravel	100–1000 m/day	Volcanic rock	almost 0–1000 m/day
Sand and gravel mixes	5–100 m/day		
Clay, sand, and gravel mixes (till)	0.001–0.1 m/day		

Another term indicative of the water yielding capacity of an aquifer is its transmissivity or transmissibility. The transmissivity (T) of an aquifer is simply the product of the hydraulic conductivity and the saturated thickness, or:

$$T = KD \tag{2.5}$$

where:

T = transmissivity of the aquifer (L^2/t)
K = hydraulic conductivity of the aquifer (L/t)
D = saturated thickness of the aquifer (L)

In many soil formations, the nature of the material and its orientation yield a preferential direction for the movement of water. In simpler terms, sometimes a formation may have a hydraulic conductivity in the horizontal direction (K_x) radically different than in the vertical direction (K_z). This phenomenon ($K_x \neq K_z$) is called anisotropy. When hydraulic conductivities are the same in all directions ($K_x = K_z$), the material is said to be isotropic.

AQUIFER TESTING

When dealing with problems concerning ground water, one is invariably interested in the water movement or the water yielding capabilities of the aquifer. In order to make calculations concerning these processes, it is necessary to have values for one or more of the aquifer properties listed earlier. A variety of techniques have been developed for obtaining these properties.

Drawdown (Pump) Tests

Steady Flow, Confined Aquifer

Darcy's Law for volumetric flow to a well can be written as

$$Q = K\ A\ S^* \tag{2.6}$$

where:

Q = volumetric flowrate (L^3/t)
A = area normal to flow (L^2)
S* = hydraulic gradient

If one assumes that radial flow to a well fully penetrating a confined aquifer is purely horizontal (Dupuit-Forchheimer assumption), then Equation 2.6 can be written as (see Figure 2.10):

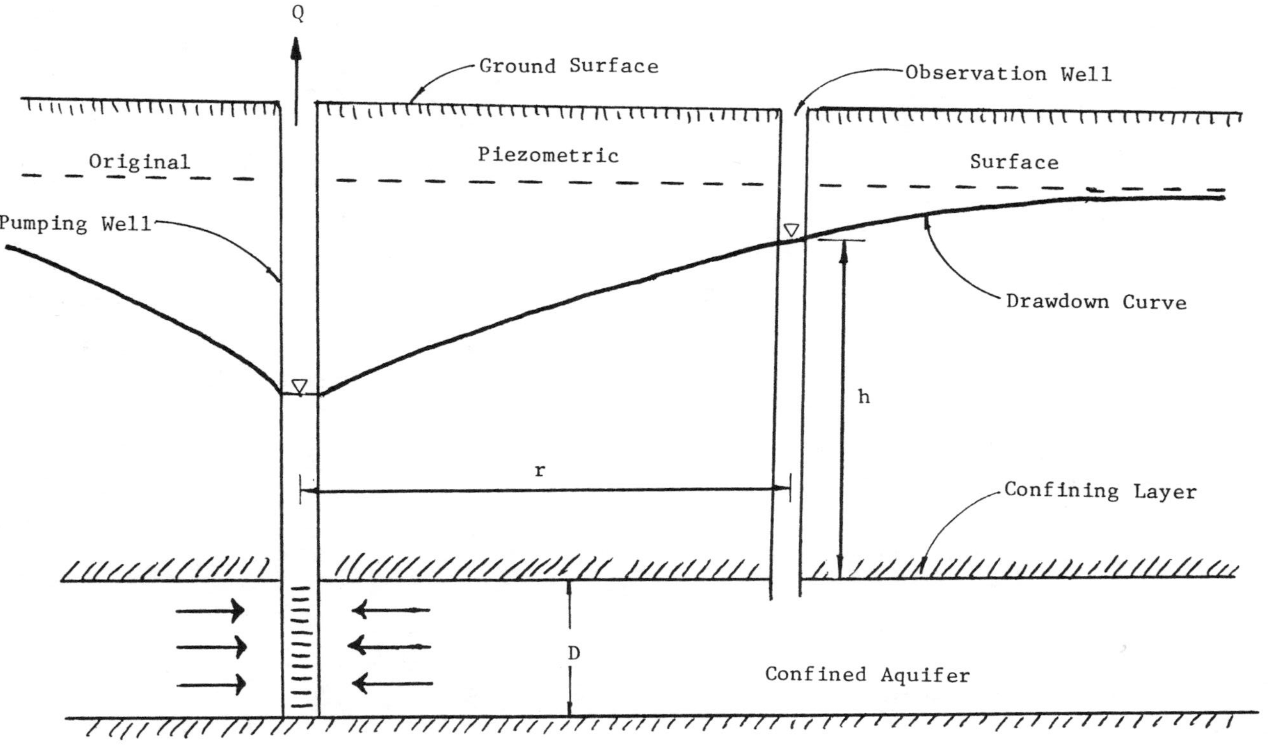

Figure 2.10. Steady radial flow in a confined aquifer.

$$Q = K(2\pi rD)\ (dh/dr) \tag{2.7}$$

where:

r = radial distance from center of well (L)

D = saturated thickness (L)

dh/dr = hydraulic gradient (or slope of piezometric surface) at distance r

Integrating Equation 2.4 yields:

$$Q = \frac{2\pi KD\ (h_2 - h_1)}{\ln\ (r_2/r_1)} \tag{2.8}$$

or:

$$Q = \frac{2\pi T\ (h_2 - h_1)}{\ln\ (r_2/r_1)} \tag{2.9}$$

Steady Flow, Unconfined Aquifer

Referring to Figure 2-11 for a fully penetrating well in an unconfined aquifer, one can utilize the Dupuit-Forchheimer assumption and develop the following equation for flow to the well:

$$Q = \frac{\pi K\ (h_2^{\ 2} - h_1^{\ 2})}{\ln\ (r_2/r_1)} \tag{2.10}$$

From Equations 2.8, 2.9, and 2.10, it is obvious that the hydraulic conductivity of an aquifer can be obtained simply by pumping one well at a constant rate and then observing the water level at two additional points (monitoring wells).

Several points concerning the phenomena associated with pumping wells need to be outlined. First, drawdown is a function of aquifer parameters and well pumping rates. If two pumping wells are spaced close to each other, the cumulative effect of their individual drawdowns would simply be the arithmetic sum of their individual drawdowns. Second, upconing of the water table due to injection of water behaves exactly the same as drawdown, only inverted. The cone of impression from an injection well will be the "mirror image" of the cone of depression of a withdrawal well.

Equations 2.8, 2.9, and 2.10 are based on steady state conditions, i.e., the cone of depression has stabilized. In actual tests, steady state is rarely achieved or is only achieved after very long periods of pumping. The more common case is for the cone of depression to change or the water level in the observation wells to drop as pumping continues. This situation is known as transient flow.

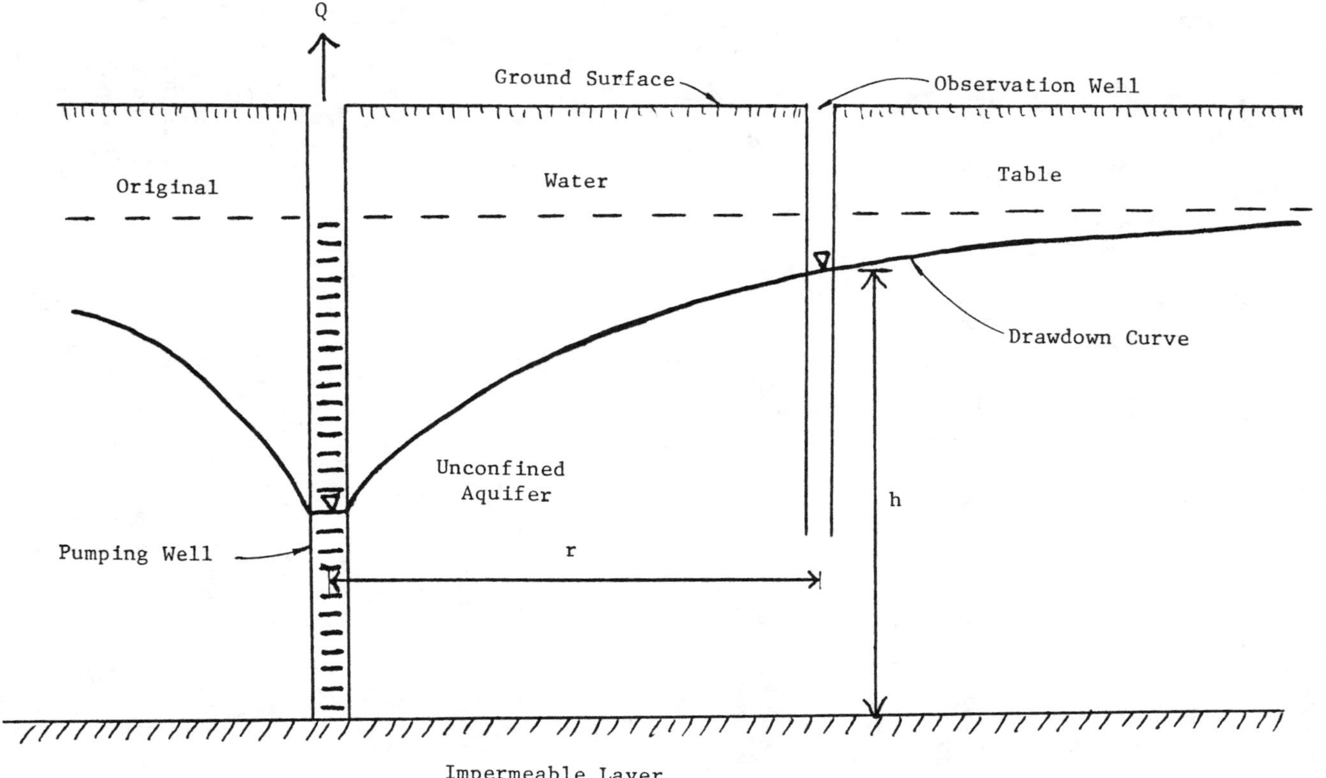

Figure 2.11. Steady radial flow in an unconfined aquifer.

Transient Flow, Confined Aquifer

The equation for transient flow to a well in a confined aquifer is:

$$\frac{1}{r}\frac{\delta h}{\delta r} + \frac{\delta^2 h}{\delta r^2} = \frac{S}{T}\frac{\delta h}{\delta r} \tag{2.11}$$

where:

S = storage coefficient

A. Theis Solution A solution to Equation 2.11 was developed by Theis in 1935 as:[5]

$$s = \frac{Q}{4\pi T} W(u) \tag{2.12}$$

where:

s = drawdown of piezometric surface
W(u) = well function

and:

$$u = \frac{r^2 S}{4Tt} \tag{2.13}$$

where:

t = time

Rearranging and taking logarithms of Equations 2.12 and 2.13 yields:

$$\log s = \log \frac{Q}{4\pi T} + \log W(u) \tag{2.14}$$

and:

$$\log r^2/t = \log \frac{4T}{S} + \log u \tag{2.15}$$

From the above relationships, it is seen that a log-log plot of s vs r^2/t and a log-log plot of W(u) vs u will be similar (Figure 2.12). By superimposing these two plots, a match point can be found. Using the four coordinates of the match point and Equations 2.12 and 2.13, the transmissivity (T) and storage coefficient (S) of the aquifer can be solved for.

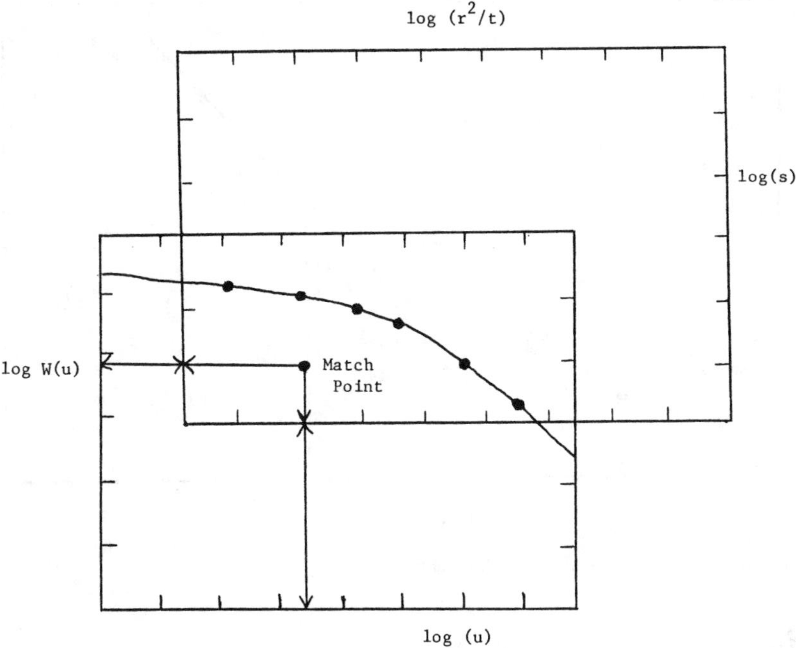

Figure 2.12. Superposition plot for Theis solution.

B. Cooper-Jacob Solution If one plots drawdown vs time (Figure 2.13) for a monitoring well on semilogarithmic paper, a significant portion of the curve will be linear. The storage coefficient (S) and transmissivity (T) of the aquifer can be obtained by the following:

$$S = \frac{2.25\ T\ to}{r^2} \tag{2.16}$$

where:

to = intersection at 0 drawdown on time axis of extended straight line portion of graph

and:

$$T = \frac{2.3\ Q}{4\pi \Delta s} \tag{2.17}$$

where:

Δs = drawdown change over 1 log cycle of the straight line portion of graph

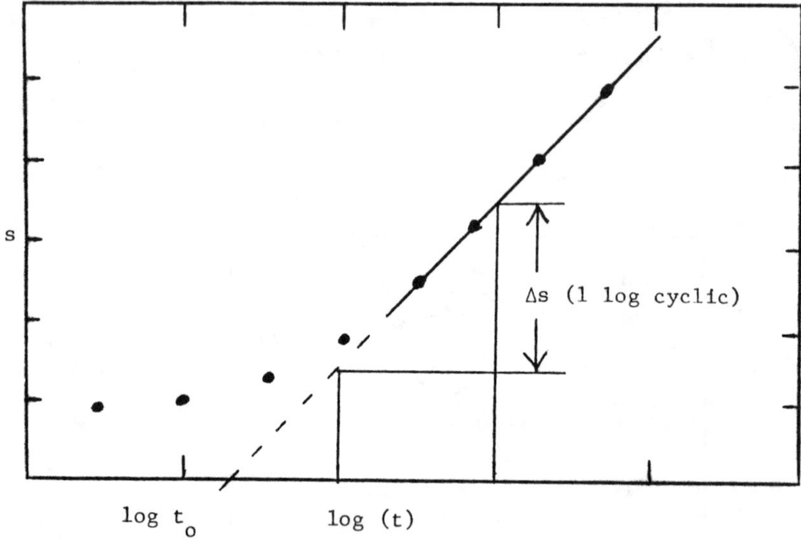

Figure 2.13. Cooper-Jacob solution plot.

Transient Flow, Unconfined Aquifer

Equations 2.11 through 2.17 are also applicable to unconfined aquifers provided that the transmissivity (T) of the aquifer can be considered constant. Most often this T is calculated based on the average height of water in the unconfined aquifer. Additionally, the S term appearing in the questions is then considered to be the specific yield of the unconfined aquifer.

Additional methods for both unconfined and confined aquifers and methods for a variety of leakage and discharge situations can be found in Bouwer.[5]

Slug Tests

All of the above described pump tests require the presence of at least two wells; one pumping well and one monitoring well. Additionally, these two wells must be screened in the same waterbearing formations, and the monitoring well must be located within the zone of influence of the pumping well. If these criteria are not met by existing wells, new wells must be drilled. Drilling and installing wells can be extremely costly, especially if the well is to be used for nothing more than observation during a pump test. It is for these reasons that several techniques have been developed for estimating the hydraulic conductivity of an aquifer based on the response of a single well to imposed stresses.

Bouwer and Rice[6] developed a procedure for estimating hydraulic conductivity based on the response of a well to the sudden removal of a slug of water.[5] Referring to Figure 2.14, the hydraulic conductivity can be calculated from:

$$K = \frac{r_c^2 \ln(R_e/r_w)}{2L_e} \frac{1}{t} \ln \frac{y_o}{y_t} \tag{2.18}$$

where:

R_e = effective radial distance over which the head difference y is dissipated
r_w = radial distance between well center and undisturbed aquifer (r_c plus thickness of gravel envelope or developed zone outside casing)
L_e = height of perforated, screened, uncased, or otherwise open section of well through which ground water enters
y_o = y at time zero
y_t = y at time t
t = time since y_o

and:

y = distance of water level in well below the water table

The effective radius R_e can be calculated by using the following equation and Figure 2.15.

$$\ln \frac{R_e}{r_w} = \frac{1}{\dfrac{1.1}{\ln (L_w/r_w)} + \dfrac{A + B \ln [(H - L_w)/r_w)]}{(L_e/r_w)}} \tag{2.19}$$

If the well penetrates to the bottom of the aquifer, Equation 2.19 cannot be used. In this instance, R_e can be calculated from:

$$\ln \frac{R_e}{r_w} = \frac{1}{\dfrac{1.1}{\ln (L_w/r_w)} + \dfrac{C}{(L_e/r_w)}} \tag{2.20}$$

The basis of the slug test is to plot y vs time on semilogarithmic paper and to utilize the straight line portion of that plot (Figure 2.16) to evaluate $1/t \ln y_o/y_t$. This is then substituted along with R_e into Equation 2.18 to calculate K.

In the field, the necessary data can be collected quite rapidly with a minimal amount of effort. However, the procedure does require the use of a fairly sensitive (hence fairly expensive) pressure transducer. The procedure involves placing the probe of the transducer just off the bottom of the well. A known volume (slug) of water is immediately withdrawn, and pressure

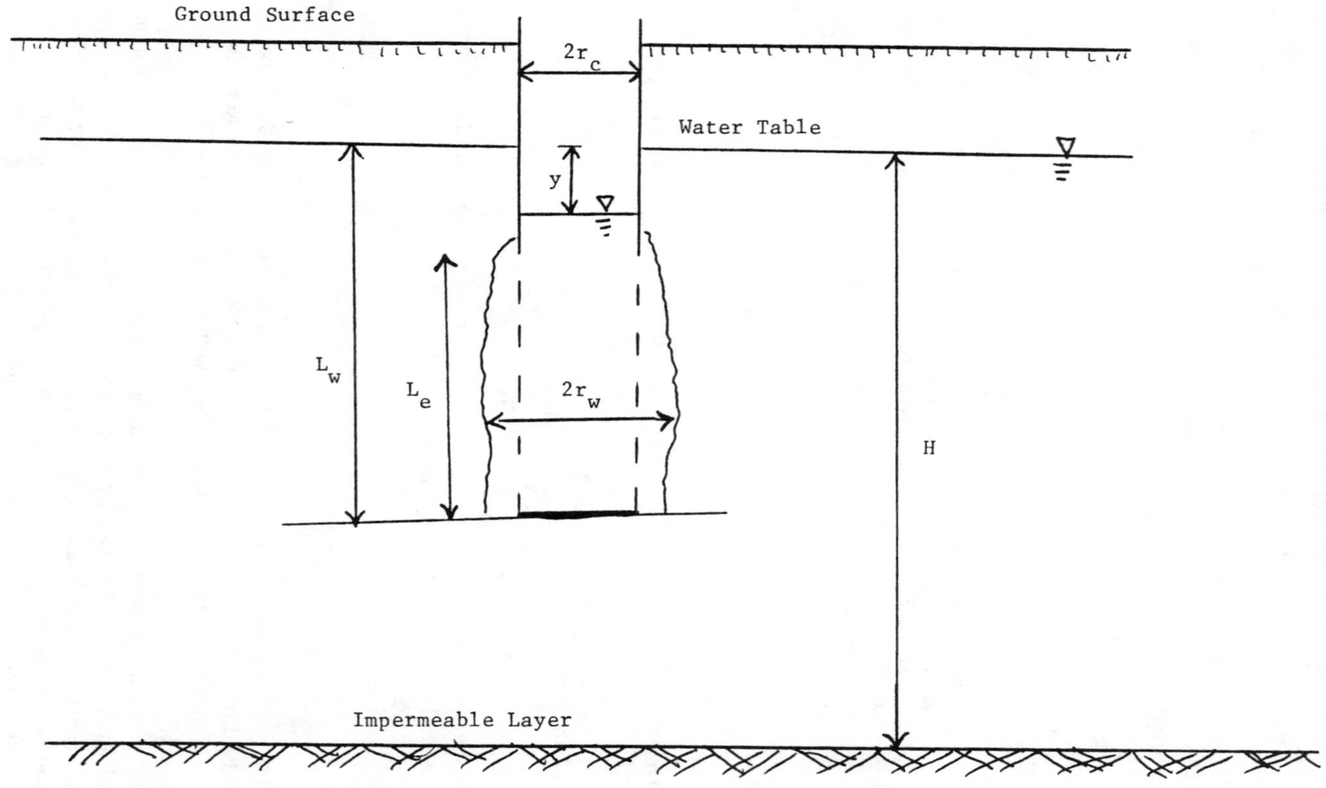

Figure 2.14. Partial penetrating well in unconfined aquifer.

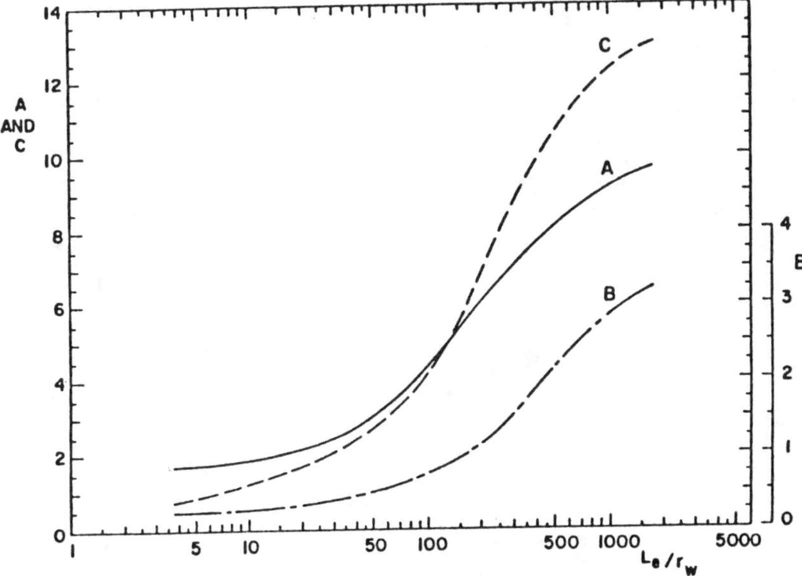

Figure 2.15. Curves relating coefficients A, B, and C to L_e/r_w.[5]

readings are recorded vs time. The slug of water can be removed by a bailing device. An alternative method is to lower a solid object of known volume into the well, let the well reach equilibrium, then instantaneously remove the object.

Other single hole techniques exist for estimating hydraulic conductivities and can be found in most elementary ground water textbooks. It should be emphasized that "rate-of-rise" or "slug test" techniques only give estimates of K for the region adjacent to the well. Additionally, these technique normally do not provide estimates for other aquifer properties such as the specific yield (S).

LABORATORY TESTS

An alternative to the field methods discussed previously is to retrieve soil or aquifer samples and take them into the laboratory for analysis. The most common type of instrument for measuring hydraulic conductivity in the laboratory is the permeameter. Basically, the permeameter involves placing a cylindrical sample in a device that measures flow rate and headloss across the sample while maintaining a flow of water through the sample.

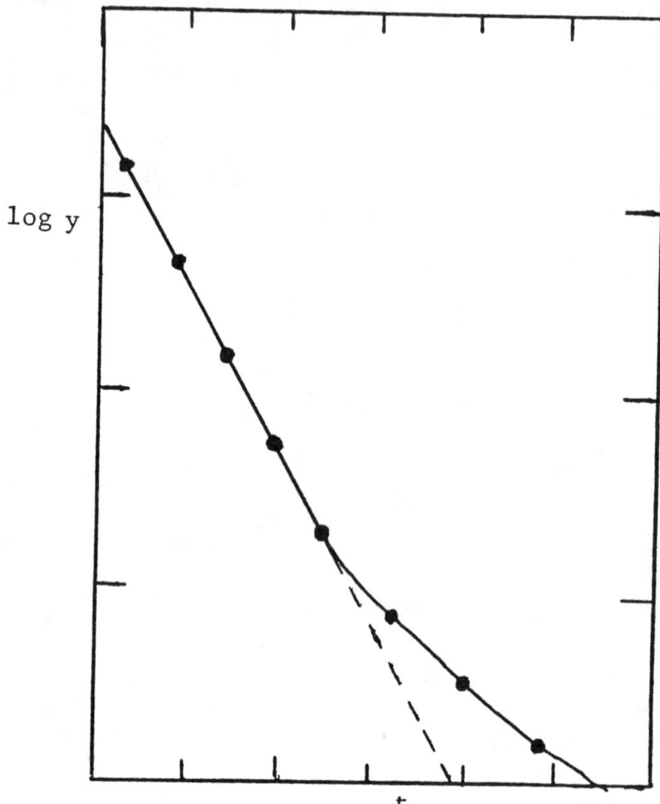

Figure 2.16. Semilogarithmic plot of depth vs time for slug test.

There are two classes of permeameters: constant head and falling head. The constant head permeameter shown in Figure 2.17 maintains a constant pressure head at both the influent and effluent sides of the sample. The flow through the sample is collected in a graduated cylinder for measurement. The hydraulic conductivity of the sample can be calculated by using Darcy's Law as:

$$K = \frac{LQ}{H\pi R^2} \tag{2.21}$$

where:

K = hydraulic conductivity of the sample
H = total head loss across sample

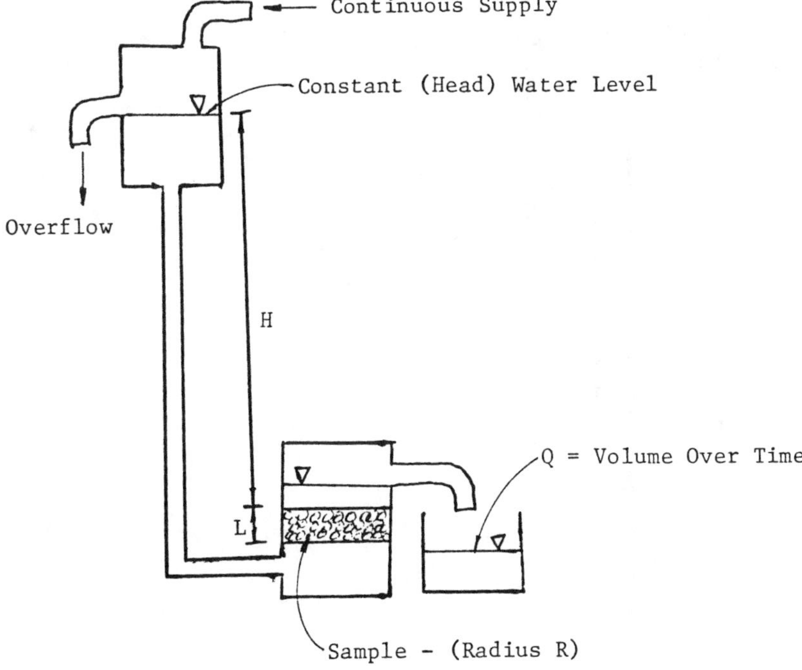

Figure 2.17. Constant head permeameter.

L = height of the sample
Q = volumetric flow rate
R = radius of the sample

The falling head permeameter shown in Figure 2.18 maintains a constant pressure head at the effluent side of the sample while letting the pressure head at the influent side decrease with time. Hydraulic conductivity can be calculated from:

$$K = \frac{Lr^2}{tR^2} \ln \frac{H_1}{H_2} \tag{2.22}$$

where:

H_1 and H_2 = H values at beginning and end of time period t
r = radius of standpipe

Falling head permeameters are used for materials thought to have relatively low hydraulic conductivities while constant head permeameters are used for more permeable materials.

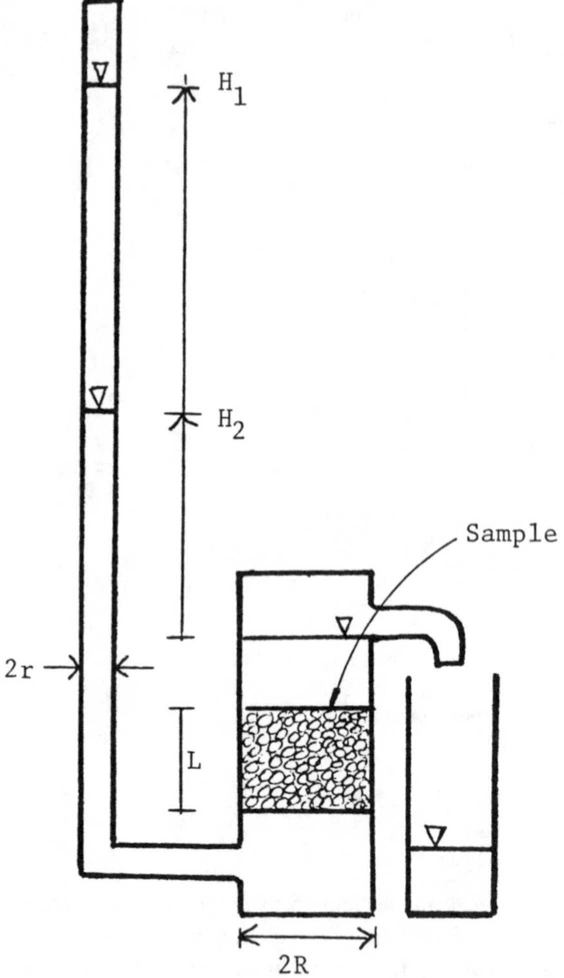

Figure 2.18. Falling head permeameter.

Laboratory methods for estimating hydraulic conductivities are possibly the most accurate methods for analyzing materials from unsaturated zones. However, the accuracy of these methods for saturated zones is much less than the in situ techniques discussed earlier. One reason for this relative inaccuracy is the fact that truly undisturbed samples for laboratory analysis are impossible to obtain. Additionally, samples can experience leakage between the container and their edges, thus leading to erroneously high hydraulic conductivities. The last disadvantage of permeameters is that they produce K values indicative of one direction only, usually the vertical direction (K_z). Obtaining deep aquifer materials for analysis of horizontal conductivities (K_x) is difficult at best.

REFERENCES

1. Wyatt, A. W., "Secondary Recovery of Capillary Water," *Proceedings of the Ogallala Aquifer Symposium II,* Texas Tech University, Lubbock, Texas, June 1984, pp. 500–513.
2. Everett, L. G., et al., "Vadose Zone Monitoring Concepts at Landfills, Impoundments, and Land Treatment Disposal Areas," *Proceedings of the National Conference on Management of Uncontrolled Hazardous Waste Sites,* 1982, Hazardous Materials Control Research Institute, Silver Spring, Maryland, pp. 100–106.
3. Lehr, J. H., "Model Analysis of Water Table Drawdown Surrounding Pumping Wells," *Journal of Soil and Water Conservation,* Vol. 18, No. 5, September–October, 1963, pp. 205–207.
4. Lehr, J. H., "Model Analysis of Groundwater Flow Around and Through Lenticular Beds," *Journal of the Water Pollution Control Federation,* Vol. 36, No. 1, January 1964, pp. 84–87.
5. Bouwer, H., *Groundwater Hydrology,* McGraw-Hill Book Company, Inc., New York, New York, 1978.
6. Bouwer, H. and Rice, R. C., "A Slug Test for Determining Hydraulic Conductivity of Unconfined Aquifers with Completely or Partially Penetrating Wells," *Water Resources Research,* Vol. 12, No. 3, 1976, pp. 423–428.

CHAPTER 3

Ground Water Information Sources

Effective protection of ground water quality is dependent upon information from a variety of substantive sources, including geology, hydrology, chemistry, microbiology, engineering, and ground water modeling. The information list could be expanded to include topical issues depending on whether the ground water quality problem was localized or regional, or over a sole source aquifer in an urban area or in a sparsely populated mountainous region.

The information which needs to be evaluated during analysis of a ground water contamination problem includes that related to the physical framework through which the ground water is moving, the hydrologic system, and present and past contamination-related operations, including construction details. Data and interpretive reports regarding the physical framework and hydrologic system would generally be available from federal, state, and/or local governmental agencies responsible for natural resources and/or environmental protection. Information may also be available from consulting firms and private industries that may have performed local specialized studies.

The type of information which might be obtained from each source category will range from fundamental data, such as well logs and water level measurements, to specialized technical reports. In some states, the local or state agency will be the principal information source, and in other states, a federal agency might have most of the necessary information.

Information on ground water quality protection and related issues has increased tremendously in the last two decades. Therefore, at times the information seeker may have a "feeling," or may have been told, that the information exists. However, finding it may be very difficult. This chapter, divided into sections on publications, computerized information systems, and associations and organizations, has been prepared to aid information seekers by identifying some starting points for sources of information.

REPORTS, PERIODICALS AND BOOKS

Publications of the U.S. Geological Survey (USGS) are available as circulars, professional papers, and water-supply papers. Most of the field work is done on a cooperative basis with individual states. Information on ground water levels in various parts of the country is published periodically. The publications may contain solutions to field problems in certain areas, the reporting of field data, or purely theoretical analyses.

The USGS national headquarters in Reston, Virginia can further explain the specific types of information available and the proper office to contact to obtain reports. One example is the Regional Aquifer System Assessment (RASA) reports which are regional aquifer studies underway or recently completed in several states. RASA projects have been selected on the basis of the following criteria: (1) the significance of the aquifer system as a present or potential supply of water—particularly its significance to the economies of more than one state; (2) the severity of the water problems facing the project area; and (3) the potential water needs in the project area—particularly those connected with energy production, irrigation, and urban development, and, to a lesser extent, water quality.

In addition, each state has a USGS Water Resources Division office with available publication lists. Pertinent addresses are shown in Table 3.1. The states produce *Water Use Data* books evey few years which provide limited information on ground water levels and quality. Each year the USGS publishes a *National Water Summary* which contains a section on ground water quality trends throughout the United States. The state USGS offices often participate in cooperative studies with state natural resources or water or environmental agencies in investigations of state-related ground water contamination problems. Publications resulting from these investigations are also available through the USGS.

Various other federal governmental agencies publish material related directly or indirectly to ground water. For example, the U.S. Army Corps of Engineers, the U.S. Bureau of Reclamation, and the U.S. Soil Conservation Service publish bulletins and manuals from time to time related to water resources in general and ground water in particular. Many of these manuals are designed for use by practicing hydrogeologists and engineers. However, many other publications are the result of extensive research.

Many recent publications of the U.S. Environmental Protection Agency (EPA) can provide useful information for ground water quality protection. These publications are typically available through the National Technical Information Service of the U.S. Department of Commerce in Springfield, Virginia. Listings of federal agency publications are also available in most university and major public libraries. Relevant ground water-related offices of

Table 3.1. Addresses of the District Offices of the U.S. Geological Survey Water Resources Division.

University of Alabama
Oil and Gas Bldg.-Room 202
P.O. Box V
Tuscaloosa, AL 35486
(205) 752-8104

218 E. Street
Anchorage, AK 99501
(907) 271-4138

Federal Building
301 W. Congress Street
Tucson, AZ 85701
(508) 378-6391

855 Oak Grove Avenue
Menlo Park, CA 94025
(405) 323-8111

Building 53
Denver Federal Center
Lakewood, CO 80225
(303) 234-5092

135 High Street-Room 235
Hartford, CT 06103
(203) 244-2528

Subdistrict-District Office/MD
Federal Building-Room 1201
Dover, DE 19901
(302) 734-2506

325 John Knox Road-Suite F-240
Tallahassee, FL 32303
(904) 386-1118

Suite B
6481 Peach Tree, Industrial Blvd.
Doraville, GA 30360
(404) 221-4858

Field Headquarters
4398D Loke St.
P.O. Box 1856
Lihue, Kauai, GUAM 96766

Subdistrict
U.S. Navy Public Works Center
FPO S.F. 96630-P.O. Box 186
Agana, GUAM 96910

P.O. Box 50166
300 Ala Moana Blvd.-Room 6110
Honolulu, HI 96850

P.O. Box 2230
Idaho Falls, ID 83401
(208) 526-2438

P.O. Box 1026
605 N. Nek Street
Champaign, IL 61820
(217) 398-5353

1819 North Meridian Street
Indianapolis, IN 46202
(317) 269-7101

Federal Building-Room 269
P.O. Box 1230
Iowa City, IA 52244
(319) 337-4191

University of Kansas
Campus West
1950 Avenue A
Lawrence, KS 66045
(913) 864-4321

Table 3.1, continued

Federal Building-Room 572
600 Federal Place
Louisville, KY 40202
(502) 582-5241

6554 Florida Boulevard
Baton Rouge, LA 70986
(504) 389-0281

District Office in Mass.
26 Ganneston Drive
Augusta, MA 04330
(207) 623-4797

208 Carroll Building
8600 Lasalle Road
Towson, MD
(301) 828-1535

150 Causeway Street
Suite 1001
Boston, MA 02114
(617) 223-2822

6520 Mercantile Way
Suite 5
Lansing, MI 48910
(517) 372-1910

702 Post Office Building
St. Paul, MN 55101
(612) 725-7841

Federal Building
Suite 710
100 West Capitol Street
Jackson, MS 39201
(601) 969-4600

Mail Stop 200
1400 Independence Road
Rolla, MO 65401
(314) 341-0824

Federal Building
Drawer 10076
Helena, MT 59601
(406) 559-5263

Federal Bldg./Courthouse-Room 406
100 Centennial Mall North
Lincoln, NE 68508
(402) 471-5082

Federal Building-Room 227
705 North Plaza Street
Carson City, NV 89701
(702) 882-1388

Subdistrict-Dist. Off./Mass
Federal Building-210
55 Pleasant Street
Concord, NH 03301

Federal Building-Room 436
402 E. State St.-P.O. Box 1238
Trenton, NJ 08607
(609) 989-2162

Western Bank Building
505 Marquette, N.W.
Albuquerque, NM 67125

236 U.S. Post Office/Courthouse
P.O. Box 1350
Albany, NY 12201
(518) 472-3107

Century Station-Room 436
Post Office Building
P.O. Box 2857
Raleigh, NC 27602
(919) 755-4510

821 E. Interstate Avenue
Bismarck, ND 58501
(701) 255-4011

975 West Third Avenue
Columbus, OH 43212
(614) 469-5553

215 N.W. 3rd-Room 621
Oklahoma City, OK 73102
(405) 231-4256

Table 3.1, continued

(Mail) P.O. Box 3202
Ship-830 NE Holladay St., 97232
Portland, OR 97208
(503) 231-5242

Federal Building-4th Floor
P.O. Box 1107
Harrisburg, PA 17108
(717) 782-4514

Building 652, Ft. Buchanan
G.P.O. Box 4424
San Juan, Puerto Rico 00936
(809) 783-4660

District Office in Mass.
Federal Bldg. and U.S. P.O.
Room 224
Providence, RI 02903
(401) 528-4655

Strom Thurmond Federal Building
1835 Assembly Street-Suite 658
Columbia, SC 29201
(803) 765-5966

Federal Building-Room 308
200 4th Street, S.W.
Huron, SD 57350
(605) 352-8651

U.S. Courthouse
U.S. Federal Building-A-413
Nashville, TN 37203
(615) 251-5424

Federal Building-649
300 East 8th Street
Austin, TX 78701
(512) 397-5766

Administration Building-1016
1745 West 1700 South
Salt Lake City, UT 84104
(801) 524-5663

District Office in Mass.
U.S. Post Office/Courthouse
Rooms 330B and 330C
Montpelier, VT 05602
(802) 229-4500

200 West Grace Street-Room 304
Richmond, VA 23220
(804) 771-2427

1201 Pacific Avenue-Suite 600
Tacoma, WA 98402
(206) 593-6510

Federal Bldg./U.S. Courthouse
500 Quarrier St., East-Room 3017
Charlestown, WV 25301
(304) 343-6181

1815 University Building
Madison, WI 53706
(608) 262-2488

P.O. Box 1125
J. C. O'Mahoney Federal Center
2120 Capitol Avenue-Room 5017
Cheyenne, WY 82001
(307) 778-2220

EPA located in Washington, DC include the Office of Ground Water Protection, Office of Water Regulations and Standards, Office of Water Program Operations, Office of Drinking Water, Office of Emergency Action and Remedial Response, Office of Pesticide Programs, and the Office of Toxic Substances.

Ground water research findings by universities and in the private sector are typically published in professional and scientific journals (national and/or international). These journal publications usually present summaries of more detailed reports by university departments and research institutes. Some of these reports may be found in the publications of local water resources research institutes (or equivalent agencies), or as separate bulletins published by universities. The reports usually contain the details of the investigations and other data that may be useful. Other sources of information are masters theses and doctoral dissertations. Although students are encouraged to adapt their results for publication in professional and scientific journals, many theses and dissertations have never been published. University Microfilms in Ann Arbor, Michigan has cumulative lists of theses and dissertations completed in the United States and Canada. Recently, scientific journals have started to summarize relevant theses and dissertations (for example, *Ground Water*, the journal of the National Water Well Association, has begun this service).

There are many professional associations and societies throughout the world that publish journals and bulletins related directly or indirectly to ground water. Many foreign scientists publish their papers in U.S. journals. Also, international conferences and congresses focused on ground water are held periodically in various locations throughout the world. National and local conferences and symposia with specific themes are also held, and the presented papers are typically compiled in proceedings.

The American Society of Civil Engineers publishes ground water-related papers in the journals from the divisions of hydraulics, irrigation and drainage, environmental engineering, and water resources planning and management. The American Geophysical Union (AGU) publishes *Water Resources Research* which includes periodic contributions related to ground water. Before 1965, AGU papers pertinent to ground water were published in the *Journal of Geophysical Research* and the *Transactions of the American Geophysical Union*. The American Water Resources Association publishes the *Water Resources Bulletin*, and this journal occasionally has papers related to ground water.

Pertinent papers also appear in journals of physics, chemical engineering, and engineering mechanics, as well as foreign journals and those of other international and U.S. associations. For example, the International Association of Hydrological Sciences (IAHS) publishes the *Journal of Hydrology*. The

American Water Works Association and other U.S. geologic, geophysical, and agricultural associations publish journals with periodic papers related to ground water quality protection.

The National Water Well Association (NWWA) in Dublin, Ohio publishes *Water Well Journal,* which is confined to water-well technology and practical aspects of well drilling and related equipment. Since 1963, the same association has published *Ground Water,* which is a scientific journal devoted exclusively to ground water. The NWWA also publishes *Ground Water Monitoring Review,* a quarterly publication. In addition to journals, the NWWA also publishes technical books and manuals, conference proceedings, brochures, films, slide shows and videotapes.

Table 3.2 lists examples of U.S. organizations which publish ground water-related information ranging from highly technical to that prepared for the general public.

Each of the 50 states has one or more agencies dealing with ground water which maintains summaries of water quality data, water levels, pollution investigations, site evaluations, and other relevant topics. This public information is usually free (or inexpensive), but may not be as easily accessible as federal agency information because there are fewer publication lists and less effort is made to disseminate the information. The technical detail of these publications varies widely from state to state. An old bibliography of state agency water resources publications is available through the Water Information Center in Syosset, New York.[1] Even though it is out of date, it provides a good indication of the types of reports available from a state health department vs a state geological survey.

There are many specialized books on hydrology, geochemistry, pollution, federal and state regulations, and other relevant topics.[2] A bibliography of approximately 4600 references, including books, reports, and articles, is available.[3]

COMPUTERIZED INFORMATION SYSTEMS

Computerized ground water information systems have been developed for data storage and retrieval, and bibliographical surveys.

Ground Water Data

In order to properly protect ground water resources, a comprehensive ground water management program is essential. Effective ground water quality management requires that relevant information be available to the de-

Table 3.2. Examples of U.S. Organizations Publishing Ground Water Journals, Magazines and Newsletters.

American Society of Civil Engineers
345 East 47th Street
New York, NY 10017

American Water Works Association Journal
6666 W. Quincy Avenue
Denver, CO 80235

EOS: Transactions of the American Geophysical Union
2000 Florida Avenue, N.W.
Washington, DC 20009

From the State Capitols—Water Supply
Box 1939
New Haven, CT 06509

Ground Water Age
National Trade Publications
8 Stanley Circle
Latham, NY 12110

Ground Water Newsletter
Water Information Center, Inc.
North Shore Atrium
6800 Jericho Turnpike
Syosset, NY 11791

National Water Well Association
6375 Riverside Dr.
Dublin, OH 43017

New England Water Works Association Journal
850 Providence Highway
Dedham, MA 02026

NWSIA Newsletter
National Water Supply Improvement Association
26 Newbury Road
Ipswich, MA 01939

Southwest and Texas Water Works Journal
306 E. Adams Avenue
Temple, TX 76501

U.S. Water News
230 Main Street
Halstead, KS 67056

Water Currents
Texas Water Resources Institute
College Station, TX 77843

Water Information News Service, Inc.
1730 M Street, N.W.
Suite 1100
Washington, DC 20036

Western Water
1007 Seventh Street, Suite 315
Sacramento, CA 95814

WRRI News Report
Water Resources Research Institute
202 Hargis Hall
Auburn, AL 36849

WSIA Journal
Water Supply Improvement Association
363 Boston Road
Box 387
Topsfield, MA 01983

cision maker in a concise, comprehensive, timely, economical, and reliable manner. A comprehensive ground water data management capability is composed of three major components: (1) maintaining the data generated by ground water surveillance; (2) indexing that data so that it can be accessed efficiently; and (3) maintaining concise citations of relevant ground water research documentation. At the federal level, these capabilities are provided mainly by the USGS and EPA.

The National Water Data Exchange (NAWDEX) provides direct, on-line access to its data bases by its members and other organizations. The program is also authorized to provide limited access to the data bases of the USGS National Water Data Storage and Retrieval System (WATSTORE) and EPA's Storage and Retrieval (STORET) System. The NAWDEX and WATSTORE data bases are maintained on the USGS's AMDAHL 470 V/7 computer system located at its national headquarters in Reston, Virginia. The STORET data base is maintained on EPA's IBM 370/168 computer located in the Research Triangle Park in Raleigh-Durham, North Carolina. Authorized users are provided with complete instructions and user manuals for use in accessing these data bases.

NAWDEX, established in 1976, is a confederation of water-oriented organizations working together on a national basis to improve access to water information by assisting users in the identification, location, and acquisition of needed data. It is a major federal information indexing system for ground water in the United States. Member organizations are linked so that their water-data holdings may be readily exchanged for maximum use. A central program office within the Water Resources Division of USGS coordinates this linkage and provides overall management of the program. The types of hydrologic data indexed include surface and ground water flows and water quality data.

Through use of local assistance at USGS district offices, the Water Data Source Directory, and the Master Water Data Index, NAWDEX can readily assist users in locating data for a specific geographical area. Listings and summary information can be provided, or the user can be referred to water-data systems, bibliographic services or data-collecting organizations that are able to provide information. Membership in NAWDEX is voluntary and open to any water-oriented organization. Members are required, however, to enter into a formal agreement with the NAWDEX program office to take an active role in NAWDEX activities.

As part of the USGS program of releasing water data to the public, a large-scale computerized system has been developed for the storage and retrieval of such data. WATSTORE was established in 1971 to modernize the USGS water-data processing procedures and techniques and to provide for more efficient data management. A minimal fee plus the actual computer cost incurred in producing a desired product is charged to the information requester/receiver. The WATSTORE system is designed to recognize the possibility that a ground water monitoring station (well) can penetrate more than one aquifer and that samples can be drawn from individual aquifers separately with the use of screen plugs. Therefore, WATSTORE allows for the storage of aquifer identifiers along with the water quality analysis data for each sample. The ground water information resides in an on-line computer data file which is maintained by a data base management system called System 2000.

Within WATSTORE, three files are pertinent to water. The Water-Quality File contains the results of chemical, physical, biological, and radio-chemical analyses for both surface and ground waters. These data are, however, stored and made available through STORET. The Groundwater-Site Inventory File contains inventory information such as site location and identification data, well construction data, geohydrologic characteristics, and other data pertinent to wells, springs, and other sources of ground water. The Water-Use File, also called the National Water-Use Data System (NWUDS), contains limited summary data on water use throughout the United States. The NWUDS is a result of a federal-state cooperative program to collect, store, and disseminate water-use data. Field activities for acquisition of the data are the responsibility of the state agency, where direct communication with the water using community can be readily established. How these responsibilities are implemented in each state ultimately is determined by the USGS Water Resources Division district chief and the state cooperator.

The Storage and Retrieval of Water Quality Data System (STORET) was developed initially by the U.S. Public Health Service and is currently operated by EPA. This system is intended to provide federal assistance to the states in the performance of surface and ground water quality management functions. Data are stored for several federal agencies and over 40 state agencies. Originally, the STORET system was only used for the storage and retrieval of water quality data from surface waterways within and contiguous to the United States. Presently, however, STORET is capable of handling chemical data from both surface and ground water, and soil and air, as well as biological data. Data may be updated, manipulated and retrieved in various formats by employing user-oriented facilities (desk-top terminals and package programs). The principal data storage file within the STORET system is the Water-Quality File (WQF) which can be used for ground water data. All users access the STORET system with data processing terminals via telephone. Both card-reading/line-printing terminals (remote or medium-speed) and keyboard terminals (low-speed, including video) can be employed by the user.

STORET maintains a full-time, user-assistance staff at EPA headquarters in Washington, DC. This group provides assistance and guidance in response to STORET-related problems. In addition, user-assistance personnel periodically conduct basic and advanced STORET training seminars at no cost to the user. This person-to-person assistance is supplemented by a set of STORET documentation manuals, provided to users, that describe both data input and information retrieval procedures.

States and other organizations can use STORET for their own local data-management programs, while those concerned with national programs, such as EPA, can review the data from broader perspectives. STORET can be responsive to the needs of state and local governments in the area of ground water data management because of its ease of use, adaptability, wide diversity of available functions and, most important, its economy.

The EPA also established the Model State Information System (MSIS) in the mid-1970s for reporting on public water supply compliance with the National Primary Drinking Water Regulations. It is a data base management system that has 78 separate programs. States reimburse EPA for use of MSIS; some states track the usage of all their drinking water supplies (community and noncommunity), and other states use it for data storage for their wells. Some ground water quality data is available through MSIS; however, well yield and capacity data is typical. These data combined with available private well data can then be used in statewide ground water planning and policy decisions as well as in general aquifer classification schemes. The MSIS has no tie-in with NWUDS.

All users of the above systems are required to sign a Memorandum of Agreement with the USGS which defines the conditions and fiscal responsibilities for use of the computer systems. In addition, STORET users must be members of NAWDEX under terms of the agreement between NAWDEX and EPA developed for this service. Nongovernmental users of STORET may also be required to coordinate their usage via governmental sponsorship. All users are charged for their actual use of the computer system plus a percentage administrative surcharge. A schedule of USGS computer charges is available upon request. Users of NAWDEX and WATSTORE are billed monthly. Users of STORET are billed quarterly. All charges are assessed by and reimbursable to the USGS. Additional information on these described data bases can be obtained from:

Program Manager
National Water Data Exchange
U.S. Geological Survey
421 National Center
Reston, VA 22092
Telephone: (703) 860-6031
FTS 928-6031

Bibliographic Information Retrieval

Two bibliographic information retrieval systems will be described: (1) the National Ground Water Information Center; and (2) the Dialog Information Retrieval Service.

National Ground Water Information Center

In 1979, the U.S. Environmental Protection Agency established, through a cooperative agreement with the National Water Well Association, the National Ground Water Information Center (NGWIC) as a repository of ground water literature for scientists, government agencies, business, and the public. It is funded in part by the EPA and managed by the National Water Well Association. Ground Water On-Line has been a natural outgrowth of the NGWIC, and it serves the international ground water communities information needs.

The NGWIC collection contains more than 50,000 documents including state publications, technical reports, government documents, maps, reference books, and other ground water quality related literature. In addition, the collection maintains more than 120 technical and trade journals and newsletter subscriptions. The literature is accessible through a telephone and computer terminal (mini, micro, or mainframe).

References within NGWIC may contain information on up to 22 different fields and may be retrieved by searching any of the fields individually or in combination. When searching a combination of fields, a topic can be limited or expanded using Boolean Logic ("or," "and," and "and not" statements). The fields that a reference may contain are:

- Accession number
- Indexer's initials
- Date entered into data base
- Author
- Title
- Source
- International Standard Serial Number (ISSN)
- International Standard Book Number (ISBN)
- Publisher
- Non-U.S. geographical area
- State abbreviation
- County name
- Aquifer region
- Publication date

- Call number
- Language
- Holding library
- Contents notes
- Chemical constituents
- Biological factors
- Author's affiliate organization
- More than 700 key hydrogeological terms

The references can be displayed either directly on a terminal or printed off-line and mailed to the user. Additional information on NGWIC can be obtained from:

National Ground Water Information Center
6375 Riverside Dr.
Dublin, OH 43017
(614) 761-1711

Dialog Information Retrieval Service

Dialog is an international computerized bibliographic information retrieval service that has been in existence since 1972. There are approximately 250 data bases available on the system containing 120 million records. Information available includes bibliographic citations and abstracts from journals, books, technical reports, yellow pages, newspapers, conference papers, dissertations, and other information sources.

There are approximately ten data bases within Dialog that provide a substantial amount of ground water information. These data bases are in categories entitled Energy and Environment and Science and Technology. Pertinent data bases within these categories include Water Resources Abstracts, National Technical Information Service, Commonwealth Agricultural Abstracts, Compendex, Conference Papers Index, Pollution Abstracts, Aqualine, Dissertation Abstracts, and Agricola. The Water Resources Abstracts data base contains the most ground water references—approximately 20,000. Table 3.3 includes detailed descriptions of the file contents of four of the most useful data bases having ground water information.

Dialog is available to users 22 hr/day and for 7 days/week. The total cost of conducting a search includes a data base cost of approximately \$30–\$120/hr, a telephone charge of \$8/hr, and a print cost of \$0.05–\$.30/line (depending on whether the desired output is printed at the user's terminal or printed at Dialog headquarters in California and mailed to the user).

A search of Dialog is conducted by selecting key terms (descriptor words)

Table 3.3. Summaries of Selected Data Bases in Dialog Containing a Large Number of Ground Water References.

WATER RESOURCES ABSTRACTS, 1968—present, 172,500 records, monthly updates (U.S. Department of the Interior, Washington, DC)

Water Resources Abstracts is prepared from materials collected by over 50 water research centers and institutes in the United States. The file covers a wide range of water resource topics including water resource economics, ground and surface water hydrology, metropolitan water resources planning and management, and water-related aspects of nuclear radiation and safety. The collection is particularly strong in the literature on water planning (demand, economics, cost allocations), water cycle (precipitation, snow, ground water, lakes, erosion, etc.), and water quality (pollution, waste treatment). The WRA covers predominantly English-language materials and includes monographs, journal articles, reports, patents, and conference proceedings.

CONFERENCE PAPERS INDEX, 1973—present, 1,086,000 records, monthly updates (Cambridge Scientific Abstracts, Bethesda, MD)

Conference Papers Index provides access to records of more than 100,000 scientific and technical papers presented at over 1,000 major regional, national, and international meetings each year. Conference Papers Index provides a centralized source of information on reports of current research and development from papers presented at conferences and meetings. It provides titles of the papers as well as the names and addresses (when available) of the authors of these papers. Also included in this data base are announcements of any publications issued from the meetings, available preprints, reprints, abstract booklets, and proceedings volumes, including dates of availability, costs, and ordering information. Primary subject areas covered include the life sciences, chemistry, physical sciences, geosciences, and engineering.

NATIONAL TECHNICAL INFORMATION SERVICE, 1964—present, 1,076,000 records, biweekly updates (U.S. Department of Commerce, Springfield, VA)

The *NTIS* data base consists of government-sponsored research, development, and engineering plus analyses prepared by federal agencies, their contractors or grantees. It is the means through which unclassified, publicly available, unlimited distribution reports are made available for sale from such agencies as NASA, DDC, DOE, HHS (formerly HEW), HUD, DOT, Department of Commerce, and some 240 other units. State and local government agencies are now beginning to contribute their reports to the file. The NTIS data base includes material from both the hard and soft sciences, as well as substantial material on technological applications, business procedures, and regulatory matters. Many topics of immediate broad interest are included, such as environmental pollution and control, energy conversion, technology transfer, behavioral/societal problems, and urban and regional planning.

COMPENDEX, 1970—present, 1,357,000 records, monthly updates (Engineering Information, Inc., New York, NY)

The *Compendex* data base is the machine-readable version of the Engineering Index (monthly/annual) which provides abstracted information from the world's significant engineering and technological literature. The Compendex data base provides worldwide coverage of approximately 3500 journals and selected government reports and books.

that will yield the desired result. Term selection is not restricted by a thesaurus—any word or group of words which may appear in the reference's title or abstract may be selected. The system searches for any combination of key terms that the user assembles using Boolean logic (and/or, intersection or union of sets).

Search selection can further be limited by journal(s), author, year(s) of publication, language, or other factors. Another useful feature within Dialog is search updating. Once a search and retrieval is complete the search can be saved within Dialog; then, six months or two years later (or whatever time period is chosen), the search can be conducted again retrieving only those references added since the previous search. In addition to providing complete bibliographic citations and abstracts, another service called Dial-Order allows the user to retrieve complete documents by mail through some data bases. Additional information on the Dialog system can be obtained from:

Dialog
3460 Hillview Avenue
Palo Alto, CA 94304
(800) 227-1927

ASSOCIATIONS AND ORGANIZATIONS

There are numerous professional and public interest associations and organizations which may have a wide variety of ground water information. Some of the areas addressed include water well drilling, modeling, hazardous waste disposal, water use and conservation, ground water research, and ground water policy issues. These associations and organizations may provide information through publications, annual and regional conferences, training courses, technical assistance programs, public education programs, trade shows, and other means of information transfer. Table 3.4 contains a listing of some ground water associations and organizations in the United States. A complete listing of organizations and associations is available elsewhere.[4]

Table 3.4. Examples of Ground Water-Related Associations or Organizations.

American Academy of Environmental Engineers
P.O. Box 269
Annapolis, MD 21404

American Public Works Association
1313 E. 60th Street
Chicago, IL 60637

American Water Resources Association
5410 Grosvenor Lane
Suite 220
Bethesda, MD 20814

Association of Ground Water Scientists and Engineers
6375 Riverside Drive
Dublin, OH 43017

Conservation Foundation
1717 Massachusetts Ave., N.W.
Washington, DC 20036

Environmental and Energy Study Institute
410 First Street, S.E.
Washington, DC 20003

Environmental and Ground Water Institute
200 Felgar Street, Room 127
The University of Oklahoma
Norman, OK 73019

Environmental Law Institute
1346 Connecticut Ave., N.W.
Suite 620
Washington, DC 20036

Georgia Ground Water Association
2360 Burnt Creek Road
Decatur, GA 30033

Groundwater Management Districts Association
1125 Maize Road
Colby, KS 67702

Hazardous Materials Control Research Institute
9300 Columbia Blvd.
Silver Spring, MD 20910

Hazardous Waste Research Information Center
Illinois State Water Survey Division
P.O. Box 5050, Station A
Champaign, IL 61820

Table 3.4, continued

Hazardous Waste Research Institute
EPA Center of Excellence
Louisiana State University
Baton Rouge, LA 70803

Holcomb Research Institute
National Ground Water Models Clearinghouse
Butler University
Indianapolis, IN 46208

Illinois Ground Water Association
Illinois Water Survey
P.O. Box 409
Warrenville, IL 60555

Iowa Ground Water Association
P.O. Box 3398
Iowa City, IA 52244

Minnesota Ground Water Association
P.O. Box 65362
St. Paul, MN 55165

National Association of Conservation Districts
1025 Vermont Avenue, N.W., Suite 730
Washington, DC 20005

National Association of Water Companies
1725 K Street, N.W., Suite 1212
Washington, DC 20006

National Center for Ground Water Research
(Consortium of Three Universities)
University of Oklahoma
Norman, OK 73019

Oklahoma State University
Stillwater, OK 74074

Rice University
Houston, TX 77005

National Water Alliance
50 E Street
Washington, DC 20003

National Water Resources Association
955 L'Enfant Plaza S.W., Suite 1202
Washington, DC 20024

Table 3.4, continued

Robert S. Kerr Environmental Research Laboratory
U.S. Environmental Protection Agency
P.O. Box 1198
Ada, OK 74820

Universities Council on Water Resources
310 Agricultural Hall
University of Nebraska
Lincoln, NE 68583-0711

Water Pollution Control Federation
2626 Pennsylvania Avenue
Washington, DC 20037

Water Resources Congress
3800 N. Fairfax Drive, Suite 7
Arlington, VA 22203

Water Supply Improvement Association
P.O. Box 387
375 Boston Road
Topsfield, MA 01983

Wisconsin Ground Water Association
P.O. Box 263
Antigo, WI 54409

SELECTED REFERENCES

1. Giefer, G. and Todd, D. *Water Publications of State Agencies*, 1976, Water Information Center, Inc., Syosset, New York.
2. Kashef, A. I. *Groundwater Engineering*, 1986, McGraw-Hill Book Company, Inc., New York, New York.
3. van der Leeden, F. *Ground Water: A Selected Bibliography*, 1983, Water Information Center, Inc., Syosset, New York.
4. Graber, C., Ed. *Encyclopedia of Associations*, 20th edition, 1986, Gale Research Company, Detroit, Michigan.

CHAPTER 4

Ground Water Pollution Sources

The sources of existing or potential contamination to ground water resources are numerous and of a wide variety. Grouping of the various sources becomes particularly difficult due to the many variables associated with the sources. One very broad grouping of the sources could be naturally occurring vs anthropogenic sources. A very specific grouping would be sources actually designed to discharge to the subsurface such as septic tank systems or injection wells. This chapter will first address the sources of ground water contamination as a whole in order to develop a perspective of the national scope of the threat to ground water. The remainder of the chapter will focus on several specific sources and the potential threat they individually pose to ground water resources.

SOURCE GROUPINGS

In 1984, the Office of Technology Assessment (OTA) of the U.S. Congress, completed a study that was designed to assess the current status of the nation's knowledge about and experience in dealing with ground water contamination problems.[1] As a result of that study, OTA identified 33 sources known to have contaminated ground water and categorized them based on the nature of their release characteristics. This list and categorization is shown in Table 4.1.

Several points need to be made relative to Table 4.1:

(1) Both wastes and nonwastes are potential sources of ground water contamination.

(2) Only a few sources are designed to actually discharge substances to the subsurface.

(3) Release of nonwastes to the subsurface is generally unplanned or results from outside influence.

Table 4.1. Sources of Ground Water Contamination.[1]

Category I—Sources designed to discharge substances
Subsurface percolation (e.g., septic tanks and cesspools)
Injection wells
  Hazardous waste
  Non-hazardous waste (e.g., brine disposal and drainage)
  Non-waste (e.g., enhanced recovery, artificial recharge, solution mining, and in-situ mining)
Land application
  Wastewater (e.g., spray irrigation)
  Wastewater byproducts (e.g., sludge)
  Hazardous waste
  Non-hazardous waste

Category II—Sources designed to store, treat, and/or dispose of substances; discharge through unplanned release
Landfills
  Industrial hazardous waste
  Industrial non-hazardous waste
  Municipal sanitary
Open dumps, including illegal dumping (waste)
Residential (or local) disposal (waste)
Surface impoundments
  Hazardous waste
  Non-hazardous waste
Waste tailings
Waste piles
  Hazardous waste
  Non-hazardous waste
Materials stockpiles (non-waste)
Graveyards
Animal burial
Aboveground storage tanks
  Hazardous waste
  Non-hazardous waste
  Non-waste
Underground storage tanks
  Hazardous waste
  Non-hazardous waste
  Non-waste
Containers
  Hazardous waste
  Non-hazardous waste
  Non-waste
Open burning and detonation sites
Radioactive disposal sites

Category III—Sources designed to retain substances during transport or transmission
Pipelines
  Hazardous waste

Table 4.1, continued

Non-hazardous waste
Non-waste
Materials transport and transfer operations
Hazardous waste
Non-hazardous waste
Non-waste
Category IV—Sources discharging substances as consequence of other planned activities
Irrigation practices (e.g., return flow)
Pesticide applications
Fertilizer applications
Animal feeding operations
De-icing salts applications
Urban runoff
Percolation of atmospheric pollutants
Mining and mine drainage
Surface mine-related
Underground mine-related
Category V—Sources providing conduit or inducing discharge through altered flow patterns
Production wells
Oil (and gas) wells
Geothermal and heat recovery wells
Water supply wells
Other wells (non-waste)
Monitoring wells
Exploration wells
Construction excavation
Category VI—Naturally occurring sources whose discharge is created and/or exacerbated by human activity
Ground water—surface water interactions
Natural leaching
Salt-water intrusion/brackish water upconing (or intrusion of other poor-quality natural water)

Several factors influence the potential for a particular source or source grouping to contaminate ground water formations. These factors include (see Table 4.2):[1]

- design, operation, and maintenance characteristics
- release characteristics
- geographic location (pervasiveness and regionality)
- number of sources and amounts of material flowing through or stored in sources
- hydrogeology

Table 4.2. Summary of Source Characteristics.[a1]

	Individual facility/activity			Aggregate of facilities/activities			
	Purpose[a]	Spatial release pattern[b]	Temporal release pattern[c]	Pervasiveness[d]	Diversity of known contaminants[e]	Numbers[f]	Amounts of material[g]
Category I							
Subsurface percolation	W	P^h	Y^i	R	High	High	High
Injection wells	W/NW	P	Y^i,S	R	Moderate	High	High
Land application	W	D,P	S	R	Moderate	Moderate	Low
Category II							
Landfills	W	P^h	S^i	W	High	High	Moderate (High?)
Open dumps	W	P^h	S^i	W	High	Moderate	Moderate
Residential disposal	W	P^h	S^i	W	High	?	?
Surface impoundments	W	P^h	S^i	W	High	High	High
Waste tailings	W	P^h	S^i	R	Moderate	?	High
Waste piles	W	P^h	S^i	R	Moderate	?	High
Materials stockpiles	NW	P^h	S^i	W	Low	?	High
Graveyards	W	P^h	S^i	W	Moderate	?	?
Animal burial	W	P^h	S^i	L	Low	?	?
Aboveground storage tanks	W/NW	P^h	R	W	Low	?	?
Underground storage tanks	W/NW	P^h	R	W	Moderate	High	Moderate
Containers	W/NW	P^h	R	W	Low	Moderate (?)	Moderate
Open burning and detonation sites	W	P	S	L	Low	Low (?)	Low
Radioactive disposal sites	W	P	Y,S,R^i	L	Low	Low	Low
Category III							
Pipelines	W/NW	P^h,F	R	W	Low	Moderate	High
Materials transport and transfer operations	W/NW	P^h,F	R	W	Moderate	Moderate	Moderate

Category IV							
Irrigation practices	NW	D	S[i]	R	Low	Moderate	Moderate
Pesticide applications	NW	D	S[i]	W,R	Low	High	Low
Fertilizer applications	NW	D	S	W,R	Moderate	High	Moderate
Animal feeding operations	W	P[h]	Y	W	Low	Moderate	Low
De-icing salts applications	NW	F	S	R	Low	?	Moderate
Urban runoff	W	P,D,F	S	W	Moderate	Moderate	?
Percolation of atmospheric pollutants	W	D	S	W	Low	?	?
Mining and mine drainage	W	P,D,F	S[i]	R	Moderate	High	Low (?)
Category V							
Production wells	NW	P	Y[i]	R	Moderate	High	Moderate
Other wells	NW	P	Y[i]	W	Low	?	?
Construction excavation	W	P,D,F	S	W	Low	?	Moderate
Category VI							
Ground water-surface water interactions	W	F	S	W	Low	NA	?
Natural leaching	NW	D,F	Y,S	L	Moderate	NA	?
Salt-water intrusion	NW	D,F	S	R	Moderate	NA	?

[a]Purpose: W = waste; NW = non-waste.
[b]Spatial Release Pattern: P = point; D = diffuse; F = frontal.
[c]Temporal Release Pattern: Y = year-round; S = seasonal; R = random.
[d]Pervasiveness: W = widespread; R = regional; L = local.
[e]Diversity of known contaminants: Low = 1–2 associated classes; Moderate = 3–5 associated classes; High = 6–8 associated classes.
[f]Number: Low = <1,000 facilities; Moderate = 1,000–25,000 facilities, 10–20,000 spills, 100,000–1,000,000 miles, or 10–100 million acres; High = 725,000 facilities or >100 million acres; ? = unable to obtain sufficient information; NA = not applicable.
[g]Amounts: Low = <10 billion gallons, <10 million tons, <10 million cubic yards, <1 billion barrels, or <100 million acres; Moderate = 10–250 billion gallons, 10–250 million tons, 10–100 million cubic yards, 1–10 billion barrels, or 10–250 million acres; High = >250 billion gallons, >250 million tons, or >10 billion barrels: ? = unable to obtain sufficient information.
[h]Point sources, but typical dense concentration leads to a diffuse problem (individual sources are not traceable).
[i]Release characteristics are also a function of age.

Design, operation, and maintenance of a source can influence the source's ground water contamination potential, usually through faulty procedures. In other words, very few sources are designed, operated, or maintained in such a manner as to contaminate ground water formations.

Release characteristics cover the geometry, temporal situation, and age of a source. Geometrical characteristics are usually divided into point and nonpoint or diffuse sources. The temporal release characteristics center on whether a source releases year-round or tends to be influenced by seasonal weather patterns. The age of a given source can affect its release characteristics due to deterioration.

The geographic location of a source can affect its contamination potential. Some sources tend to be nationwide such as landfills, while others tend to be localized such as saltwater intrusion in coastal areas. Some sources correlate with heavily populated areas such as impoundments, and some correlate with economic activities such as fertilizers and pesticides in the west and midwest regions of the country.

NUMBER OF SOURCES

Attempts have been made to estimate the magnitude of the ground water contamination problem in the United States.[2] Listed in Table 4.3 are current estimates of the number of sources and the amounts of material flowing through or stored in these sources.

At least four limitations to Table 4.3 can be cited:[1]

(1) The estimates are specifically for the amounts of material flowing through or stored in the source and are not estimates of the amounts of material actually reaching the ground water (unless otherwise indicated). Thus the estimates suggest only the maximum potential for ground water contamination.

(2) An estimate of the amount reveals nothing about the nature and concentration of substances in that material. Industrial and municipal sludge provides an example. The amount of industrial sludge used in land applications is roughly 7% of that used from municipal systems, yet often the chemical compounds or their concentrations in industrial sludge (e.g., inorganic acids and higher concentrations of hydrocarbons) pose greater health threats than the chemical compounds found in municipal sludge.

(3) Accuracy of the quantitative estimates varies considerably from source to source, depending on the underlying assumptions and completeness of the data. The OTA study has attempted to address this problem by indicating the range of values within which the true value

Table 4.3. Numbers of Sources and Amounts of Material Flowing Through or Stored in Sources.[a1]

Source	OTA Update			1977 Report	
	Approximate number of facilities	Approximate amount of material[b]	Possible uncertainty in number estimate[c]	Possible uncertainty in amount estimate[c]	Approximate amount of material
Category I					
Subsurface percolation					
Domestic	16.6–19.5 million	820–1,460 bgy	<2x	<2x	800 bgy
Industrial	25,000	1–2 bgy	>10x	>10x	1.2 bgy
Injection wells					
Hazardous waste	87	8.6 bgy[d]	<10x	<10x	—
Drainage, etc.	350,000	?	<10x	?	—
Brine	140,000 (Brine and Non-waste (enhanced oil recovery))	525 bgy	<10x	<10x	460 bgy
Non-waste (enhanced oil recovery)		24.5 bgy	?	<10x	—
Non-waste (solution, in-situ)	12,000	?	?	<10x	0.3 mt
Land application					
Municipal sludge	2,500	3–4 mty (dry)	<10x	<10x	4 mty
Industrial hazardous waste	70	0.10 bgy[d]	<10x	<10x	—
Spray irrigation	485	?	>10x	?	—
Category II					
Landfills					
Industrial hazardous waste	199	0.81 bgy[d]	<10x	>10x	50 bgy
Industrial non-hazardous waste	75,700	40–140 mty (wet)	<10x	>10x	—
Utility	?	30 mty (wet)	?	>10x	—
Municipal	15–20,000	138 mty	<2x	<2x	90 bgy
Open dumps	2,400	10 bgy	>10x	>10x	—

Table 4.3, continued

Source	OTA Update			1977 Report	
	Approximate number of facilities	Approximate amount of material[b]	Possible uncertainty in number estimate[c]	Possible uncertainty in amount estimate[c]	Approximate amount of material
Category II, continued					
Residential disposal sites	?	?	?	?	—
Surface impoundments					
Hazardous waste	1,078	35.8 bgy[d]	<10x	<10x	—
Non-hazardous waste	180,000	1,800 bgy[e]	<2x	<10x	161 bgy
Waste tailings	?	580 mty	?	<2x	—
Waste piles					
Hazardous waste	174	0.4 bgy	>10x	>10x	—
Non-hazardous waste	?	1,730 mty	?	<2x	—
Materials stockpiles	?	700 mty	?	<10x	—
Graveyards	?	?	?	?	—
Animal burial	?	?	?	?	—
Aboveground storage tanks	?	?	?	?	—
Underground storage tanks					
Hazardous waste	2,031	13.8 bgy	<10x	<10x	—
Non-hazardous waste	2.5 million	25 bg	<2x	<10x	—
Non-waste	?	?	?	?	—
Containers					
Hazardous waste	3,577	0.16 bgy[d]	>10x	>10x	—
Non-hazardous waste	?	?	?	?	—
Non-waste	?	?	?	?	—
Open burning and detonation sites	?	?	?	?	—
Radioactive disposal sites	31[f]	3.7 million cubic yards	<2x	<2x	—

Category III					
Pipelines					
Hazardous waste	?	?	?	?	—
Non-hazardous waste	700,000 miles	280 bgy[e]	<10x	>10x	250 bgy
Non-waste	175,000 miles	10 billion barrels	?	?	—
Materials transport and transfer operations					
Hazardous waste	16,000 spills (hazardous and non-hazardous waste)	14 mty	<10x	>10x	—
Non-hazardous waste					
Non-waste	?	?	?	?	—
Category IV					
Irrigation practices	50–60 million acres	169 million acre-feet	<2x	<2x	—
Pesticide applications	280 million acre-treatments	0.26 mty active	<2x	<2x	—
Fertilizer applications	229 million acre-treatments	42 mty	<2x	<2x	—
Animal feeding operations	1,935	8 mty	<2x	<10x	—
De-icing salts applications	?	10–12 mty	?	<2x	—
Urban runoff	21.2–32.6 million acres	?	<2x	?	—
Percolation of atmospheric pollutants	NA	?	?	?	—
Mining and mine drainage					
Surface	15,000 active	4 million acres; 0.36–1.0 mty acid (surface and underground)	<10x	<10x	108 billion gallons
Underground	67,000 inactive				

Table 4.3, continued

Source	OTA Update			1977 Report	
	Approximate number of facilities	Approximate amount of material[b]	Possible uncertainty in number estimate[c]	Possible uncertainty in amount estimate[c]	Approximate amount of material
Category V					
Production wells					
Oil wells	548,000 activity 2 million abandoned	g	<10x	<10x	—
Geothermal, heat recovery	32	?	?	?	—
Water supply	350,000	?	?	?	—
Other wells (non-waste)					
Monitoring	?	?	?	?	—
Exploration	?	?	?	?	—
Construction excavation	?	45 mty	?	>10x	—
Category VI					
Ground water-surface water interactions	NA	?	NA	?	—
Natural leaching	NA	?	NA	?	—
Salt-water intrusion	NA	?	NA	?	—

[a]? = OTA unable to obtain sufficient information to develop estimate.
— = No estimate presented in 1977 report (EPA, 1977).
NA = Not applicable.
[b]mty = million tons per year.
bgy = billion gallons per year.
bg = billion gallons.

[c]Confidence in estimates is defined as follows:
<2x = estimate considered correct within 100%.
<10x = estimate considered correct within one order of magnitude.
>10x = estimate *could* be incorrect by more than one order of magnitude.
[d]Note that this figure refers to hazardous wastes regulated under RCRA.
[e]Estimate of actual amount of leachate.
[f]Excludes nuclear reactors.

probably falls, but even this approach is arbitrary. It is important to remember that there is a high degree of uncertainty underlying the estimates and that they are best used to indicate the most numerous and most material-intensive sources.

(4) Comparing estimates is difficult because they are expressed in different units of measurement. The units cannot be converted into a common base unit; thus only simple categorizations of large vs small numbers or amounts can be made.

Despite the above limitations, Table 4.3 can be quite informative. First, the table identifies those sources which are numerous or have large amounts of materials, thus posing a significant threat through sheer numbers. Second, the table points out that some sources associated with nonwastes and nonhazardous wastes are at least as important, in terms of numbers, as those sources that have received much notoriety, such as the hazardous waste sources.

A final outcome of the OTA study was a listing of the important sources of ground water contamination (Table 4.4). It should be noted that Table 4.4 was developed on a set of specific criteria and that different lists could be developed based on different criteria. However, it could be expected that a majority of the sources in Table 4.4 would be found in any listing of "important" ground water contamination sources.

POINT SOURCES

Presented below is an overview discussion of seven point sources of ground water contamination. The sources are called point sources because they are confined or generally release contaminants from a single, isolated geographical area. The sources are not arranged in any particular order and are not meant to show any potential significance over other sources. These sources simply represent those for which substantial overview information is available.

Septic Tank Systems

Septic tank systems were introduced in the United States in 1884 and have since become the most widely used method of onsite sewage disposal. Estimates on the number of septic tanks in the United States range up to 20 million with more than 70 million people utilizing septic tanks. Approximately one third of all existing housing units and about 25% of all new homes being constructed are using septic tanks. Intensive septic tank usage

Table 4.4. "Important" Sources of Ground Water Contamination Based on Selected Sets of Criteria.[a1]

Sources/criteria[b]	1 H diversity	2 H amounts	3 H numbers	4 H numbers M–H diversity	5 M–H numbers M–H amounts H diversity	6 M–H numbers M–H amounts M–H diversity Widespread	7 M–H numbers M–H amounts M–H diversity Widespread or Regional	8 Same as 7 but H diversity only	9 Same as 7 but at least one H ranking	10 Toxicity
Subsurface percolation	X	X	X	X	X		X	X	X	X
Injection wells		X	X	X			X		X	X
Land application										X
Landfills	X	X	X	X	X	X	X	X	X	X
Open dumps	X				X	X	X	X	X	X
Residential disposal	X									X
Surface impoundments	X	X	X	X	X	X	X	X	X	X
Waste tailings		X								
Waste piles		X								X
Materials stockpiles		X								
Aboveground storage tanks										X
Underground storage tanks			X	X		X	X		X	X
Containers										X
Pipelines		X								X
Materials transport						X	X			X
Pesticides applications			X							X
Fertilizer applications			X	X		X	X		X	
Mining			X	X						
Production wells (oil)			X	X			X		X	

[a]Abbreviations: H = High.
M–H = Moderate to High.

[b]Other sources listed in Table 4.1 do not meet any of the selected criteria.

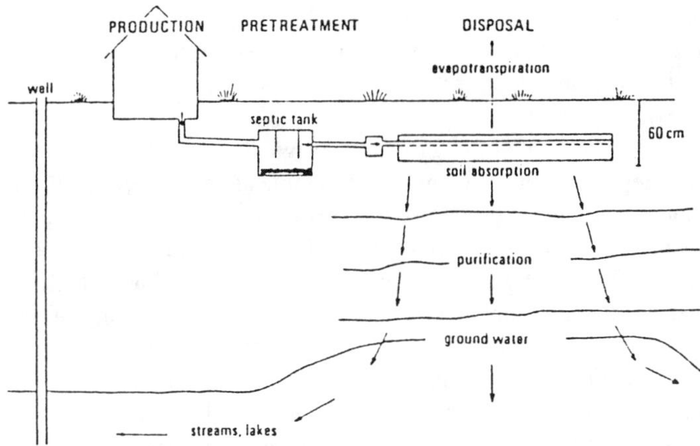

Figure 4.1. Schematic cross-section through a conventional septic tank soil disposal system for onsite disposal and treatment of domestic liquid waste.[3]

occurs in the east and southeast as well as the northern tier and northwest portions of the United States. Early estimates of wastewater discharged by septic tanks to the subsurface were about 800 billion gallons annually;[2] however, recent updates of this figure range up to 1,460 billion gallons annually.[1]

The basic septic tank system consists of a buried tank and a subsurface soil absorption system as depicted in Figure 4.1. Wastewater is directed first to the buried tank where scum, grease, and settleable solids are removed from the liquid by gravity separation. The clarified effluent then proceeds to the subsurface drain system where it percolates into the soil. The performance of the system depends on the system design, construction techniques, waste characteristics, hydraulic loading, climate, hydrogeology and soil composition, and proper maintenance scheduling.

The effluent from the tank portion of the system that is allowed to percolate into the soil represents a potential source of ground water contamination. This effluent will be typical of domestic wastewater, unless the system has received industrial wastes. Canter and Knox (1985) present the following physical and chemical parameter concentrations as being representative of typical septic tank effluent: suspended solids—75 mg/l; BOD_5—140 mg/l; COD—300 mg/l; total nitrogen—40 mg/l; and total phosphorus—15 mg/l. Based on reported efficiencies of soil absorption systems, Canter and Knox[4] report the following typical concentrations entering the ground water: suspended solids—18 to 53 mg/l; BOD—28 to 84 mg/l; COD—57 to 142 mg/l; ammonia nitrogen—10 to 78 mg/l; and total phosphates—6 to 9 mg/l.

Other wastewater constituents of concern include bacteria, viruses, nitrates, synthetic organics, metals (lead, tin, zinc, copper, iron, cadmium, and arsenic), and inorganics (sodium, chlorides, potassium, calcium, magnesium, and sulfates).

Septic tank systems that are properly designed, constructed, maintained, and located represent an efficient and economical sewage disposal alternative without threatening ground water resources. However, in many instances poor system design, improper construction and maintenance, and bad system locations have led to ground water pollution. Another concern in many locations is septic tank density exceeding the natural ability of the subsurface environment to absorb and purify system effluents. Additionally, most of the septic tank systems installed during the rapid growth of the 1960s are now exceeding their design life, usually 10 to 15 years. Hence, deterioration of these systems represents a threat to underlying ground water formations. A recent development regarding septic tanks is the use of synthetic organic chemical degreasers (such as trichloroethylene) and tank cleaners. The use of these chemicals has resulted in organic contamination of ground water.[5]

Excessive septic tank densities in many areas has degraded ground water quality with high concentrations of nitrates, bacteria, and organic contaminants. One reason for this degradation is pollutants moving too rapidly through soils. Soils with high permeabilities can be overloaded resulting in rapid downward migration of the organic and inorganic chemicals and microorganisms. These high seepage rates do not allow the soil's physical, chemical, and biological removal mechanisms to operate on the percolating effluent.

The transport and fate of pollutants from the soil absorption system through underlying soils to ground water formations is an important consideration relative to the pollution potential of septic tank systems. The transport and fate issues must be addressed for the biological contaminants (bacteria and viruses), inorganic contaminants (phosphorus, nitrogen, and metals), and organic contaminants (synthetic organics and pesticides).

Biological contaminants show wide ranges in size, shape, surface properties, and die-away rates. Bacteria, the larger of the microorganisms, tend to be removed from percolating water through the physical process of straining and/or the chemical process of adsorption. The most important removal mechanism for the smaller viruses is adsorption onto soil particles.

Phosphorus movement through soils is limited due to chemical changes and adsorption. Ammonia-nitrogen is primarily removed by adsorption but can also be subject to cation exchange or incorporation into microbial biomass. Nitrate-nitrogen tends to be highly mobile, moving essentially with the ground water. However, nitrates can be utilized by plants or crops, or subjected to microbiological denitrification.

Metals in soils can be rendered immobile through adsorption, ion exchange, chemical precipitation, or complexation with organic substances. Adsorption seems to be the major fixation process for metals. All four of the fixation processes can be influenced by the soil composition and characteristics such as pH and redox potential.

The transport and fate of organic contaminants in the subsurface could be influenced by volatilization of the organics, retention due to adsorption, incorporation into plant or microbial biomass, and bacterial degradation. The relative importance of these possibilities in a given situation is dependent upon the characteristics of the organic, the soil types and characteristics, and the subsurface environmental conditions.[4]

Underground Storage Tanks

A source of potential ground water contamination that has seen a meteoric rise in level of concern is leaking underground storage tanks (USTs). USTs became such an important concern that the U.S. Environmental Protection Agency (EPA) created the Office of Underground Storage Tanks (OUST) in September 1985. The OUST is a separate program office within the Office of Solid Waste and Emergency Response and was established in response to the 1984 Hazardous and Solid Waste Amendments to the Resource Conservation and Recovery Act. These amendments require EPA to promulgate regulations to prevent, detect, and correct the leaking of USTs.

One major category of USTs includes those that contain petroleum hydrocarbons. The number of gasoline storage tanks in the United States is estimated to be between 1.5 and 2 million, not including abandoned tanks or tanks used to store other liquids. Tejada[6] reported about 1.2 million tanks in use are made of steel and only about 16,000 of these have corrosion protection. Predpall, Rogers, and Lamont[7] estimate that as much as 23% of all tanks leak; and Robbins and Nicholas[8] estimate the actual number of leaking tanks between 75,000 and 100,000. The Office of Technology Assessment[1] has put estimates on the number of hazardous and nonhazardous underground storage tanks at 2031 and 2.5 million, respectively.

The main cause of leakage of steel USTs is corrosion. The American Petroleum Institute[9] found that corrosion was the cause of leakage in 92% of steel tanks and 64% of steel pipes. Conversely, Tejada[6] reports that breakage is the major cause of leaks in fiberglass tanks.

Buried gasoline tanks and service stations are the most common source of leaks investigated. Listed in Table 4.5 is summary information regarding leakage around the country. From this table it can be noted that service stations are the dominant source throughout the country.

Table 4.5. The Extent, Amount, and Nature of Pollution from Gasoline in a Few Example Incidents.

Place	Year	Identified By	Comments
Michigan[1]	1982	Assessment of Ground Water Contamination	Among 897 known and suspected cases of ground water contamination, 100 were caused by leaking underground storage tanks.
California, Death Valley National Monument[11]	1982	U.S. Geological Survey	A leak in a service station storage tank, probably totaling more than 19,000 gallons, caused the formation of a gasoline layer overlying the water table, creating the potential for ground water contamination.
Pennsylvania[1]	1982	Water Quality Inventory	Out of 249 cases of ground water contamination by toxic materials, 75% were caused by gasoline and finished petroleum products. The majority of these cases involved leaking underground storage tanks.
Vermont[1]	1982	Congressional Research Service (CRS)	Survey identified leaking underground gasoline and fuel oil storage tanks and pipelines as second leading cause of ground water contamination.
Tennessee[1]	1981	Profile of Existing Ground Water Problem	Gasoline leaks from underground storage tanks and pipelines were a common problem.
Connecticut[1]	1981–1982	Annual Oil and Chemical Spill Summary for FY 1981–1982	Identified 45 cases of ground water contamination by gasoline fuel oil, waste oil, or kerosene. Almost all of them caused by leaks from in-ground storage tanks and pipelines.
New Jersey[10]	1978	—	More than 1,400 chemical spills and leaks were detected involving 1.1 million gallons of petroleum compounds spilled. The number of incidents is increasing to as much as 2,000 per year.
New Mexico[1]	—	—	Cases of ground water contamination by gasoline leaking from tanks totaled up to 28.
Maine[1]	—	—	A 10,000-gallon leak has rendered one quarter of the town's supply undrinkable.
Wyoming[1]	—	—	Sixteen out of 40 homes in subdivision have contaminated water.
Colorado	—	—	About 30,000 gallons of gasoline were lost over a 3- or 4-year period before leak was discovered.

Petroleum compounds in USTs are composed of hydrogen and carbon atoms, hence the term hydrocarbons. Bruell and Hoag[12] found more than 70 separate hydrocarbon compounds in regular gasoline using gas chromatography and mass spectrometry. Included in gasoline are the aliphatic hydrocarbons and the aromatic hydrocarbons. The aliphatic compounds, such as pentane and butane, are characterized by carbon atoms linked in an open chain. The aromatic compounds, such as benzene and toluene, are characterized by a ring structure of carbon atoms.

Gasoline that has seeped down to a ground water formation will tend to float on top of the water table. However, this free product is just one phase in which gasoline can exist in the subsurface. Volatile components can exist in a gaseous phase and actually escape as fumes or odors. Other components, such as benzene, toluene, and various xylenes (or BTX) can attach to soil and exist in an adsorbed phase. Finally, some compounds can exist in the soluble phase within the ground water.

BTXs are of concern due to their possible adverse health effects. Minor constituents which also could be of concern include ethylene dibromide (EDB) and ethylene dichloride.[13] Bruell and Hoag[13] identify other health affecting components of gasoline to be ethylbenzene, naphthalene, ethers, and alcohols.

Table 4.6 shows that benzene is classified as a human carcinogen by the International Agency for Research on Cancer (IARC) and as a hazardous substance, a hazardous waste, and a priority pollutant by the U.S. Environmental Protection Agency. Many other petroleum hydrocarbons including toluene, xylene, EDB, and ethylene dichloride are also classified either as priority pollutants or hazardous substances by the U.S. Environmental Protection Agency. However, these compounds are yet to be tested for human carcinogenity.

The U.S. Environmental Protection Agency has set up Suggested No Adverse Response Levels (SNARLS) advising health effects concerning unregulated contaminants found in drinking water supplies. Permissible concentrations in water to protect human health is zero in the case of benzene, with consumption of 6.6 micrograms per liter (parts per billion) causing an additional lifetime cancer risk of 1 in 100,000. In other words, an estimated 1 person in 100,000 will get cancer from ingesting that concentration over a lifetime. Permissible concentrations of toluene in water to protect human health is 14.3 milligrams per liter (parts per million). For xylenes, however, permissible concentrations to protect human health have not been set, but the U.S. Environmental Protection Agency suggests a permissible ambient goal of 6,000 micrograms per liter (parts per billion) based on health effects.

Benzene, toluene, and xylenes are proven to systematically affect the blood, central nervous system (CNS), skin and bone marrow, and through

Table 4.6. The Extent of Toxic Organic Chemicals Found in Drinking Water Wells.[14]

Chemical	Concentration (ppb)	State	Highest Surface Water Concentration Reported (ppb)	Evidence of Carcinogenicity
Benzene	330	New Jersey		
	230	New York		
	70	Connecticut		
	30	New York	4.4	H[a]
Toluene	6,400	New Jersey		
	260	New Jersey		
	55	New Jersey	6.1	NA[b]
Xylene	300	New Jersey		
	69	New York	24	NTA[c]
Ethylene dibromide (EDB)	300	Hawaii		
	100	Hawaii		
	35	California	NT[d]	NTA[c]
1,1-Dichloroethylene	280	New Jersey		
	118	Massachusetts		
	70	Maine	0.5	NTA[c]
1,2-Dichloroethylene	323	Massachusetts		
	294	Massachusetts		
	91	New York	9.8	NTA[c]

[a]H—Confirmed human carcinogen.
[b]NA—Negative evidence of carcinogenicity from animal bioassay.
[c]NTA—Not tested in animal bioassay.
[d]NT—Not investigated.

vapor exposure can affect the eyes and the respiratory system. Symptoms from ingestion range from headaches, dizziness, and nausea to convulsions, coma, and death from extreme exposure or ingestion. Chronic exposure can lead to hypo- or hyperactive bone marrow and possibly leukemia.

Hazardous Waste Sites

Hazardous waste sites appear in a variety of forms ranging from waste piles to impoundments to landfills. The figures in Table 4.3 show the number of facilities containing hazardous waste to be above 7,000. This number goes well over 20,000 if spills and pipeline breaks are also included.

The one type of hazardous waste site that has received much notoriety and increased attention is the uncontrolled hazardous waste site. These types of sites are addressed under the provisions of the Comprehensive Environmental Response, Compensation and Liability Act (CERCLA) of 1980. The act, nicknamed Superfund, provides for " . . . liability, compensation, cleanup, and emergency response for hazardous substances released into the environment and the cleanup of inactive hazardous waste disposal sites."[15] The 1980 Superfund Act set up a $1.6 billion fund to pay for cleanup of abandoned hazardous waste sites and spills of toxic chemicals for a five-year period. The funds were developed through taxes on petroleum, certain chemical substances and federal appropriations. In 1986, President Reagan signed a $9 billion extension of the Superfund program.

The staggering number of abandoned waste sites in the United States is attributable to growth of the synthetic chemical industry (Figure 4.2). In 1940, approximately one billion pounds of synthetic organic chemicals were produced. By 1965, this quantity had increased by two orders of magnitude and currently exceeds 300 billion pounds per year. The exponential growth of the industry has led to the massive problem of toxic waste disposal.[17]

The major concern with hazardous and toxic waste disposal is the possibility of human contact and subsequent health effects. Very real examples have surfaced over the past decade. Shown in Table 4.7 is a list of recent situations involving potential human exposure to hazardous or toxic waste materials. In each case, the threat posed by the toxic material is due, in most part, to careless handling or improper disposal of the material.

Probably the best means of depicting the threat that hazardous waste sites pose to ground water formations is through consideration of a case study. The case study presented below is one of the first Superfund sites to be addressed to the point that remedial actions have been implemented.

The Gilson Road hazardous waste dump site is located in Nashua, New Hampshire, near the New Hampshire-Massachusetts border. The six-acre disposal site is located within one quarter mile of several family residences

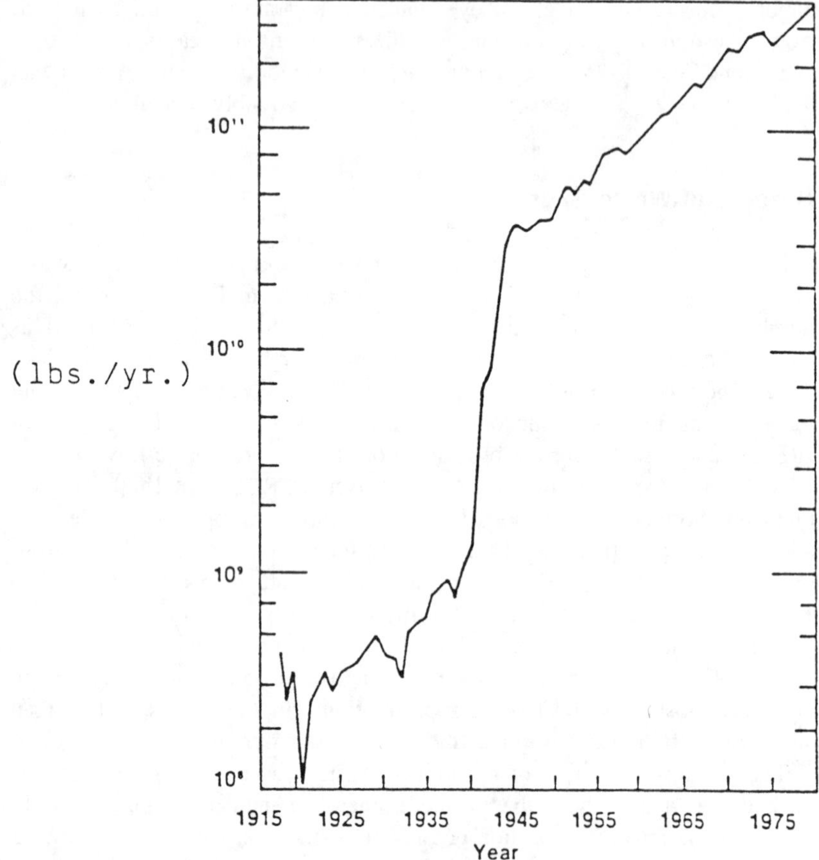

Figure 4.2. Synthetic organic chemical production.[16]

and adjacent to a small stream named Lyle Reed Brook. The site was originally developed as a sand borrow pit. The extensive sand removal operations extended many times into the underlying ground water formation at the site.

At an unknown time in the late 1960s, the owner discontinued the sand mining operations and began operating an unapproved, illegal refuse dump. Initially, the site received only household refuse and demolition materials. Eventually, the site began receiving chemical sludges and aqueous chemical wastes.

The received wastes were handled in a generally haphazard fashion including dumping of wastes into a makeshift leaching field and allowing them to percolate into the ground. The total amount of waste materials disposed

Table 4.7. Recent Toxic Waste Contamination.[16]

Location of Site	Toxic Materials	Physical Conditions	Principal Routes of Potential Human Exposure
Love Canal dump, Niagara Falls, New York	Largely hydrocarbon residue from pesticide production	Inactive landfill in residential area	Direct, airborne, and waterborne contacts
Melvin Wade dump, Chester, Pennsylvania	Diverse organic chemicals	Surface collection of waste in drums in urban setting	Direct contact, explosion, and fire
Woburn, Massachusetts	Arsenic compounds, heavy metals, organic chemicals	Abandoned waste lagoon with multiple surface dumps dumps	Direct and waterborne contacts
Triana, Alabama	DDT and related compounds	Industrial waste dumped in a rural stream	Food chain (fish)
Bloomington, Indiana	Polychlorinated biphenyls (PCBs)	Industrial waste contaminating municipal sewage used for garden manure	Direct contact and possibly food chain
Tristate Mining District, Oklahoma, Kansas, and Missouri	Heavy metals, acidic aquifer conditions	Mine tailings and acidic aquifer recharge	Airborne and irreparably contaminated aquifer
Montgomery County, Alabama	Trichlorethylene	Industrial waste contaminating aquifer; underground storage tank rupture	Direct and waterborne contacts
Pittston, Pennsylvania	Chlorinated solvents, HCN and heavy metals	Millions of gallons of waste dumped into abandoned mine shafts	Food chain and waterborne exposure

of at the site is unknown and not obtainable. However, in the last four months of operation alone, the site received 800,000 gallons of aqueous waste.

Cleanup activities at the site began in 1980 with removal of 1,314 drums. At the end of 1982, a three-foot wide slurry wall had been installed around the 20 acres of contaminated soil and ground water. By 1985, a ground water extraction, treatment, and reinjection system was in operation.

The contaminants found in the ground water at the Gilson Road site are listed in Table 4.8. The treatment system to be employed will utilize chemical precipitation for metals removal, steam stripping to remove volatile organics, and biological treatment for the extractable organics. The effluent from the treatment process will be reinjected to increase the "flushing" rate of the contaminated ground water. The treatment system alone was bid at a total cost of $5,375,000.

Although each Superfund site is unique in terms of the source of contamination, hydrogeology, etc., the above example does show some general aspects that are typical of many sites. Such features as the wide range of contaminants, unknown quantities of material disposed, lack of responsible parties, and multimillion dollar cost figures can be expected at most all sites.

Table 4.8. Primary Contaminants of Concern in Ground Water at the Gilson Road Site.[18]

Heavy Metals	Volatile Organics	Extractable Organics
Arsenic	Tetrahydrofuran	1,4-Dichlorobenzene
Barium	1,1,1-Trichloroethane	1,2-Dichlorobenzene
Cadmium	Benzene	Naphthalene
Chromium	Trichloroethylene	2-Chlorophenol
Lead	Methyl isobutyl ketone	2-Nitrophenol
Mercury	Xylenes	Phenol
Selenium	Toluene	2,4-Dimethylphenol
Silver	Ethylbenzene	O-Cresol
Copper		M-Cresol
Iron		Benzoic acid
Manganese		Pentachlorophenol
Nickel		
Zinc		

The most interesting aspect of the above case study is one which is probably common to all sites. This aspect is subtle in presentation, but devastating in terms of impact. The aspect involves the size of the problem, i.e., although the disposal site consists of six surface acres, the underlying subsurface contamination has grown to encompass 20 acres.

Landfills

Landfills represent a significant threat to ground water resources due to their nature of operation and shear numbers. As noted in Table 4.3, the number of industrial and municipal landfills could exceed 100,000, including several thousand open dumps.

Landfills and other land disposal sites can threaten fresh ground water formations through the production of low-quality leachate. Simplistically, leachate is generated by precipitation (or other moisture sources) percolating through a landfill and removing soluble components of the disposed waste. The chemical composition of landfill leachate will depend on the nature of the refuse (municipal or industrial), on the leaching rate, and on the age of fill.[19] The chemical composition of landfill leachate reported by the American Society of Civil Engineers is shown in Table 4.9. Table 4.9 applies mainly to domestic waste. Vesiland and Rimer[20] suggested an alternative leachate composition as shown in Table 4.10.

Estimating the amount of leachate generated by a landfill centers on an accounting procedure called the Water Balance Method.[21] The calculations involve a one-dimensional analysis of the various moisture components at a landfill as shown in Figure 4.3. The mass balance equation used in the calculations is:

$$C = P(1 - R) - S - E \quad (4.1)$$

where:

C = total percolation into the top soil layer, mm
P = precipitation, mm
R = runoff coefficient
E = evapotranspiration, mm
S = storage within the soil or refuse, mm

In the above equation, some fraction of the precipitation actually percolates into the landfill while the remainder is lost to evapotranspiration, runoff, or storage within the soil or refuse. If the percolation exceeds the evapotranspiration for a sufficient period of time, the field capacities of the soil, and subsequently the refuse, will be exceeded. When the field capacities are exceeded, excess moisture (leachate) is released at the bottom of the refuse layer.

Table 4.9. Chemical Composition (mg/l) of Landfill Leachate.[22]

	Normal Range	Upper Limits
Calcium	240–2330	4080
Magnesium	64–410	15600
Sodium	85–3800	7700
Potassium	28–1700	3770
Iron	0.1–1700	5500
Manganese		1400
Zinc	0.03–135	1000
Nickel	0.01–0.8	
Copper	0.1–9	9.9
Lead		5.0
Chloride	47–2400	2800
Sulfate	20–730	1826
Orthophosphate	0.3–130	472
Total nitrogen	2.6–945	1416
BOD	21700–30300	54610
COD	100–51000	89520
pH (units)	3.7–8.5	8.5
Hardness ($CaCO_3$)	200–7600	22800
Alkalinity ($CaCO_3$)	730–9500	20850
Total residue	1000–45000	
TDS		42276

Table 4.10. Composition of a Typical Leachate.[20]

Component	Typical Value (mg/l)	Range[a]
BOD	20,000	0.01x–2x
COD	30,000	0.01x–3x
Specific conductance	6,000	0.5x–1.5x
Ammonia nitrogen	500	0.01x–1.5x
Chloride	2,000	0.05x–1.5x
Total iron	500	0.5x–5x
Zinc	50	0.5x–5x
Lead	2	0.1x–5x
pH	6.0	0.7x–1.3x

[a]For example, the range of BOD commonly reported is $0.01 \times 20,000 = 200$ mg/l to $2 \times 20,000 = 40,000$ mg/l.

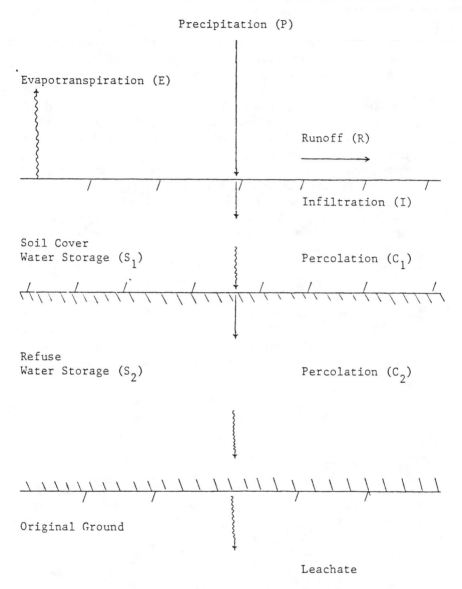

Precipitation (P)

Evapotranspiration (E)

Runoff (R)

Infiltration (I)

Soil Cover
Water Storage (S_1)

Percolation (C_1)

Refuse
Water Storage (S_2)

Percolation (C_2)

Original Ground

Leachate

Figure 4.3. Percolation of water and leachate in a landfill.[21]

In humid areas, where precipitation exceeds evaporation, the amounts of leachate generated can be high. EPA's Office of Solid Waste has estimated that an average land disposal site, seventeen acres in size, with an annual infiltration of ten inches of water, can generate 4.6 million gallons of leachate a year, and can maintain this production for 50 to 100 years.[23]

Bennett, Feates, and Wilder[24] report on a 1977 study of 50 landfill sites in which organic chemical contamination was discovered at forty sites and migration of at least one hazardous chemical was detected at 43 sites. Additionally, 26 of the sites exceeded EPA safe drinking water levels for heavy metals.

Bouwer[22] notes that landfills behave essentially as point sources of pollution. When leachate from a landfill reaches the ground water, it travels in the general direction of the ground water movement. The chloride ion concentration is usually the best indicator for detecting the presence of a leachate in ground water.[25] The contours of a leachate contaminant plume for a thoroughly characterized landfill are shown in Figure 4.4.

Current state of the art for solid waste disposal is a far cry from the indiscriminate dumping of previous years. Sanitary landfills are now the accepted form of solid waste disposal. Brunner and Keller[26] describe sanitary landfilling as an engineered method of disposing of solid wastes on land by spreading, compacting to the smallest practical volume, and covering with soil each working day in a manner that protects the environment.

Methods of ground water protection at sanitary landfills will include surface water control, capping, and liners. In fact, the Hazardous and Solid Waste Amendments of 1984 require that each new landfill (or addition to existing landfills) be constructed with two or more liners, and include leachate collection systems above and between the liners as shown in Figure 4.5.[27]

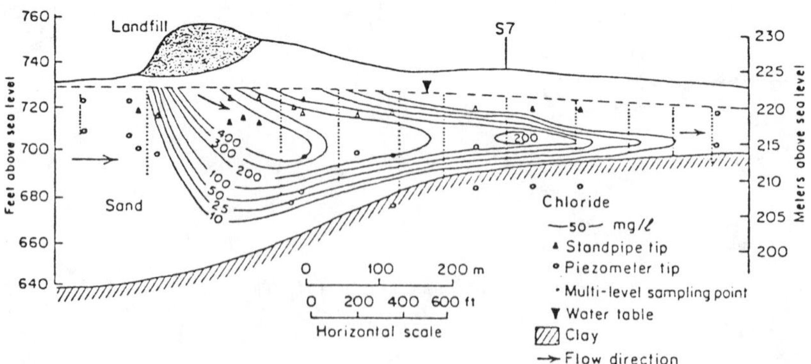

Figure 4.4. Plume of leachate migrating from a sanitary landfill on a sandy aquifer; contaminated zone is represented by contours of Cl^- concentration in ground water.[25]

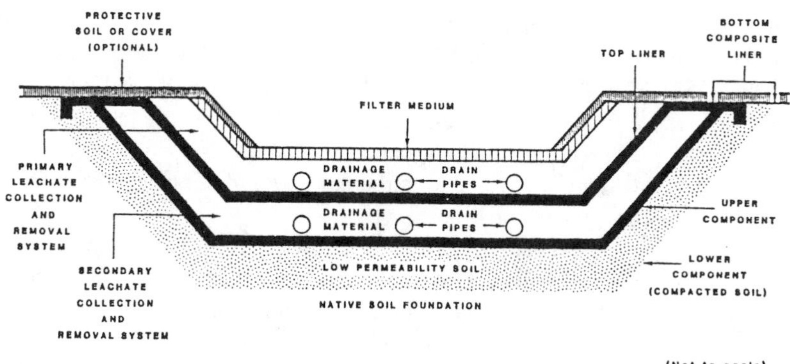

Figure 4.5. Liner leachate collection system for landfills.[27]

Surface Impoundments

Domestic and industrial surface impoundments are often referred to as pits, ponds, lagoons and basins. The terms are used interchangeably and they refer to natural or man-made depressions, lined or unlined, and from a few feet in diameter to hundreds of acres in size.[2]

Municipal impoundments include all of the facilities handling, storing, or treating municipal wastewater. This would include all waste stabilization ponds and lagoons (aerobic, anaerobic, and facultative). The total volume of sewage generated for the United States is approximately 15 billion gallons per day. More than 5,000 of the 22,000 municipal treatment plants in the nation have waste stabilization ponds.[28] The volume of wastewater sewage entering soils from unlined stabilization ponds is quite large. As noted in Table 4.3, the amount of leachate generated from nonhazardous waste impoundments was estimated to be 161 billion gallons per year, but has been updated to a current estimate of 1,800 billion gallons per year.

Surface impoundments are often used for treatment, storage, or disposal of wastewaters from a variety of industries including food and poultry processing; refining, petrochemical and chemical production; and mining and oil and gas production. A 1980 report of a study on industrial impoundments covered 8,163 industrial sites containing 25,779 impoundments. The report estimated that 50 billion gallons of liquid waste are annually placed in those impoundments. Rail[29] reports the following facts related to the study: 18,200 impoundments (70%) are unlined; 9,100 impoundments (35%) may contain hazardous constituents; 7,800 impoundments (30%) are unlined and sitting above ground water sources; and 2,600 impoundments (10%) are unlined and sitting above ground water sources within one mile of potential water supply wells. Additionally, 90% of the impoundments were unmonitored; meaning that no monitoring wells existed for assessing the potential effects of the impoundment on local ground water quality.

The 1984 OTA data summarized in Table 4.3 shows a total of 1,078 hazardous waste impoundments regulated under RCRA. These facilities handle a total of 35.8 billion gallons per year. Estimates on the amount of leachate generated are not available, although it would not be expected to be high for RCRA regulated facilities.

The potential contaminants entering ground water formations from leaking surface impoundments will depend on the type of effluent disposed, whether domestic or industrial. The major pollutants possible from domestic impoundments could include ammonia, nitrates, nitrites, phosphates, and biological parameters such as bacteria and viruses.

The liquids stored in industrial ponds may contain brines, arsenic compounds, heavy metals, acids, gasoline products, phenols, radioactive substances, and a long list of miscellaneous chemicals. Table 4.11 shows some of the significant industrial pollutants posing a threat to ground water quality. Table 4.12 shows a summary of 57 cases of contamination from surface impoundments. Each case involves a separate location where leakage of contaminants from a surface impoundment has entered the ground water.

In a related study, Ghassemi, Haro, and Fargo[30] examined nine industrial waste impoundments (see Table 4.13). The study developed a number of conclusions regarding ground water protection from industrial impoundments; two of the obvious conclusions were: (1) siting in suitable geologic formations is the best defense against ground water contamination; and (2) subdrain leak detectors allow for early warning of potential ground water contamination. A third conclusion drawn from the study was that ingradient design, where the facility is intentionally located in the saturated zone and the high ground water table provides a positive pressure, can prevent migration of leachate or waste in the event of failure.

Abandoned Wells

Abandonment of oil and gas wells usually results when drilling operations are unsuccessful (economically recoverable oil or gas supplies are not found), or the production rate of the well drops too low. Historically, well abandonment and plugging procedures have not been stringent. Abandoned and improperly plugged wells can serve as avenues of access for contaminants to reach drinking water aquifers. Locating abandoned wells represents an area of increasing concern. The Underground Injection Control (UIC) regulations of the Safe Drinking Water Act require that areas around injection wells be surveyed to locate abandoned wells.

Table 4.11. Industrial Wastewater Parameters Having or Indicating Significant Ground Water Contamination Potential.[2]

CHEMICALS AND ALLIED PRODUCTS		
	Pulp and Paper Industry	
COD	Phenols	Nutrients (nitrogen and phosphorus)
TOC	Sulfite	Total dissolved solids
pH	Color	
Ammonia	Heavy metals	
PETROLEUM AND COAL PRODUCTS		
	Petroleum Refining Industry	
Ammonia	Chloride	Nitrogen
Chromium	Color	Odor
COD	Copper	Total phosphorus
pH	Cyanide	Sulfate
Phenols	Iron	TOC
Sulfide	Lead	Turbidity
Total dissolved solids	Mercaptans	Zinc
PRIMARY METALS		
	Steel Industries	
pH	Cyanide	Tin
Chloride	Phenols	Chromium
Sulfate	Iron	Zinc
Ammonia		
CHEMICALS AND ALLIED PRODUCTS		
	Organic Chemicals Industry	
COD	TOC	Phenols
pH	Total phosphorus	Cyanide
Total dissolved solids	Heavy metals	Total nitrogen

Table 4.11, continued

CHEMICALS AND ALLIED PRODUCTS (Continued)		
	Inorganic Chemicals, Alkalies and Chlorine Industry	
Acidity/alkalinity	Chlorinated benzenoids and polynuclear aromatics	Chromium
Total dissolved solids		Lead
Chloride	Phenols	Titanium
Sulfate	Fluoride	Iron
COD	Total phosphorus	Aluminum
TOC	Cyanide	Boron
	Mercury	Arsenic
	Plastic Materials and Synthetics Industry	
COD	Phosphorus	Ammonia
pH	Nitrate	Cyanide
Phenols	Organic nitrogen	Zinc
Total dissolved solids	Chlorinated benzenoids and polynuclear aromatics	Mercaptans
Sulfate		
	Nitrogen Fertilizer Industry	
Ammonia	Sulfate	COD
Chloride	Organic nitrogen compounds	Iron, total
Chromium	Zinc	pH
Total dissolved solids	Calcium	Phosphate
Nitrate		Sodium
	Phosphate Fertilizer Industry	
Calcium	Acidity	Mercury
Dissolved solids	Aluminum	Nitrogen
Fluoride	Arsenic	Sulfate
pH	Iron	Uranium
Phosphorus		

Table 4.12. Origins and Contaminants in 57 Cases of Ground Water Contamination in the Northeast Caused by Leakage of Wastewater from Surface Impoundments.[2]

Type of Industry or Activity	Number of Cases	Principal Contaminant(s) Reported
Chemical	13	Ammonia Barium Chloride Chromium Iron Manganese Mercury Organic chemicals Phenols Solvents Sulfate Zinc
Metal processing and plating	9	Cadmium Chromium Copper Fluoride Nitrate Phenols
Electronics	4	Aluminum Chloride Fluoride Iron Solvent
Laboratories (manufacturing and processing)	4	Arsenic Phenols Radioactive materials Sulfate
Paper	3	Sulfate
Plastics	3	Ammonia Detergent Fluoride
Sewage treatment	3	Detergents Nitrate
Aircraft manufacturing	2	Chromium Sulfate
Food processing	2	Chloride Nitrate
Mining sand and gravel	2	Chloride
Oil well drilling	2	Chloride Oil

Table 4.12, continued

Type of Industry or Activity	Number of Cases	Principal Contaminant(s) Reported
Oil refining	2	Oil
Battery and cable	1	Acid Lead
Electrical utility	1	Iron Manganese
Highway construction	1	Turbidity
Mineral processing	1	Lithium
Paint	1	Chromium
Recycling	1	Copper
Steel	1	Acid Ammonia
Textiles	1	Chloride

The first petroleum-related well in the United States was drilled in 1859, and since then over 2.2 million oil and gas wells have been drilled, with some 30,000 new wells being drilled on an annual basis.[32] Table 4.14 gives oil and gas production figures for the top ten producing states.

In addition to oil and gas wells, there are also millions of municipal and private water wells. There are also numerous waste injection and mineral resource extraction wells.

Table 4.15 outlines the five classes of wells addressed by the UIC program.[33] Although examples of abandonment and improper plugging exist for all classes of wells, the problem is most prevalent for wells associated with oil and gas activities.

An oil or gas well can be considered as having several phases in its life cycle, including: (1) drilling, (2) production, (3) stimulation, (4) brine disposal, and (5) abandonment. All wells would at least include phases (1) and (5); some would include phases (1), (2), and (5); and others can include still other groupings of the phases. A key point to note is that ground water pollution could occur from any of these phases if they are not properly planned, designed, and implemented. Of importance herein is the abandonment phase.

Figure 4.6 illustrates how abandoned wells can serve as conduits for ground water pollution when they are in close proximity to a brine disposal well.[34] Rusted casing or the absence of cement can provide openings for pollutant migration through abandoned wells with casing. Abandoned wells with no plugs, or which have been improperly plugged, can also allow pol-

Table 4.13. Case Study Facilities—General Features and Waste Characteristics.[30]

Case Study No.	Type of Facility	Type of SI (No. and Function)	Year Placed in Service	SI Size (Acres)	Waste Type	Waste Quantity
1	Electrolytic metal refining plant	Small disposal pond	1972	0.4	Acidic process liquor and sludge waste (pH <2) high in heavy metal content	A total of 843,750 gal in 1982
		Disposal pond	1979	1.1	Acidic process liquor and sludge waste (pH <2) high in heavy metal content	
2	Pesticide formulation and distribution plant	Pesticide washdown evaporation disposal pond	1979	<0.1	Pesticide rinsewater	Batch operation (400 gal/day maximum)
		Pesticide rinsewater evaporation disposal pond	1982	<0.1	Pesticide rinsewater	Batch operation (400 gal/day maximum)
3	Commercial hazardous waste disposal facility	Site A: 8 impoundments used for settling, storage, and sludge disposal	1951	15	Oily water and brines, alkaline and acid wastes, heavy metals, paint sludge, tank bottom sediments, cyanide, pesticides, and other chemical wastes	A total of 53 million gal in 1982
4	Agricultural fertilizer manufacturing plant	11 settling ponds used to remove gypsum	1965	14	Production water for ammonium phosphate/phosphoric acid plant with pH <2 and high radionuclides content	20,000 gal/day

Table 4.13, continued

Case Study No.	Type of Facility	Type of SI (No. and Function)	Year Placed in Service	SI Size (Acres)	Waste Type	Waste Quantity
		One evaporation pond (treatment)	1976	8	Wastewaters from plant boilers, water treaters, and nitric and sulfuric acid plants (pH <2, high in radionuclides)	130,000 gal/day
		One cooling pond (treatment)	1976	38	Same as gypsum SIs	10,000 gal/day
5	Mineral ore mining/ manufacturing plant	5 low-head solar ponds (treatment and storage)	1972	90	Mineral liquor tailings with high arsenic and boron content	A total of 50 million gal/month to Ponds A–E, 4, and 5
		High-head evaporation pond (treatment and storage)	1975	80	Mineral liquor tailings with high arsenic and boron content	
		High-head evaporation pond (treatment and storage)	1976	100	Mineral liquor tailings with high arsenic and boron content	
		Evaporation pond (treatment and storage) high arsenic and boron content	1980	120	Acid plant wastewater with high arsenic and boron content	12 million gal/month
6	Commercial hazardous waste disposal facility	Evaporation pond (disposal)	1980	5	Geothermal muds and brines, wastewater treatment sludge, tank bottom sediments, cooling tower blowdown sludge and oil drilling muds	9.85 million gal in 1982

		Disposal pond (currently used for land treatment)	1980	5	Geothermal muds and brines, wastewater treatment sludge, tank bottom sediments, cooling tower blowdown sludge and oil drilling muds	
7	Agricultural fertilizer manufacturing plant	Cooling pond	1974	100	Process water from phosphoric acid plant with pH <2 and high fluoride content	40,000 gal/min (maximum)
		Initial gypsum pond (disposal)	1974	150	Gypsum slurry with pH <2 and fluoride and phosphorus content	No data available
		Expansion gypsum pond (disposal)	1980	200	Gypsum slurry with pH <2 and fluoride and phosphorus content	No data available
8	Chemical production plant	2 equalization/retention basins (treatment)	1976	3.5	Wastes high in organic nitrogen content and varying pH, resulting from synthetic fiber production	3,000 gal/min
9	Uranium mining/milling	Tailings pond (disposal)	1980	64	Acidic tailings slurry containing kerosene and radium-226	No data available

Table 4.14. Oil and Gas Production by State in 1979.[32]

State	Crude Oil (million bbls)	Natural Gas (billion ft^3)	State	Crude Oil (million bbls)	Natural Gas (billion ft^3)
Alabama	19	86	Mississippi	38	144
Alaska	512	221	Montana	30	54
Arkansas	19	110	Nebraska	6	3
California	352	248	New Mexico	79	1,181
Colorado	32	191	North Dakota	31	19
Illinois	22	2	Ohio	13	123
Kansas	57	798	Oklahoma	144	1,835
Kentucky	5	60	Texas	1,013	7,175
Louisiana	494	7,266	Utah	27	59
Michigan	34	160	Wyoming	125	414

Total U.S. Oil Production— 3,052 million bbls.

Gas Production—20,149 billion $ft.^3$

Table 4.15. Wells Addressed by Underground Injection Program.[33]

Wells	Comments
Class I	Used to inject industrial, nuclear, and municipal wastes beneath the deepest stratum containing an underground drinking water source.
Class II	Used to dispose of fluids which are brought to the surface in connection with oil and gas production, to inject fluids for the enhanced recovery of oil or gas, or to store hydrocarbons.
Class III	Used to inject fluids for the solution mining of minerals, for in situ gasification of oil shale and coal, and to recover geothermal energy.
Class IV	Used by generators of hazardous wastes or by owners and operators of hazardous waste management facilities to inject into or above strata that contain underground drinking water sources.
Class V	Includes all wells not incorporated in Classes I–IV; typical examples of such wells are recharge wells and air conditioning return flow wells.

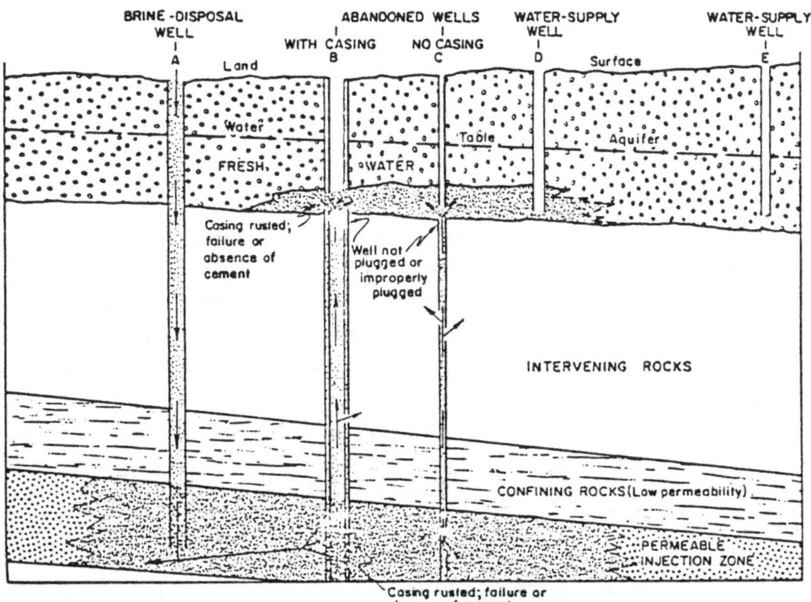

Figure 4.6. Potential pathways of ground water pollution from abandoned wells.[34]

lutant migration from the permeable injection zone to the fresh water aquifer.

The leakage of contaminated or highly mineralized water through abandoned wells and unplugged exploration holes has led to insidious ground water pollution problems.[35] Improperly abandoned wells and unplugged exploration holes are potential avenues for salt water to infiltrate fresh water aquifers under normal conditions. The pressure injection conditions for disposal and secondary recovery increase the likelihood of salt water contamination. The presence of unrecorded abandoned wells in areas subject to flooding (natural or permanent) may introduce surface water into fresh water aquifers. The addition of a surface water reservoir can cause high water inflows with the danger of continuous ground water contamination.[36]

During 1957 to 1961, a field study of salt water disposal in three oil fields in Limestone County, Texas was conducted. Brines were disposed of in large surface pits, and the resulting degradation of ground water quality was compounded by 600 abandoned oil and gas wells which were improperly plugged.[37] In 1962, a study in Fisher County, Texas revealed serious soil damage problems on 25 farms due to rising ground water levels and a general increase in ground water salinity. In some areas, the presence of sodium chloride in ground water suggested brine contamination from improperly plugged wells.[38] Unplugged wells and artesian brines have resulted in flowing salt water wells in Know, Hopkins, and Young Counties, Texas. In Hopkins County, one abandoned oil test well flowed between 100 and 125 barrels per day of brine water before it was plugged.[34]

A classic example of problems that can arise from abandoned wells has occurred in Colorado. In 1915, an oil test hole was drilled in west-central Colorado to a depth of 1,837 ft and it encountered warm, mineralized water. In 1968, this well was discharging 1,350 gpm of brackish water with a TDS concentration of 19,200 mg/l. It was estimated that this flow contributed 57,000 tons of dissolved solids per year to the White River. The well was subsequently plugged, after which the hydrostatic pressure built up, causing other nonflowing wells in the area to flow, and thus creating saline seeps in the area.[34]

Leakage of acid mine drainage through old oil and gas test wells and other open holes has caused extensive ground water pollution in the coal fields of Pennsylvania and elsewhere in Appalachia.[34] In the Pocatalico River Basin of West Virginia, brine, oil, or gas was present throughout the shallow salt sands of the region. Increased brine disposal well drilling threatened upward movement of these pollutants to fresh ground water unless all wells tapping the salt sands were permanently and properly cased.[38] Also in West Virginia, serious ground water contamination (measured by area and thickness of aquifer) occurred because over 50,000 oil and gas wells penetrated artesian salt water. Most of the wells had pulled casings but were never plugged, and over half of the wells allowed salt water to flow upwards to fresh water aquifers.[39]

Abandoned, unplugged wells from frantic wildcat drilling in Taylor

County, Kentucky have hampered brine disposal through injection wells. When brines were injected under pressure, the abandoned wells in the area also began to flow brine because the formation had been previously penetrated by several abandoned gas and oil test wells drilled in the 1920s and 1930s. The brines changed the potable ground water from a calcium bicarbonate type containing moderate amounts of magnesium and sulfate, to a sodium chloride type. Chloride concentrations prior to oil production generally were less than 60 mg/l; after oil production, chloride concentrations were as high as 51,000 mg/l.[34]

A 1960 and 1961 field study near Greensburg, Kentucky showed that when deep well injection of oil field brines was carried out, abandoned oil and gas test wells allowed the brines to move upward and contaminate fresh ground water. The contamination declined as the oil production decreased, but secondary recovery methods threatened even greater pollution if the abandoned wells were not plugged. It was recommended that residual brines be injected into a permeable zone separated by impermeable material from a fresh ground water zone, and that several wells and springs be sampled periodically for chemical analyses and water level measurements.[38]

An abandoned, unplugged oil test well in Glynn County, Georgia yielded salt water with a chloride content of 7,780 mg/l. The 4,615-foot well penetrated both fresh water and salt water aquifers, and the well bore served as a conduit between them. Because the pressure was greater in the saltwater aquifers, it flowed upward into the freshwater aquifer. Elevated chloride contents appeared to extend 1.5 miles along the hydraulic gradient, and any additional pumping in the area threatened contamination of nearby wells. The study stressed the need for better well construction, plugging of abandoned wells, and the location of well fields away from the test well.[40]

During 1979 and 1980, the Oklahoma Corporation Commission received eleven pollution complaints from five counties for cases of improperly plugged or abandoned wells.[41] Despite the number of ground water quality problems which have resulted from abandoned wells, the information is disaggregated and difficult to find in routinely published literature. Probably the majority of the data on abandoned wells is associated with complaints and lawsuits, and it may be found in the offices of the legal staffs of state regulatory agencies. No comprehensive and systematic nationwide study has been conducted to establish information on locations of abandoned wells and the likelihood that they were properly or improperly plugged at the time of abandonment. Additionally, no studies have been made of the effectiveness of various plugging practices in terms of preventing ground water pollution. Finally, there are numerous legal issues related to the question of who is responsible for plugging old wells which have been abandoned and the drilling companies are no longer in existence.[39]

Drilling Mud Disposal

Rotary drilling is probably the most popular drilling method utilized in oil and gas exploration activities. A basic component of the rotary method is the drilling fluid. The fluid, usually a slurry composed of water, clay, and other additives, serves a variety of essential purposes, including:

(1) cooling and lubricating the drillstring and bit
(2) building a filter (impermeable) cake on the well bore
(3) controlling subsurface pressure
(4) suspending cuttings and weighting material when circulation is interrupted
(5) releasing sand and cuttings at the surface
(6) supporting part of the weight of the drillstring and collars
(7) minimizing damage to the geological formation
(8) ensuring maximum information about the formations penetrated
(9) transmitting hydraulic horsepower to the drill bit

During drilling operations the drilling fluids are usually stored onsite in tanks or open pits. Upon completion of drilling operations, the drilling fluids must be properly disposed. Disposal is usually accomplished by allowing the liquid to evaporate and burying the mud onsite or hauling the fluid to an approved offsite disposal location. The chief concerns related to either onsite or offsite disposal are associated with potential surface or ground water pollution which might result from pit overflows or seepage.

The basic components of drilling muds include clays to increase viscosity and create a gel; barium sulfate (barite), a weighting agent; and lime and caustic soda to increase the pH and control viscosity.[42] Additional conditioning constituents include polymers, starches, lignitic material, and various other chemicals. Clay-based muds are formulated using naturally occurring clays such as bentonite and attapulgite, and a variety of organic and inorganic additives to achieve the desired consistency, lubricity, or density. Fresh or salt water is the liquid phase used for these muds. Additives are also used for pH control, corrosion inhibition, and emulsification. The circumstances surrounding the drilling determines the type of water-based drilling fluid that should be used for a given situation. The number of additives, weighting agents, deflocculants, and treating chemicals now on the market provide the basis of a trend toward "tailor-made" drilling fluids. The annual list of drilling fluid additives published in World Oil has over 1,400 tradename additives. Nearly 100,000 tons of common inorganic chemicals are added to drilling muds annually.[43]

Four products comprise about 90% of the total volume of all additives used in drilling mud.[44] These products (barite, bentonite, lignosulfonate, and lignite) are natural materials obtained from sources predominately located in the western and midwestern areas of the United States. Typical chemical analyses for each additive will be discussed.

Of the various functions of drilling muds, controlling subsurface pressures is probably the most crucial. When high pressure zones are penetrated, the hydrostatic head of the mud column must be sufficient to offset the varying subsurface pressures. Consequently, drilling fluid densities must vary from 8.3 to over 22.0 pounds per gallon. These weights are attained by adding finely ground high-density materials such as barium sulfate, calcium carbonate, and iron carbonate.[45] Barite (barium sulfate) is the most common additive used to increase the density of the drilling fluid for control of formation pressures. It accounts for 98% of weighting agents used. Barite is inert in alkaline, acid, and neutral solutions. The physical and chemical properties of oil field barite must meet specifications adopted by the American Petroleum Institute concerning density, particle size, and chemical hardness. The primary contaminants in finished barite are silica and iron compounds. Barite concentrations ranging from 50 to 100 lb per 42-gallon barrel may be used in well drilling operations.

Clays comprise the second largest volume of products used in drilling fluids. The commonly used clays include sodium bentonite, processed calcium bentonite, attapulgite, and sepiolite. Attapulgite and sepiolite clays are used as viscosifiers in saltwater fluids in concentrations of 10 to 30 lb per 42-gallon barrel. These clays are described as hydrous magnesium silicates. Of these four clays, sodium bentonite is the most widely used.[46]

Bentonite is a soft, moisture-absorbing rock composed of clayey minerals which are highly colloidal and plastic. The primary clay mineral is sodium montmorillonite; however, calcium montmorillonite may be the predominant species in some deposits. Silica, shale, calcite, mica, and feldspar are the most common impurities associated with various bentonite deposits.[46]

Lignosulfonates are the third highest volume chemical additive used in drilling muds. Lignosulfonates are considered the best all-purpose deflocculant for water-based drilling muds for control of the rheology of the system. In the drilling operation, heavy solids-laden drilling fluids become gelled from effects of drill solids, contaminants, and temperature; therefore, a deflocculant is required to maintain the mud in a fluid state. Muds containing high concentrations of barite are not generally "thinned" by dilution with water. Chemical deflocculants such as lignosulfonates aid in minimizing the total volume of mud and barite consumed. To obtain this necessary deflocculating action, concentrations of 1 to 15 lb of lignosulfonates per 42-gallon barrel are normally used. Chrome lignosulfonates are the most widely used deflocculants in drilling fluids. These deflocculants perform over a wide alkaline pH range, are resistant to common mud contaminants, are temperature stable to about 325 to 350° F, and will function in high concentrations

Table 4.16. High Values Reported for Merkle Pits in Prestudy Samples.[47]

Parameter	Pit Water	Pit Sediment
Chloride	1,103 mg/l	2,500 mg/kg
Arsenic	860 μg/l	7.5 mg/kg
Barium	195,000 μg/l	780 mg/kg
Chromium	9,600 μg/l	39 mg/kg
pH low	4.5	8.0
pH high	7.5	9.5
Zinc	24,200 μg/l	
Lead	3,700 μg/l	
TOC	82 mg/l	
H Scan (crude oil)	300,000 mg/l	
Conductivity	32,200 μmhos	
Salinity	22.2%	
Temperature	38°C	

of soluble salts. The additives of potential concern from a pollution standpoint include ferochrome lignosulfonate and lead compounds. Ferochrome lignosulfonate contains 2.6% iron, 5.5% sulfur, and 3% chromium.[46]

Fairchild and Knox[47] report the results of an in-depth study of the environmental implications of offsite disposal pits for waste drilling fluids at a site in Oklahoma. In that study, samples of pit liquids and pit sediments were analyzed for a variety of constituents and the extreme values are reported in Table 4.16. Additionally, seven ground water monitoring wells were drilled and sampled. The results of this sampling program are presented in Table 4.17.

Results of chemical analysis of the liquid portion of each pit indicate high levels of pollutants in the pits. Some of these constituents exceed Oklahoma Corporation Commission (OCC) discharge water maximum effluent concentrations (chemical oxygen demand, chlorides, total dissolved solids, and conductivity). The constituents in the liquid portion are of particular concern because of their mobility relative to disposal site permeability. Species such as chlorides will move with any infiltration or seepage water into ground water or surface water.[47]

The sediments consist of solids and interstitial waters of very high ionic strength. Contributing constituents include many metals of concern; for

Table 4.17. Summary of Ground Water Data.[a]

	pH	Cond (μmhos)	Salinity (%)	Temp (°C)	Alkalinity (mg/l $CaCO_3$ to pH 8.3)	Alkalinity (mg/l $CaCO_3$ to pH 4.5)
Wells M-1 & 4	6.7	158	0	17.6	0	84.2
	6.75	169	0	17.6	0	82.9
Wells M-2 & 5	6.77	184	0	17.2	0	89.0
Wells M-3 & 6	6.66	157	0	17.0	0	72.3
Well M7	6.36	91	0	19.7	0	34.0
Edgin Well	6.43	180	0	—	0	93.6

	Nitrate (mg/l NO_3-N)	Chloride (mg/l)	Total Phosphate (mgP/l)	COD (mg/l)	TOC (mg/l)	Sulfate (mg/l)	TDS (mg/l)
Wells M-1 & 4	0.32	132	0.034	37	13	14.8	189
	0.42	108	0.037	45	19	14.2	199
Wells M-2 & 5	0.91	158	0.033	7	12	15.8	181
Wells M-3 & 6	0.38	109	0.064	7	12	18.0	165
Well M7	2.62	398	0.02	45	14	7.6	110
Edgin Well	1.42	190	0.015	4	4	9.8	198

[a]Well M7 and Edgin Well are removed from pit locations. Well nests M-1 and 4, M-2 and 5, and M-3 and 6 are adjacent to disposal pits.

example, chromium, barium, lead, and arsenic. This reservoir of potential pollutants is subject to chemical changes which may release metals out of the pits. Factors such as dilution or concentration of the aqueous phase, pH shifts, or biological activity could trigger such a release. Once in the aqueous phase, the metals would be a direct threat such as the pit liquids are at the present time.[47] Analysis of the ground water quality data indicated contamination due to elevated levels of nitrates, chlorides, phosphates, sulfates, and organics.

DIFFUSE SOURCES

Presented below is general information on two diffuse sources of ground water contamination: agriculture and acid precipitation. Diffuse sources are those activities that encompass a large areal extent and/or possess the potential to cause widespread ground water contamination. Potential contaminants from diffuse sources will approach ground water formations along a wide front.

Agriculture

Agriculture is an industry that encompasses a variety of activities, many of which represent potential sources of ground water contamination. Gratto et al.[48] grouped the five major sources of agricultural pollutants as: animal wastes, fertilizers, irrigation residues, pesticides, and sedimentation. Robbins and Kritz[49] cite the same first four sources of agricultural contamination, but list the fifth source as plant residues.

Animal feeding operations (mostly beef and poultry) represent an agricultural activity that can affect both surface and ground water. Contaminants of concern from these sources would include nitrogen compounds, phosphates, bacteria, chlorides, and, in some cases, heavy metals. Miller[37] identifies those factors influencing the susceptibility of ground water to feedlots as: stocking rate, manure removal, depth to water table, soil texture and structure, and infiltration rates.

Use of fertilizers containing nitrogen, phosphorus, and potassium compounds can also threaten ground water supplies. Duke, Smika, and Heerman[50] describe how excessive irrigation resulted in the leaching of significant amounts of nitrogen fertilizer as nitrate to underlying ground water. Crabtree[51] found a large percentage of nitrate contamination in rural Wisconsin wells and hypothesized the source to be organic fertilizer (manure) that had been ammoniafied and then nitrified. Exner and Spalding[52] found that up to 50 percent of applied nitrogen fertilizer infiltrates to ground water reservoirs.

The practice of irrigation has increased crop productivity in arid areas, but it also presents a potential source of ground water contamination. Irrigation water usually has high dissolved solids. Plants will take up the irrigation water and concentrate the dissolved minerals (salts) in the root zone. When the soils become too saline to be productive, they are flushed by overirrigating. This practice tends to flush the salts and the applied agricultural chemicals down to ground water systems.

Regardless of the various practices used or the type of agricultural industry considered, the contaminants of most concern in relation to agriculture are nitrates, herbicides, pesticides, and fertilizers. These contaminants are grouped as nitrates and agricultural chemicals and are discussed below.

Nitrates

The agricultural sources of nitrate in the environment are usually animal feeding operations or fertilizer applications. Cech and Harrist[53] describe an exhaustive study done in Texas to determine relationships between nitrate concentrations in ground water and the presence of gardens or fields (fertilizers), animal wastes (feedlots, barnyards), and domestic wastes (septic tanks, cesspools). Young[54] discusses the changes in influence of various nitrate contamination sources over a 34-year period. The biggest growth in relative importance was inorganic fertilizers which jumped from 2% (1938) to 23% (1972).

Nitrate is the most common environmental form of nitrogen because it is the end product of the aerobic biological process called nitrification (see Figure 4.7). This oxidation of ammonia or organic nitrogen to nitrate nitrogen is a two-phase reaction commonly occurring in nature. The complete reaction involves conversion of ammonia to nitrite by nitrifying bacteria (*Nitrosomonas* group); then conversion of nitrite to nitrate by bacteria from the *Nitrobacter* group.

The main concern over nitrate contamination of waters is the health effects related to consuming nitrate contaminated drinking water. Comley[55] first initiated concern over nitrates when he reported methemoglobinemia in infants was related to nitrates. More recently, concern has risen over the possible compounds nitrates may form in the subsurface. Magee and Barnes[56] and Bogovski[57] both express concern about nitrates combining with amine compounds to form carcinogenic nitrosamine compounds. Ginocchio[58] says that high nitrate intake can cause cyanosis in small children and can lead to gastric cancer in adults. The U.S. Environmental Protection Agency has set the maximum contaminant level in drinking water supplies at 10 mg/l nitrate in order to prevent infantile methemoglobinemia. It is estimated that approximately 1% of the U.S. population uses public water supplies, derived primarily from ground water formations, having excessive nitrate levels.[59]

Several technologies have been promoted for removal of nitrates from water supplies. Rautenbach and Henne[60] discuss a hybrid reverse osmosis system that was able to reduce nitrate levels to less than 50 ppm. Bilidt[61]

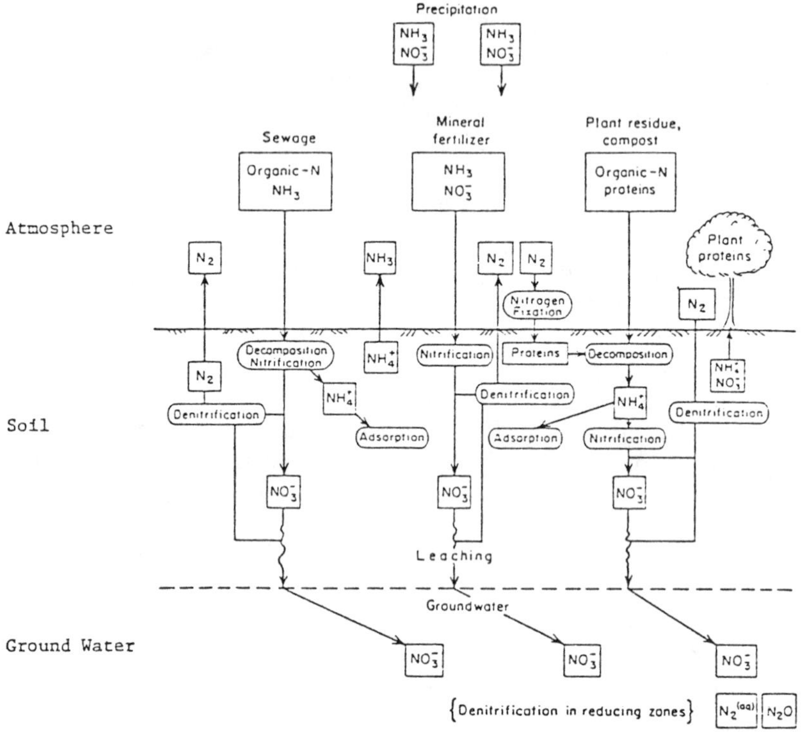

Figure 4.7. Sources and pathways of Nitrogen in the Subsurface Environment.[25]

discusses an application of reverse osmosis for treating nitrate-contaminated ground water at the surface. Contrary to these two studies is the work done by Sorg and Love[62] who reported limited success of reverse osmosis for nitrate removal.

Another traditional wastewater treatment technology being touted for application to nitrates is ion exchange. In an early study, Sheinker and Codoluto[63] describe the use of a synthetic resin as an exchange medium for denitrification. In a comparative study, Sorg[64] reported ion exchange to be more selective for sulfate removal, but still the most economical means of removing nitrates. Musterman et al.[65] report good removal efficiencies for nitrate, sulfate, and hardness by using strong base resins. Hoell and Feuerstein[66] describe the CARIX process for simultaneous softening and nitrate and sulfate removal. The process is a combination of both anionic and cationic exchange processes. Horng[67] describes a study aimed at modeling the ion-exchange process for nitrate removal from ground water. Musterman et al.[65] developed two models for their ion-exchange process; one for regeneration requirements of the strong base resins; and one for the length of time for exhaustion of the resins. Rautenbach et al.[68] discuss the advantage of high water recovery ratios when using electrodialysis.

Other technologies for controlling nitrate contamination of ground waters have been reported and center mostly on sound management practices. Martin and Watts[69] discuss how irrigation management can be used for nitrate control. Swoboda[70] describes how timing of nitrogen applications (mainly in the spring) can reduce nitrate leaching to ground water. Finally, Timmons[71] describes the successful use of "nitrapyrin" which is a nitrification inhibitor that can be applied to soils.

Interest has been generated concerning biological treatment of nitrates through denitrification processes. Four parameters tend to influence both the rate and degree of denitrification in soil and biological systems; oxygen content, moisture content, pH, and carbon source. Denitrification is an anaerobic process which requires a lack of oxygen to occur.[59] Increased moisture content of a soil is conducive to an anaerobic environment; hence, increased denitrification. In general, denitrification processes are retarded in acidic soils. The most important factor for denitrification to occur is the presence of an adequate carbon (food) source. Lance[59] reports a rule of thumb to be a "1 to 1 ratio (carbon to nitrogen) to achieve 80 to 90% denitrification." Davenport, Lembke, and Jones[72] were one of the first groups to propose the idea of in situ denitrification by simply discing or ploughing in a supplemental carbon source. Sontheimer[73] stated that underground treatment of nitrates showed promise. In the only real application of the in situ approach to denitrification, Trudel, Gillham, and Cherry[74] injected bromide and nitrate into the subsurface and found preferential nitrate removal. Additionally, bicarbonate production, organism growth, and removal rates were all found to be comparable to previously published figures.

Agricultural Chemicals

Chemicals used in the agriculture industry can be grouped as those compounds that promote growth and those that do not. The compounds promoting growth are referred to as fertilizers; the compounds inhibiting growth are referred to as pesticides. More specifically, pesticides would include fungicides, herbicides, insecticides, and nematicides. A listing of some of the common pesticides in use is found in Table 4.18.

Concern over pesticides centers on the fact that many of the pesticides are complex organic chemicals that could cause adverse health effects. Soukup[76] examined the poisoning symptoms, toxic levels, and synergistic potential of several pesticides. The conclusion is that all of the above effects are species-specific.

Table 4.18. Common Names, Trade Names, and Use Classes for Some Pesticides.[75]

Common Name	Trade Name(s)[a]	Use Class[b]
alachlor	Lasso	H
aldicarb	Temik	I,N
atrazine	AAtrex	H
benomyl	Benlate	F
bromacil	Hyvar-X	H
butylate	Sutan	H
carbaryl	Sevin	I
carbofuran	Furadan	I,N
chloramben	Amiben	H
cypermethrin	Ammo, Cymbush	I
2,4-D	(numerous)	H
DBCP	Fumazone, Nemagon	N
DCPA	Dacthal	H
diazinon	Spectracide	I
dicamba	Banvel II, Banvel D	H
dinoseb	Premerge 3	H
disulfoton	Di-Syston	I
diuron	Karmex	H
EDB	Bromofume, Soilbrom 40	I,N
EPTC	Eptam, Eradicane	H
fenvalerate	Pydrin	I
lindane (γ-BHC)	Gammexane, Kwell	I
linuron	Lorox	H
metolachlor	Dual	H
oxamyl	Vydate	I,N
paraquat	Gramoxone, Dextrone, Ortho Paraquat CL	H
permethrin	Pounce, Ambush	I
picloram	Tordon	H
propachlor	Ramrod	H
propanil	Stam M-4, Stampede 3E, Rogue	H

Table 4.18, continued

Common Name	Trade Name(s)[a]	Use Class[b]
propazine	Milogard	H
simazine	Princep, Primatol	H
2,4,5-T	Weedar 2,4,5-T, Esteron 245 BE	H
trifluralin	Treflan	H
vernolate	Vernam	H

[a]Most pesticides have additional trade names. Different trade names commonly are given to different formulations of a given pesticide and to formulations including two or more pesticides. All trade names are registered.
[b]F = fungicide, H = herbicide, I = insecticide, and N = nematicide.

Significant work has been done on the transport and fate of pesticides in soil and ground water systems. Junk, Spalding, and Richard[77] studied organics from agriculture in ground water and found partial removal from solution during vertical transport within the saturated zone. Weidner[78] conducted a study on herbicide degradation and found degradation to occur in ground water, but to a much lesser degree than in soils. Rothschild, Manser, and Anderson[79] studied the pesticide aldicarb and determined it to be a good indicator of potential ground water contamination due to its high solubility and poor adsorption.

Many factors can affect the potential of a given compound to contaminate ground water formations. Of particular concern with agricultural chemicals is their persistence in nature. Rao and Davidson[80] reviewed the literature for information on sorption partition coefficients, degradation coefficients, and half-lives of several organic pesticides. They concluded that the octanol-water partition coefficient was a good predictor of pesticide adsorption parameters. Figure 4.8 gives a graphical comparison of the persistence of a number of agricultural chemicals. Table 4.19 lists properties of pesticides which indicate their ground water contamination potential.

Although sound management practices and prudent use are the best means of preventing agricultural contamination of ground water, certain site-specific factors can be beneficial. LeGrand[82] notes that the chances of ground water contamination can be reduced at those sites that have: (1) a deep water table, (2) high clay content, (3) hydraulic gradient moving contaminants away from water supply wells, and (4) large distances between wells and contamination sources.

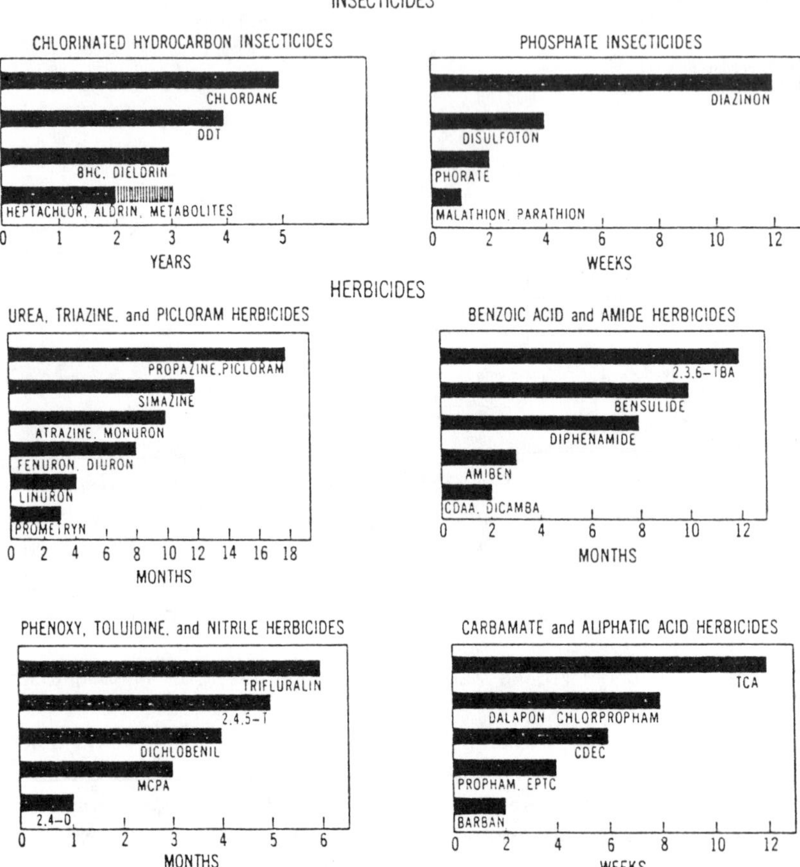

Figure 4.8. Persistence of pesticides.[75]

Acid Precipitation and Dry Deposition

As the awareness of acidic precipitation becomes more acute and interest in ground water quality grows, concerns have risen over the effects of "acid rain" and other airborne pollutants on soils and ground water quality. Because the path of the potential contaminant is sequential through soils and then into ground water formations, it is important to note the processes operating in both the soils (unsaturated zone) and the ground water formations (saturated zone).

Two processes of concern regarding acid precipitation on soils are the changes in leaching patterns of the soils and the movement of chemical con-

Table 4.19. Chemical and Physical Properties of Pesticides: Values Which Indicate Potential for Ground Water Contamination.[81]

Water solubility	—	greater than 30 ppm
K_d	—	less than 5, usually less than 1
K_{oc}	—	less than 300–500
Henry's Law Constant	—	less than 10^{-2} atm-m^{-3} mol
Speciation	—	negatively charged, fully or partially at ambient pH
Hydrolysis half-life	—	greater than 25 weeks
Photolysis half-life	—	greater than 1 week
Field dissipation half-life	—	greater than 3 weeks

stituents of precipitation through the soils. Gorham and McFee[83] have summarized information concerning chemical constituents capable of being leached from soils or retained by soils when exposed to acid rain.

Hutchinson[84] studied Canadian soils and found that mobilization and leaching of heavy metals increase with increases in acid precipitation. Cronan[85] examined the effects of acid precipitation on forest soils and watershed biogeochemistry in New England. Key findings included the following:

(1) Acid precipitation may cause increased aluminum mobilization and leaching from soils to sensitive aquatic systems.

(2) Acid deposition may shift the historic carbonic acid/organic acid leaching regime in forest soils to one dominated by atmospheric H_2SO_4.

(3) Acid precipitation may accelerate nutrient cation leaching from forest soils and may pose a particular threat to the potassium resources of northeastern forested ecosystems.

(4) Progressive acid dissolution of soils in the laboratory may provide an important tool for predicting the patterns of aluminum leaching from soils exposed to acid deposition.

Sulfates are a common constituent of acid precipitation. Johnson[86] reviewed the mechanisms of sulfate adsorption in soils, and soil properties in relation to sulfate adsorption and susceptibility to leaching by H_2SO_4. The study found that cation leaching from soil caused by atmospheric sulfuric acid only occurs if the sulfate is mobile in the soil or can displace other mobile anions. Soils rich in sesquioxides are likely to be resistant to leaching by sulfuric acid.

The kinetics of elemental release of three Delaware soils were investigated by Sparks and Curtis.[87] In all soils and at all acid rain pH levels, the order of quantities of element released was Ca, K, Mg, and Na. The amounts of silicon and heavy metals released were low. However, as the pH of the acid rain decreased, the total quantity of each released element increased. In all cases, the release pattern consisted of a rapid initial release followed by slower release rates.

Canter[88] notes that ground water can potentially be influenced by changes in soil chemistry from direct precipitation and/or dry deposition in recharge areas. Tschupp[89] notes that most atmospheric pollutants, both gases and particulates, are soluble in precipitation and are carried into the soil with the precipitation that does not run off. The major effect of these pollutants is to increase the acidity of the precipitation. This leads to increased weathering and chemical reactions on the soils and rocks, with the net effect being a reduction in ground water quality due to increased mineralization.

Acid ground water has been detected both in the United States and abroad. Taylor and Symons[90] studied the health implications of acid precipitation in the New England states. Although no adverse human health effects were detected, the highly corrosive nature of the precipitation is partly attributable to the acidic aquifers. Jacks and Knutsson[91] studied ground water supplies in three Swedish counties, and the wells in one of these counties showed definite decreases in alkalinity with time, as well as increases in carbonic acid. Henriksen and Kirkhusmo[92] found a correlation between acidic surface waters and low pH ground waters in Norway.

A common phenomenon accompanying acid ground water is an increased aluminum content. Eriksson[93] hypothesizes the process as follows. When precipitation turns acidic, it attacks the cation exchange capacity of the soil. When this capacity is depleted, aluminum hydroxides are attacked, and when sulfate concentrations are high enough, the aluminum oxides are transformed into basic (soluble) sulfates. These sulfates eventually leach down to underlying ground water formations.

To summarize, the impacts of acid precipitation and dry deposition on ground water quality are now being identified. In general, the impacts consist of changes in soil leaching patterns, decreases in ground water pH levels, and increases in soluble inorganic content. Most notably, sulfate and aluminum content increases in acidic ground water. Concerns also arise regarding the increased mobility of heavy metals in the soil system.

SELECTED REFERENCES

1. Office of Technology Assessment, "Protecting the Nation's Groundwater from Contamination," OTA-O-233, October 1984, U.S. Congress, Washington, DC.
2. U.S. Environmental Protection Agency, "The Report to Congress: Waste Disposal Practices and Their Effects on Ground Water," EPA 570/9-77-001, June 1977, Washington, DC.

3. Bouma, J., "Use of Soil Survey Data for Preliminary Design of Land Treatment Systems and Regional Planning," *Simulating Nutrient Transformation and Transport During Land Treatment of Wastewater,* August 1980, John Wiley & Sons, Inc., New York, New York.
4. Canter, L. W. and Knox, R. C., "Ground Water Pollution from Septic Tank Systems," *Septic Tank System Effects on Ground Water Quality,* 1985, Lewis Publishers, Chelsea, Michigan.
5. U.S. Environmental Protection Agency, "Design Manual—Onsite Wastewater Treatment and Disposal Systems," EPA 62511-80-012, October 1980, Cincinnati, Ohio.
6. Tejada, S., "Underground Tanks Contaminate Groundwater," *EPA Journal,* Vol. 10, No. 1, January–February 1984, pp. 20–22.
7. Predpall, D. F., Rogers, W. and Lamont, A., "An Underground Tank Spill Prevention Program," *Conference and Exposition on Petroleum Hydrocarbon and Organic Chemicals in Ground Water—Prevention, Detection and Restoration,* November, 5–7, 1984, National Water Well Association, Worthington, Ohio, pp. 17–32.
8. Robbins, R. J. and Nicholas, D. G., "Electrical Leak Detection System for Underground Stored Chemicals and Fuels," *Seventh National Groundwater Quality Symposium,* Las Vegas, Nevada, September, 26–28, 1984, pp. 40–45.
9. American Petroleum Institute, "Underground Spill Cleanup Manual," API Pub. No. 1628, June, 1980, Washington, DC.
10. Kramer, W. H., "Groundwater Pollution from Gasoline," *Ground Water Monitoring Review,* Vol. 2, No. 2, Spring 1982, pp. 18–22.
11. Buono, A. and Packard, E. M., "Delineation and Hydrologic Effects of a Gasoline Leak at Stovepipe Wells Hotel, Death Valley National Monument, California," Water-Resources Investigations 82–45, July 1982, U.S. Geological Survey, Washington, DC.
12. Bruell, C. J. and Hoag, G. E., "Capillary and Packed-Column Gas Chromatography of Gasoline Hydrocarbons and EDB," *Conference and Exposition on Petroleum Hydrocarbon and Organic Chemicals in Ground Water—Prevention, Detection and Restoration,* November 5–7, 1984, National Water Well Association, Worthington, Ohio.
13. U.S. Environmental Protection Agency, "A Groundwater Protection Strategy for Environmental Protection Agency—Draft," January 1984, Washington, DC.
14. Crump, K. S. and Guess, H. A., "Drinking Water and Cancer: Review of Recent Findings and Assessment of Risks." In *Groundwater Pollution—Environmental and Legal Problems,* American Association for the Advancement of Science Selected Symposium 95, 1984, Westview Press, Inc., Boulder, Colorado, p. 82.
15. Davis, J., "Environmentalists Hold Edge as Laws Come Up for Renewal," *Congressional Quarterly,* January 1985, pp. 81–83.
16. Long, F. A. and Schweitzer, G. E., *Risk Assessment at Hazardous Waste Sites,* 1982, American Chemical Society, Washington, DC.
17. Ingle, F. W., *TSCA's Impact on Society and Chemical Industry,* 1983, American Chemical Society, Washington, DC.

18. Stover, E. L. and Kincannon, D. F., "Treatability Studies for Aquifer Restoration," Presented at the 1982 Joint Annual Conference, Southwest and Texas Sections of American Water Works Association, October 1982, Oklahoma City, Oklahoma.
19. Liu, Yeong-Ming, "The Effects of Acid Rain and Cover Soil on Leachate Quality from Simulated Solid Waste Disposal Sites," Master's Thesis, 1983, University of Oklahoma, Norman, Oklahoma.
20. Vesiland, P. A. and Rimer, A. E., *Unit Operations in Resource Recovery Engineering,* 1981, Prentice-Hall, Englewood Cliffs, New Jersey.
21. Fenn, D. G., Hanley, K. J., and DeGeare, T. V., "Use of the Water Balance Method for Predicting Leachate Generation from Solid Waste Disposal Sites," EPA-SW-168, October 1975, U.S. Environmental Protection Agency, Cincinnati, Ohio.
22. Bouwer, H., *Groundwater Hydrology,* 1978, McGraw Hill Book Company, Inc., New York, New York.
23. Brown, K. W., Evans, G. B., and Greentrup, B. D., *Hazardous Waste Land Treatment,* 1983, Butterworth Publishers, Boston, Massachusetts.
24. Bennett, G. F., Feates, F. S., and Wilder, I., Hazardous Materials Spills Handbook, 1982, McGraw-Hill Book Company, Inc., New York, New York.
25. Freeze, R. A. and Cherry, J. A., *Groundwater,* Prentice-Hall, 1979, Englewood Cliffs, New Jersey.
26. Brunner, D. R. and Keller, D. J., "Sanitary Landfill Design and Operation," EPA-SW-65, 1972, U.S. Environmental Protection Agency, Washington, DC.
27. Environmental Institute for Waste Management Studies, "Disposal of Solvents and Solvent-Contaminated Wastes to Land—A Position Paper," December 1985, University of Alabama, University, Alabama.
28. Sammy, G. K. and Canter, L. W., "Effects of Wastewater Ponds on Ground Water Quality," NCGWR 80-6, October 1980, National Center for Ground Water Research, University of Oklahoma, Norman, Oklahoma.
29. Rail, C. D., "Groundwater Monitoring Within an Aquifer: A Protocol," *Journal of Environmental Health,* Vol. 48, No. 3, November-December 1986, pp. 128–132.
30. Ghassemi, M., Haro, M., and Fargo, L., "Assessment of Hazardous Waste Surface Impoundment Technology: Case Studies and Perspectives of Experts," EPA-600/52-84-173, January 1985, U.S. Environmental Protection Agency, Cincinnati, Ohio.
31. U.S. Environmental Protection Agency, "Salt Water Detection in the Cimarron Terrace, Oklahoma," April 1975, EPA 660/3-74-033, Corvallis, Oregon.
32. U.S. Department of Commerce, "Statistical Abstracts of the United States—National Data Book and Guide to Sources," 1980, 101st Edition, Washington, DC.
33. U.S. Environmental Protection Agency, "A Guide to the Underground Injection Control Program," June 1979, Washington, DC.
34. U.S. Environmental Protection Agency, "Impact of Abandoned Wells on Ground Water," March 1977, EPA-600/3-77-095, Washington, DC.
35. Gass, T. E., Lehr, J. H., and Heiss, H. W., Jr., "Impact of Abandoned Wells on Ground Water," EPA/600/3-77-05, August 1977, National Water Well Association, Worthington, Ohio.

36. Campbell, M. D. and Lehr, J. H., *Water Well Technology,* 1973, McGraw-Hill Book Company, Inc., New York, New York.
37. Miller, D. W., *Waste Disposal Effects on Ground Water,* 1980, Premier Press, Berkeley, California, pp. 294–321.
38. U.S. Environmental Protection Agency, "Saline Groundwaters Produced with Oil and Gas," April 1974, EPA-660/2-74-010, Washington, DC.
39. Canter, L. W., "Problems of Abandoned Wells," *Proceedings of the First National Conference on Abandoned Wells: Problems and Solutions,* May 1984, University of Oklahoma, Norman, Oklahoma, pp. 1–16.
40. Wait, R. L. and McCollum, M. J., "Contamination of Fresh Water Aquifers through an Unplugged Oil Test Well in Glynn County, Georgia," *Georgia Geological Survey Mineral Newsletter,* 1963, Vol. 16, No. 3–4, pp. 74–80.
41. Fairchild, D. M., Hall, B. J., and Canter, L. W., "Prioritization of the Ground Water Pollution Potential of Oil and Gas Field Activities in the Garber-Wellington Area," Report No. 81-4, September 1981, National Center for Ground Water Research, University of Oklahoma, Norman, Oklahoma.
42. Sittig, M., *Petroleum Transportation and Production: Oil Spill and Pollution Control,* 1981, Noyes Data Corporation, Park Ridge, New Jersey, pp. 3–20.
43. Wright, T. R., "Drilling Fluids File," *World Oil,* Vol. 180, No. 1, January 1977, pp. 37–71.
44. Perricone, A. C. and Browning, W. C., "Clay Chemistry and Drilling Fluids," *Texas Conference on Drilling Rock Mechanics,* 1979.
45. Grantham, C. K. and Sloan, J. P., "Toxicity Study—Drilling Fluid Chemicals on Aquatic Life," In: "Environmental Aspects of Chemical Use in Well Drilling Operations," EPA 560/1-75-004, 1975, U.S. Environmental Protection Agency, Washington, DC.
46. Canter, L. W., "Drilling Waste Disposal: Environmental Problems and Issues," *Proceedings of a National Conference on Disposal of Drilling Wastes,* May 1985, University of Oklahoma, Norman, Oklahoma, pp. 1–12.
47. Fairchild, D. F. and Knox, R. C., "A Case Study of Off-Site Disposal Pits in McLoud, Oklahoma," *Proceedings of a National Conference on Disposal of Drilling Wastes,* May 1985, University of Oklahoma, Norman, Oklahoma, pp. 47–66.
48. Gratto, C. P. et al., "Primer on Agricultural Pollution," *Journal of Soil and Water Conservation,* Vol. 26, March–April 1971, pp. 44–65.
49. Robbins, J. W. D. and Kritz, G. J., "Groundwater Pollution by Agriculture," *Groundwater Pollution: Proceedings of Groundwater Pollution Conference,* October 1971, Underwater Research Institute, St. Louis, Missouri, pp. 91–114.
50. Duke, H. R., Smika, D. E., and Heerman, D. F., "Ground-Water Contamination by Fertilizer Nitrogen," *Journal of the Irrigation and Drainage Division, American Society of Civil Engineers,* Vol. 104, No. IR3, September 1978, pp. 283–291.
51. Crabtree, K. T., "Nitrite and Nitrate Variation in Ground Water," Technical Bulletin No. 58, 1972, Wisconsin Department of Natural Resources, Madison, Wisconsin.

52. Exner, M. E. and Spalding, R. F., "Evolution of Contaminated Groundwater in Holt County, Nebraska," *Water Resources Research,* Vol. 15, No. 1, February 1979, pp. 139–147.
53. Cech, I. and Harrist, R., "Ground Water Contamination: Data Analysis and Modeling," *Groundwater Pollution Microbiology,* 1st ed., John Wiley, 1984, pp. 261–302.
54. Young, C. P., "Distribution and Movement of Solutes Derived from Agricultural Land in the Principal Aquifers of the United Kingdom, with Particular Reference to Nitrate," *Water Science and Technology,* Vol. 13, No. 4–5, 1981.
55. Comley, H., *Journal of the American Medical Association,* Vol. 129, 1945.
56. Magee, P. and Barnes, J., "Carcinogenic Nitroso Compounds," *Advances in Cancer Research,* Vol. 10, Academic Press, New York, 1967.
57. Bogovski, P., "The Importance of the Analysis of N-nitroso Compounds in International Research," *N-Nitroso Compounds Analysis and Formation,* 1972, International Agency for Research and Cancer, Lyon, France.
58. Ginocchio, J., "Nitrate Levels in Drinking Water are Becoming too High," *Water Services,* Vol. 88, No. 1058, April 1984, pp. 143–147.
59. Lance, J. C., "Land Disposal of Sewage Effluents and Residues," *Groundwater Pollution Microbiology,* 1st ed., 1984, John Wiley, New York, New York, pp. 197–224.
60. Rautenbach, R. and Henne, K. H., "Removal of Nitrates from Groundwater by Reverse Osmosis," RFP-TRANS-409, August 1983, U.S. Department of Energy, Washington, DC.
61. Bilidt, H., "Use of Reverse Osmosis for Removal of Nitrate in Drinking Water," *Desalination,* Vol. 53, No. 1–3, September 1985.
62. Sorg, T. J. and Love, O. T., "Reverse Osmosis Treatment to Control Inorganic and Volatile Organic Contamination," EPA-600/D-84-198, July 1984, U.S. Environmental Protection Agency, Municipal Environmental Research Laboratory, Cincinnati, Ohio.
63. Sheinker, M. and Codoluto, J. P., "Making Water Supply Nitrate Removal Practicable," *Public Works,* Vol. 108, No. 6, June 1977, pp. 71–73.
64. Sorg, T. J., "Compare Nitrate Removal Methods: For Some Communities, Ion Exchange May be the Most Economical, Practical Way of Handling Nitrate-Contaminated Groundwater," EPA-600/J-80-279, December 1980, U.S. Environmental Protection Agency, Cincinnati, Ohio.
65. Musterman, J. L., et al., "Removal of Nitrate, Sulfate, and Hardness from Groundwater by Ion Exchange," OWRT-A-075-IA(1), February 1983, Office of Water Research and Technology, U.S. Department of the Interior, Washington, DC.
66. Hoell, W. and Feuerstein, W., "Combined Hardness and Nitrate/Sulfate Removal from Groundwater by the CARIX Ion Exchange Process," *Proceedings of the Specialty Conference of the IWSA,* 1985, pp. 99–109.
67. Horng, L. L., "Modelling of Ion-Exchange Process," *Journal of the Chinese Institute of Chemical Engineers,* Vol. 16, No. 2, April 1985, pp. 91–101.
68. Rautenbach, R., et al., "Electrodialysis for Nitrate Removal from Groundwater," *Wasser Abwasser,* Vol. 126, No. 7, July 1985, pp. 349–355.

69. Martin, D. L. and Watts, D. G., "Potential Nitrogen Purification of Groundwater Through Irrigation Management," *ASAE Summer Meeting,* 1980.
70. Swoboda, A. R., "The Control of Nitrate as a Water Pollutant," EPA/600 12-77/158, August 1977, U.S. Environmental Protection Agency, Ada, Oklahoma.
71. Timmons, D. R., "Nitrate Leaching as Influenced by Water Application Level and Nitrification Inhibiters," *Journal of Environmental Quality,* Vol. 13, No. 2, April–June 1984, pp. 305–309.
72. Davenport, L. A., Lembke, W. D., and Jones, B. A., "Denitrification in Laboratory Sandy Columns," *Transactions of the American Society of Agricultural Engineers,* Vol. 18, No. 1, January–February 1975, pp. 95–105.
73. Sontheimer, H., and Rohmann, U., "Groundwater Pollution by Nitrates-Causes, Significance and Solutions," *Wasser Abwasser,* Vol. 125, No. 12, December 1984, pp. 599–608.
74. Trudell, M. R., Gillham, R. W., and Cherry, J. A., "In-Situ Study of the Occurrence and Rate of Denitrification in a Shallow Unconfined Sand Aquifer," *Journal of Hydrology,* Vol. 83, No. 3–4, March 1986, pp. 251–268.
75. Council for Agricultural Science and Technology, "Agriculture and Ground Water Quality," May 1985, Ames, Iowa.
76. Soukup, A. V., "Trace Elements in Water," *Environmental Chemicals: Human and Animal Health (Proceedings),* EPA-540/9-72-015, August 1972, Fort Collins, Colorado, pp. 11–12.
77. Junk, G. A., Spalding, R. F. and Richard, J. J., "Areal, Vertical and Temporal Differences in Ground Water Chemistry: II. Organic Constituents," *Journal of Environmental Quality,* Vol. 9, No. 3, July–September 1980, pp. 479–483.
78. Weidner, C. W., "Degradation in Groundwater and Mobility of Herbicides," OWRT-A-024-NEB(2), June 1974, Office of Water Research and Technology, U.S. Department of the Interior, Washington, DC.
79. Rothschild, E. R., Manser, R. J., and Anderson, M. P., "Investigation of Aldicarb in Ground Water in Selected Areas of the Central Sand Plain of Wisconsin," *Ground Wcter,* Vol. 20, No. 4, July–August 1982, pp. 437–445.
80. Rao, P.S.C. and Davidson, J. M., "Retention and Transformation of Selected Pesticides and Phosphorus in Soil—Water Systems: A Critical Review," EPA-600/3-82-060, May 1982, Environmental Research Laboratory, U.S. Environmental Protection Agency, Athens, Georgia.
81. U.S. Environmental Protection Agency, "Safe Drinking Water Act: 1986 Amendments," EPA 570/9-86-002, August 1986, Washington, DC.
82. LeGrand, H. E., "Movement of Agricultural Pollutants with Groundwater," *Agricultural Practices and Water Quality,* 1970, Iowa State University Press, Ames, Iowa, pp. 303–313.
83. Gorham, E. and McFee, W. W., "Effects of Acid Deposition Upon Outputs from Terrestrial to Aquatic Ecosystems," *Effects of Acid Rain Precipitation on Terrestrial Ecosystems,* 1980, Plenum Press, New York, New York, pp. 465–480.
84. Hutchinson, T. C., "Effects of Acid Leaching on Cation Loss from Soils," *Effects of Acid Rain Precipitation on Terrestrial Ecosystems,* 1980, Plenum Press, New York, New York, pp. 481–497.

85. Cronan, C. S., "Effects of Acid Precipitation on Cation Transport in New Hampshire Forest Soils," DOE/EV/04498-1, July 1981, U.S. Department of Energy, Washington, DC.
86. Johnson, D. W., "Site Susceptibility to Leaching by H_2SO_4 in Acid Rainfall," *Effects of Acid Rain Precipitation on Terrestrial Ecosystems,* 1980, Plenum Press, New York, New York, pp. 525–535.
87. Sparks, D. L. and Curtis, C. R., "An Assessment of Acid Rain on Leaching of Elements from Delaware Soils into Groundwater," OWRI-A-053-DEL(1), 1983, College of Agricultural Sciences, University of Delaware, Newark, Delaware.
88. Canter, L. W., *Acid Rain and Dry Deposition,* 1986, Lewis Publishers, Chelsea, Michigan, pp. 73–124.
89. Tschupp, E. J., "Effect on Groundwater Resources of Precipitation Modified by Air Pollution," *55th Annual Meeting of the American Geophysical Union,* April 1974, Washington, DC.
90. Taylor, F. B. and Symons, G. E., "Effects of Acid Rain on Water Supplies in the Northeast," *Journal of the American Water Works Association,* Vol. 76, No. 3, March, 1984, pp. 34–41.
91. Jacks, G. and Knutsson, G., "Susceptibility to Acidification of Ground Water in Different Parts of Sweden," KHM-TR-11, October 1981, Statens Vattenfallsverk, Stockholm, Sweden.
92. Henriksen, A. and Kirkhusmo, L. A., "Acidification of Groundwater in Norway," *Nordic Hydrology,* Vol. 13, No. 3, 1982, pp. 183–192.
93. Eriksson, E., "Aluminum in Groundwater: Possible Solution Equilibria," *Nordic Hydrology,* Vol. 12, No. 1, 1981, pp. 43–50.

CHAPTER 5

Pollutant Transport and Fate Considerations

The availability and use of information on pollutant transport and fate in the subsurface environment is vital to successful ground water quality management efforts. The central issue is whether the contaminants will move with the water phase through the unsaturated and saturated zone, or whether they will be adsorbed onto subsurface materials or be subjected to chemical reactions or biological degradation. This information is important in pollutant and source evaluations, permit reviews, and assessment of potential remedial action measures. In addition, it is necessary for inclusion in ground water flow and solute transport models.

Pollutant transport and fate can be considered in terms of hydrodynamic, abiotic, and biotic processes. The first three sections of this chapter provide basic information on these processes. The use of microcosms for the simulation of these processes is described in the fourth section. The next major section summarizes field and laboratory studies on the major categories of pollutants, including bacteria and viruses, nitrogen, phosphorus, metals, organics, pesticides, and radionuclides. Field and laboratory studies oriented to septic tank systems, solid waste disposal operations, and the land application of municipal wastewater and sludges are addressed in the sixth major section. The final section highlights other relevant issues such as leachate testing, the use of tracers, and facilitated transport.

HYDRODYNAMIC PROCESSES

The hydrodynamic processes of advection and dispersion are involved in the subsurface transport of contaminants. Advection refers to the transport of a solute at a velocity equivalent to that of the ground water movement.[1] This transport occurs via the bulk motion of the flowing ground water. Advection is also referred to as convection. Dispersion refers to the spreading of a solute concentration front as a result of spatial variation in aquifer permeability, fluid mixing, and molecular diffusion. This spreading results in the dilution of the solute concentration. Microscale dispersion can result from hydraulic drag in pore channels, differences in pore size, and pore channel tortuosity, branching, and interfingering.

The following transport equation can be used to describe one-dimensional, horizontal, single-phase flow in a saturated, unconsolidated, homogeneous medium:[1]

$$-u\frac{\partial C}{\partial x} + D\frac{\partial^2 C}{\partial x^2} - \frac{\rho_b}{\Sigma}\frac{\partial S}{\partial t} + \left(\frac{\partial C}{\partial t}\right)_{rn} = \frac{\partial C}{\partial t} \quad (5.1)$$

where:

u = average fluid velocity (m/sec)
C = solute concentration in aqueous phase (gm/m^3)
x = distance in flow direction (m)
D = dispersion coefficient (m^2/sec)
ρ_b = bulk density of soil (gm/m^3)
Σ = soil void fraction (unitless)
S = mass of solute adsorbed per unit dry mass of soil (gm/gm)
t = time (sec)
rn = chemical reactions and/or biological degradation

The first term in Equation 5.1 relates to advection or convection, and the second term addresses dispersion. Adsorption is incorporated in the third term, and other abiotic reactions and biological degradation are accounted for in the fourth term. The influence of convection, convection plus dispersion, and convection plus dispersion plus retardation from adsorption is shown in Figure 5.1.[2]

The dispersion coefficient, D, is determined primarily by the spatial variation of aquifer permeability, thus it must be determined by tracer measurements in the field.[1] The importance of dispersion in an aquifer system can be approximated by the dimensionless Peclet number (Pe) as follows:

$$Pe = \frac{ux}{D} \quad (5.2)$$

The smaller the value of Pe, the greater the extent of dispersion. For example, when Pe is greater than 1,000, dispersion can be neglected; however, when Pe is less than 5, the flow regime approaches complete mixing.[1] In ground water systems, the dispersion coefficient, D, has been found to be approximately proportional to the velocity as follows:

$$D = \alpha u \quad (5.3)$$

where:

α = dispersivity (m)

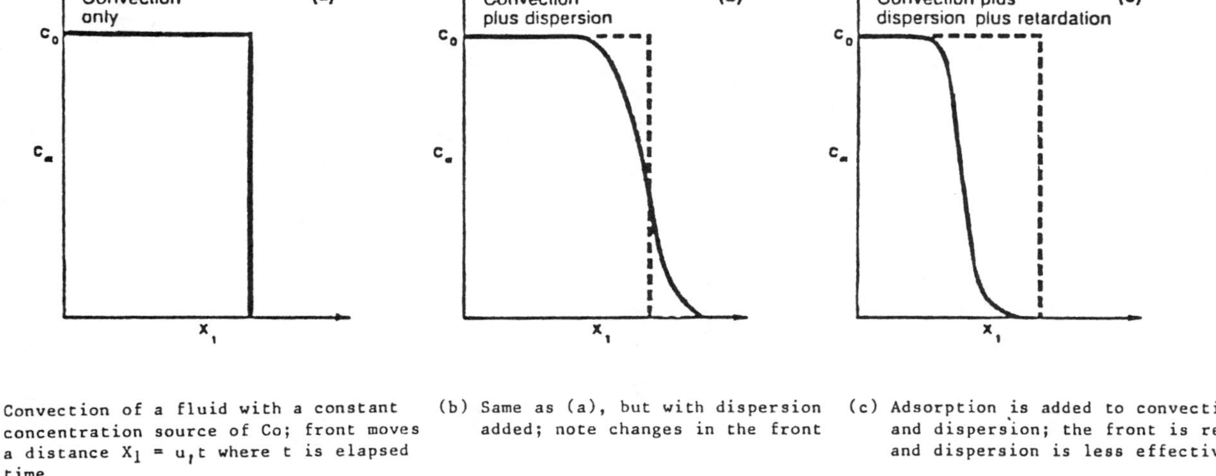

(a) Convection of a fluid with a constant concentration source of Co; front moves a distance $X_1 = u_1 t$ where t is elapsed time

(b) Same as (a), but with dispersion added; note changes in the front

(c) Adsorption is added to convection and dispersion; the front is retarded and dispersion is less effective

Figure 5.1. Influence of subsurface processes on pollutant transport.[2]

Values of dispersivity calculated from field measurements range from 0.1 to 100 m.[1]

The above discussion primarily relates to subsurface flow through homogeneous unconsolidated materials. Hydrodynamic processes in fractured materials may be characterized by channeling through macropores.[3]

ABIOTIC PROCESSES

Abiotic processes in the subsurface environment can include adsorption/desorption of solutes onto/from subsurface materials, ion exchange reactions, ion pairing or complexation, hydrolysis, solution-precipitation reactions, acid-base reactions, and oxidation-reduction reactions. For example, Scrivner, et al.[4] identified the following potential reactions in the subsurface at a hazardous waste injection well: carbonate dissolution, sand dissolution, clay dissolution, hydrolysis, coprecipitation, and ion exchange. Examples of these reactions are presented in the subsections related to types of pollutants.

Some important physical properties of chemicals which influence the occurrence of various abiotic properties are solubility, equilibrium vapor pressure, and partition coefficient. Solubility refers to the tendency of a chemical to move from the pure solid into solution and is usually expressed as the concentration of a saturated solution in equilibrium with excess solid. This equilibrium process is dependent on the balance between those forces holding the molecules or ions in the solid and the solvating ability of the particular solvent. The equilibrium vapor pressure of a gas can be conceived as the solubility of the material in air. The vapor pressure of a liquid or solid is the pressure of the gas in equilibrium with the liquid or solid at a given temperature. This property is important in the case of volatile organic chemicals in the subsurface environment.

The concentration of any singular molecular species in two phases that are in equilibrium with one another will bear a constant ratio to each other. This equilibrium system is defined as follows:

$$K = \frac{C_2}{C_1} \qquad (5.4)$$

This relationship assumes that there are no significant solute-solute interactions and no strong, specific, solute-solvent interactions that would influence the distribution process. Concentrations are expressed as mass/unit volume and usually C_1 refers to an aqueous phase and C_2 to the nonaqueous phase. The equilibrium constant defining this system, K, is usually referred to as the partition coefficient or, on some occasions, a distribution ratio. A commonly used partition coefficient in subsurface pollutant transport and fate studies is the octanol-water partition coefficient (K_{ow}); Table 5.1 contains examples for various organic compounds.[1] The relationship between the oc-

Table 5.1. Octanol-water Partition Data for Organic Compounds.[1]

Compound	log K_{ow} Average ± Standard Deviation
Aliphatic alcohols	
methanol	−0.71 ± 0.077
ethanol	−0.22 ± 0.17
n-propanol	0.275 ± 0.035
n-butanol	0.885 ± 0.007
n-pentanol	1.41 ± 0.14
n-hexanol	2.03
n-dodecanol	5.13
Alkanes and alkenes	
methane	1.09
ethane	1.81
ethylene	1.13
propane	2.36
pentane	3.39
Aromatic hydrocarbons	
benzene	2.01 ± 0.23
naphthalene	3.32 ± 0.18
Halogenated aliphatics	
chloroform	1.95 ± 0.02
carbon tetrachlorida	2.72 ± 0.15
dichlorodifluoromethane	2.16
trichloroethylene	2.29
tetrachloroethylene	2.60
1,1,1-trichloroethane	2.49
Halogenated aromatics	
chlorobenzene	2.49 ± 0.33
o-dichlorobenzene	3.38
m-dichlorobenzene	3.38
p-dichlorobenzene	3.39
hexachlorobenzene	4.13
2,4,5,2′,4′,5′-PCB	6.72
2,4,5,2′,5′-PCB	6.11
4,4′-PCB	5.58
DDT	4.98 ± 1.16
Phenols and substituted phenols	
phenol	1.49 ± 0.02
resorcinol	0.77 ± 0.01
p-nitrophenol	1.70 ± 0.47
o-chlorophenol	2.16 ± 0.03
m-chlorophenol	2.50 ± 0.03
p-chlorophenol	2.40 ± 0.04
2,4-dichlorophenol	3.15 ± 0.13
Carboxylic acids	
benzoic acid	1.95 ± 0.11
Nitrogenous compounds	
adenine	−0.12
pyridine	0.64
caffeine	−0.02 ± 0.05

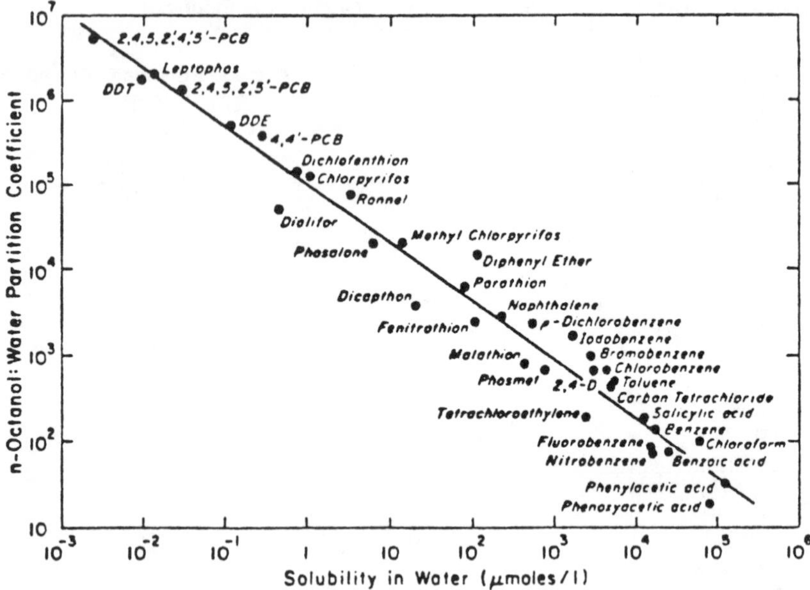

Figure 5.2. Relationship between octanol-water partition coefficient and aqueous solubility of organic compounds.[5]

tanol-water partition coefficient and the aqueous solubility of various organics is shown in Figure 5.2.[5] These two parameters are significantly correlated with the following regression equation:

$$\text{low } K_{ow} = 5.00 - 0.670 \log S \quad (5.5)$$

where:

S = solubility (μmol/L)

This regression extends over six orders of magnitude for solubility and eight orders of magnitude for partition coefficient, and includes values for a diverse group of compounds.

Adsorption is perhaps the most significant abiotic process in the subsurface environment, particularly when considering organic contaminants. This process refers to the retention of a solute in the soil phase by means of partitioning between the aqueous phase and solids.[1] Solutes that adsorb strongly onto subsurface materials are retarded in their movement through the unsaturated and saturated zones. A number of factors influence adsorption, including physical and chemical characteristics of both the adsorbent (adsorbing surface) and adsorbate (solute).[6]

If it is assumed that the mass transfer of the solute to adsorption sites is rapid, and the adsorption equilibrium is linear, then the following relationship can be written:

$$\frac{dS}{dC} = K_d \tag{5.6}$$

where

S = mass of solute sorbed per unit dry mass of soil (gm/gm)
C = solute concentration in the aqueous phase (gm/m^3)
K_d = distribution coefficient (m^3/gm)

If it is assumed that no dispersion occurs and that no biodegradation or chemical reactions occur, then Equation 5.1 can be written as:

$$-u\frac{\partial C}{\partial x} + \frac{\rho_b}{\Sigma}\frac{\partial S}{\partial t} = \frac{\partial C}{\partial t} \tag{5.7}$$

Substitution of Equation 5.6 into Equation 5.7 yields:

$$-u\frac{\partial C}{\partial x} = \left(\frac{1 + \rho_b K_d}{\Sigma}\right)\frac{\partial C}{\partial t} \tag{5.8}$$

The term in parenthesis in Equation 5.8 is known as the retardation factor. Thus, an advancing front of a sorbing solute in the subsurface environment moves at a linear velocity that is smaller than the velocity of ground water movement by the following factor:

$$R_f = \left(\frac{1 + \rho_b K_d}{\Sigma}\right) \tag{5.9}$$

where

R_f = velocity of movement of a front of a conservative tracer divided by the velocity of movement of a concentration front of an adsorbing solute

For a given subsurface environment, R_f is influenced primarily by K_d. The value K_d is determined by the strength of solute-soil interactions; the greater the affinity of the solute for the soil phase, relative to its affinity for water, the more strongly the solute will adsorb, and the greater the value of K_d.[7] Therefore, solutes that exhibit strongly hydrophobic behavior will adsorb onto soil materials. Karickhoff, Brown, and Scott[8] demonstrated that the larger the organic carbon fraction of a soil or sediment, the greater the value of K_d. In experiments with 15 soil materials having organic carbon contents in the range of 0.1 to 3.3%, Karickhoff, Brown, and Scott[8] found that the value of K_d for a given solute was proportional to the organic carbon content of the soil. In addition, they found that the value of K_d was approximately proportional to the degree of hydrophobicity of the solute, as measured by the octanol/water partition coefficient, K_{ow}. Therefore K_d for a given organic compound can be estimated by:

$$K_d = (6.3 \times 10^{-7})(f_{oc})(K_{ow}) \tag{5.10}$$

where:

f_{oc} = fraction of organic carbon in the soil (gm organic carbon per gm dry soil)

Equation 5.10 was developed as a best fit regression line for experiments with 10 solutes having K_{ow} values from 100 to 1,000,000.[8] To serve as an example of the use of Equations 5.9 and 5.10, if an aquifer material contains 2% organic carbon, has a bulk density of 2.0×10^{-6} gm/m^3, and has a void fraction of 0.2, chloroform ($K_{ow} = 93$) would have an R_f value of about 13. Thus, the chloroform would move only 1/13 as fast as a conservative tracer or the water front through the system.

BIOTIC PROCESSES

Biological degradation of organic contaminants can occur in the subsurface environment. This degradation is generally beneficial in terms of ground water contamination in that contaminant concentrations are decreased; however, intermediate products produced during degradation can be of concern. The subsequent subsections on organics and pesticides contain examples of degradation and intermediate products under both aerobic and anaerobic conditions.

The subsurface environment, including the saturated zone, has been found to contain naturally occurring bacteria capable of carrying out biodegradation of organics. For example, Ghiorse and Balkwill[9] described preliminary microscopic studies of aseptically procured subsurface material from two locations in Oklahoma. Total cell counts done with epifluorescence microscopy indicated that bacterial numbers in one sample remained constant during six weeks of storage at 4°C. In this sample the mean number of bacteria was 5.2 $(\pm 3.1) \times 10^{-6}$ cells per gm of oven-dried material. Viable cell counts on nutrient-rich media were 10^3 lower than on nutrient-poor media, and the highest viable cell counts were a factor of two lower than epifluorescence counts. These results suggested that at least one half of the epifluorescence-counted cells were viable. Electron microscopic examination confirmed the abundance of bacteria and scarcity of eukaryotic microorganisms in these samples. Small bacterial forms (<0.8 μm in size) were abundant. Both Gram-positive and Gram-negative cell wall types were detected, but Gram-positive types were observed more frequently.

Ghiorse and Balkwill[10] further described enumeration and morphological characterization studies of bacteria indigenous in the unsaturated and satu-

Table 5.2. Numbers of Bacteria in Subsurface Samples from Fort Polk, Louisiana.[10]

		AO-Fluorescent Cells (S.D.)/g dry wt		
Borehole	**Depth, m**	**Fixed[a]**	**Stored[b]**	**CFU/g dry wt[b,c]**
6B	3.7 (U)[d]	$1.6(5.1) \times 10^6$	$3.4(1.7) \times 10^6$	$<10^2$
6B	5.5 (U)	$2.2(0.8) \times 10^6$	$1.3(1.0) \times 10^6$	$<10^2$
6B	6.7 (S)[e]	$2.7(2.8) \times 10^6$	$3.0(2.8) \times 10^6$	$<10^2$
7	1.2 (U)	$1.5(0.3) \times 10^7$	$7.0(7.8) \times 10^6$	1×10^5
7	4.0 (U)	No data	$1.2(1.7) \times 10^6$	$<10^2$
7	5.1 (S)	$8.9(7.6) \times 10^6$	$9.8(0.7) \times 10^6$	$<10^2$

[a]Samples were fixed with glutaraldehyde at time of sampling then shipped to the laboratory for acridine orange (AO)-fluorescent cell counts.

[b]Samples were cooled immediately at 0 to 4°C then shipped to the laboratory. They were stored at 4°C until plate counts or AO-fluorescent cell counts were made.

[c]Only the colony counts of aerobically incubated SEA from a single plating are reported. Similar results were obtained when PYG and 1% media were used. A second plating after 6 months of storage at 4°C gave similar results. Colony counts on plates incubated anaerobically were always lower than on those incubated aerobically.

[d]U = unsaturated zone.

[e]S = saturated zone.

rated zones of a site at Fort Polk, Louisiana. Table 5.2 contains a summary of the results.[10] Like the Oklahoma samples reported earlier,[9] those from Fort Polk contained 1 to 10 million acridine orange (AO)-fluorescent bacteria per gm dry wt; however, the Fort Polk samples also contained forms that may have been eukaryotic microorganisms, and plate counts were very low, usually less than 100 colony forming units (CFU) per gm dry wt. Since total organic carbon (TOC) was invariably less than 0.1% in the samples, attempts were made to cultivate the indigenous bacteria on low-nutrient media; but very few isolates were obtained. Indigenous bacteria were observed by AO-fluorescence microscopy, however. They were morphologically diverse and frequently were found in groups (microcolonies). Transmission electron microscopy of the indigenous microflora revealed a greater proportion of Gram-positive than Gram-negative bacteria.

MICROCOSM STUDIES

Microcosms refers to laboratory or field-scale model ecosystems, designed to simulate the processes and interactions of a larger ecosystem, which can be used to study the environmental transport and fate of chemicals. Microcosms have been used to study surface water, terrestrial, and subsurface ecosystems. The emphasis herein will be on the use of microcosms to study pollutant transport and fate in the subsurface environment.

Table 5.3. Advantages and Limitations of the Usage of Microcosms.[11]

Advantages

1. Their small size permits replication and the use of appropriate controls.
2. The chemical composition of the media within the systems and physical parameters (pH, salinity, temperature, etc.) can all be varied within an experimental regime and the subsequent effects monitored.
3. The systems can be perturbed in a variety of ways.
4. The trophic structure of the systems can be manipulated, within limits, to provide greater similarity to natural situations.
5. The imports and exports to the systems can be controlled.
6. The systems circumvent large-scale environmental contamination for the examination of pollutant effects and are cost- and time-effective approaches to assessments of the environmental risks of chemical and biological pollution.
7. The difficulty in accessibility and containment of a field study can be supplanted in certain situations by a microcosm study.

Limitations

1. Containerization of environmental components will, on a pragmatic laboratory scale, lead to the exclusion of certain higher trophic levels (i.e., their existence in a laboratory system cannot naturally be sustained).
2. Containerization of environmental components can lead to structural and functional changes that are more related to system design rather than to the environment from which the samples were taken.
3. The cost and time expenditure for a microcosm study are often initially high but tends to even itself out with continued study and application.
4. Biological recruitment into microcosms must be carefully modeled and controlled because natural systems depend heavily on recruitment for maintenance of organic diversity under stressed conditions.
5. Laboratory microcosms invariably have unnaturally high surface-to-volume ratios, due primarily to large vessel wall surfaces.

Pritchard[11] listed the generic advantages and limitations of microcosms as shown in Table 5.3. The primary advantage of microcosms for subsurface studies is that information can be procured on the relative influence of hydrodynamic, abiotic, and biotic processes in chemical transport and fate.

Microcosms for studying subsurface transport and fate can have a variety of sizes and design features. Pritchard[11] identified the following operational criteria for microcosm studies and design:

(1) A microcosm study must involve a laboratory test system in which a component, in as natural a state as possible, of a real world system is brought into the laboratory and kept under standard conditions of light, temperature, humidity, aeration, pH, Eh, and other similar factors.

(2) Microcosms are ecosystems unto themselves that attempt to simulate a natural environmental situation. They must accordingly be calibrated with the environmental situation to be simulated. Extensive efforts to calibrate or parameterize the effects of scaling and containerization should be considered in initiating any microcosm program.

(3) The integrity of a microcosm study cannot stand solely on the merits of the microcosm system. The microcosm system (that is, the actual container and accessories used to study the environmental sample) should be considered only as a tool in a research undertaking, not as the research undertaking itself.

(4) A microcosm study should be considered initially as a verification tool, not as a screening tool. Its results are the integration of a variety of processes associated with the natural complexities of the environment. It will be the experimental focus for other investigations to describe and elucidate the individual processes and interactions occurring in a real world setting. From such information, either new screening tests will be developed or current screening tests will be validated.

Table 5.4 contains brief comments on 12 references involving the use of microcosms for transport and fate studies. A detailed review of the use of model ecosystems for environmental transport and fate studies has been prepared by Pritchard.[11] Van Voris, et al.[12] described a detailed study on the development and validation of a terrestrial microcosm test system for assessing the potential ecological effects of the disposal of utility wastes. Microcosms have been used in subsurface transport and fate studies to determine in-situ and onsite biodegradation of industrial landfill leachate,[13] the influence of the herbicide atrazine on denitrification processes,[14] leachates from dredged material upland disposal,[15] cadmium sorption onto two mineral subsoils,[16] and the movement of the miticide acarol and the herbicide terbacil through fine sand.[17]

Microcosms have been used in the study of hydrodynamic, abiotic, and biotic processes. For example, Saxena, Lindstrom, and Boersma[18] measured dispersion profiles of 2,4-D in order to evaluate a theoretical dispersion model and the effect of pore size on dispersion. The 2,4-D was introduced to a column comprising glass beads. Four porous media systems consisting of glass beads with particle diameters of less than 30, 28–52, 105–149, and 149–210 nm obtained by dry sieving were used. Corresponding predominant pore radii were 4.5, 8.2, 19.5, and 36.2 millimicrons. The corresponding dispersion cooefficients were found to be 0.53, 0.69, 1.41, and 2.05 sq cm/day, respectively. Using these values, theoretical dispersion profiles were calculated. The theoretical distributions of the chemical in the columns were found to be in good agreement with the experimentally determined distributions. The values of the dispersion coefficients could be related to pore size, diffusion coefficient, and flow rate.

Miller[22] studied the adsorption or partitioning between the solid (soil) and aqueous (ground water) phases of two hydrophobic organic contaminants. The study focused on evaluation of the nature and importance of rate and

Table 5.4. Summary of Twelve References Involving the Use of Microcosms for Transport and Fate Studies.

Author(s)	Comments
Ahlert and Kosson[13]	Use of large columns in a field installation to study biodegradation of industrial landfill leachate.
Cervelli and Rolston[14]	Use of soil columns to study influence of atrazine on denitrification.
Chen, et al.[15]	Use of laboratory column in dredged material leachate studies.
Christensen[16]	Use of small laboratory columns to study cadmium sorption onto two mineral soils.
Gile, Collins, and Gillett[19]	Use of soil core microcosms to assess effects of chemicals on soil ecosystems, and the fate of chemicals within soil core microcosms.
Huff, et al.[20]	Use of laboratory columns to study transport and fate of chemicals associated with petroleum sludges.
Hutchins[21]	Use of soil columns to study microbial involvement in trace organic removal in rapid infiltration recharge operations.
Mansell, et al.[17]	Use of sand columns to study subsurface movement of acarol and terbicil pesticides.
Miller[22]	Use of batch reactors and soil column experiments to study sorption and desorption phenomena for hydrophobic organic contaminants.
Pritchard[11]	Review of the use of model ecosystems for transport and fate studies.
Saxena, Lindstrom and Boersma[18]	Development of dispersion profiles of 2,4-D in a nonsorbing media using laboratory columns.
Van Voris, et al.[12]	Detailed study on the development and validation of a terrestrial microcosm test system for assessing the ecological effects of utility wastes.

equilibrium processes in organic solute/soil partitioning and on the development of submodels of the adsorption process for subsequent use in predictive fate and transport models. Adsorption rates were investigated for four aquifer soils and two representative hydrophobic organic contaminants, the pesticide lindane and nitrobenzene. Extended kinetic studies of several days' duration were carried out in completely mixed batch reactors consisting of either tumbled bottle or stirred reactors. Long-term (14-day) isotherm measurements were made to provide a basis for evaluating the character of the equilibrium partitioning of solutes between the aqueous phase and the soils tested. Both batch reactor and soil column experiments were conducted to assess rates of approach to partitioning equilibrium. The experimental investigations indicated that conventional linear, local-equilibrium assumptions may not be appropriate for hydrophobic contaminants such as lindane and nitrobenzene. Studies of the kinetic approach to adsorptive equilibrium with soils indicated that an initial rapid adsorption process is followed by a slower adsorption which continues over periods of several days. A number of different modeling approaches were examined for description of the observed rate data, including first order, plus parallel instantaneous equilibrium, second order, and coupled film and diffusion transport models. The coupled film transport and diffusion model was found to provide the mathematical framework best suited to an accurate description of the observed data.

Hutchins[21] conducted a series of soil column tests and field experiments to evaluate microbial removal of trace organics during rapid infiltration recharge of ground water. Field experiments demonstrated that operation of rapid infiltration systems using either primary or secondary wastewater can contribute trace organics to associated ground waters although concentrations of individual compounds can be reduced by the treatment process. Most of the removal occurred in the upper meter of the soil. Soil column tests demonstrated that trace organic removal from either primary or secondary wastewater was minimal when nonacclimated soil was conditioned by continuous flooding to eliminate adsorption as a removal mechanism. Batch tests indicated that most of the removal occurred in the wastewater prior to infiltration. Conversely, column tests using acclimated soil from an operational system demonstrated good removal of trace organics. Increased concentrations of target compounds in the feed did not always result in corresponding increases in the column effluent. Microbial adaptation was evident for some compounds. Other compounds appeared to exhibit a minimum concentration below which biodegradation did not proceed.

Gile, Collins, and Gillett[19] described the usage of a soil core microcosm (SCM) for subsurface studies. They investigated the effects of ^{14}C-labeled dieldrin, methyl parathion, 2,4,5-T, and HCB on a soil microcosm, as well as the fate of the chemicals within the microcosm. Figure 5.3 is a diagram of the microcosm used. In this case, the cores were 5 cm in diameter and 10 cm deep. After the cores were set up as shown in Figure 5.3, they were leached with 50 ml of standard reference rainwater, and cores which leached in less than 10 minutes or more than 12 hours were rejected. The remaining cores were then allowed to equilibrate for 28 days, with weekly leaching with enough standard rainwater to allow collection of 30 ml of leachate. The SCMs were then treated with the chemicals, and leached with standard reference rainwater at weekly intervals for 4 more weeks, and the experiments were terminated on day 58. The samples taken as well as the soil core itself

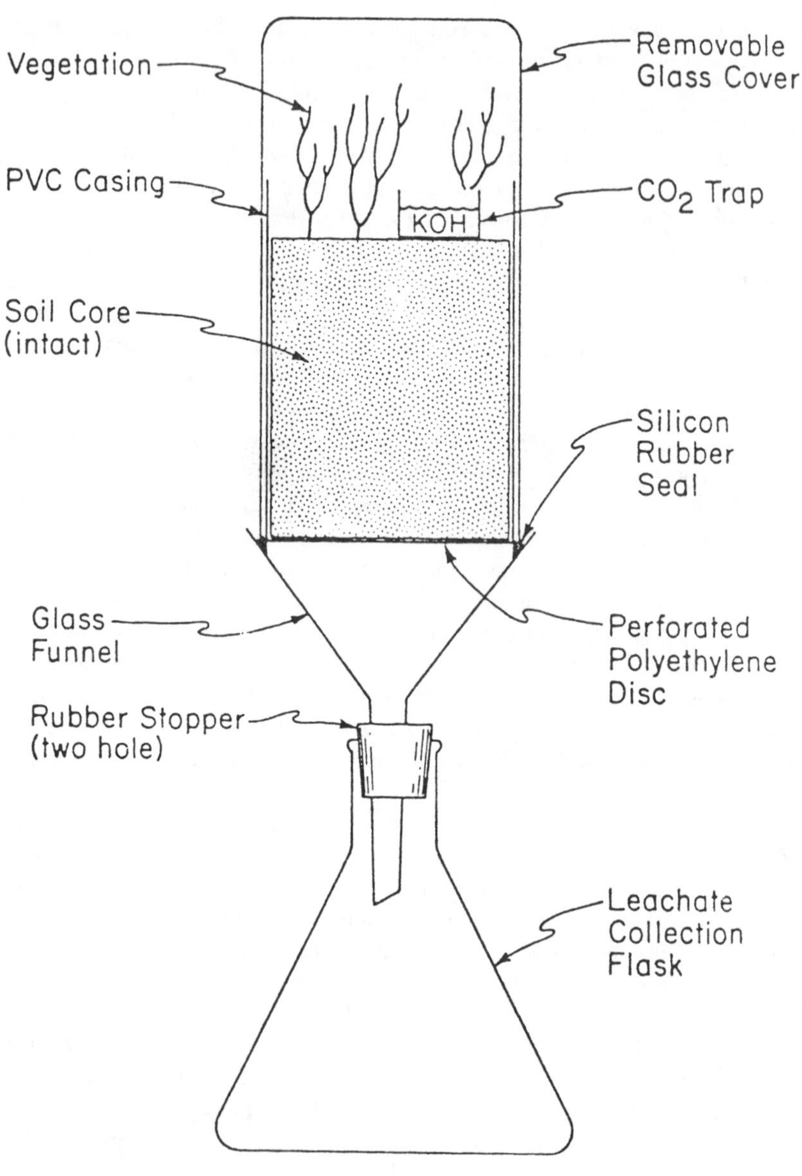

Figure 5.3. Soil core microcosm.[19]

were analyzed for total ^{14}C mass balance per SCM; fraction of radioactivity in each SCM section, metabolites, and bound residue; total nutrient content, weekly elution rate, and terminal nutrient residue for nitrate, ammonia, phosphate, cadmium, and DOC; and CO_2 evolution rate as a weekly 8-hr value. Analysis of the results indicated that the majority of the ^{14}C was in the soil, irrespective of which chemical was used, with the next highest amount being in the plants. Gile, Collins, and Gillett[19] pointed out that although the overall recovery of ^{14}C was lower than in their experiments performed with other microcosms, the lower amounts were probably not due to the different techniques used, since the coefficients of variation were typical for analysis of biological materials and the respective experimental conditions. They also noted that for results to have statistical significance, at least 3 or 4 SCMs must be analyzed to overcome the variability in the soil cores.

Huff, et al.[20] described the use of vadose zone soil columns for determining the subsurface transport and fate of contaminants leached from the land treatment and disposal of hazardous wastes. The column design is shown in Figure 5.4. All surfaces in contact with the feed water, soil, and/or effluent were either glass or Teflon.™ The soil used was a mixed, thermic Typic Ustifluent of the Lincoln series. The soil was separated into 10-cm depth increments when collected. Soil from these 10-cm increments was packed into the columns in the same relative position that it occupied in the original profile. The columns were wetted from the bottom up using a standing head reservoir connected to the bottom cap port. The water table was raised slowly (20 to 30 cm/day), allowing water to first enter the soil by capillary action. The water table was kept below the wetting front until the entire column was wet. Wetting the column slowly in this way allowed for complete saturation of the soil pore spaces. Tritiated water tracer tests were conducted to determine the saturated pore volume of each column. After the tritium tracer tests were completed for each saturated column, the water table was allowed to fall freely until the entire column was no longer saturated (yet still wet; not allowed to dry). A flow approximately equal to 10% of the saturated hydraulic conductivity was then maintained in each column for the remainder of the study.

Huff, et al.[20] used a total of 14 unsaturated columns divided into two sets of seven each. One column from each set remained undosed and served as an analytical control. Six columns of one set received a slug dose of five target organic compounds. The target organic compounds were benzene, o-xylene, 2-methylnaphthalene, phenanthrene, and pyrene. The other six columns received a slug dose of a complex petroleum refinery API separator sludge waste. The same five target organic compounds were added to the refinery sludge waste such that the amount of each target compound received by all the columns was approximately the same. The findings after about 20

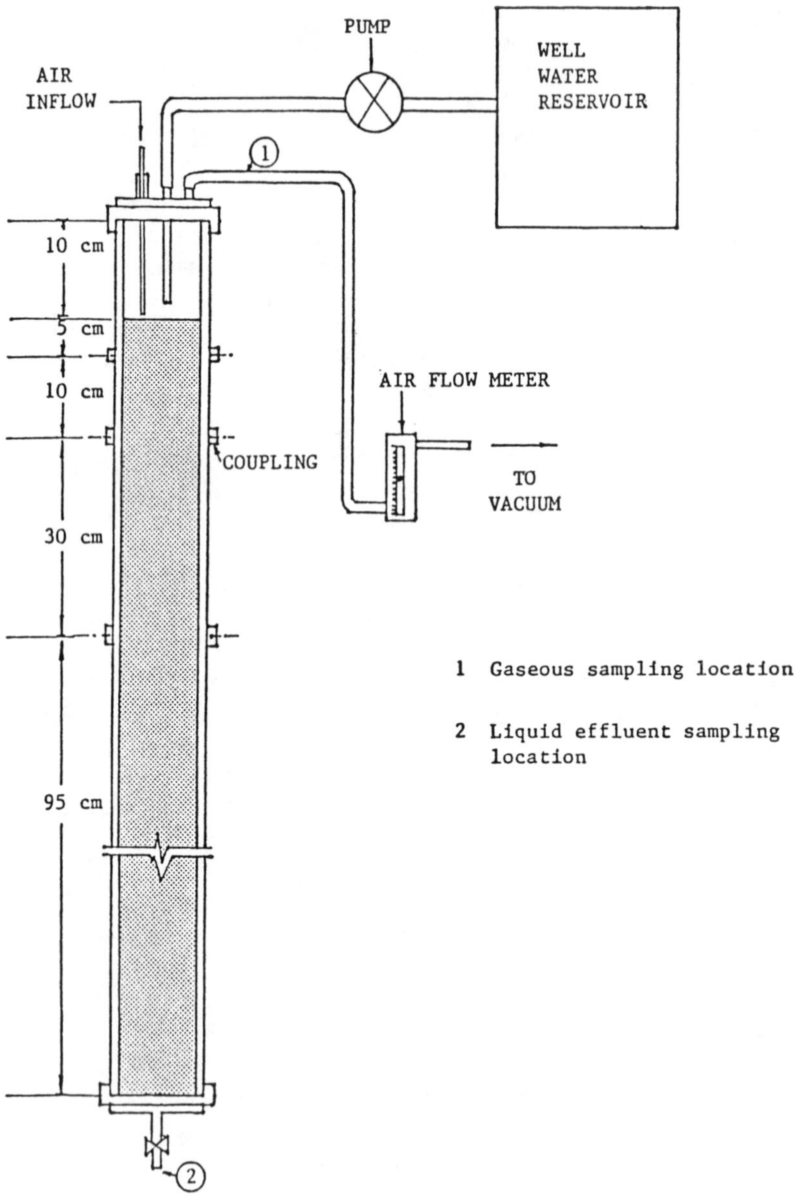

Figure 5.4. Design of vadose zone soil column.[20]

pore volumes of flow through the columns were that: benzene and o-xylene were released from the vadose zone soil columns in the liquid and gaseous phases; pyrene, phenanthrene, and 2-methylnaphthalene were not detected in liquid effluent samples; and the transport of benzene and o-xylene was differentially affected by API separator sludge.

Additional examples of the use of microcosms in subsurface transport and fate studies will be given in subsequent sections of this chapter.

TYPES OF POLLUTANTS

Ground water pollutants can be categorized by type, including bacteria, viruses, nitrogen, phosphorus, metals, organics, pesticides, and radionuclides. This section includes information on the subsurface transport and fate of these pollutant types.

Bacteria

Pathogenic bacteria can be introduced into the subsurface environment from septic tank systems, land application of municipal wastewaters and sludges, and seepage from municipal waste stabilization ponds. Bacteria in the subsurface environment may undergo natural die-away, or they may be retained in the soil or transported to ground water. Table 5.5 contains brief comments on nine studies of factors influencing the survival and transport and fate of bacteria in the subsurface environment.

Table 5.6 delineates several environmental factors that affect survival of enteric bacteria in soil. Gerba[23] reported that under adverse conditions survival of enteric bacteria seldom exceeded 10 days; under favorable field conditions survival may extend up to approximately 100 days. The principal factor determining the survival of bacteria in soil is moisture. Temperature, pH, and the availability of organic matter can also influence enteric bacteria survival. Temperature changes, the presence of oxygen, a reduction in readily available food supply, and predation by native soil organisms can create unsuitable conditions for bacterial growth. Periodic or partial drying of the soil increases the death rate. Also, bacteria seem to survive longer in cool soils than in warm soils, while low pH, low organics, and low moisture content increase the death rate. It has been surmised that low pH could not only act to adversely affect the viability of the organisms but also the availability of nutrients; pH could also interfere with the action of inhibiting agents.

Several mechanisms combine to remove bacteria from water percolating through the soil.[32] The physical process of straining (chance contact) and the

Table 5.5. Summary of Nine Studies on the Transport and Fate of Bacteria in the Subsurface Environment.

Author(s)	Comments
Brown, et al.[24]	Lysimetric study involving 3 undisturbed soils to investigate the movement of fecal coliforms and coliphages from septic tank systems.
Gerba[23]	Study of the fate of wastewater bacteria and viruses in soil.
Gerba and Lance[25]	Review of removal mechanisms for pathogenic bacteria and viruses from wastewater during ground water recharge.
Hagedorn, Hansen, and Simonsson[26]	Study of the potential of antibiotic-resistant fecal bacteria as monitors of subsurface flow under saturated conditions.
Matthess and Pekdeger[27]	Description of a survival and transport model of pathogenic bacteria and viruses in ground water.
Rahe, et al.[28]	Field experiments on the transport of *E. coli* from septic-tank drainfields under conditions of saturated flow.
Reneau and Pettry[29]	Field study of the movement of total and fecal coliform bacteria from septic tank effluent through 3 coastal plain soils.
Viraraghavan[30]	Field study of the horizontal travel of indicator microorganisms from a septic tank system tile field.
Weaver[31]	Laboratory study of the survival and leaching of salmonellae in soil.

Table 5.6. Factors Affecting Survival of Enteric Bacteria in Soil.[23]

Factor	Remarks
Moisture content	Greater survival time in moist soils and during times of high rainfall
Moisture holding capacity	Survival time is less in sandy soils than in soils with greater water-holding capacity
Temperature	Longer survival at low temperatures; longer survival in winter than in summer
pH	Shorter survival time in acid soils (pH 3–5) than in alkaline soils
Sunlight	Shorter survival time at soil surface
Organic matter	Increased survival and possible regrowth when sufficient amounts of organic matter are present
Antagonism from soil microflora	Increased survival time in sterile soil

chemical process of adsorption (bonding and chemical interaction) appear to be the most significant. Additional mechanisms include competition for nutrients and the production of antibiotics by high populations of actinomycetes. Physical straining occurs when the bacteria are larger than the pore openings in the soil. Adsorption is the other major mechanism in the removal of bacteria by soil. The process of adsorption appears to be significant in soils having pore openings several times larger than typical sizes of bacteria. Since most soils also carry a net negative charge, one might expect rejection rather than attraction of bacteria on soils. This adsorption takes place in spite of the fact that bacteria are hydrophilic colloids which possess a net negative charge at the surface. Adsorption will occur in water with high ionic strength and neutral or slightly acidic pH. Cations (Ca^{++}, Na^{+}, H^{+}) in water neutralize and sometimes supersaturate the surface of the bacteria, thus making them susceptible to adsorption by negatively charged soil particles.

Three studies will be highlighted to illustrate the survival and transport and fate of bacteria in the subsurface environment. For example, Weaver[31] investigated the survival of salmonellae in soil and their tendency to leach through soil under controlled conditions in the laboratory. The time required for 99.9% reduction in population was dependent on soil type, incubation temperature, and soil moisture. Soil temperature above 35°C was consistently detrimental to salmonellae survival and, generally, less than 2 weeks were required for a 99% reduction in population. Dry soil was not as suitable for salmonellae survival as moist soil. Leachability of salmonellae through soil was also dependent on the soil. Soils containing more than 35% clay were able to reduce the populations of salmonellae by 99.9% by passage through only 5 cm of soil. The combined effects of poor survival of salmonellae in soil, adsorption, and filtration greatly reduces the likelihood of salmonellae reaching ground water.

Hagedorn, Hansen, and Simonsson[26] investigated the potential of antibiotic-resistant fecal bacteria as monitors of subsurface flow under saturated conditions. Two pits of different depth were constructed to simulate septic tank drainfield beds, and ground water samples were removed during 32-day sampling intervals from sampling wells installed at set distances from each inoculation pit. The bacteria added to the deep pit were released into a B2t horizon which contained a higher clay content than the A horizon in which the shallower pit was installed. Streptomycin-resistant strains of *Escherichia coli* and *Streptococcus faecalis* amended to each pit site moved in a directional manner, required more time to reach sampling wells when inoculated into the deeper of the two pits, and moved relatively long distances when considering that the area where the sites were located had only a 2% slope. Bacterial numbers peaked in the sampling wells in association with major rainfall patterns, and the populations required longer periods to peak in the wells farthest from the inoculation pits. The results indicated that antibiotic-resistant bacteria eliminated the problem of differentiating between the amended bacteria and those nonresistant strains already in the soil, and the potential is excellent for including this type of microbiological procedure for assessing the suitability of a soil site for a septic tank system drainfield installation.

The movement of total and fecal coliform bacteria from septic tank system effluent through three Virginia Coastal Plain soils was monitored in situ over a 2-year period by Reneau and Pettry.[29] These soils were considered to be marginally suited for sanitary disposal of domestic wastes because of fluctuating seasonal water tables and/or restricting layers. Since septic effluent moved predominantly in a horizontal direction in these soils as a result of slowly permeable subsurface horizons, a series of piezometers were installed to collect samples at selected distances and depths from the source (subsurface drainfield) in each of the soils studied. Generally, the most probable numbers (MPN) of both total and fecal coliforms decreased significantly with horizontal distance and depth. At the Varina location, total and fecal coliform counts were observed to decrease from 11 and 1.3×10^6/100 ml in the distribution box to 11 and 2.5×10^3/100 ml at 6.1 m, respectively, above a relatively impermeable plinthic horizon. At the Goldsboro and Beltsville sites, large reductions in both total and fecal coliform counts were noted within 13.5 m. Few detectable fecal coliform were present below the restricting layers at all three sites.

Viruses

Viruses can be introduced into the subsurface environment from septic tank systems, land application of municipal wastewaters and sludges, and municipal waste stabilization ponds. A review of sources and survival of viruses in ground water is available.[33] Viruses in the subsurface environment may undergo natural die-away, or they may be retained in the soil or transported to ground water. The most important retention mechanism is via adsorption on soil particles. Table 5.7 contains brief comments on 12 studies of factors influencing the survival and transport and fate of viruses in the subsurface environment.

The transport and fate of viruses in soil is a function of factors influencing virus survival and factors influencing transport and retention. Table 5.8 lists important factors which influence virus survival in soils.[34] A major factor influencing virus survival is temperature, with generally increased survival at lower temperatures. Soil moisture also influences virus survival, with de-

Table 5.7. Summary of Twelve Studies on the Transport and Fate of Viruses in the Subsurface Environment.

Author(s)	Comments
Bitton, Davidson, and Farrah[35]	Critical assessment of the value of soil columns for assessing the transport pattern of viruses through soil.
Funderberg, et al.[36]	Laboratory study of the movement of three types of viruses in 8 soil types under saturated flow conditions.
Gerba[23]	Study of the fate of wastewater bacteria and viruses in soil.
Keswick and Gerba[33]	Literature review on viruses in ground water.
Lance and Gerba[37]	Laboratory study of the movement of polio viruses during saturated and unsaturated flow of sewage through soil columns.
Lance, Gerba, and Wang[38]	Laboratory study of the comparative movement of three enteroviruses in sand and sandy loam soils.
Moore, Sagik, and Sorber[39]	Field study of the survival and transport of human enteric viruses through soil at a wastewater irrigation site.
Schaub, Bausam, and Taylor[40]	Field study of the fate of viruses in wastewater applied by spray irrigation to sandy loam and silt loam soils.
Sobsey[34]	Literature review on the transport and fate of viruses in soils.
Sobsey, et al.[41]	Laboratory studies of virus removal from septic tank effluent applied to pilot scale soil absorption systems.
Sobsey, et al.[42]	Laboratory study of the abilities of 8 different soil materials to remove and retain viruses from settled sewage.
Vilker, et al.[43]	Development of ion exchange/adsorption models for predicting virus transport in percolating beds.

Table 5.8. Factors Influencing Virus Survival in Soils.[34]

Temperature
Microbial activity and related chemical activity
Moisture content
pH
Salt species and concentrations
Virus association with soil and other particulate matter
Virus aggregation
Soil type (textural, chemical and mineralogical properties)
Virus type

creasing survival in dry or drying soils. However, information is not entirely consistent regarding the moisture levels at which maximum virus inactivation occurs, and more studies are needed in this area with a variety of viruses and soils. Aerobic microbial activity also contributes to virus inactivation in soils at moderate to high temperatures. Although pH and ionic effects may play a role in virus inactivation in soils, these properties are related to the chemical and mineralogical characteristics of soils. Information is not entirely consistent on which chemical, mineralogical, and textural properties of soils are the most influential upon virus survival or inactivation. The role of organic matter in virus survival in soils is uncertain. Although interaction with humic and fulvic acids may cause reversible loss of virus infectivity, there is also some evidence that such virus complexation with organics may actually protect viruses from inactivation by preventing their adsorption to soil particles. The latter effects may result in greater virus survival and mobility in soils.[34]

As noted earlier, the primary mechanism of virus removal in soil is by adsorption of viruses onto soil particles. Virus adsorption is greatly affected by the pH of the soil-water system. This effect is due primarily to the amphoteric nature of the protein shell of the virus particles. At pH values below 7.4, virus adsorption by soils is rapid and effective. Higher pH values considerably decrease the effectiveness of virus adsorption by soils because of increased ionization of the carboxyl groups of the virus protein and the increasing negative charge on the soil particles. Virus adsorption by some soils is greatly enhanced by increasing the cation concentration of the liquid phase of the soil-water system. The cations in the water neutralize or reduce the repulsive electrostatic potential (the negative charge) on either the virus particles or the soil particles, or both, and allow adsorption to proceed. Adsorption of virus particles by soils increases with increasing clay content, silt content, and ion-exchange capacity. Finally, adsorption also differs as a function of virus type.[32]

Table 5.9 lists factors that may influence the removal efficiency of viruses by soil,[23] and Table 5.10 lists important factors influencing virus transport and retention in soils.[34] Virus retention by soils is also influenced by a variety of factors, some of which also influence virus survival. Virus association with particulate matter may enhance retention in soils but may also result in longer survival. Soil type also influences virus retention since adsorption is an important mechanism of virus removal in soils. Ionic conditions and pH also influence virus retention by soils, and retention is generally better at lower pH levels and at high salt concentrations or with salts having higher cation valencies. The effects of these factors are also related to the characteristics of the soil material and other environmental factors such as hydraulic conditions and rainfall. Increased hydraulic loads and flow rates may

Table 5.9. Factors that May Influence Removal Efficiency of Viruses by Soil.[23]

Factor	Remarks
Flow rate	Low flow rates (less than 1/64 gpm/sq ft) result in very efficient removal of viruses (greater than 99%) in clean waters. As flow rate increases, virus retention decreases proportionally.
Cations	Cations, especially divalent cations, can act to neutralize or reduce repulsive electrostatic potential between negatively charged virus and soil particles, allowing adsorption to proceed.
Clays	This is the active fraction of the soil. High virus retention by clays results from their high ion exchange capacity and large surface area per volume.
Soluble organics	Soluble organic matter has been shown to compete with viruses for adsorption sites on the soil particles, resulting in decreased adsorption or elution of an already adsorbed virus.
pH	The hydrogen ion concentration has a strong influence on virus stability as well as adsorption and elution. Generally, a low pH favors virus adsorption while a high pH results in elution of adsorbed virus.
Isoelectric point of virus	The most optimum pH for virus adsorption is expected to occur at or below its isoelectric point, where the virus possesses no charge or a positive charge. A corresponding negative charge on a soil particle at the same pH would be expected to favor adsorption.
Chemical composition of soil	Certain metal complexes such as magnetic iron oxide have been found to readily adsorb viruses to their surfaces.

Table 5.10. Factors Influencing Virus Transport and Retention in Soil.[34]

Virus association with particulate matter and virus aggregation
Soil composition
Virus type
pH
Salt species and concentrations
Organic matter
Hydraulic conditions and moisture content

increase virus transport in soils, especially when the soils become saturated.[34]

A number of laboratory studies on the survival and movement of viruses through soil columns have been conducted. A critical review of the merits and disadvantages of using soil columns for assessing virus transport through soils has been prepared.[35] Funderburg, et al.[36] found that poliovirus retention was greatest in soils with high cation exchange capacity values and low organic matter and clay content. Retention of the bacteriophage OX174 was greatest in soils with high organic carbon, and high residence time in combination with low soil pH or percent clay. Cation exchange capacity in combination with either soil pH or specific surface area was important in reovirus retention.

Lance and Gerba[37] compared virus movement in soil during saturated and unsaturated flow by adding poliovirus to sewage water and applying the water at different rates to a 250-cm-long soil column equipped with ceramic samplers at different depths. Movement of viruses during unsaturated flow of sewage through soil columns was much less than during saturated flow. Viruses did not move below the 40-cm level when sewage water was applied at less than the maximum infiltration rate; virus penetration in columns flooded with sewage was at least 160 cm.

Sobsey, et al.[41] studied the extent of virus removal from septic tank effluent applied to pilot scale soil absorption systems. Each soil absorption system consisted of a soil column which was 5 feet long and 6 inches in diameter and had an unsaturated upper zone of 3 feet and a saturated bottom zone of 2 feet. The soil composition of the upper zone was either fine, loamy sand or a 1:1 mixture of fine, loamy sand and medium sand. The saturated bottom zone was composed of either gravel or a 1:2 mixture of organic soil (27% organic matter) and medium sand. There were a total of four different columns; one for each possible combination of upper and lower zones. The results of these column studies with both poliovirus and reovirus indicate that adequately designed and properly operated soil absorption systems can potentially achieve extensive virus removals from septic tank effluent.

In terms of modeling, Vilker, et al.[43] used ion exchange and adsorption equations for the prediction of the breakthrough of low levels of virus from percolating columns under conditions of both adsorption (application of wastewater to uncontaminated bed) and elution. The effects of external mass transfer and nonlinear adsorption isotherms were included. The predictions were in qualitative agreement with reported observations from experiments which measured virus uptake by columns packed with activated carbon or a silty soil.

Nitrogen

The transport and fate of nitrogen in the subsurface environment is dependent upon the form of the entering nitrogen and various biological conversions which may take place. Figure 5.5 displays the forms and fate of nitrogen in the subsurface environment.[44] The four primary forms are organic nitrogen, ammonia nitrogen, nitrites, and nitrates. Organic nitrogen can be converted to the ammonium form through biological decomposition processes. Nitrates (NO_3^-) can be formed by nitrification involving ammonium ion conversion to nitrites and then to nitrates. Nitrification ($NH_4^+ \rightarrow NO_2^- \rightarrow NO_3^-$) is an aerobic reaction performed primarily by obligate autotrophic organisms and NO_3^- is the predominant end product.

Denitrification is another important nitrogen transformation in the subsurface environment (soils and ground water) underlying septic tank systems. It is the only mechanism by which the NO_3^- concentration in the percolating (and oxidized) water flow can be decreased. Denitrification, or the re-

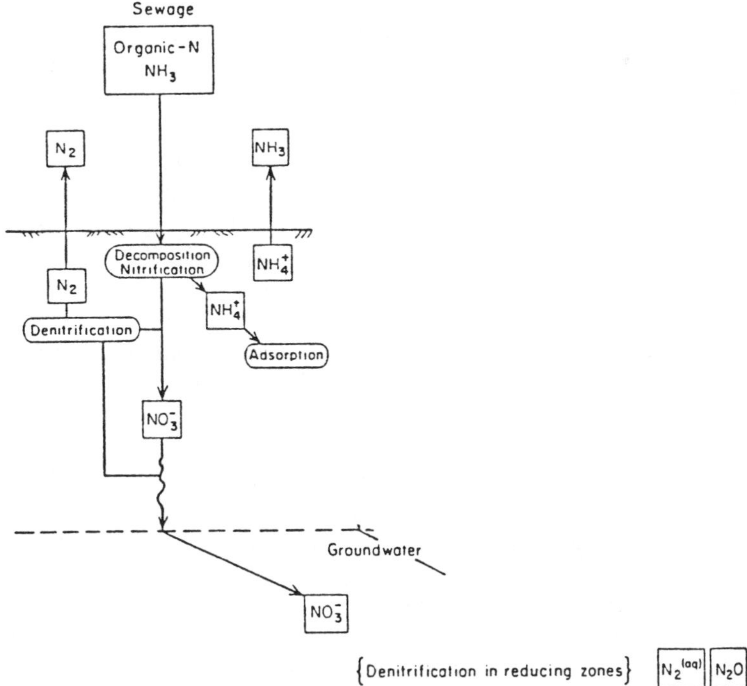

Figure 5.5. Form and fate of nitrogen in the subsurface environment.[44]

duction of NO_e^- to N_2O or N_2, is a biological process performed primarily by ubiquitous facultative heterotrophs. In the absence of O_2, NO_3^- acts as an acceptor of electrons generated in the microbial decomposition of an energy source. However, in order for the denitrification to occur in soils subject to nitrogen input, the nitrogen must usually be in the NO_3^- form, and an energy source must be available.

Based upon the forms of nitrogen which can be introduced into the subsurface environment from either point or diffuse sources, and based on the biological transformations which can occur, there are two forms of major concern relative to ground water pollution—ammonium ion (NH_4^+) and nitrates (NO_3^-). Ammonium ions can be discharged directly into the subsurface environment, or they can be generated within the upper layers of soil from the ammonification process (conversion of organic nitrogen to ammonia nitrogen). The transport and fate of ammonium ions may involve adsorption, cation exchange, incorporation into microbial biomass, or release to the atmosphere in the gaseous form. Adsorption is probably the major mechanism of removal in the subsurface environment.[32]

Under anaerobic conditions in the subsurface environment, positively charged ammonium ions (NH_4^+) are readily adsorbed onto negatively charged soil particles. Since anaerobic conditions in soils are usually associated with saturated soils, some movement of ammonia with ground water can occur. However, this movement will be slow since adsorption will continue to occur on soil particles in the aquifer.

Cation exchange may be involved along with adsorption in the retention of ammonium ions in soils underneath septic tank systems.[32] However, just as the adsorption capacity of a soil can be exceeded, the cation exchange capacity can also be exhausted. Under these conditions, the cation exchange sites in the soil beneath a seepage bed would become equilibrated with the cations in the effluent. The effluent would then move to the ground water with its cation composition essentially unchanged. Ammonia nitrogen can be incorporated into microbial or plant biomass in the subsurface environment; however, this is probably not a major removal mechanism relative to nitrogen in the subsurface. Finally, ammonia gas can be released to the atmosphere as a function of the soil-liquid pH conditions. When the pH is neutral or below, most of the nitrogen is in the ammonium ion form. As the pH becomes basic, the NH_4^+ is transformed into ammonia and can be released from the soil as a gas.

An important aspect of the subsurface transport and fate of nitrogen is the rate of ammonium nitrification. Corey, McWhorter, and Smith[45] utilized laboratory experiments to determine the time required for nitrification of ammonium and subsequent leaching of nitrates through soils. The transport and fate of nitrate ions may involve movement with the water phase, uptake in plants or crops, or denitrification. Of importance herein is movement with the water phase. Since nitrate ions (NO_3^-) have a negative charge, they are not attracted to soils which also possess negative charges. Accordingly, nitrates are more mobile than ammonium ions in both unsaturated and saturated soils. Two illustrations of nitrate behavior in the subsurface environment from diffuse sources will be cited.[46,47]

Substantial increases in ground water nitrate-nitrogen are of particular concern on intensively cropped irrigated sandy soils. A study was conducted by Hubbard, Asmussen, and Allison[46] to compare shallow ground water nitrate-nitrogen concentrations under intensive multiple cropping systems with those from under nearby nonagricultural sites. The intensive multiple cropping systems involved center pivot irrigation and sprinkler-applied fertilizer. The study was conducted on a sandy, coastal plain soil by weekly measurement and sampling of observation wells located under a center pivot area and at two forest sites. Nitrate-nitrogen concentrations ranged from less than 0.1 mg/L to about 133 mg/L, with a mean of 20 mg/L. In contrast, samples from the forest sites had nitrate-nitrogen concentrations ranging from less than 0.1 mg/L to just over 1 mg/L. Mean nitrate-nitrogen concentrations under the center pivot area were found to vary seasonally according to cropping and hydrologic patterns such that the mean values for March–May, June–August, September–November, and December–February were 7, 21, 27, and 21 mg/L, respectively. The lower value for March–May indicated that winter rains leached most of the root zone nitrate-nitrogen beneath the wells by March, and there was a two- to three-month lag between spring-applied nitrogen and its appearance in shallow ground water.

Walter[47] developed a mathematical model of manurial nitrogen movement through soil profiles under typical early spring conditions. Nitrogen transformation studies were conducted with batch systems of plainfield sand and anaerobic dairy waste. Laboratory soil columns were used to investigate flow and transport processes. The specific conditions studied were temperatures from 0 to 20°C, soil moisture from 5 to 20% by dry soil weight, soil pH from 6 to 8, and an aerobic soil environment. The quantity of nitrate in an incremental volume of soil depended upon its movement in or out of the soil volume due to mass flow of water and to the net production of nitrate within the volume of soil due to mineralization of organic nitrogen and nitrification of ammonium. Nitrate accumulation as predicted by the computer model was based on nitrification of added manurial ammonium and soil nitrogen mineralization. Estimates of solute dispersion were made based on the movement of the soil water after infiltration.

When nitrogen in the form of nitrate reaches ground water, it becomes very mobile because of its solubility and anionic form. Nitrates can move with ground water with minimal transformation. They can migrate long distances from input areas if there are highly permeable subsurface materials which contain dissolved oxygen. The only condition which can affect this process is a decline in the redox potential of the ground water. In this case, the denitrification process can occur.

Phosphorus

Phosphorus is typically present as phosphate ions in septic tank system effluents, leachates from sanitary landfills, land application systems for wastewater, and surface impoundments. While phosphorus can be a ground water pollutant of concern, it is typically not a major contaminant since it can be easily retained in surface and subsurface soils due to chemical changes and adsorption.

Phosphate ions become chemisorbed on the surfaces of Fe and Al minerals in strongly acid to neutral systems, and on Ca minerals in neutral to alkaline systems. As the concentration in the soil solution is raised, there comes a point above which one or more phosphate precipitates may form. For example, in the pH range encountered in septic tank seepage fields, hydroxyapatite is the stable calcium phosphate precipitate. At relatively high phosphorus concentrations similar to those found in septic tank effluents, dicalcium phosphate or octacalcium phosphate are formed initially, followed by a slow conversion to hydroxyapatite.[48] Therefore, both chemical precipitation as well as chemisorption is involved in phosphorus retention in soils. These removal mechanisms have been found by many investigators, including Enfield, et al.[49] who determined that the ability of soil to remove wastewater orthophosphate from a solution passing through a soil matrix is primarily related to the formation of relatively insoluble phosphate compounds of aluminum, iron, and calcium.

Sawhney[50] used laboratory soil columns to assess the ground water pollution potential of phosphorus from septic tank system drain fields. The sorption capacities of various soils were determined over an extended period of time and related to phosphorus movement through the soil columns using solutions having phosphorus concentrations similar to wastewaters. The amounts of phosphorus adsorbed by fine sandy loam and silt loam soil columns before breakthrough occurred were approximately equal to the adsorption capacities determined from isotherms obtained over a sufficiently long reaction time of about 200 hours. In Merrimac fine sandy loam, breakthrough occurred after about 50 pore volumes of wastewater had passed through the column while about 100 pore volumes passed through Buxton silt loam before the breakthrough occurred.

Mansell, et al.[51] found that reversible equilibrium adsorption-desorption relationships were inadequate for describing the transport of orthophosphate through water-saturated and water-unsaturated cores from surface A_1 and subsurface A_2 horizons of Oldsmar fine sand. Using a kinetic model with nonlinear reversible adsorption-desorption improved descriptions of phosphorus transport through these soils. Phosphorus effluent concentrations were described best by using an irreversible sink for chemical immobilization by precipitation with nonlinear reversible kinetic adsorption-desorption equation.

Stuanes and Enfield[52] noted that the capacity of a soil to react with phosphorus is often underestimated when equilibrium isotherms are used to describe phosphorus sorption. An analytical one-dimensional convective dispersive solute transport model, assuming linear adsorption and first-order irreversible precipitation, was evaluated to describe the movement of phosphorus in small laboratory columns. The model was calibrated to experimental data using nonlinear least squares analysis. The length of the column did not make a significant difference in the model coefficients. This finding suggests that the approximation of a linear adsorption isotherm with first-order irreversible precipitation is adequate for making predictions of phosphate movement through soil. Stuanes and Enfield[52] also found that experimental data fit to a Langmuir equation using the same soils seriously underestimated the capacity of the soils to react with phosphorus. The same data fit to a Freundlich or other equilibrium isotherm would yield similar results. It was concluded that use of equilibrium isotherms without considering the kinetics of the reactions will lead to overly conservative designs of land application of wastewater treatment systems.

Metals

Table 5.11 contains brief comments on 11 studies related to the transport and fate of metals in the subsurface environment. A review of the transport and fate of heavy metals in the subsurface environment has been prepared by Bates.[53] The four major reactions that metals may be involved in with soils are adsorption, ion exchange, chemical precipitation, and complexation with organic substances. Of these four, adsorption seems to be the most important for the fixation of heavy metals and their retention in the subsurface environment. Laboratory studies of various influence factors on the four major reactions for various metals and soil types have been conducted by Alesii and Fuller,[54] Frost and Griffin,[55] Fuller, et al.,[56] Gambrell, et al.,[57] Giordano and Mortvedt,[58] Patel,[60] and Wangen and Stallings.[63] A field study on trace metals in ground water at a fly ash disposal site is addressed by Theis.[61] Finally, mathematical models for the transport and fate of metals in the subsurface have been developed by Jurinak and Santillan-Medrano[59] and by Theis, Kirkner, and Jennings.[62]

Table 5.11. Summary of Eleven Studies on the Transport and Fate of Metals in the Subsurface Environment.

Author(s)	Comments
Alesii and Fuller[54]	Laboratory study of the movement of three forms of cyanide through 5 soil types.
Bates[53]	Literature review on the transport and fate of heavy metals in the subsurface environment.
Frost and Griffin[55]	Laboratory study of the effect of pH on clay adsorption of arsenic and selenium in landfill leachate.
Fuller, et al.[56]	Laboratory study of the influence of municipal sanitary landfill leachate quality on the movement of Cd, Fe, Ni, and Zn through soils.
Gambrell, et al.[57]	Laboratory studies of the effects of pH and redox potential on metal mobility and availability in sludge-amended soils.
Giordano and Mortvedt[58]	Laboratory study of the effects of nitrogen on the mobility of some heavy metals in soil.
Jurinak and Santillan-Medrano[59]	One-dimensional transport model for the movement of lead and cadmium through soils under steady state saturated flow conditions.
Patel[60]	Laboratory study of the adsorption of barium and chromium in five soil types.
Theis[61]	Trace metals in ground water at a fly ash disposal site for a coal-fired power plant.
Theis, Kirkner, and Jennings[62]	Finite element/chemical equilibrium subsurface flow model for heavy metals in fly ash pond leachates.
Wangen and Stallings[63]	Laboratory studies on the transport and fate of several trace elements in selected soil types.

As noted earlier, adsorption seems to be the most important attenuation mechanism for metals in the subsurface environment. Ion exchange is thought to provide only a temporary or transitory mechanism for the retention of trace and heavy metals. The competing effects exhibited by more common metal ions such as Ca^{+2}, Na^{+}, H^{+}, and K^{+} limit the cation exchange sites available for heavy metal removal. Precipitation reactions as a mechanism of metal fixation in soils are greatly influenced by pH and concentration, with precipitation predominantly occurring at neutral to high pH values and in macroconcentrations. Organic materials in soils may immobilize metals by complexation reactions or cation exchange. Organic materials have a very high cation exchange capacity, therefore providing more available exchange sites than most clays. Complexation reactions between metals and organic substances, although definitely serving to fix the metals, may only provide for temporary immobilization. If the organic complex is biodegradable, the metal may be subsequently released back to the soil environment. Conversely, organics-metal complexes may facilitate the movement of metals through the subsurface environment.[64] In summary, relative to the four major reactions involved in soil attenuation of metals, it is important to note that these reactions are dependent on a number of factors, including soil composition, soil texture, pH, and the oxidation-reduction potential of the soil and associated ions.[53]

Soil type or composition is a very important factor in all heavy metal fixation reactions. Clays are extremely important in adsorption reactions because of their high cation exchange capacity. In addition, soils high in humus or other organic matter also exhibit good exchange capacity. The type of clay mineral present is, in addition, an important factor. Many sorption reactions take place at the surface of iron and aluminum hydroxides and hydroxy oxides and, therefore, the iron and aluminum content of soils becomes an essential factor governing the ability of a soil for heavy metal immobilization. A number of studies have been conducted on the retention of zinc, copper, cadmium, lead, arsenic, mercury, and molybdenum by various soil types.[53]

The influence of soil type on the adsorption of barium and chromium was studied by Patel.[60] Muck, Hoytville, Nebraska, Norfolk, and Troup soils, which differ in genesis, organic matter content, pH, and cation exchange status, were used. The observed order of total adsorbed barium was Muck>Hoytville>Nebraska>Troup>Norfolk. The order of total adsorbed chromium was the same as for barium. The Muck soil had a higher tenacity to adsorbed metallic cations than the other four soils because it contains a higher percentage of organic matter, is more acidic, and has a higher cation exchange capacity than the other four soils. The magnitude of the average stability constants for the soil-metal complexes are in the order of Muck>Hoytville>Nebraska>Norfolk>Troup soil. This indicates that the relative stability of the resulting soil-metal complexes increase with an increase in organic matter. The results of these studies may be used to explain the high leachability of metal ions from certain soils or to predict which of the metal ions would be adsorbed to the most or least extent by a given soil.

Soil texture or soil particle size is another factor that can influence the fixation of metals by soils. In general, finely textured soils immobilize trace and heavy metals to a greater extent than coarsely textured soils. Also, finely textured soils usually have a greater cation exchange capacity which is an important factor in heavy metal fixation. Soil texture has been found to influence the transport of mercury, lead, nickel, and zinc.[53]

Soil pH plays a very important role in the retention and mobility of metals in soil. The pH is a controlling factor in adsorption-desorption reactions and precipitation-solubilization reactions. In addition, the cation exchange capacity of soils generally increases with an increase in pH. Even with a soil that has a high affinity for a specific metal, the degree to which the metal is fixed is a function of pH.[53] For example, Frost and Griffin[55] studied the effect of pH on the adsorption of arsenic and selenium anions from municipal sanitary landfill leachate. The results suggested that disposal of arsenic and selenium wastes in municipal landfills, especially under alkaline conditions, could produce ground water contamination by these two metals. Soil pH has been determined to be a major factor along with cation exchange capacity for the fixation of lead by soils. Soil pH also influences the retention of zinc, molybdenum, mercury, and copper.[53]

The oxidation-reduction or redox potential of a soil is very important in determining which species of an element is available for adsorption, precipitation, or complexation. For example, Gambrell, et al.[57] conducted laboratory studies on the effects of pH and redox potential on the chemical mobility and plant availability of copper, zinc, cadmium, lead, chromium, nickel, and arsenic in sludge-amended soils. In general, the reduced forms of a metal are more soluble than the oxidized forms. The redox potential of a soil system can be altered through biological activity, and a change in redox potential is many times correlated with changes in pH. Reducing conditions may be associated with a low pH resulting from the formation of CO_2 and organic acids from the microbial degradation of organic matter. A reducing environment typically exists in saturated soils underneath septic tank systems and waste disposal sites. The anaerobic conditions would enhance the mobility of metals in the subsurface environment. Iron is a good example of a metal which readily undergoes redox reactions. In the oxidized or ferric state, iron may form insoluble compounds of $Fe(OH)_3$ or $FePO_4$. However, when iron is reduced under anoxic conditions, the ferrous form, which is more soluble, predominates.[53]

Another factor affecting the retention or mobility of metal ions is competing ions. The presence of phosphate affects the retention of both arsenic and zinc. Arsenic tends to become more mobile in the presence of phosphate, and zinc is more highly retained. The effects of chlorides on the mobility of several heavy metals have also been investigated.[53] Two examples of the influence of competing ions will be cited.[54,56] Fuller, et al.[56] found that readily measurable parameters in municipal sanitary landfill leachates could influence the movement of cadmium, iron, nickel, and zinc through soils. The influencing parameters were total organic carbon, pH, and concentration of soluble common salts.

Alesii and Fuller[54] leached three solutions of cyanide, KCN in deionized water (simple form), $K_3Fe(CN)_6$ in deionized water (complex form), and

KCN in natural landfill leachate (mixed form) through five soils of varying physical and chemical properties to evaluate which soil characteristics govern the movement of the various cyanide forms in soils. The effluent from each soil column was collected daily and analyzed for total cyanide. In general, KCN and $K_3Fe(CN)_6$ in water were both found to be very mobile in soils, while KCN in landfill leachate was found to be less mobile. Soil properties such as low pH, presence of free iron oxide and kaolin, chlorite, and gibbsite-type clay (high positive charges) tended to increase attenuation of cyanide in the three forms. High pH, the presence of free $CaCO_3$ (high negative charges), low clay content, and montmorillonite clay tended to increase the mobility of the three cyanide forms.

Models for predicting the subsurface transport and fate of metals have been developed. Jurinak and Santillan-Medrano[59] developed a one-dimensional transport model for the movement of lead and cadmium ions through soils under steady state saturated moisture flow conditions. The chemical processes considered were: precipitation and dissolution, ion-pair formation, pH flux, and adsorption or cation exchange. The transport model was tested by data from laboratory soil columns using three soil types. Initially, both chemical equilibrium and kinetic studies were conducted to obtain data necessary for interpretive aspects of the column studies. The solubility of cadmium in soils is considerably greater than lead in the pH range of 5 to 9. At low concentrations of cadmium, adsorption by soil is an important mechanism of retention. At higher concentrations, precipitation of $Cd_3(PO_4)_2$ appears to regulate cadmium solubility. The transport model was effective in predicting cadmium and lead movement in soils after the initial 20 to 25 pore volumes of effluent passed through the soil column.

Theis, Kirkner, and Jennings[62] described the development of a finite element/chemical equilibrium subsurface flow model for application to metal transport around waste disposal sites. The tasks in the model development included soil column studies for use with the one-dimensional model, laboratory batch studies to define the necessary thermodynamic constants, soil chemistry characterization studies, organic studies, leachate composition evaluation, and assessment of available field data.

Organics

The transport and fate of organic contaminants in the subsurface environment is a relatively new topical area of concern. A variety of possibilities exist for the movement of organics, including transport with the water phase, volatilization and loss from the soil system, retention on the soil due to adsorption, incorporation into microbial or plant biomass, and bacterial degradation. The relative importance of these possibilities in a given situation is dependent upon the characteristics of the organic, the soil types and characteristics, and the subsurface environmental conditions. This very complicated topical area is being actively researched at this time.[32]

Information on volatilization, adsorption, and biodegradation will be presented in the following subsections. It is important to note that some studies listed in these subsections have two or even all three of these processes associated with pollutant transport and fate; their simple division by process was not possible. Finally, a subsection on immiscible organics is included.

Volatilization

Volatile organic compounds can be emitted to the atmosphere from soil subjected to planned waste disposal activities or accidental spills. For example, Wetherold, Randall, and Williams[65] conducted laboratory studies to determine the volatile organic emissions resulting from oily petroleum sludges from refineries being incorporated into soils. The sludges tested included three of the five listed hazardous wastes from the refining industry: dissolved air flotation float, slop oil emulsion solids, and API separator sludge. The volatile components of the sludges were first identified. Then the effects of air temperature and humidity, wind speed, soil type, temperature and moisture, sludge loading and volatility, and method of waste application were studied. The volatile components identified in the sludge were also present in the emissions from the soil-waste mixtures. The quantity of emissions was most affected by the sludge volatility, sludge loading, application method, and atmospheric humidity; the cumulative emissions at a given period of time could be correlated with these operational variables. The emission rate or level was not significantly affected by soil type or soil moisture.

Enfield, et al.[66] developed a methodology for estimating volatilization of toxic organic chemicals from unsaturated soils. Projections were compared with laboratory data for simulated rapid infiltration wastewater treatment systems receiving primary municipal wastewater spiked with a suite of 18 toxic organic compounds, nine of which were considered volatile. Predictions within a factor of two were possible under the quasi-steady state conditions of the experiment for most compounds applied. Volatilization was calculated by applying Fick's second law under the experimental conditions.

One of the issues in volatilization is whether or not soil adsorption of vapors will decrease the atmospheric emissions. Manos, et al.[67] conducted a study to determine if adsorption should be considered in volatile organic compound (VOC) vapor emission models of gas movement through soils. Liquid n-pentane was placed into the bottom of a soil column containing incorporations of silt plus clay into sand, and the resulting n-pentane gaseous emission concentrations measured over a 24-hour period. N-pentane emission concentrations were reduced by 1% for every percent incorporation into sand of the undisturbed silt plus clay fraction containing clay mineral-organic matter complexes. N-pentane emission concentrations were reduced by 4% for every 1% incorporation into sand of the 400°C fired (organic matter removed) silt plus clay fraction. Incorporation of the "less than 0.063 mm" (silt plus clay) fraction, taken from a natural soil (Houston Black Series), into Ottawa sand matrix soil columns resulted in decreased emissions of n-pentane over a 24-hour period of monitoring. Similar results were obtained by incorporation of 400°C fired silt plus clay fractions into Ottawa sand. Thus, it was found that both soil organic matter and clay mineral content of soils significantly affect VOC gaseous emissions from soils.

Adsorption

Adsorption refers to an abiotic process wherein solutes are adsorbed onto subsurface materials; therefore, adsorption retards the movement of contaminant chemicals through the unsaturated and saturated zones. Table 5.12 contains brief comments on 8 studies wherein adsorption was a dominant factor in the subsurface transport and fate of organics. Adsorption processes during ground water recharge with treated domestic wastewater were noted by Hutchins, et al.,[68] McCarty, Reinhard, and Rittman,[69] and Roberts, et al.[70] The adsorption of phenol on silt loam soils was addressed by Scott, Wolf, and Lavy.[71]

Griffin and Chian[72] reported on laboratory studies of the adsorption, biodegradation, and volatility of polychlorinated biphenyls (PCBs) when contacted with five soils. A simple linear relation described the adsorption of water-soluble PCBs by the five earth materials. An adsorption constant (K) unique to each adsorbent was obtained. The adsorption was strongly correlated to the total organic carbon (TOC) content and surface area of the earth materials. TOC was the dominant of these two earth material properties by a ratio greater than three to one.

The adsorption, mobility, and degradation of polybrominated biphenyls (PBBs) and hexachlorobenzene (HCB) in soil materials and in a carbonaceous adsorbent were also studied by Griffin and Chou.[73] Freundlich adsorption isotherm plots of PBBs and HCB sorption on soils and the carbonaceous adsorbent yielded straight and nearly parallel lines. PBBs and HCB were found to be strongly adsorbed by the carbonaceous adsorbent and by soil materials, with HCB being adsorbed to a greater extent than PBBs. The adsorption capacity and mobility of PBBs and HCB were highly correlated

Table 5.12. Summary of Eight Studies Involving the Adsorption of Organics in the Subsurface Environment.

Author(s)	Comments
Griffin and Chian[72]	Laboratory study of the adsorption of polychorinated biphenyls by earth materials.
Griffin and Chou[73]	Laboratory study of the adsorption of polybrominated biphenyls and hexachlorobenzene by soil materials and a carbonaceous adsorbent.
Hutchins, et al.[68]	Field study of fate of trace organics during rapid infiltration of primary wastewater.
McCarty, Reinhard, and Rittman[69]	Study of relative retention times of chloroform, chlorobenzene, and 1,4-dichlorobenzene on subsurface media.
Roberts, et al.[70]	Field study of organic contaminant behavior during artificial recharge, particularly for chlorobenzene and chloroform.
Schwarzenbach and Westall[74]	Laboratory sorption studies of the transport of nonpolar organic compounds from surface water to ground water.
Scott, Wolf, and Lavy[71]	Study of adsorption and degradation of phenol at low concentrations in silt loam soils.
Wilson, et al.[75]	Laboratory study of the transport and fate of 13 organic pollutants in a sandy soil with low organic matter content.

with the organic carbon content of the soil materials. In a soil incubation study, it was found that PBBs and HCB persisted for six months in soil with no significant microbial degradation. Because of their low water solubilities, strong adsorption, and persistence in soils, these two compounds are highly resistant to aqueous phase mobility through earth materials; however, they are highly mobile in the presence of organic solvents.

Laboratory sorption studies of nonpolar organic compounds have been conducted by Schwarzenbach and Westall.[74] The experiments were conducted using toluene; di-, tri-, and tetramethylbenzenes; n-butylbenzene, tetrachloroethylene, chlorobenzene, and di-, tri-, and tetrachlorobenzenes with several adsorbents (natural river and lake sediments, kaolin, silica, and activated sewage sludge). For a variety of adsorbents, the partition coefficient for a particular compound can be estimated from its octanol/water partition coefficient and from the organic carbon content of the adsorbents if the amount of organic carbon is greater than 0.1%. These findings are similar to the results of other researchers.

Wilson, et al.,[75] studied the transport and fate of 13 organic pollutants in a sandy soil with low organic matter content (0.087% organic C). Glass columns were packed with soil in a manner that preserved the original soil profile. Water containing 1.0 or 0.2 mg per liter of the organic compound was applied at the rate of 14 cm per day for 45 days. Concentrations of compounds in the effluents and the amounts volatilized were determined and the mass ballances calculated. Chemicals which percolated rapidly through the soil without degradation were chloroform, 1,2-dibromo-3-chloropropane, dichlorobromomethane, 1,2-dichloroethane, tetrachloroethene, 1,1,2-trichloroethane, and trichloroethene; retardation factors (velocity of tritiated water through the soil divided by the apparent velocity of pollutant through soil) were 2.5 or less. Little or no biodegradation was observed. From 19 to 65% of chemical applied to the surface reached the 140-cm depth, and the rest volatilized. Retardation of chlorinated benzenes generally increased with decreasing water solubility. Some degradation occurred in this group of compounds. Bis(2-chloroethyl) ether did not degrade, and its retardation factor was less than 1.5. A method was developed to predict the retardation factors on the basis of water solubility of the chemical and soil organic C content. In general, it could be concluded that soils containing low organic matter are more pollution vulnerable.

Biodegradation

Organic contaminants can be biologically degraded in the subsurface environment under either aerobic or anaerobic conditions. Biodegradation is an active research area, and Table 5.13 includes brief comments on 13 pertinent studies. Biotransformation information can be useful for understanding subsurface transport and fate and for planning and evaluating in-situ remedial action programs.

Larson and Ventullo[76] discussed techniques for measuring the density, heterotrophic activity, and biodegradation potential of ground water bacteria. Based on the Canadian ground water used, it was determined that bacteria were present at significant levels in the ground water samples tested and were the predominant microbial species. However, only 1 to 10% of the total bacteria present appeared to be metabolically active and capable of growth on dilute complex nutrient media. The activity of ground water bacteria, based on turnover time and specific activity indices for several natural substrates, approximated that typical of microorganisms in oligotrophic surface waters. Finally, the biodegradation potential of ground water bacteria on natural and xenobiotic substrates was variable and related to suitability of the substrate as a carbon and energy source. For degradable substrates, rates of biodegradation were fairly rapid, even in the oligotrophic ground water tested.

The fate of six organic compounds during rapid infiltration of primary

Table 5.13. Summary of Thirteen Studies Involving Biodegradation of Organics in the Subsurface Environment.

Author(s)	Comments
Barker and Patrick[77]	Field study of the processes controlling the rate of movement and persistence of dissolved benzene, toluene, and xylene in a shallow, unconfined sand aquifer.
Bouwer[78]	Laboratory column studies of biofilm processes for the transformation of organic contaminants in ground water.
Bouwer and McCarty[79]	First-order kinetic model for describing subsurface microbial processes.
Bouwer, Rittman, and McCarty[80]	Laboratory studies of the anaerobic degradation of trihalomethanes, trichloroethylene, and tetrachloroethylene.
Enfield, et al.[81]	Compartmental screening model for describing movement of volatile and transformable organic chemicals in rapid infiltration wastewater treatment systems.
Godsy, Goerlitz, and Ehrlich[82]	Field study of methanogenesis of phenolic compounds in a contaminated aquifer.
Hutchins, et al.[83]	Laboratory study of the fate of six organic compounds during rapid infiltration of primary wastewater through soil columns.
Larson and Ventullo[76]	Techniques for measuring the number, activity, and biodegradation potential of heterotrophic bacteria in ground water.
Piwoni, et al.[84]	Use of microcosms to study the behavior of organic pollutants during rapid infiltration of municipal wastewater into soil.
Tsentas and Supkow[85]	Natural subsurface biotransformations of a tetrachloroethylene spill.
White, et al.[86]	Study of the subsurface biodegradation rates of methanol and tertiary butyl alcohol.
Wilson and Rees[87]	Laboratory study of the biotransformation of gasoline hydrocarbons in methanogenic aquifer material.
Wilson and Wilson[88]	Aerobic biotransformation of trichloroethylene in soil under enhanced conditions.

wastewater through soil columns was studied by Hutchins, et al.[83] Feed solutions were prepared which contained all six compounds in individual concentrations ranging from 1 to 1,000 μg/L and were applied to separate soil columns on a flooding and drying schedule. Breakthrough profiles of o-phenylphenol were relatively consistent during the test, with fractional breakthrough (mass output/mass input) being independent of input concentration. Consistent profiles were also observed for 2-(methylthio)benzothiazole, although fractional breakthroughs were higher at lower input concentrations, indicating that removal processes were operating less efficiently at these levels. The behavior of p-dichlorobenzene was similar to that of 2-(methylthio)benzothiazole after the first inundation cycle, with the exception that increased fractional breakthroughs were observed at the highest input concentration as well. Microbial adaptation was evident for benzophenone, 2-methylnaphthalene, and p-(1,1,3,3-tetramethylbutyl)phenol, as indicated by increased removal efficiencies during successive inundation cycles, especially at the higher input concentrations. The data demonstrate that microbial activity is important in the removal of several trace organic compounds during rapid infiltration of wastewater and that this removal is generally enhanced by the presence of adapted microflora.

Piwoni, et al.[84] used laboratory microcosms to study the behavior of organic pollutants during rapid infiltration of municipal wastewater into soil. The microcosms received 4.4 ± 0.2 cm of wastewater each day; an application was made every four hours. The wastewater was amended with selected organic compounds at individual average concentrations of 0.2 to 32 μmol/L. Chloroform, 1,1-dichloroethane, 1,1,1-trichloroethane, trichloroethene, and tetrachloroethene volatilized extensively. Individual average concentrations in the microcosm effluent ranged from $9 \pm 5\%$ to $27 \pm 11\%$ of the applied concentrations. There was observable degradation of 1,1-dichloroethane, but there was no detectable degradation of the other volatile compounds. Small proportions of chlorobenzene, 1,2,-dichlorobenzene, and toluene volatilized; the remaining portion was extensively degraded, producing concentrations in the column effluent of $9 \pm 10\%$ of the influent concentration or less. The concentration of 1,2,4-trichlorobenzene in the column effluent was less than 0.7% of that applied. The proportion that volatilized was not measured. Degradation was extensive for nitrobenzene, phenol, 2-chlorophenol, 2,4-dichlorophenol, and 2,4,6-trichlorophenol; concentrations in the column effluent were less than 0.2% of applied concentrations.

Enfield, et al.,[81] developed a compartmental model to describe the movement of volatile and transformable organic chemicals in rapid-infiltration wastewater treatment systems, as studied by Piwoni, et al.[84] The first compartment describes losses of the chemical from the infiltration basin when the basin is flooded with wastewater. The second compartment considers losses due to volatilization and transformation in near surface soils during periods of drying. The third compartment describes the transport and transformation of the remaining chemical to ground water. The model was developed to consider an accidental spill of chemical into a treatment system or a constant input of chemical into a system. Laboratory data for 18 organic compounds from microcosms designed to simulate a rapid-infiltration system were used to evaluate the model.[84] Data used in the evaluation were obtained after quasi-steady state had been achieved. Projections of volatilization were within a factor of two for slowly transformable compounds. The model overestimates volatilization for degradable compounds if one assumes no transformation. When transformation is considered, accounting can be made for chemical losses. A major limitation to implementation of the model in design practice is lack of knowledge on actual field transformation rates.

One of the concerns relative to biodegradation of trace organics in the subsurface is that a sufficient concentration of substrate may not be present. However, Bouwer[78] noted that in the presence of one compound at a relatively high concentration, termed the primary substrate, the contaminant present at trace concentrations, termed the secondary substrate, can be biotransformed as well. This concept was evaluated through laboratory column studies with several petrochemicals of importance in ground water. Trace concentrations of chlorinated benzenes and halogenated aliphatics were biotransformed by acetate-supported biofilms; the former under aerobic conditions and the latter under methanogenic conditions. The laboratory results compare favorably with field evidence for biotransformations. These results have implications for subsurface transport and fate studies as well as in the planning of remedial action programs. For example, injection of a primary substrate and nutrients into the subsurface to stimulate microbial activity may be an important tool for cleaning up aquifers that have become contaminated with organic micropollutants. Bouwer and McCarty[79] developed a simplified batch model with first-order kinetics for describing subsurface microbial processes when low active organism and pollutant concentrations exist over a large scale. This model can be used to describe this secondary utilization phenomena by biofilm processes.

As noted earlier, biodegradation of organics in the subsurface environment can also occur under anaerobic conditions. For example, methanogenesis of phenolic compounds by a bacterial consortium in a contaminated aquifer has been described by Godsy, Goerlitz, and Ehrlich.[82] The ground water was contaminated by the operation of a coal tar distillation and wood treatment facility.

Bouwer, Rittman, and McCarty[80] incubated trihalomethanes, trichloroethylene, and tetrachloroethylene at concentrations found in ground water. The incubations were conducted aerobically in the presence of primary sewage bacterial cultures and anaerobically in the presence of mixed methanogenic bacterial cultures. No aerobic conditions were found under which these compounds could be degraded. Anaerobic degradation was observed for the trihalomethanes, but the 2-carbon chlorine-substituted aliphatic compounds remained unchanged or were degraded only slightly. The brominated trihalomethanes were degraded rapidly in both anaerobic sterile controls and seeded cultures, indicating a chemical mechanism. However, the rate of decomposition increased in the presence of microbial activity. Chloroform degradation was much slower, occurred only in anaerobically seeded cultures, and appeared to be biologically mediated.

Natural anaerobic biodegradation has been suggested as a cause of the occurrence of nonspilled organic compounds at the site of a tetrachloroethylene spill. For example, Tsentas and Supkow[85] described a spill of an unknown volume of tetrachloroethylene over an eight-year period, thus contaminating the ground water in a shallow bedrock aquifer in central New Jersey. In addition to tetrachloroethylene, several other organic compounds were detected in ground water samples. These compounds may have been present as impurities in the spilled tetrachloroethylene or may be the products of the biodegradation of tetrachloroethylene in the subsurface. Figure 5.6 suggests a possible biodegradation scheme for tetrachloroethylene.[85]

Subsurface biodegradation of dissolved gasoline products has also been noted in field and laboratory studies. In a field study, Barker and Patrick[77] examined the processes controlling the rate of movement and the persistence of dissolved benzene, toluene, and xylene (BTX) in ground waters in a shallow, unconfined sand aquifer. BTX-spiked ground water was introduced below the water table, and this migration of contaminants through a dense sampling network was monitored. BTX components migrated slightly slower than the ground water due to adsorptive retardation. Essentially all the injected mass of BTX was lost within 434 days due to biodegradation. Rates of mass loss were highest for m- and p-xylene, lower for o-xylene and toluene, and lowest for benzene which was the only component to persist beyond 270 days. Laboratory biodegradation experiments produced similar rates under aerobic conditions.

Wilson and Rees[87] studied the behavior of four compounds characteristic of a weathered gasoline spill in authentic aquifer material that receives municipal landfill leachate and is known to support methanogenesis. The compounds were: benzene, methylbenzene (toluene), ethylbenzene, and 1,2-dimethylbenzene (o-xylene). All manipulations were done in an anaerobic glovebox to ensure the maintenance of methanogenic conditions. The treatments were: (1) aquifer material plus alkylbenzenes, (2) aquifer material plus alkylbenzenes and nutrients, (3) aquifer material, autoclaved plus alkylbenzenes, (4) autoclaved water plus alkylbenzenes, and (5) aquifer material without alkylbenzenes. Initial concentrations were approximately 600 μg/L for benzene and toluene and 250 μg/L for ethylbenzene and o-xylene. All of the compounds were degraded in the anaerobic subsurface material with or without nutrients. Toluene degradation was apparent after six weeks. At the end of twenty weeks of incubation, the concentration of toluene was reduced

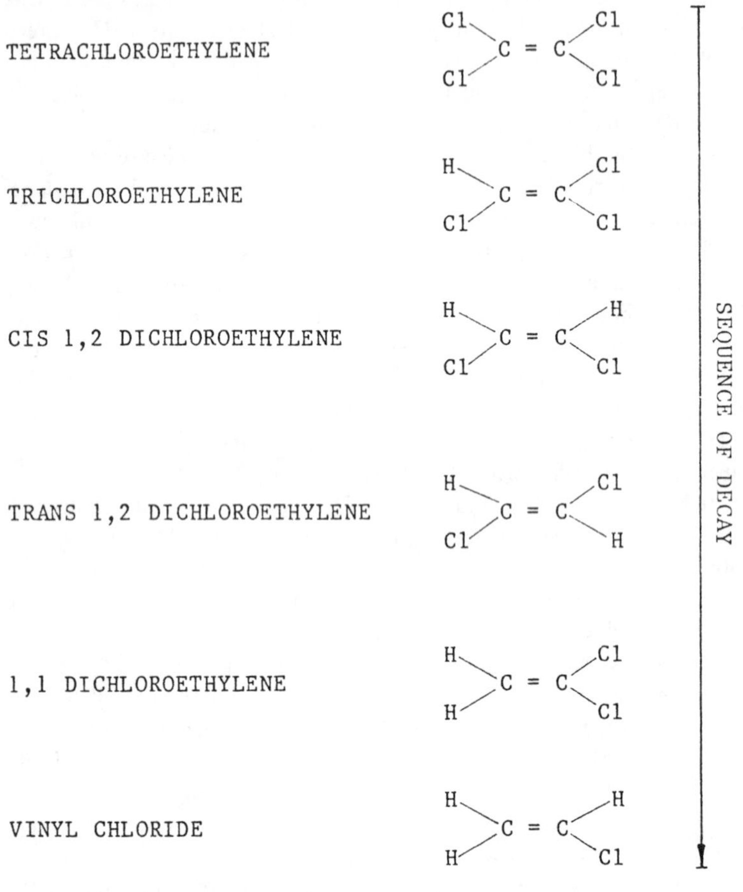

```
                                     Cl\       /Cl
TETRACHLOROETHYLENE                     C = C
                                     Cl/       \Cl

                                     H\        /Cl
TRICHLOROETHYLENE                       C = C
                                     Cl/       \Cl

                                     H\        /H
CIS 1,2 DICHLOROETHYLENE                C = C
                                     Cl/       \Cl

                                     H\        /Cl
TRANS 1,2 DICHLOROETHYLENE              C = C
                                     Cl/       \H

                                     H\        /Cl
1,1 DICHLOROETHYLENE                    C = C
                                     H/        \Cl

                                     H\        /H
VINYL CHLORIDE                          C = C
                                     H/        \Cl
```

SEQUENCE OF DECAY

Figure 5.6. Biodegradation sequence of tetrachloroethylene.[85]

by 80% of the original amount. Benzene and o-xylene were reduced by 20%, and ethylbenzene was reduced by 12%. Significant degradation of all compounds occurred after 40 weeks of incubation. The compounds did not degrade in autoclaved aquifer material or water, thus implicating a biological process.

The subsurface biodegradation rates of methanol and tertiary butyl alcohol (TBA), two compounds used commercially as gasoline additives, were studied by White, et al.[86] Four locations were chosen for subsurface soil and ground water sampling. One of the sampling sites had been previously contaminated with gasoline containing TBA for several years. Significant bacterial populations were found to exist in all subsurface samples, down to a depth of 100 feet. In an aerobic aquifer, both methanol and TBA were found to biodegrade readily. Pristine, anaerobic ground water systems also exhibited high methanol utilization rates. However, under these ground water conditions, TBA was only slowly biodegraded. Complete utilization of TBA never occurred in pristine, anaerobic aquifers. Contamination of an aquifer by gasoline had no observed effect on methanol biodegradation, but seemed to cause an enhancement of TBA biodegradation. In this anaerobic contaminated ground water system, TBA utilization proceeded slowly initially, similar to uncontaminated systems, then increased dramatically. Anaerobic TBA utilization rates were found to be first order with respect to initial substrate concentration in both contaminated and pristine systems. Kinetic analysis revealed that TBA biodegradation in the contaminated system could be predicted from laboratory data and extrapolated to lower temperatures.

One of the most commonly detected organic contaminants in ground water is trichloroethylene. Trichloroethylene is resistant to biodegradation in aerobic subsurface environments; however, it can be biotransformed under anaerobic conditions as shown in Figure 5.6. Wilson and Wilson[88] have conducted laboratory studies to explore the possibility of enhancing aerobic biodegradation of trichloroethylene through the addition of natural gas to a soil system. Trichloroethylene was shown to degrade aerobically to carbon dioxide in unsaturated soil columns exposed to a mixture of 0.6% natural gas in air. This finding has important implications for in-situ remedial action measures for contaminated ground water.

Immiscible Organics

In terms of ground water protection, fluids immiscible with water are chiefly organic substances and for the most part mineral oil products and products of the chemical industry. Mineral oil products include medium distillates, jet fuel, and gasoline.[89] Immiscible organics are not absolutely insoluble in water; they may exhibit a weak solubility of several mg/L up to several hundred mg/L. Oil products tend to migrate in the subsurface along the surface of the water table as shown in Figure 5.7.[89] An oil phase and a phase involving oil components dissolved in water are shown.

Therefore, the transport of immiscible organics in the subsurface environment occurs in several phases, including bulk liquid, dissolved, and vapor phases.[90] Mechanisms that influence transport include the physicochemical properties of the specific compounds present such as density, vapor pressure, viscosity, and hydrophobicity, as well as the physical and chemical

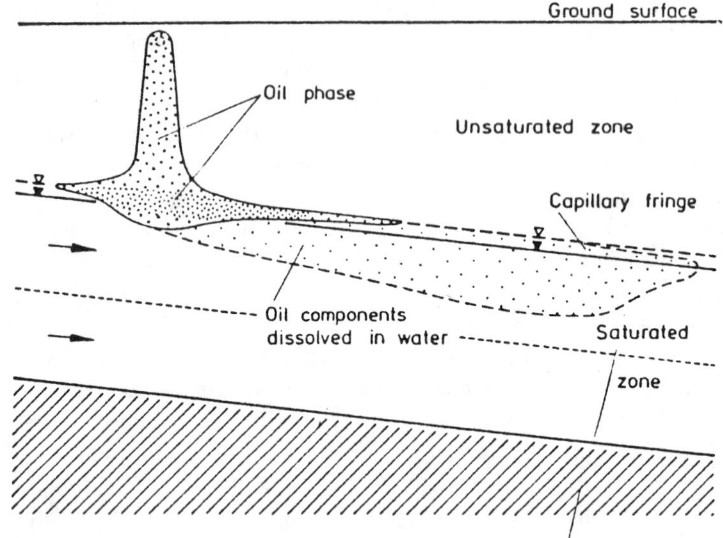

Figure 5.7. Oil migration pattern in the subsurface.[89]

properties of the subsurface environment, including geology and ground water hydrology. Hydrocarbon liquids are typically complex mixtures composed of numerous compounds, each with its own individual physicochemical and, therefore, transport properties. Ground water transport results in relative enrichment by more soluble, less hydrophobic hydrocarbon compounds as a function of distance from a spill. Vapor phase transport typically results in relative enrichment in more volatile hydrocarbon compounds.

Adsorption and biodegradation can be associated with the transport and fate of immiscible organics in the subsurface environment. For example, Van Der Waarden, Groenewoud, and Bridie[91] investigated the adsorption of oil components on lime, clay, and organic soil components. They found that adsorption may reduce the total amount of oil transported to ground water, reduce oil concentrations in ground water, and delay transport.

Kappeler[92,93] conducted experiments on the microbial degradation of the water-soluble fraction associated with an oil spill on a sandy soil. The rate of degradation of 47 water-soluble components of gas-oil by the mixed autochthonous microflora of clean ground water and by four isolates of the genus *Pseudomonas* was measured in batch experiments. In assays with the natural microbial flora of ground water, the aqueous solution (2 ppm hydrocarbons) was incubated directly. In tests with pure strains, the solution was sterilized prior to inoculation. Control experiments on degradation of single compounds were performed in sterilized ground water containing 10 ppm of the compound to be investigated. Identification of some intermediate degradation products gave insight into the primary reactions of the oxidative breakdown of some of the aromatic components of gas oil. The two operative degradation pathways were oxidation of the alkyl substituents followed by ring opening and direct oxidation of substituted or unsubstituted mono- or polynuclear aromatics. Kappeler[93] found that available nitrogen and dissolved oxygen are limiting factors in the microbial purification of hydrocarbon-contaminated ground water.

An understanding of the transport and fate mechanisms for immiscible organics is necessary in assessing contamination, source identification, predicting contaminant fate, and design of an appropriate remedial program.[88] One approach for gaining this understanding is to use a model, and Khaleel[94] has described the development of a two-dimensional, two-phase numerical model for the flow and transport of organic contaminants resulting from gasoline spills and leakage.

Pesticides

The transport and fate of pesticides in the subsurface environment will be addressed in terms of field studies, laboratory studies, modeling, and summary information.

Field Studies

Table 5.14 contains brief comments on 11 field studies related to the transport and fate of pesticides in the subsurface environment. Factors associated with the type and characteristics of the pesticide and the subsurface environment influence pesticide migration to ground water. For example, pesticides with significant adsorptive properties in the soil structure and/or short-lived persistence may be of minimal concern.[95] LeGrand[96] identified the following subsurface characteristics which would limit concern about pesticides in ground water: (1) a deep water table which allows for sorption of pollutants on earth materials, slows the subsurface movement of pollutants, and facilitates oxidation or other beneficial retention or removal mechanisms; (2) sufficient clay in the path of pollutants so that retention or sorption of pollutants is favorable; (3) a water table gradient beneath an application area away from nearby wells; and (4) a large distance between wells and the application area. Specific pesticides addressed in the field studies listed in Table 5.14 include atrazine;[97–99] 2,4-D;[100,101] bromacil;[102] toxaphene and fluonethuron;[103] and DDT.[104]

Table 5.14. Summary of Some Field Studies on the Transport and Fate of Pesticides in the Subsurface Environment.

Author(s)	Comments
Dregne, Gomez, and Harris[100]	Adsorption and degradation of 2,4-D herbicide in three soil systems.
Garrett, Maxey, and Katz[95]	Influence of subsurface materials on pesticide migration to ground water.
Hall and Hartwig[97]	Conventional tillage effects on atrazine mobility in two Pennsylvania soils.
Hebb and Wheeler[102]	Field study of bromacil transport to ground water in a sandy zone.
Lafluer, Wojeck, and McCaskill[103]	Transport of toxaphene and fluonethuron through Dunbar soil.
Le Grand[96]	Natural environmental factors tending to mitigate pesticide transport to ground water.
Mansell, et al.[101]	Influence of tillage and soil management on herbicide and acaricide movement from citrus groves in a Florida sandy flatwood soil.
Reece, et al.[105]	Mobility of organic arsenic herbicides at an abandoned herbicide manufacturing plant.
Scalf, et al.[104]	Movement of DDT from a recharge well in the Ogallala aquifer.
Schneider, Wiese, and Jones[98]	Field study of the transport of three herbicices into a sand aquifer.
Wehtje, et al.[99]	Fate of atrazine in ground water underlying Nebraska corn producing areas.

Atrazine is a premergence herbicide effective in controlling broadleaf weeds and some grasses. Hall and Hartwig[97] studied the mobility of atrazine in two Pennsylvania soils subjected to conventional tillage. Atrazine application rates ranging from 1.0 to 4.5 kg/ha were used, and only very low levels were detected in lysimeter leachates collected at 122 cm. Therefore, it appeared from this study that application of atrazine to fine-textured, conventionally tilled soils at rates ranging from 1.0 to 4.5 kg/ha would not seriously affect ground water supplies. Schneider, Wiese, and Jones[98] studied the movement of three herbicides (picloram, atrazine, and trifluralin) and a sodium nitrate tracer in an aquifer recharged by injection for 10 days at an average rate of 81.8 cu m/hour. After a 10-day pause, the well was pumped for 12 days to determine if the herbicides and tracer could be recovered. Water samples from observation wells located 9, 20, and 45 m from the dual-purpose well were collected, and herbicides were detected in the 9- and 20-m distant wells, but none of the herbicides or the tracer was detected in the 45-m distant well.

Ground water under large portions of corn producing areas in the Platte Valley of Nebraska has been found to be contaminated with trace levels of atrazine ranging from 0.2 to 0.8 ppb.[99] The widespread occurrence of these low concentrations result from leaching of normal agricultural usage of atrazine. Field studies found that the usage of atrazine on sandy soil will result in a small percentage (<0.1%) of the applied amount reaching the underlying aquifer, assuming irrigation is sufficient to result in deep percolation and thus provide recharge to the aquifer. Long-term buildup of atrazine in ground water would be expected in areas with continuous corn production under irrigation with yearly applications of the herbicide; however, this buildup has not occurred due to dilution, dispersion, and chemical hydrolysis. The rate of hydrolysis is conservatively estimated at 5% per year. Therefore, annual additions of atrazine into ground water, followed by some hydrolysis, will result in either a slow increase in concentration, or an equilibrium situation.

The movement of the herbicide 2,4-dichlorophenoxyacetic acid (2,4-D) in three soils was studied by Dregne, Gomez, and Harris.[100] They found that 2,4-D in the acid or salt form is only slightly adsorbed in permeable sandy loam soil; however, adsorption occurred in silty clay loam soil. Some biodegradation of 2,4-D was also noted. Mansell, et al.[101] determined the concentrations and discharge amounts of nitrates, phosphates, 2,4-D, terbacil herbicide, and chlorobenzilate acaricide in surface and subsurface drainage waters from a citrus grove located in an acid, sandy flatwood soil of southern Florida. Citrus growing in three soil management treatments was examined: ST (shallow-tilled plowed to 15 cm); DT (deep-tilled and soil mixed within the top 105 cm); and DTL (deep-tilled to 105 cm and 56 mt/ha of dolomitic limestone mixed with the soil). Deep tillage decreased the leaching losses of terbacil and 2,4-D. Discharges of these herbicides in subsurface drainage were usually in the order: ST>DTL>DT. Chlorobenzilate pesticide was not detected in drainage water from any of the three soil treatments.

Hebb and Wheeler[102] evaluated the leaching of a persistent and mobile herbicide, bromacil, into a shallow ground water with a set of environmental conditions highly conducive to leaching. These conditions included a sandy soil low in organic matter and plentiful rainfall. Bromacil was applied in a single application at a high rate of 22 kg/ha to a lakeland sand-bearing scrub vegetation of small oaks and poor grasses. Ground water (at depths ranging from 4.5 to 6 m) was sampled for bromacil residue at weekly intervals for two years. Residue was first found in the ground water three months after application and was highest (1.25 mg/l) one month later. Thereafter, the amount declined to less than 0.1 mg/l in about a year and less than 0.001 mg/l in two years. Peaks in residues generally followed periods of increased rainfall by about two years. Residues (0.24 mg/l) were still detected in the surface soil two years after application.

Toxaphene and fluonethuron were applied at 100 and 40 kg/ha, respectively, to topsoil by Lafluer, Wojeck, and McCaskill.[103] Losses from soil and accumulations in the underlying ground water were monitored for one year. Toxaphene loss from topsoil occurred in two stages; the second (major) stage was approximately linear on a log residue vs log time plot. The half residence time was less than 100 days. Fluonethuron loss from topsoil occurred in one continuous approximately linear episode. The half residence time was less than 120 days. Both compounds were found in the underlying ground water within two months after application, and they persisted in the ground water during the entire study year.

The movement of DDT and nitrates from an artificial recharge well in the Ogallala aquifer was studied by Scalf, et al.[104] Information was assembled on the distance and manner in which DDT and nitrates move from a well being recharged and how DDT and nitrates are released from the aquifer when pumping the well that has been recharged for a specified time with water containing measured concentrations of these substances.

Reece, et al.[105] addressed the mobility of organic arsenic herbicides in the subsurface environment at an abandoned herbicide manufacturing plant. Organic arsenic herbicides such as monosodium methanearsonate, along with several phenolic amine-based herbicides, were produced from 1968 through 1981 at a now abandoned plant located in Houston, Texas. Contamination of the first water-bearing zone, called the 35-foot sand, has been recently found. Organic and inorganic arsenicals have migrated from now closed waste ponds and from spill sources to the ground water of the 35-foot sand.[105] This sand, which is of limited extent and is not presently utilized as a drinking water or industrial water source, contains water having up to 900 ppm total arsenic. Trimethylarsine oxide, methanearsonic acid, cacodylic acid and arsenate are present in the ground water of the 35-foot sand. The chemical forms of organic arsenic present in the soil and ground water and potential chemical transformations are a critical consideration in assessing the mobility and attenuation potential of the arsenic species.

There is some concern related to further movement of organic arsenic herbicides from the abandoned Houston, Texas plant into deeper sands which are water-bearing zones. The most probable attenuation mechanism in the clay aquitard at the site is adsorption to the soil. The presence of iron, manganese, and aluminum oxides and hydroxides in the soil will tend to maximize attenuation of arsenic by adsorption and precipitation. Organic species (particularly trimethyl and dimethyl forms) of arsenic will reduce the rate of attenuation since the organics are adsorbed to a lower degree than the inorganic species. Laboratory studies of adsorption on clays from a depth of about 50 feet confirmed the applicability of Langmuir isotherms. Based upon these results and various calculations, it has been found that migration of arsenic through the aquitard to the 100-foot sand by direct flow through the aquitard should require a time period on the order of 2,000 to 6,000 years, if the adsorption properties of the clay are relatively homogeneous and consistent with test results. If the source of recharge of contaminants to the 35-foot sand is eliminated (leaching of overlying contaminated soils), and only the contamination currently present in the ground water of the 35-foot sand is considered, the aquitard has sufficient adsorptive capacity to prevent migration of the contaminants through the aquitard to the 100-foot sand.[105]

Laboratory Studies

Table 5.15 contains brief comments on 8 laboratory studies dealing with the transport and fate of pesticides in the subsurface environment. The

Table 5.15. Summary of Some Laboratory Studies on the Transport and Fate of Pesticides in the Subsurface Environment.

Author(s)	Comments
Awad, Kilgore, and Winterlin[108]	Laboratory studies of the movement of aldicarb in eight different soil types.
Biggar, Nielsen, and Tillotson[109]	Movement of the nematocide DBCP in Panoche clay loam soil.
Conley and Mikucki[110]	Determination of whether DDT will leach from or migrate through a simulated sanitary landfill.
Davidson, Ou, and Rao[106]	Study of factors affecting pesticide mobility from hazardous waste disposal sites.
Davidson, Rao, and Ou[107]	Laboratory studies of the adsorption, movement, and biological degradation of selected pesticides in four major soil orders.
Robertson and Kahn[111]	Movement of aldrin through laboratory soil columns containing Ottawa sand.
Van Genuchten, Davidson, and Nieranga[112]	Movement of picloram through a water-saturated Norge loam soil.
You[113]	Microbial metabolism of the herbicide metabolite, 3,4-dichloroaniline (DCA) in soil.

studies listed in Table 5.15 address 10 pesticides or their metabolites as follows: 2,4-D;[106,107] atrazine, methyl parathion, terbacil, and trifluralin;[107] aldicarb;[108] DBCP;[109] DDT;[110] aldrin;[111] picloram;[112] and DCA.[113]

Factors affecting the subsurface mobility of 2,4-D from hazardous waste disposal sites were examined by Davidson, Ou, and Rao.[106] Major attention was given to the role of adsorption based on an equilibrium adsorption isotherm developed for a dimethyamine salt of 2,4-D on Webster silty clay loam. The 2,4-D concentrations ranged up to 5000 μg/ml. The observed adsorption isotherm was nonlinear in shape with the exponent in the Freundlich equation equal to 0.71. The adsorption sites for 2,4-D on the Webster soil were not saturated even in the presence of 5000 μg/ml of 2,4-D. The mobility of 2,4-D in the Webster soil at various 2,4-D concentrations was then simulated with a numerical solution to a solute transport model. These simulations revealed that pesticide mobility increased as solution concentration increased when the Freundlich exponent N was less than 1.0. However, an increase in solution concentration resulted in a decreased mobility when N was greater than 1.0. The pesticide solution concentration did not influence the mobility when N was equal to 1.0. Therefore, significant errors may be introduced by assuming a linear adsorption isotherm ($N = 1$) when predicting pesticide transport under waste disposal sites where high pesticide concentrations exist. A procedure for estimating the arrival time of a selected pesticide concentration at various soil depths below a disposal site was also presented by Davidson, Ou, and Rao.[106]

Davidson, Rao, and Ou[107] evaluated the adsorption and bacterial degradation of 2,4-D, atrazine, methyl parathion, terbacil, and trifluralin in soils representing four major soil orders in the United States. Solution concentrations ranged from zero to the aqueous solubility limit for each pesticide. The mobility of each pesticide increased as its concentration in the soil solution phase increased. These results were in agreement with the adsorption isotherm data. Pesticide degradation rates and soil microbial populations generally declined as the pesticide concentration in soil increased; however, some soils were able to degrade a pesticide at all concentrations studied, while others remained essentially sterile throughout the incubation period (60 to 80 days).

The influence of adsorption and/or physical binding of aldicarb to soil particles was investigated by Awad, Kilgore, and Winterlin.[108] Thirty-five grams of eight different but well-defined soils were packed in individual glass columns and tapped lightly. The soil types included sand, loam, clay, sandy loam, clay loam, silt loam, muck, and peat. Twenty-five mg of aldicarb active ingredient in formulated form were added to the top of each column and mixed thoroughly with the top 3 cm of soil. Three 50-ml aliquots of distilled water were then added to the top of the soil columns.

Clay soil retained the highest amount of aldicarb residues followed by loam soil. The apparent retention of aldicarb residues by loam and clay soils may be a function of water flow as well as some type of physical binding to the soil particles. The lowest amounts of aldicarb residues were found in sandy, sandy loam, and peat soils.

The nematocide 1,2-dibromo-3-chloropropane (DBCP) has been found in California ground waters.[109] The solubility of DBCP in water is 1,230 ppm, and it can be subject to leaching, adsorption and/or microbial degradation in the subsurface environment. Due to its low vapor pressure, losses from the subsurface due to volatilization is expected to be minimal. Biggar, Nielsen, and Tillotson[109] developed adsorption isotherms of DBCP on surface soil samples, and they conducted both column studies and field experiments to determine DBCP mobility characteristics and influencing processes. Panoche clay loam, an alluvial soil, was used in the adsorption and column studies as well as the field experiments. A Freundlich isotherm with an N of unity was observed. The adsorption partition coefficient (K_D) was 0.20 cm^3/gm for equilibrium solution concentrations (C) ranging from 0.5 to 95 $\mu g/ml$. The small K_D value is indicative of the mobility of DBCP in Panoche clay loam soil. Some microbial degradation was noted in the column studies and field experiment; however, leaching below the root zone could be expected in many soils.

Conley and Mikucki[110] determined whether DDT will leach from or migrate through a simulated sanitary landfill. Leachate samples were tested, and DDT and two of its degradation products were not recovered. However, analyses of soil core samples showed some vertical migration and only minimal horizontal migration.

Robertson and Kahn[111] indicated that four factors influencing the mobility of chlorinated hydrocarbon insecticides through soils are: the type of formulation applied, the frequency of application, soil characteristics, and the frequency and rate of rainfall or irrigation. These factors were studied via the infiltration of aldrin through columns of Ottawa sand in four experiments. Most of the eluted aldrin in organic solvent systems was eluted during the passage of the first liter of water. The aldrin distributed itself in a characteristic manner through each column, depending upon the composition and stability of the emulsion and rate of water infiltration. A commercial emulsible concentrate mixture transmitted more aldrin to the effluent than did a synthetic solution containing no emulsifier.

The influence of the frequency and rate of rainfall or irrigation is reflected by the average pore water velocity, and Van Genuchten, Davidson, and Nieranga[112] examined this factor in studies of the movement of picloram through a water-saturated Norge loam soil. The equilibrium adsorption and desorption isotherms for picloram and Norge loam soil were not single-valued relations. Picloram mobility was reduced significantly when the average pore-water velocity was decreased from 145 to 14.2 cm/day. Predictions were made by using two kinetic rate equations and an equilibrium Freundlich equation. The two kinetic models and the equilibrium model each satisfactorily described the observed effluent concentration distributions at low pore-water velocities provided the nonsingle-valued character of the adsorption-desorption process was included in the calculations. At high pore-water velocities, the kinetic adsorption models were found inadequate to predict the picloram movement. An empirical model was then developed, based on the assumption that equilibrium existed during displacement and that only a fraction of the soil participated in the adsorption process. This fraction was found to be a function of the average pore-water velocity. With the empirical model, a reasonable fit between data and calculated effluent curves was obtained for all pore-water velocities.

The fate of the herbicide metabolite, 3,4-dichloroaniline (DCA), was investigated in culture solution and in soil by You.[113] The addition of aniline to soil containing 0.2 to 100 μg DCA in free and/or humus-bound form per gm of soil increased the microbial degradation rates of DCA severalfold. Identification of the key biodegradation intermediates revealed that DCA biodegradation occurred through 4,5-dichlorocatechol, 3,4-dichloromuconate, 3-chlorobutenolide, 3-chloromaleylacetate, 3-chloro-4-ketoadipate to succinate plus acetate. Absolute degradation of DCA per unit time was positively correlated with both increasing DCA and increasing aniline concentrations. The enrichment of specific microbial populations and/or the induction of pathways that can cometabolize DCA were found to be the most plausible explanations for the effect of aniline. The observed phenomenon shows a potential for eliminating xenobiotic pollutants from contaminated soils.[113]

Mathematical Models

Several types of mathematical models are available for predicting the transport and fate of pesticides in the subsurface environment. For example, Mansell et al.[114] developed a mathematical model to simulate the transport and physical and chemical reactions for potassium, phosphorus, and 2,4-D in two representative Florida soils: Wauchula sand and Troup sand. They found that reactions such as adsorption-desorption and chemical precipitation greatly influenced the movement and thus potential leaching of these constituents through the soil.

Enfield, Carsel, and Phan[115] compared a one-dimensional, steady-state hydraulic model with a two-dimensional, transient hydraulic model for aldicarb data from Long Island, New York. They found that both models adequately projected chemical transport velocity and concentration through the vertical soil profile at the relatively homogeneous field site evaluated. The two-dimensional solution also projects transport through the ground water and was able to describe three successive years of aldicarb application. In addition, they noted that the one-dimensional approach is computationally efficient and requires minimal input data, while the two-dimensional approach requires complex input data to utilize its ability to describe the natural system hydrogeologically, biologically, and chemically. The one-dimensional approach is limited to relatively homogeneous systems while the two-dimensional numerical approach permits describing nonhomogeneous (layered) systems. The one-dimensional approach does not evaluate transient hydraulics and assumes constant water content and flux while the two-dimensional system describes transient water flow and calculates water content and water flux as a function of time and space.

Enfield et al.[116] described three models for estimating the transport of organic chemicals through soils to ground water. The models consider mobility and first-order degradation, and they have been compared to field data for the pesticides aldicarb and DDT. The first model calculates linear sorption/desorption of the pollutant and first-order degradation without considering dispersion. The second is similar but also considers dispersion. The third considers nonlinear adsorption following a Freundlich equation and first-order degradation, but does not consider dispersion.

Dean, Jowise, and Donigian[117] have developed a methodology to assess potential pesticide leaching from the crop root zones in major (corn, soybean, wheat, and cotton) crop growing areas of the United States. Use of the Leaching Evaluation of Agricultural Chemicals (LEACH) methodology provides an indication of the presence or absence of leaching past the rooting depth and, if such leaching is indicated, its severity. LEACH was developed through the use of long-term simulation (i.e., 15 years) of annual pesticide leaching via a pesticide root zone model. The user must evaluate key parameters for a pesticide-site-crop-management scenario to locate pesticide leaching cumulative frequency distributions. Each scenario has a unique distribution associated with it. The distribution functions indicate the chance that the annual quantity of pesticide leached past the crop rooting depth will exceed a given value.

Finally, Javandel, Doughty, and Tsang[118] have developed a handbook which introduces the reader to various mathematical methods for estimating solute transport in ground water systems. It contains tables, figures, and simple computer programs that can be directly used for field studies. Three levels of mathematical methods are covered: (1) analytical, (2) semianalytical, and (3) numerical. The first two levels require relatively small amounts of data. At the third level, numerical approaches are discussed, and a number of currently available numerical models are listed, indicating code capabilities and code developers to be contacted for further information. An example of the use of one such model is presented. A discussion on method selection and data requirements is also included in the handbook.

Summary

Whether or not pesticides represent a threat to ground water is dependent upon factors related to the pesticides themselves, the soil, and application factors. Important physical and chemical characteristics of pesticides include water solubility, soil adsorption, volatility, and soil dissipation. Values for these characteristics which indicate increased potential for ground water contamination are summarized in Table 5.16.[119] Descriptive summary information for pesticide characteristics is in Table 5.17.[119]

A number of factors related to the physical and chemical characteristics of the soils receiving pesticide application also influence the potential for

Table 5.16. Chemical and Physical Properties of Pesticides: Values Which Indicate Potential for Ground Water Contamination.[119]

Water solubility	—	greater than 30 ppm
K_d	—	less than 5, usually less than 1
K_{oc}	—	less than 300–500
Henry's Law Constant	—	less than 10^{-2} atm-m^{-3} mol
Speciation	—	negatively charged, fully or partially at ambient pH
Hydrolysis half-life	—	greater than 25 weeks
Photolysis half-life	—	greater than 1 week
Field dissipation half-life	—	greater than 3 weeks

Table 5.17. Descriptive Information on Key Physical and Chemical Characteristics of Pesticides Leachers.[119]

Water Solubility— the propensity for a pesticide to dissolve in water. The higher a pesticide's water solubility, the greater the amount of pesticide that can be carried in solution to ground water. Water solubility of greater than 30 ppm has been identified as a "flag" for a possible pesticide "leacher."

Soil Adsorption— the propensity of a pesticide to "stick" to soil particles. It is defined as the ratio of the pesticide concentration in soil (C_s) to the pesticide concentration in water (K_d; C_s/C_w). There are different mechanisms for pesticide adsorption in soils, with particularly importantly differences occurring in clays as opposed to organic soil

Table 5.17, continued

	matter. A second measure, K_{oc}, is used to help characterize the mechanism of adsorption. K_{oc} is a measure of the pesticide adsorption to the organic part of the soil. The lower a pesticide's K_d and K_{oc} values, the more likely these chemicals will not be adsorbed to soil particles but leach to ground water. Of the pesticides found in ground water to date, most have had K_d values of less than 5, and usually less than 1.0. These ground water contaminants have also generally been shown to have K_{oc}values of less than 300.
Volatility—	the propensity for a pesticide to disperse into the air. It is primarily a function of the vapor pressure of the chemical and is strongly influenced by environmental conditions (e.g., temperature, wind speed, etc.). Nonvolatile pesticides such as DDT have low vapor pressures which will increase their persistence on and in the soil. Pesticides with high vapor pressures have not been considered a threat to ground water because they rapidly volatilize from the soil surface. However, the major ground water problems caused by the very volatile pesticides EDB, DBCP, and DCP have dispelled this notion. Contamination by these volatile pesticides has been blamed on their mode of application which is direct injection into the soil. It has also become apparent that the actual volatility of these pesticides when present in water is critically changed. This aqueous volatility is determined by dividing a chemical's vapor pressure by its solubility; this value is termed Henry's Law Constant (H). A compound such as DBCP has a very high vapor pressure but is also very water soluble. High water solubility can cause high vapor pressure chemicals to remain in the soil, particularly when these pesticides are applied just prior to irrigation or rainfall.
Soil Dissipation—	a simplified, general measure of a pesticide's persistence in soil. It is usually measured as the length of time required for dissipation of one-half the concentration of a pesticide and often referred to as a pesticide's "soil half-life." The soil half-life of a pesticide can be derived from either laboratory or field studies, but care must be taken in recording the conditions of the test including temperature, type of soil, soil moisture, etc. Soil dissipation is dependent on a number of environmental processes including vaporization and several decomposition processes that cause chemical breakdown, particularly hydrolysis, photolysis and microbial transformation. Hydrolysis is the reaction of a chemical with water. Photolysis is the breakdown of a chemical from exposure to the energy of the sun. And, microbial transformations result from the metabolic activities of microorganisms within the soil. When a pesticide resists these decomposition processes and does not readily evaporate, it will have a long soil half-life, increasing its potential as a threat to ground water. This is particularly true if the same pesticide is highly soluble and does not readily adsorb to the soil particles. Pesticides with half-lives greater than 2 or 3 weeks should be carefully assessed.

ground water contamination. Table 5.18 provides descriptive information on seven important soil factors.[119] Finally, the methods and conditions of soil application of pesticides can influence the potential for pesticide transport to ground water. Table 5.19 contains descriptive information on a number of pertinent methods (practices) and conditions.

Table 5.18. Descriptive Information on Key Soil Factors Affecting Pesticide Transport to Ground Water.[119]

Clay Content—	refers to the presence of clay minerals. Clay minerals contribute to cation exchange capacity, or the ability of the soil to adsorb positively charged molecules (i.e., cations). Positively charged pesticides will thus be adsorbed to soil containing negatively charged clay particles. Clay soils also have a high surface area which further contributes to adsorption capacity. Adsorption onto clay colloids leads to chemical degradation and inactivation of some pesticides, but it inhibits degradation of others.
Organic Matter Content—	also contributes to adsorption of pesticides in soil. Organic matter content affects bioactivity, bioaccumulation, biodegradability, leachability, and volatility of pesticides. Soils with high organic content adsorb pesticides and, therefore, inhibit their movement into ground water. However, pesticides which are highly adsorbed to organic soil will often be applied at higher rates by a farmer in order to compensate for the adsorbed portion. There is evidence that pesticide residues adsorbed to high organic (humus) soils may eventually be released, intact, to ground water when microbial degradation of the humus occurs.
Soil Texture—	also influences pesticide leaching. Texture refers to the percent sand, silt, and clay. Leaching is more rapid and deeper in coarse or light-textured sandy soils than in fine or heavy-textured clayey soils.
Soil Structure—	refers to the way soil grains are grouped together into larger pieces or aggregates—platy, prismatic, blocky, or granular. Structure is affected by texture and percent organic matter. Pesticides and water can seep, unimpeded, through seams between the aggregates.
Porosity—	is a function of total pore space, pore size, and pore size distribution, and is determined by soil texture, structure, and particle shape. Pesticide transport is more rapid through porous soils.
Soil Moisture—	refers to the presence of water in soil. The water in soil ultimately transports pesticides that are not adsorbed into the water table below. Upward movement may also occur through capillary action and by a process termed evapotranspiration in which water in the soil is lost to the air.
Depth to Ground Water—	the distance a pesticide must travel through the soil or underlying foundation material to reach ground water is, of course, a key determinant in whether contamination will occur at a particular site.

Table 5.19. Descriptive Information on Key Application Methods and Conditions Affecting Pesticide Transport to Ground Water.[119]

Local Climatic Conditions—	the degree of pesticide leaching at a particular site can be directly dependent on the amount of local rainfall. The temperature of the soil and surrounding air at a site can also greatly affect the processes that result in a pesticide's movement and degradation in the environment.
Rate of Application—	how much and how often a pesticide is applied to the soil can be the critical determinant in ground water contamination.
Timing of Application—	when a pesticide is applied can be a major factor depending on local environmental conditions and temperature and rainfall.
Method of Application—	a pesticide can be applied to crops by aerial spraying, topsoil application (granular, dust, or liquid formulations), soil injection, soil incorporation, or through irrigation. Soil injection and incorporation are generally considered to pose the greatest likelihood for ground water contamination problems. The application of pesticides through irrigation, often referred to as chemigation, can also be a significant source of ground water contamination. An irrigation pump may shut down due to a mechanical or electrical failure while the pesticide-adding equipment continues to operate. This malfunction can cause a backflow of pesticides into the well or cause highly concentrated pesticide levels to be applied to a field.
Irrigation Practices—	increase the soil moisture content and flow through the soil, raising the potential for chemical leaching. Irrigation can decrease the amount of volatilization of some pesticides from the soil. Excess irrigation can also carry pesticides down the well casings of abandoned or poorly constructed wells, directly injecting contaminants into an aquifer. The use of drainage tiles can also lead to direct input of pesticides into ground water regardless of their leaching potential.
Cultivation Practices—	conservation tillage or no-till practices used to decrease soil erosion and pollutant runoff into streams will increase water infiltration and hence potential for pesticides to leach. Furthermore, these practices usually require increased use of herbicides which may leach. Other soil conservation practices designed to inhibit runoff may also increase infiltration.
Spillage/ Disposal—	Can result in high concentrations of pesticides in soil. These "slugs" can overwhelm normal decomposition processes and soil adsorption capacity, resulting in high potential for ground water contamina-

Table 5.19, continued

	tion. Spillage, in particular, can be a common problem where pesticide mixing and loading take place. Handling of unwanted pesticides and empty containers may also pose problems. Rinse water from the cleaning of spray equipment may also be washed into the soil; the large amounts of contaminated water associated with this practice can increase pesticide leaching.

Radionuclides

The subsurface transport and fate of radionuclides at radioactive waste disposal sites is of concern due to the potential, albeit low, for the occurrence of ground water pollution. Monitoring of ground water quality at commercial disposal sites is routinely accomplished.[120] Basic studies on radionuclide retention mechanisms and influencing factors have also been done, and two will be highlighted. The general findings of these studies can also be applied to the behavior of stable isotopic forms of the radionuclides.

The transport of radionuclides through geologic materials has been predicted using the assumption that adsorption along the transport path occurs instantaneously. This assumption is valid if equilibrium adsorption is reached rapidly relative to ground water velocity. To test this assumption, Fenske[121] conducted a series of laboratory experiments with varying particle sizes and experimental conditions. The findings indicated that equilibrium adsorption may be reached too slowly under some conditions for the assumption of instantaneous adsorption to be valid. An empirical equation was devised from the laboratory experiments to describe the time-dependent sorption process for cesium-137 on breccia and andesite. The hypothetical example was then extended for flow through granular materials (pebble and cobble size) and flow through fractured rocks considering a time-dependent adsorption. Dispersion was not considered in the analysis. The hypothetical prediction showed that for usual ground water velocities, transport considering time-dependent adsorption is not significantly greater than transport assuming equilibrium adsorption. The analysis also showed that adsorption during transport through granular materials is less, with less retardation, than adsorption during transport through fractured rocks.

One of the concerns in radioactive waste disposal is related to the facilitated transport of metals due to complexation with organics. Jones, et al.[122] investigated the ability of the organic complexant EDTA to enhance the mobility of cobalt-60 in both laboratory and field experiments. Laboratory tests consisted of short-term (approximately 7-day) column and batch adsorption tests using soil from the Hanford, Washington, site as well as long-term (approximately 70-day) batch tests with Hanford soil and soils from Oak Ridge, Tennessee, and Savannah River, South Carolina. In addition, two large scale tracer tests were conducted using Hanford soil. One used a large (1.6 m) laboratory column, spiked with cobalt-60/EDTA, and the other was a field test conducted in an 8-m deep lysimeter. Enhanced mobility decreased adsorption was observed in both column and batch tests when the cobalt-60/EDTA solutions contacted Hanford and Oak Ridge soil for only a few days. When long contact times were allowed (months) the Hanford soil showed large increases in adsorption with time. The low adsorption exhibited initially by the Oak Ridge soil increased slightly over time; however, the high adsorption observed with the Savannah River soil remained constant with time. The reduced mobility, with time, observed in Hanford soils was confirmed in the large-scale laboratory experiments when contacted with Hanford and Savannah River soil, and to a lesser extent, in the Oak Ridge soil. The implication to waste management is that the potential for transport of cobalt by EDTA complexation may not be as serious as once thought.

CATEGORIES OF POLLUTANT SOURCES

As discussed in Chapter 4, there are numerous categories of ground water pollution sources. Of concern in source evaluation is the subsurface transport and fate of released pollutants. Numerous laboratory and field studies have been made regarding potential and actual ground water pollution from specific source categories. This section will summarize selected studies on septic tank systems, solid waste disposal operations, and the land application of municipal wastewater sludges. Detailed information on the subsurface transport and fate of pollutants from other source categories is available from other recent works. For example, the ground water quality effects from the artificial recharge of municipal wastewater are addressed by O'Hare, et al.,[123] and comparable effects from salt water intrusion are discussed by Atkinson, et al.[124]

Septic Tank Systems

Detailed information on the ground water quality effects of septic tank systems is contained in a recent book by Canter and Knox.[32] Three laboratory studies will be briefly described herein. Magdoff, Bouma, and Keeney[125] designed some laboratory columns to represent the vertical dimensions of a mound-type disposal system for receiving septic tank effluent on problem soils. The columns were filled with gravel (representing creviced bedrock), silt loam (representing the original topsoil), a sand or sandy loam till (fill material), gravel (the seepage bed), and another layer of silt loam (the mound cover). The columns were dosed with septic tank effluent at the rate of 2 cm every 6 hours. Until crusting caused permanent ponding at the fill-gravel interface, the fill was aerobic and the silt loam at the bottom of the column was anaerobic. Redox potentials were higher in the fill (350 to 600 mV) than in the silt loam (200 to 410 mV). Moisture tension fluctuations after a 2-cm addition were greatest near the fill-gravel interface and decreased with depth in the column. After continuous ponding, tension fluctuations almost ceased, the subcrustal soil became anaerobic, and the redox potentials decreased to around minus 250 mV. These columns were found to be useful in evaluating the ground water pollution potential of mound-type septic tank systems.

Lotse[126] also used laboratory soil columns to study the rate and extent of phosphorus movement in selected Maine soils under continuous and intermittent loading. For intermittently operated columns, there was no breakthrough of phosphorus when 8.0, 12.4, and 15.0 pore volumes of effluent, respectively, had been collected. For continuously operated columns, however, breakthrough occurred at 10.8 and 11.2 pore volumes, respectively. The greater the hydraulic loading, the greater was the rate of phosphorus movement through a given soil. Based on these findings, it was determined that septic tank absorption field systems should have several trenches and a large total length of trench in order to minimize the movement of phosphorus to ground waters.

The effects of soil moisture content upon the movement and adsorption of phosphorus was studied by Brutsaert, Hedstrom, and McNeice.[127] Three plexiglass laboratory soil columns containing Adams soil were loaded with a phosphorus solution (25 mg/liter as P) at rates to maintain constant saturation levels throughout the column (1.0, i.e., fully saturated; 0.86; and 0.64). Reducing the degree of saturation, by controlling the infiltration rate in otherwise physically similar columns, increased the soil adsorptive capacity and, hence, greatly reduced the extent of phosphorus penetration into the column per pore volume of influent applied. At saturation levels of 0.86 and 0.64, phosphorus adsorption increased five and seven times, respectively. The 24-hour phosphorus retention capacity of 476 micrograms P/gm soil as determined by batch study was exceeded in the partially saturated columns. These results can be used to optimize domestic leach field design by maximizing partially saturated flow conditions and thereby substantially extending the life of these systems.

Absorption field design for septic tank systems was studied by Mitchell.[128] Laboratory columns (lysimeters) were used to determine the biodegradation capacity of various sizes and depths of washed river gravel. The results of the laboratory analysis of influent and effluent wastewater were discussed as they relate to the establishment of the most efficient biodegradation size and depth of the column. The biodegradation capacity of each lysimeter was evaluated in terms of COD removal, total inorganic nitrogen removal and phosphate removal in relation to dosage rates, loading rates, and pH. In lysimeters containing river gravel passing a 3/8 inch sieve, degradation as measured by COD, phosphate, and nitrogen removal varied directly with increasing depth. Based upon these findings, it was determined that an absorption field with an addition of twelve inches of gravel passing a number 4 sieve with a coefficient of uniformity of 30 and an effective size of 0.4 micrometers would provide a COD removal of 90% or above at minimum dosage rates.

Solid Waste Disposal Operations

Solid waste disposal operations can involve either municipal or industrial solid wastes placed in sanitary or hazardous waste landfills. Sanitary landfills may also be used for the disposal of residues from municipal solid waste-to-energy conversion facilities. A number of theoretical and laboratory studies have been conducted on the generation and fate of sanitary landfill leachates, and three will be highlighted.

Burchinal and Qasim[129] presented a procedure for theoretically estimating the chemical quality of leachates from sanitary landfills. A theory of column chromatography was employed to determine the leaching of chloride during the movement of water through sanitary landfill cells. Generalized breakthrough curves were used to obtain the theoretical chloride concentrations. Families of curves were then derived from experimental results representing the ratios of various leaching materials to the theoretical results. The method can be applied to the field study of existing landfills to determine the chemical composition of polluting liquid during the leaching process if the cumulative precipitation, evaporation, evapotranspiration, runoff coefficients, and the depth of the landfill cells are known.

The movement of sanitary landfill leachate metals through 11 soils from 7 major soil orders in the United States was studied in the laboratory by Fuller.[130] The movement and retention of As, Be, Cd, CN, Cr, Cu, Hg, Ni, Pb, Se, V, and Zn was influenced by the individual properties of the elements, by the permeability of the soil, and by the amounts of clay, lime, and hydrous iron oxides present in the soil. The movement of iron was also studied; its movement and retention in soil were related most strongly to the content of clay and hydrous iron oxides. Total organic carbon and COD in the municipal solid waste leachate were only slightly retained by soils. A simulation model for predicting solute concentration changes during leachate flow through soils was developed and partially validated using data from the study.

Griffin and Shimp[131] conducted both laboratory column and batch adsorption studies related to the attenuation of pollutants in municipal landfill leachates by mixtures of sand and calcium-saturated clays. Based on the column studies, they found that K, NH_4, Mg, Si, and Fe were moderately attenuated; and the heavy metals Pb, Cd, Hg, and Zn were strongly attenuated even in columns with small amounts of clay. Precipitation was the principal attenuation mechanism for the heavy metals; cation exchange was responsible for any attenuation of the other elements. The clays, in order of increasing attenuation capacity, were Kaolinite, Illite, and Montmorillonite. The second part of the project involved batch studies of adsorption of Cr, Cu, Pb, Cd, Hg, and Zn by Montmorillonite and Kaolinite from water solutions and from landfill leachate. Adsorption of the cations Cr (III), Cu, Pb, Cd, Hg, and Zn increased with increasing pH; adsorption of the anions Cr (VI), As, and Se decreased with increasing pH. Above a pH of about 5.3, precipitation of the cations was an important mechanism, while adsorption was the principal mechanism for the anions over the pH range studied. Griffin and Shimp[131] noted that the mobility of any element was affected by other solutes in the leachate; adsorption information on one leachate may not be directly applied to predicting adsorption of the same element at the same concentration in another leachate.

The chemical characteristics of leachates from municipal solid waste-to-energy conversion facilities were evaluated by Rademaker and Young.[132] Leachate samples were collected daily and analyzed weekly to see if heavy metals and organic priority pollutants could be leached in sufficient quantities to pose a potential environmental hazard. The effects of soil on the character of the resulting leachate produced from the selected waste residues were also examined. Bench scale column tests revealed that levels of effluent constituents declined gradually from a peak early in the leaching period to essentially zero at 5 to 7 weeks. Although no organic priority pollutants were found in ash residues, they were present in sludges from the biological treatment of pyrolysis wastewaters. Soil columns were successful in removing most base-neutral extractable organic compounds, but not acidic extractable compounds.

Miller, Braids, and Walker[133] reported on a study to determine the prevalence of subsurface migration of hazardous chemical constituents at 50 land disposal sites that had received large volumes of industrial wastes. The facilities included landfills, lagoons, and combinations of the two, both active and abandoned. They were located in 11 states in the humid region, east of the Mississippi River. Hazardous substances which were monitored included: (1) heavy metals other than iron and manganese, (2) cyanide, arsenic, and selenium, and (3) various organic substances. At 43 of the 50 sites, migration of one or more hazardous constituents was confirmed. Migration of heavy metals was confirmed at 40 sites; selenium, arsenic and/or cyanide at 30 sites; and organic chemicals at 27 sites. A total of 86 wells and springs used for monitoring yielded water containing one or more hazardous substances with concentrations above background.

Land Application of Municipal Wastewater Sludges

Municipal wastewater treatment plant sludges are frequently applied to land via either land spreading or disposal in landfills. Sommers, et al.[134] evaluated the chemical composition of sewage sludges, the mobility of sludge constituents in laboratory soil columns, and the use of synthetic sludges as a model system in studying the dynamics of carbon and nitrogen in soils amended with sewage sludges. For the majority of the soils and sludges examined, the amount of nitrate leached from sludge-amended columns was less than or equal to the amount of nitrate from columns not treated with sludge. A model was developed to depict the chemical and physical factors responsible for the movement of nitrogen and metals into ground water following sludge disposal on land.

An assessment of ground water quality in the vicinity of eight landfill sites accepting various quantities of municipal sewage sludge was performed by Lofty, et al.[135] An appraisal was also made of the operations and economics at each location. The study consisted of two phases. The first included a preliminary assessment of the environmental impacts of sewage sludge burial practices at the eight sites. In the second phase, additional data on leachate and ground water quality trends over a 12-month period were obtained. Two sites were selected for more intensive ground water quality impact studies. The results showed that landfilling of sewage sludge released significant contaminants in the form of leachate. Among the contaminants analyzed, iron, lead, and TOC in excess of EPA Drinking Water Standards were found most frequently. Contaminants at the sites typically traveled at least several hundred meters beyond the immediate disposal areas despite the existence of soil strata with low permeability.

RELATED ISSUES

Three general issues related to pollutant transport and fate in the subsurface environment will be addressed in this section. These issues include leachate testing, ground water tracers, and facilitated transport.

Leachate Testing

Evaluation of the ground water pollution potential of various sludges and waste materials requires the conduction of testing protocols to determine constituent concentrations in produced leachates. Leaching tests have been used for dredged materials and municipal and industrial solid wastes and sludges. The basic concept of a leachate test is that it should simulate the conditions in the field; when exact simulation is not possible, more stringent test conditions are used to achieve worst case (maximum release) conditions. Lee and Jones[136] identified the following factors for consideration in designing a leaching test for solid waste:

(1) The leaching solution must bear some relationship to that actually encountered at a disposal site
(2) The oxidation-reduction conditions expected to be present in the field should be simulated in the test
(3) The importance of the liquid-to-solid ratio, that is, the ratio of the leaching solution to the material to be leached in the disposal area (the use of several liquid-to-solid ratios that cover the range of expectations may be desirable)
(4) The contact time between the leaching solution and the material to be leached
(5) The degree and method of agitation during the leaching procedure
(6) The leaching solution pH and temperature
(7) The organic content of the material to be leached
(8) The method of separation of the leachate from the material which has been leached

The two most frequently used leachate tests relative to determining the ground water pollution potential of tested materials are the Extraction Procedure Toxicity Characteristic (EP Toxicity) test and the Toxicity Characteristic Leaching Procedure (TCLP) test.

Under section 3001 of the Resource Conservation and Recovery Act (RCRA), the U.S. Environmental Protection Agency was charged with identifying those wastes which pose a hazard to human health and the environment if improperly managed. One approach for this identification is through determining certain characteristics of hazardous wastes. One of these characteristics, the Extraction Procedure Toxicity Characteristic (EPTC), was intended to identify wastes which pose a hazard due to their potential to leach significant concentrations of specific toxic species.[137] Determination of the EPTC entails the use of a leaching test, called the Extraction Procedure (EP) Toxicity test, for determining if an unacceptably high level of ground water contamination might result from improper waste management. The EP Toxicity test, adopted in 1980 by the U.S. Environmental Protection Agency, results in a liquid extract (leachate) which is analyzed for eight metals (arsenic, barium, cadmium, chromium, lead, mercury, selenium, and silver), four insecticides (endrin, lindane, methoxychlor, and toxaphene), and two herbicides (2,4-D and 2,4,5-TP). Regulatory thresholds as shown in Table 5.20 were established for these 14 species taking into account the attenuation

Table 5.20. Maximum Concentration of Contaminants for Characteristics of EP Toxicity.[138]

Contaminant	Maximum Concentration (mg/l)
Arsenic	5.0
Barium	100.0
Cadmium	1.0
Chromium	5.0
Lead	5.0
Mercury	0.2
Selenium	1.0
Silver	5.0
Endrin	0.02
Lindane	0.4
Methoxychlor	10.0
Toxaphene	0.5
2,4-D (2,4-Dichlorophenoxyacetic acid)	10.0
2,4,5-TP Silvex	1.0

and dilution expected to occur during migration of the leachate to the ground water, through use of a generic dilution/attenuation factor of 100. The policy is that a solid waste exhibits the characteristic of EP toxicity if, following application of the test, the leachate from a representative sample of the waste contains any of the contaminants listed in Table 5.20 at a concentration equal to or greater than the respective concentration listed.

Detailed information on the EP Toxicity test is available elsewhere.[138] The two major shortcomings of this test have been identified as:[137]

(1) Dependence upon the National Interim Primary Drinking Water Standards for constituent evaluation
(2) The focus of the test on leaching of elemental rather than organic constituents

Due to these shortcomings, in 1986 the U.S. Environmental Protection Agency adopted the Toxicity Characteristic Leaching Procedure (TCLP) test as an amended version of the EP Toxicity test.[137] The TCLP test expands the EPTC by including 38 additional organic compounds. Also, the TCLP test applies compound-specific dilution/attenuation factors generated from a ground water transport model rather than the uniformly used factor of 100 in the EPTC. The TCLP test is designed to determine the mobility of both organic and inorganic contaminants present in liquid, solid, and multiphasic wastes.[137]

The policy regarding the TCLP test is that a solid waste exhibits the characteristic of toxicity if, following application of the test, the leachate (extract) from a representative sample of the waste contains any of the 52 compounds listed in Table 5.21 at a concentration equal to or greater than

Table 5.21. Toxicity Characteristic Contaminants and Regulatory Levels.[137]

Contaminant	Regulatory level (mg/l)
Acrylonitrile	5.0
Arsenic	5.0
Barium	100.0
Benzene	0.07
Bis (2-chloroethyl) ether	0.05
Cadmium	1.0
Carbon disulfide	14.4
Carbon tetrachloride	0.07
Chlordane	0.03
Chlorobenzene	1.4
Chloroform	0.07
Chromium	5.0
o-Cresol[a]	10.0
m-Cresol[a]	10.0
p-Cresol[a]	10.0
2,4-D	1.4
1,2-Dichlorobenzene	4.3
1,4-Dichlorobenzene	10.8
1,2-Dichloroethane	0.40
1,1-Dichloroethylene	0.1
2,4-Dinitrotoluene	0.13
Endrin	0.003
Heptachlor (and its hydroxide)	0.001
Hexachlorobenzene	0.13
Hexachlorobutadiene	0.72
Hexachloroethane	4.3
Isobutanol	36.0
Lead	5.0
Lindane	0.06
Mercury	0.2
Methoxychlor	1.4
Methylene chloride	6.6
Methyl ethyl ketone	7.2
Nitrobenzene	0.13
Pentachlorophenol	3.6
Phenol	14.4
Pyridine	5.0
Selenium	1.0
Silver	5.0
1,1,1,2-Tetrachloroethane	10.0
1,1,2,2-Tetrachloroethane	1.3
Tetrachloroethylene	0.1
2,3,4,6-Tetrachlorophenol	1.5
Toluene	14.4
Toxaphene	0.07
1,1,1-Trichloroethane	30.0

Table 5.21, continued

Contaminant	Regulatory level (mg/l)
1,1,2-Trichloroethane	1.2
Trichloroethylene	0.07
2,4,5-Trichlorophenol	5.8
2,4,6-Trichlorophenol	0.30
2,4,5-TP (Silvex)	0.14
Vinyl chloride	0.05

[a]o-, m-, and p-Cresol concentrations are added together and compared to a threshold of 10.0 mg/l.

the respective concentration listed.[137] If a total analysis of the waste demonstrates that individual contaminants are not present in the waste, or that they are present but at such low concentrations that the appropriate regulatory thresholds could not possibly be exceeded, the TCLP need not be run. Detailed information on the TCLP test is available elsewhere.[137]

The specific environment modeled by both the EP Toxicity test and the TCLP test is codisposal of industrial waste with refuse in an sanitary landfill. This model assumes that the landfill is composed of 5% industrial solid waste and 95% municipal waste, and that the character of the leaching fluid that the waste will be exposed to is predominantly a function of the decomposing refuse in the landfill.[137]

Ground Water Tracers

Information on the subsurface transport and fate of chemicals and microbial species can be useful in selecting tracers of ground water flow or solute movement. Table 5.22 contains a detailed listing of factors to consider in selecting an appropriate tracer for meeting a given need.[139] The most important selection criterion is that the potential chemical and physical behavior of the tracer in ground water must be clearly understood. Table 5.23 summarizes some of the most important tracers presently available.[139] A manual on ground water tracers has been prepared for use by engineers, hydrogeologists, and other ground water scientists. The manual is specifically concerned with the selection of tracers, their field application, collection of samples containing tracers, sample analysis, and interpretation of the results.[139]

Table 5.22. Factors to Consider in Tracer Selection.[139]

PURPOSE OF STUDY		
		Tracer Type to be Used
Determination of:	flow path	
	velocity (solute)	Nonconservative
	velocity (water)	Conservative
	porosity	Conservative
	dispersion coefficient	Conservative
	distribution coefficient	Nonconservative
Delineation of contaminant plume		Constituent of plume
Recharge		Environmental isotope or anthropogenic compound
Dating		Radioactive isotopes

AVAILABLE FUNDS

Manpower and equipment to run tests to completion (e.g., drilling, tracer cost, sampling, analysis)

TYPE OF MEDIUM	
	Tracer Type
Karst	Fluorescent dyes, spores, tritium, as well as other tracers
Porous media (alluvium, sandstone, soil)	Wide range of choices Dyes and particulate material are rarely useful
Fractured rock	Wide range of choices Dyes and particulate material only occasionally are useful

STABILITY OF TRACER	
	Tracer Type
Distance from injection to sampling point	Must be stable for length of test and analysis
Approximate velocity of water and approximate estimate of time required for test, given: distance from injection to sampling point, porosity, thickness of aquifer	

DETECTABILITY OF TRACER

Background level

Dilution expected in test (function of distance, dispersion, porosity, and hydraulic conductivity)

Detection limit of tracer (ppm, ppb, ppt)

Table 5.22, continued

Interference due to other tracers, water chemistry	
DIFFICULTY OF SAMPLING AND ANALYSIS	
Factors to Consider	**Example of Difficult Tracer**
Availability of tracer	Radioactive (must have special permits)
Ease of sampling	Gases (will escape easily from poorly sealed container)
Availability of technology for and ease of analysis	Cl-36 (only one or two laboratories in the world can do analyses)
PHYSICAL/CHEMICAL/BIOLOGICAL PROPERTIES OF TRACER	
Density, viscosity	May affect flow (e.g., high concentrations of Cl)
Solubility in water	Affects mobility
Sorptive properties	Affects mobility
Stability in water	Affects mobility
Physical radioactive decay; **Chemical** decomposition and precipitation; **Biological** degradation	
PUBLIC HEALTH CONSIDERATIONS	
Toxicity	
Dilution expected	
Maximum permissible level—determined by federal, state, provincial, and county agencies	
Proximity to drinking water	

Table 5.23. Summary of Most Important Tracers.[139]

Tracer	Characteristics
A. Particulates	
Spores	Used in karst tracing; inexpensive Detection: high, multiple tests possible by dying spores different colors Low background Moderately difficult sampling and analysis (trapping on plankton, then microscopic identification and counting) No chemical sorption May float on water, travels faster than mean flow rate

Table 5.23, continued

Tracer	Characteristics
Bacteria	Most useful for studying transport of microorganisms Detection: highly sensitive Sampling: filtration, then incubation and colony counting No diffusion, slight sorption
Viruses	Detection: highly sensitive Sampling: culturing, colony counting Some sorption Smallest particulate
B. Ions (nonradioactive, excludes dyes)	
Chloride	Conservative Inexpensive Stable Detection: 1 ppm by titration, electrical conductivity, or selective ion electrode High background may be problematic In large quantities, affects density which distorts flow No sorption
Bromide	Inexpensive Stable Detection: 0.5 ppm by selective ion electrode Low background No sorption
C. Dyes	
Rhodamine WT	Used in karst and highly permeable sands and gravels Inexpensive Moderate stability Detection: 0.1 ppb by fluorimetry Low background fluorescence Moderate sorption
Fluorescein	Properties similar to Rhodamine WT, except: Degraded by sun *Chlorella* bacteria interferes High sorption
D. Radioactive Tracers	
Tritium	High stability Detection: >1 ppt by weak β radiation Varying background Complex analysis (expensive field and lab equipment) Half-life $= 12.3$ years Radiation hazard Handling and administrative problems No sorption

Table 5.23, continued

Tracer	Characteristics
^{131}I	High stability Detection: high sensitivity by measuring β and α emission Background negligible Complex analysis Half-life = 8.2 days Radiation hazard Sorption on organic material
EDTA-^{51}Cr	Moderately stable (affected by cations) Detection: highly sensitive, by radiation or post-sampling neutron activation analysis No background Half-life = 28 days Radiation hazard Little sorption
^{82}Br	High stability Detection: high sensitivity by measuring β emission No background Half-life = 35 hours Radiation hazard No sorption
E. Other Tracers	
Fluorocarbons	Expensive High stability Detection: 1 ppt by gas chromatography with electron capture detection Low background Difficult to maintain integrity of samples Nondegradable, volatile, low solubility, strong sorption by organic materials Low toxicity
Organic anions	Detection: few ppb by HPLC Low background Expensive analysis Very low sorption Low toxicity

Facilitated Transport

Relatively immobile or highly attenuated chemicals have been observed to move more rapidly through the subsurface environment under certain conditions. This facilitated transport has implications for problem analysis in terms of solute transport modeling; and it could be utilized in remedial action program design. Facilitated transport can occur as a result of complexation between organics and metals, the presence of a multiphase flow system as characterized by a high concentration of organics, or the enlargement of subsurface macropores in the presence of certain chemicals.

Complexation between organics and metals has been addressed by Jones, et al.,[122] and Loch, Lagas, and Haring.[64] Jones, et al.[122] studied the facilitated transport of cobalt-60 by the organic complexant EDTA, and the results are summarized in the section on radionuclides. Loch, Lagas, and Haring[64] used laboratory column experiments to study the behavior of heavy metals from landfill leachate in soil. Special attention was given to the influence of fatty acids in the leachate on the mobility of lead, copper, nickel, and zinc. The fatty acids which were examined included acetic, propionic, butyric, valeric, and ascorbic acids. In addition, the influence of the kind of soil material was investigated. Of the six columns used in the experiments, three were percolated with synthetic landfill leachate and three with the same leachate without fatty acids. In both cases, the three columns contained sand, sand mixed with fine clay particles, and sand mixed with large clay aggregates. All experiments were carried out under anaerobic conditions. The soil solution of each column was periodically analysed. At the end of the experiments 7 months later, the soil was analysed for precipitates and adsorption of metals.

As a result of the laboratory column experiments, Loch, Lagas, and Haring[64] found that in the presence of fatty acids from landfill leachate, mobilization of zinc, nickel, and lead in the soil takes place. Complexation of the metal cations and the absence of sulphides explain this behavior. They also found that the presence of clay increases retention of the metals, even when they are largely complexed by fatty acids. The breakthrough sequence of the metals in the landfill leachate was, respectively, nickel, zinc, lead, and copper. This agrees well with known adsorption properties of these metals. Loch, Lagas, and Haring[64] also found that it makes a substantial difference whether the clay material in the sand is finely divided or aggregated. Because of slow but prolonged diffusion of ions toward and into the aggregates, before adsorption can take place, breakthrough started relatively early, but it took a long time before the maximum effluent concentration was reached. In the absence of fatty acids, most of the Pb, Ni, and Zn was retained by sulphide precipitation. The behavior of copper deviated substantially from that of the other metals. Copper was retained in the upper layer of the soil, no matter whether fatty acids were present or absent. This was probably due to preferential adsorption or complexation of copper on soil colloids.

Enfield[140] developed a theoretical approach for addressing chemical transport facilitated by multiphase flow systems. The subsurface environment can be subject to multiphase flow; for example, in certain waste disposal situations, a separate organic fluid phase has been observed flowing through soil in addition to the water phase. Enfield[140] solved the convective dispersive transport equation utilizing a transformation of variables which permits the use of existing solutions covering a wide variety of boundary conditions. This equation was then utilized for predicting the transport and transformation of pollutants in cases where there are up to three separate mobile phases. It was determined, based upon this theoretical perspective, that the presence of a second mobile phase can facilitate the transport of hydrophobic chemicals through ground water. The impact of the facilitated transport was demonstrated for one case where the soil organic carbon was 10%. If 2% of the fluid phase was an organic fraction, the theory developed projected that hydrophobic theory may underestimate mobility by more than 100 times. At concentrations of dissolved organic carbon normally observed in nature (5 to 10 mg/L) a measurable increased mobility is anticipated for the very immobile compounds like dioxins.

Barbee and Brown[141] conducted a lysimeter study to determine the rapidity and extent of xylene movement after simulated spills on undisturbed soils of diverse texture and structure. The soils used were loamy sand, silt loam, and clay. Xylene has no net charge, low water solubility, and little polarity with which to compete with water for adsorption sites on clay minerals and organic matter. A volume equivalent to a depth of 5 cm of dye-labeled xylene was spilled on one plot of each soil contained in large lysimeters. Split applications of 1.25-cm depth equivalent were also made 12 days apart on one lysimeter of each of the soils. Significant quantities of xylene were detected at the 0.61-m depth approximately one day, one hour, and one half hour after the 5-cm application in the loamy sand, silt loam, and clay, respectively. Xylene movement after the first 1.25 cm of the split xylene application was almost as rapid as after the 5-cm application. However, with the second 1.25-cm application of xylene, breakthrough of free xylene was collected in the leachate at the 0.61-m depth for all three soils. Dye patterns observed upon excavation of the soils indicated that the xylene moved as a relatively uniform front in the sand but moved through isolated macropores between structural units in the silt loam and clay soils. Barbee and Brown[141] noted that xylene causes the removal of water from clay mineral surfaces, thus causing shrinkage and the formation of channels. Xylene has been shown to enhance soil aggregation and, thus, could enlarge existing macropores in a structured soil. In addition, xylene is hydrophobic and tends to remain concentrated in the large pores and does not easily diffuse through water films into the soil matrix. These factors all contribute to the rapid movement of xylene through silt loam and clay soils.

SELECTED REFERENCES

1. Roberts, P. V., Reinhard, M., and Valocchi, A. J., "Movement of Organic Contaminants in Groundwater: Implications for Water Supply," *Journal of the American Water Works Association,* August 1982, pp. 408–413.
2. Pinder, G. F., "Groundwater Contaminant Transport Modeling," *Environmental Science and Technology,* Vol. 18, No. 4, 1984, pp. 108A–114A.
3. Bouma, J., et al., "The Function of Different Types of Macropores During Saturated Flow Through Four Swelling Horizons," *Soil Science Society of America Journal,* Vol. 41, 1977, pp. 945–950.
4. Scrivner, N. C., et al., "Chemical Fate of Injected Wastes," *Ground Water Monitoring Review,* Vol. 6, No. 3, Summer 1986, pp. 53–58.
5. Chiou, C. T., et al., "Partition Coefficient and Bioaccumulation of Selected Organic Chemicals," *Environmental Science and Technology,* Vol. 11, 1977, pp. 475–478.
6. Hounslow, A. W., "Adsorption and Movement of Organic Pollutants," *Proceedings of the Third National Symposium on Aquifer Restoration and Ground Water Monitoring,* May 1983, National Water Well Association, Worthington, Ohio, pp. 334–346.
7. Roberts, P. V., and Valocchi, J., "Principles of Organic Contaminant Behavior During Artificial Recharge," *Quality of Groundwater,* van Duijvenbooden, W., Glasbergen, P., and van Lelyveld, H., Editors, 1981, Elsevier Scientific Publishing Company, Amsterdam, The Netherlands, pp. 439–450.
8. Karickhoff, S. W., Brown, D. S., and Scott, T. A., "Sorption of Hydrophobic Pollutants on Natural Sediments," *Water Research,* Vol. 13, 1979, pp. 241–248.
9. Ghiorse, W. C., and Balkwill, D. L., "Microbiological Characterization of Subsurface Environments," *Proceedings of the First International Conference on Ground Water Quality Research,* October 1981, Rice University, Houston, Texas.
10. Ghiorse, W. C., and Balkwill, D. L., "Enumeration and Morphological Characterization of Bacteria Indigenous to Subsurface Environments" *Developments in Industrial Microbiology,* Vol. 24, 1983, pp. 213–224.
11. Pritchard, P. H., "Model Ecosystems," *Environmental Risk Analysis for Chemicals,* Conway, R. A., Editor, 1982, Van Nostrand Reinhold Company, New York, New York, pp. 257–353.
12. Van Voris, P., et al., "Development and Validation of a Terrestrial Microcosm Test System for Assessing Ecological Effects of Utility Wastes," August 1984, Battelle Columbus Laboratories, Columbus, Ohio.
13. Ahlert, R. C., and Kosson, D. S., "In-situ and On-site Biodegradation of Industrial Landfill Leachate," June 1983, Center for Coastal and Environmental Studies, Rutgers University, New Brunswick, New Jersey.
14. Cervelli, S., and Rolston, D. E., "Influence of Atrazine on Denitrification in Soil Columns," *Journal of Environmental Quality,* Vol. 12, No. 4, October–December 1983, pp. 482–486.

15. Chen, K. Y., et al., "Confined Disposal Area Effluent and Leachate Control," Technical Report DS-78-7, October 1978, U.S. Army Engineer Waterways Experiment Station, Vicksburg, Mississippi.
16. Christensen, T. H., "Cadmium Sorption onto Two Mineral Soils," PhD dissertation, 1980, University of Washington, Seattle, Washington.
17. Mansell, R. S., et al., "Movement of Acarol and Terbacil Pesticides During Displacement Through Columns of Wabasso Fine Sand," *Proceedings of Soil and Crop Science Society of Florida,* Vol. 31, December 1971, pp. 239–243.
18. Saxena, S. K., Lindstrom, F. T., and Boersma, L., "Experimental Evaluation of Chemical Transport in Water-Saturated Porous Media: 1. Nonsorbing Media," *Soil Science,* Vol. 118, No. 2, August 1974, pp. 120–126.
19. Gile, J. D., Collins, J. C., and Gillett, J. W., "The Soil Core Microcosm—A Potential Screening Tool," EPA 600/3-79-089, August 1979, Corvallis Environmental Research Laboratory, U.S. Environmental Protection Agency, Corvallis, Oregon.
20. Huff, D., et al., "Use of Microcosms for Evaluation of the Land Applicability of Hazardous Wastes," September 1986, University of Oklahoma, Norman, Oklahoma.
21. Hutchins, S. R., "Microbial Involvement in Trace Organic Removal from Ground Water Recharge During Rapid Infiltration," PhD dissertation, 1984, Rice University, Houston, Texas.
22. Miller, C. T., "Modeling of Sorption and Desorption Phenomena for Hydrophobic Organic Contaminants in Saturated Soil Environments," PhD dissertation, 1984, University of Michigan, Ann Arbor, Michigan.
23. Gerba, C. P., "Fate of Waste Water Bacteria and Viruses in Soil," *Journal of the Irrigation and Drainage Division, ASCE,* Vol. 101, No. IR3, September 1975, pp. 157–173.
24. Brown, K. W., et al., "Movement of Fecal Coliforms and Coliphages Below Septic Lines," *Journal of Environment Quality,* Vol. 8, No. 1, January–March 1979, pp. 121–125.
25. Gerba, C. P., and Lance, J. C., "Pathogen Removal from Wastewater During Groundwater Recharge," *Proceedings of the Symposium on Wastewater Reuse for Groundwater Recharge,* May 1980, California State Water Resources Control Board, Sacramento, California, pp. 137–144.
26. Hagedorn, C., Hansen, D. T., and Simonsson, G. H., "Survival and Movement of Fecal Indicator Bacteria in Soil Under Conditions of Saturated Flow," *Journal of Environmental Quality,* Vol. 7, No. 1, January–March 1978, pp. 55–60.
27. Matthess, G., and Pekdeger, A., "Concepts of a Survival and Transport Model of Pathogenic Bacteria and Viruses in Groundwater," *Quality of Groundwater,* van Duijvenbooden, W., Glasbergen, P., and van Lelyveld, H., Editors, 1981, Elsevier Scientific Publishing Company, Amsterdam, The Netherlands, pp. 427–437.
28. Rahe, T. M., et al., "Transport of Antibiotic-Resistant *Escherichia Coli* Through Western Oregon Hillslope Soils Under Conditions of Saturated Flow." *Journal of Environmental Quality,* Vol. 7, No. 4, October–December 1978, pp. 487–494.

29. Reneau, R. B., Jr., and Pettry, D. E., "Movement of Coliform Bacteria from Septic Tank Effluent Through Selected Coastal Plain Soils of Virginia," *Journal of Environmental Quality,* Vol. 4, 1975, pp. 41–44.
30. Viraraghavan, T., "Travel of Microorganisms from a Septic Tile," *Water, Air, and Soil Pollution,* Vol. 9, No. 3, April 1978, pp. 355–362.
31. Weaver, R. W., "Transport and Fate—Bacterial Pathogens in Soil," *Proceedings of Microbial Health Considerations of Soil Disposal of Domestic Wastewaters,* EPA-600/9-83-017, September 1983, U.S. Environmental Protection Agency, Cincinnati, Ohio, pp. 121–147.
32. Canter, L. W., and Knox, R. C., *Septic Tank System Effects on Ground Water Quality,* 1985, Lewis Publishers, Inc., Chelsea, Michigan.
33. Keswick, B. H., and Gerba, C. P., "Viruses in Ground Water," *Environmental Science and Technology,* Vol. 14, No. 11, November 1980, pp. 1290–1297.
34. Sobsey, M. D., "Transport and Fate of Viruses in Soil," *Proceedings of Microbial Health Considerations of Soil Disposal of Domestic Wastewaters,* EPA-600/9-83-017, September 1983, U.S. Environmental Protection Agency, Cincinnati, Ohio, pp. 174–197.
35. Bitton, G., Davidson, J. M., and Farrah, S. R., "On the Value of Soil Columns for Assessing the Transport Pattern of Viruses Through Soils: A Critical Outlook," *Water, Air, and Soil Pollution,* Vol. 12, No. 4, November 1979, pp. 449–457.
36. Funderburg, S. W., et al., "Viral Transport Through Soil Columns Under Conditions of Saturated Flow," *Water Research,* Vol. 15, No. 6, 1981, pp. 703–711.
37. Lance, J. C., and Gerba, C. P., "Virus Movement in Soil During Saturated and Unsaturated Flow," *Applied and Environmental Microbiology,* Vol. 47, No. 2, February 1984, pp. 335–337.
38. Lance, J. C., Gerba, C. P., and Wang, D. S., "Comparative Movement of Different Enteroviruses in Soil Columns," *Journal of Environmental Quality,* Vol. 11, No. 3, 1982, pp. 347–351.
39. Moore, B. E., Sagik, B. P., and Sorber, C. A., "Viral Transport to Ground Water at a Wastewater Land Application Site," *Journal of the Water Pollution Control Federation,* Vol. 53, No. 10, October 1981, pp. 1492–1502.
40. Schaub, S. A., Bausam, H. T., and Taylor, G. W., "Fate of Virus in Wastewater Applied to Slow-Infiltration Land Treatment Systems," *Applied and Environmental Microbiology,* Vol. 44, No. 2, August 1982, pp. 383–394.
41. Sobsey, M. D., et al., "Enteric Virus Removal from Septic Tank Effluent by Pilot Scale Soil Absorption Systems," Paper at 3rd National Conference on Environmental Engineering Research, Development, and Design, July 1976, North Carolina University at Chapel Hill, Chapel Hill, North Carolina.
42. Sobsey, M. D., et al., "Interactions and Survival of Enteric Viruses in Soil Materials," *Applied and Environmental Microbiology,* Vol. 40, No. 1, July 1980, pp. 92–101.
43. Vilker, V. L., et al., "Application of Ion Exchange/Adsorption Models to Virus Transport in Percolating Beds," *American Institute of Chemical Engineers Symposium Series,* Vol. 74, No. 178, 1978, pp. 84–92.

44. Freeze, R. A., and Cherry, J. A., *Groundwater,* 1979, Prentice-Hall, Inc., Englewood Cliffs, New Jersey.
45. Corey, P. R., McWhorter, D. B., and Smith, J. L., "Rate of Ammonium Nitrification and Nitrate Leaching in Soil Columns," NSF/RA-760505, November 1976, National Science Foundation, Washington, DC.
46. Hubbard, R. K., Asmussen, L. E. and Allison, H. D., "Shallow Groundwater Quality Beneath an Intensive Multiple-Cropping System Using Center Pivot Irrigation," *Journal of Environmental Quality,* Vol. 13, No. 1, 1984, pp. 156–161.
47. Walter, M. F., "Nitrate Movement in Soil Under Early Spring Conditions," PhD dissertation, 1974, University of Wisconsin, Madison, Wisconsin.
48. Bouma, J., "Subsurface Applications of Sewage Effluent," in *Planning the Uses and Management of Land,* 1979, ASA-CSSA-SSSA, Madison, Wisconsin, pp. 665–703.
49. Enfield, C. G., et al., "Fate of Waste Water Phosphorus in Soil," *Journal of the Irrigation and Drainage Division,* American Society of Civil Engineers, Vol. 101, No. IR3, September 1975, pp. 145–155.
50. Sawhney, B. L., "Predicting Phosphate Movement Through Soil Columns," *Journal of Environmental Quality,* Vol. 6, No. 1, January–March 1977, pp. 86–89.
51. Mansell, R. S. et al., "Experimental and Simulated Transport of Phosphorus Through Sandy Soils," *Water Resources Research,* Vol. 13, No. 1, February 1977, pp. 189–194.
52. Stuanes, A. O., and Enfield, C. G., "Prediction of Phosphate Movement Through Some Selected Soils," *Journal of Environmental Quality,* Vol. 13, No. 2, April–June 1984, pp. 317–320.
53. Bates, M. H., "Fate and Transport of Heavy Metals," *Proceedings of Seminar on Ground Water Quality,* July 1980, University of Oklahoma, Norman, Oklahoma, pp. 213–229.
54. Alesii, B. A., and Fuller, W. H., "The Mobility of Three Cyanide Forms in Soils," *Hazardous Wastes Research Symposium on Residual Management by Land Disposal,* University of Arizona, Tucson, Arizona, 1976, pp. 213–223.
55. Frost, R. R., and Griffin, R. A., "Effect of pH on Adsorption of Arsenic and Selenium Landfill by Clay Minerals," *Soil Science Society of America Journal,* Vol. 41, No. 1, January–February 1977, pp. 53–57.
56. Fuller, W. H., et al., "Influence of Leachate Quality on Soil Attenuation of Metals," *Disposal of Hazardous Wastes—Proceedings of 6th Annual Symposium,* EPA 600/9-80-010, March 1980, U.S. Environmental Protection Agency, Cincinnati, Ohio, pp. 108–117.
57. Gambrell, R. P., et al., "Soil Physicochemical Parameters Affecting Metal Availability in Sludge-Amended Soils," EPA-600/S2-85/123, January 1986, U.S. Environmental Protection Agency, Cincinnati, Ohio.
58. Giordano, P. M., and Mortvedt, J. J., "Nitrogen Effects on Mobility and Plant Uptake of Heavy Metals in Sewage Sludge Applied to Soil Columns," *Journal of Environmental Quality,* Vol. 5, No. 2, April–June 1976, pp. 165–168.

59. Jurinak, J. J., and Santillan-Medrano, J., "The Chemistry and Transport of Lead and Cadmium in Soils," PB-237 497/3SL, June 1974, National Technical Information Service, U.S. Department of Commerce, Springfield, Virginia.
60. Patel, N. C., "The Release of Metal Ions to Ground Water by Soils," OWRI-B-028-ALA(6), August 1974, Tuskegee Institute, Department of Plant and Soil Science, Tuskegee, Alabama.
61. Theis, T. L., "Contamination of Groundwater by Heavy Metals from the Land Disposal of Fly Ash," Technical Progress Report, September 1, 1976–May 31, 1977, June 1977, Department of Civil Engineering, Notre Dame University, South Bend, Indiana.
62. Theis, T. L., Kirkner, D. J., and Jennings, A. A., "Hydrodynamic and Chemical Modeling of Heavy Metals in Ash Pond Leachates," March 1980, Notre Dame University, South Bend, Indiana.
63. Wangen, L. E., and Stallings, E. A., "Subsurface Transport of Contaminants from Energy Process Waste Leachates," LA-10011-PR, February 1984, Los Alamos National Laboratory, Los Alamos, New Mexico.
64. Loch, J. P., Lagas, P., and Haring, B. J., "Behavior of Heavy Metals in Soil Beneath a Landfill: Results of Model Experiments," *Quality of Groundwater,* van Duijvenbooden, W., Glasbergen, P., and van Lelyveld, H., Editors, 1981, Elsevier Scientific Publishing Company, Amsterdam, The Netherlands, pp. 545–555.
65. Wetherold, R. G., Randall, J. L., and Williams, K. R., "Laboratory Assessment of Potential Hydrocarbon Emissions from Land Treatment of Refinery Oily Sludges," EPA-600/S2-84-108, July 1984, U.S. Environmental Protection Agency, Ada, Oklahoma.
66. Enfield, C. G., et al., "Toxic Organic Volatilization from Land Treatment Systems," EPA/600/D-85/031, February 1985, U.S. Environmental Protection Agency, Ada, Oklahoma.
67. Manos, C. G., et al., "Effects of Clay Mineral-Organic Matter Complexes on Gaseous Hydrocarbon Emissions from Soils," *Proceedings of the NWWA/API Conference and Petroleum Hydrocarbons and Organic Chemicals in Ground Water—Prevention, Detection and Restoration,* November 1984, National Water Well Association, Worthington, Ohio, pp. 189–206.
68. Hutchins, S. R., et al., "Fate of Trace Organics During Rapid Infiltration of Primary Wastewater at Fort Devens, Massachusetts," *Water Research,* Vol. 18, No. 8, 1984, pp. 1025–1036.
69. McCarty, P. L., Reinhard, M., and Rittman, B. E., "Trace Organics in Ground Water," *Environmental Science and Technology,* Vol. 15, No. 1, January 1981, pp. 40–51.
70. Roberts, P. V., et al., "Organic Contaminant Behavior During Ground Water Recharge," *Journal Water Pollution Control Federation,* Vol. 52, No. 1, January 1980, pp. 161–171.
71. Scott, H. D., Wolf, D. C., and Lavy, T. L., "Adsorption and Degradation of Phenol at Low Concentrations in Soil," *Journal of Environmental Quality,* Vol. 12, No. 1, January–March 1983, pp. 91–95.

72. Griffin, R. A., and Chian, E. S. K., "Attenuation of Water-Soluble Polychlorinated Biphenyls by Earth Materials," EPA-600/2-80-027, May 1980, Municipal Environmental Research Laboratory, U.S. Environmental Protection Agency, Cincinnati, Ohio.
73. Griffin, R. A., and Chou, S. J., "Disposal and Removal of Halogenated Hydrocarbons in Soils," *Disposal of Hazardous Wastes: Proceedings of the Sixth Annual Symposium,* EPA-600/9-80-010, March 1980, U.S. Environmental Protection Agency, Cincinnati, Ohio, pp. 82–92.
74. Schwarzenbach, R. P., and Westall, J., "Transport of Nonpolar Organic Compounds from Surface Water to Groundwater—Laboratory Sorption Studies," *Environmental Science and Technology,* Vol. 15, No. 11, November 1981, pp. 1360–1367.
75. Wilson, J. T., et al., "Transport and Fate of Selected Organic Pollutants in a Sandy Soil," *Journal of Environmental Quality,* Vol. 10, No. 4, October–December 1981, pp. 501–506.
76. Larson, R. J., and Ventullo, R. M., "Biodegradation Potential of Ground Water Bacteria," *Proceedings of the Third National Symposium on Aquifer Restoration and Ground Water Monitoring,* May 1983, National Water Well Association, Worthington, Ohio, pp. 402–409.
77. Barker, J. F., and Patrick, G. C., "Natural Attenuation of Aromatic Hydrocarbons in a Shallow Sand Aquifer," *Proceedings of the NWWA/API Conference and Petroleum Hydrocarbons and Organic Chemicals in Ground Water—Prevention, Detection and Restoration,* November 1984, National Water Well Association, Worthington, Ohio, pp. 160–177.
78. Bouwer, E. J., "Biotransformation of Organic Micropollutants in the Subsurface," *Proceedings of the NWWA/API Conference and Petroleum Hydrocarbons and Organic Chemicals in Ground Water—Prevention, Detection and Restoration,* November 1984, National Water Well Association, Worthington, Ohio, pp. 66–81.
79. Bouwer, E. J., and McCarty, P. L., "Modeling of Trace Organics Biotransformation in the Subsurface," *Ground Water,* Vol. 22, No. 4, July–August 1984, pp. 433–440.
80. Bouwer, E. J., Rittmann, B. E., and McCarty, P. L., "Anaerobic Degradation of Halogenated 1- and 2-Carbon Organic Compounds," *Environmental Science and Technology,* Vol. 15, No. 5, May 1981, pp. 596–599.
81. Enfield, C. G., et al., "Behavior of Organic Pollutants During Rapid Infiltration of Wastewater into Soil: Part II, Mathematical Description of Transport and Transformation," 1984, R. S. Kerr Environmental Research Laboratory, U.S. Environmental Protection Agency, Ada, Oklahoma.
82. Godsy, E. M., Goerlitz, D. F., and Ehrlich, G. G., "Methanogenesis of Phenolic Compounds by a Bacterial Consortium from a Contaminated Aquifer in St. Louis Park, Minnesota," *Bulletin of Environmental Contamination and Toxicology,* Vol. 30, No. 3, March 1983, pp. 261–268.
83. Hutchins, S. R., et al., "Microbial Removal of Wastewater Organic Compounds as a Function of Input Concentrations in Soil Columns," *Applied and Environmental Microbiology,* Vol. 48, No. 5, November 1984, pp. 1039–1045.

84. Piwoni, M. D., et al., "Behavior of Organic Pollutants During Rapid Infiltration of Wastewater into Soil: Part I, Process Definition and Characterization Using A Microcosm," 1984, R. S. Kerr Environmental Research Laboratory, U.S. Environmental Protection Agency, Ada, Oklahoma.
85. Tsentas, C., and Supkow, D. J. "Migration and Apparent Subsurface Biodegradation of Organic Compounds in a Fractured Bedrock Aquifer," *Proceedings of the NWWA/API Conference on Petroleum Hydrocarbons and Organic Chemicals in Ground Water—Prevention, Detection and Restoration,* November 1985, National Water Well Association, Worthington, Ohio, pp. 77–89.
86. White, K. D., et al., "Microbial Degradation Kinetics of Alcohols in Subsurface Systems," *Proceedings of the NWWA/API Conference on Petroleum Hydrocarbons and Organic Chemicals in Ground Water—Prevention, Detection and Restoration,* November 1985, National Water Well Association, Worthington, Ohio, pp. 140–159.
87. Wilson, B. H., and Rees, J. F., "Biotransformation of Gasoline Hydrocarbons in Methanogenic Aquifer Material," *Proceedings of the NWWA/API Conference on Petroleum Hydrocarbons and Organic Chemicals in Ground Water—Prevention, Detection and Restoration,* November 1985, National Water Well Association, Worthington, Ohio, pp. 128–139.
88. Wilson, J. T., and Wilson, B. H., "Biotransformation of Trichloroethylene in Soil," *Applied and Environmental Microbiology,* Vol. 49, No. 1, January 1985, pp. 242–243.
89. Schwille, F., "Groundwater Pollution in Porous Media by Fluids Immiscible with Water," *Quality of Groundwater,* van Duijvenbooden, W., Glasbergen, P., and van Lelyveld, H., Editors, 1981, Elsevier Scientific Publishing Company, Amsterdam, The Netherlands, pp. 451–463.
90. Hinchee, R. E., and Reisinger, H. J., "Multi-phase Transport of Petroleum Hydrocarbons in the Subsurface Environment: Theory and Practical Application," *Proceedings of the NWWA/API Conference on Petroleum Hydrocarbons and Organic Chemicals in Ground Water—Prevention, Detection and Restoration,* November 1984, National Water Well Association, Worthington, Ohio, pp. 58–75.
91. Van Der Waarden, M., Groenewoud, W. M., and Bridie, A. L., "Transport of Mineral Oil Components to Ground Water—II. Influence of Lime, Clay and Organic Soil Components on the Rate of Transport," *Water Research,* Vol. 11, No. 4, 1977, pp. 359–366.
92. Kappeler, T., "Microbial Degradation of the Water-Soluble Fraction of Gas Oil—I," *Water Research,* Vol. 12, No. 5, May 1978, pp. 327–333.
93. Kappeler, T., "Microbial Degradation of the Water-Soluble Fraction of Gas Oil—II. Bioassays with Pure Strains," *Water Research,* Vol. 12, No. 5, May 1978, pp. 335–342.
94. Khaleel, R., "Transport of Organic Contaminants in Ground Water," *Proceedings of the Seventh National Ground Water Quality Symposium,* September 1984, National Water Well Association, Worthington, Ohio, pp. 367–387.
95. Garrett, D., Maxey, F. P., and Katz, H., "The Impact of Intensive Application of Pesticides and Fertilizers on Underground Water Recharge Areas Which

May Contribute to Drinking Water Problems," EPA 560/3-75-006, January 1976, U.S. Environmental Protection Agency, Washington, DC.

96. LeGrand, H. E., "Movement of Agricultural Pollutants with Groundwater," *Agricultural Practices and Water Quality,* 1970, Iowa State University Press, Ames, Iowa, pp. 303–313.
97. Hall, J. K., and Hartwig, N. L., "Atrazine Mobility in Two Soils Under Conventional Tillage," *Journal of Environmental Quality,* Vol. 7, No. 1, January–March 1978, pp. 63–68.
98. Schneider, A. D., Wiese, A. F., and Jones, O. R., "Movement of Three Herbicides in a Fine Sand Aquifer," *Agronomy Journal,* Vol. 69, No. 3, May–June 1977, pp. 432–436.
99. Wehtje, J. R., et al., "Atrazine Contamination of Groundwater in the Platt Valley of Nebraska from Non-point Sources," *Quality of Groundwater,* van Duijvenbooden, W., Glasbergen, P., and van Lelyveld, H., Editors, 1981, Elsevier Scientific Publishing Company, Amsterdam, The Netherlands, pp. 141–145.
100. Dregne, H. E., Gomez, S. and Harris, W., "Movement of 2,4-D in Soils," New Mexico Agricultural Experiment Station Western Regional Research Project, Progress Report, November 1969, New Mexico State University, Las Cruces, New Mexico.
101. Mansell, R. S., et al., "Fertilizer and Pesticide Movement from Citrus Groves in Florida Flatwood Soils," EPA 600/2-77/177, August 1977, National Technical Information Service, U.S. Department of Commerce, Springfield, Virginia.
102. Hebb, E. A., and Wheeler, W. B., "Bromacil in Lakeland Soil Ground Water," *Journal of Environmental Quality,* Vol. 7, No. 4, October–December 1978, pp. 598–601.
103. Lafleur, K. S., Wojeck, G. A., and McCaskill, W. R., "Movement of Toxaphene and Fluonethuron Through Dunbar Soil to Underlying Ground Water," *Journal of Environmental Quality,* Vol. 2, No. 4, 1973, pp. 515–518.
104. Scalf, M. R., et al., "Movement of DDT and Nitrates During Ground Water Recharge," *Water Resources Research,* Vol. 5, No. 5, October 1969, pp. 1041–1052.
105. Reece, D. E., et al., "Influence of Chemical Species on Mobility of Sursurface Organic Arsenic Contamination," *Proceedings of the NWWA/API Conference on Petroleum Hydrocarbons and Organic Chemicals in Ground Water–Prevention, Detection and Restoration,* November 1984, National Water Well Association, Worthington, Ohio, pp. 99–110.
106. Davidson, J. M., Ou, L. T., and Rao, P. S., "Behavior of High Pesticide Concentrations in Soil Water Systems," *Hazardous Waste Research Symposium on Residential Management by Land Disposal,* 1976, University of Arizona, Tucson, Arizona, pp. 206–212.
107. Davidson, J. M., Rao, P. S., and Ou, L. T., "Movement and Biological Degradation of Large Concentrations of Selected Pesticides in Soils," *Disposal of Hazardous Wastes—Proceedings of the Sixth Annual Symposium,* EPA-600/9-80-010, March 1980, U.S. Environmental Protection Agency, Cincinnati, Ohio, pp. 93–107.

108. Awad, T. M., Kilgore, W. W., and Winterlin, W., "Movement of Aldicarb in Different Soil Types," *Bulletin of Environmental Contamination and Toxicology,* Vol. 32, No. 4, April 1984, pp. 377–382.
109. Biggar, J. W., Nielsen, D. R., and Tillotson, W. R., "Movement of DBCP in Laboratory Soil Columns and Field Soils to Groundwater," *Environmental Geology,* Vol. 5, No. 3, 1984, pp. 127–131.
110. Conley, K. A., and Mikucki, W. J., "Migration of Explosives and Chlorinated Pesticides in a Simulated Sanitary Landfill," Report No. AD-A030 453/5WP, September 1976, National Technical Information Service, U.S. Department of Commerce, Springfield, Virginia.
111. Robertson, J. B., and Kahn, L., "The Infiltration of Aldrin Through Ottawa Sand Columns," Professional Paper 650-C, 1969, U.S. Geological Survey, Idaho Falls, Idaho, pp. C219–C223.
112. Van Genuchten, M. T., Davidson, J. M., and Nierenga, P. J., "An Evaluation of Kinetic and Equilibrium Equations for the Prediction of Pesticide Movement Through Porous Media," *Proceedings of the Soil Society of America,* 1974, pp. 29–35.
113. You, I. S., "Binding to Humus and Microbial Metabolism of the Herbicide Residue 3,4-Dichloroaniline," PhD dissertation, 1982, Rutgers University, New Brunswick, New Jersey.
114. Mansell, R. S., et al., "Movement of Fertilizer and Herbicide Through Irrigated Sands," Publication No. 38, September 1976, Florida Water Resources Research Center, University of Florida, Gainesville, Florida.
115. Enfield, C. G., Carsel, R. F., and Phan, T., "Comparison of a One-Dimensional, Steady-State Hydraulic Model with a Two-Dimensional, Transient Hydraulic Model for Aldicarb Transport Through Soil," *Quality of Groundwater,* van Duijvenbooden, W., Glasbergen, P., and van Lelyveld, H., Editors, 1981, Elsevier Scientific Publishing Company, Amsterdam, The Netherlands, pp. 507–510.
116. Enfield, C. G., et al., "Approximating Pollutant Transport to Ground Water," *Ground Water,* Vol. 20, No. 6, November–December 1982, pp. 711–722.
117. Dean, J. D., Jowise, P. P., and Donigian, A. S., Jr., "Leaching Evaluation of Agricultural Chemicals (LEACH) Handbook," EPA-600/3-84-068, June 1984, Environmental Research Laboratory, U.S. Environmental Protection Agency, Athens, Georgia.
118. Javandel, I., Doughty, C., and Tsang, C. F., "Groundwater Transport: Handbook of Mathematical Models," Water Resources Monograph 10, 1984, American Geophysical Union.
119. U.S. Environmental Protection Agency, "Pesticides in Ground Water: Background Document," May 1986, Office of Ground Water Protection, Washington, DC.
120. Oblath, S. B., and Hawkins, R. H., "Performance of Special Wasteform Lysimeters at a Humid Site," DP-MS-84-83, 1984, Savannah River Laboratory, DuPont de Nemours (E. I.) and Co., Aiken, South Carolina.
121. Fenske, P. R., "Time-Dependent Sorption on Geological-Materials," *Journal of Hydrology,* Vol. 43, October 1979, pp. 415–425.

122. Jones, T. L., et al., "Field and Laboratory Evaluation of the Mobility of Cobalt-60/EDTA in an Arid Environment," PNL-SA-10780, February 1983, Battelle Pacific Northwest Laboratories, Richland, Washington.
123. O'Hare, M. P., et al., *Artificial Recharge of Ground Water—Status and Potential in the Contiguous United States,* 1986, Lewis Publishers, Inc., Chelsea, Michigan.
124. Atkinson, S. F., et al., *Salt Water Intrusion—Status and Potential in the Contiguous United States,* 1986, Lewis Publishers, Inc., Chelsea, Michigan.
125. Magdoff, F. R., Bouma, J., and Keeney, D. R., "Columns Representing Mound-type Disposal Systems for Septic Tank Effluent: 1. Soil-Water and Gas Relations," *Journal of Environmental Quality,* Vol. 3, No. 3, July–September 1974, pp. 223–228.
126. Lotse, E. G., "Septic Tank Effluent Movement Through Soil," PB-261 368/5ST, June 1976, National Technical Information Service, U.S. Department of Commerce, Springfield, Virginia.
127. Brutsaert, W. F., Hedstrom, W. E., and McNiece, T. G., "Effect of Soil Moisture Content Upon Adsorption and Movement of Phosphorus from Leachates of Domestic Waste Disposal Systems," OWRT-A-044-ME(2), November 1980, Land and Water Resources Center, University of Maine, Orono, Maine.
128. Mitchell, D., "Improving Design Criteria for Septic Tank Systems," W77-02621, August 1976, Water Resources Research Center, University of Arkansas, Fayetteville, Arkansas.
129. Burchinal, J. C., and Qasim, S. R., "Leaching of Pollutants from Refuse Beds," *Journal of Sanitary Engineering Division,* ASCE, Vol. 96, No. SA1, February 1970, pp. 49–58.
130. Fuller, W. H., "Investigation of Landfill Leachate Pollutant Attenuation by Soils," EPA-60012-78/158, April 1977, U.S. Environmental Protection Agency, Cincinnati, Ohio.
131. Griffin, R. A., and Shimp, N. F., "Attenuation of Pollutants in Municipal Landfill Leachate by Clay Minerals," PB-287 140/8ST, August 1978, National Technical Information Service, U.S. Department of Commerce, Springfield, Virginia.
132. Rademaker, A. D., and Young, J. C., "Leachates from Solid Waste Recovery Operations," *Journal of the Energy Division, ASCE,* Vol. 107, No. EY 1, May 1981, pp. 17–29.
133. Miller, D. W., Braids, O. C., and Walker, W. H., "The Prevalence of Subsurface Migration of Hazardous Chemical Substances at Selected Industrial Waste Land Disposal Sites," PB-272 973, September 1977, National Technical Information Service, U.S. Department of Commerce, Springfield, Virginia.
134. Sommers, L. E., et al., "Nitrogen and Metal Contamination of Natural Waters from Sewage Sludge Disposal on Land," OWRT-B-066-IND, December 1976, Office of Water Research and Technology, U.S. Department of the Interior, Washington, DC.
135. Lofty, R. J., et al., "Environmental Assessment of Subsurface Disposal of Municipal Waste-Water Treatment Sludge," EPA/530/SW-167C, June 1978, SCS Engineers, Long Beach, California.

136. Lee, G. F., and Jones, R. A., "A Risk Assessment Approach for Evaluating the Environmental Significance of Chemical Contaminants in Solid Wastes," *Environmental Risk Analysis for Chemicals,* Conway, R. A., Editor, 1982, Van Nostrand Reinhold Company, New York, New York, pp. 529–549.
137. U.S. Environmental Protection Agency, "Hazardous Waste Management System," *Federal Register,* Vol. 51, No. 114, June 13, 1986, pp. 21648–21693.
138. U.S. Environmental Protection Agency, "Extraction Procedure Toxicity Characteristic," Code of Federal Regulations, Vol. 40, No. 261.24, May 19, 1980.
139. Davis, S. N., et al., "An Introduction to Ground Water Tracers," EPA/600/S2-85/022, February 1986, U.S. Environmental Protection Agency, Ada, Oklahoma.
140. Enfield, C. G., "Chemical Transport Facilitated by Multiphase Flow Systems," 1984, U.S. Environmental Protection Agency, Ada, Oklahoma.
141. Barbee, G. C., and Brown, K. W., "Movement of Xylene Through Unsaturated Soils Following Simulated Spills," 1985, Soil and Crop Sciences Department, Texas A and M University, College Station, Texas.

CHAPTER 6

Flow and Solute Transport Modeling

Ground water modeling is a general term that encompasses several different aspects of the behavior of subterranean water systems. Four processes of potential relevance in any ground water modeling study include ground water flow, solute transport, heat transport, and deformation. Ground water flow modeling studies are usually undertaken to determine the responses of an aquifer to pumping, injection, or recharge stresses. Responses would include flow velocities and drawdowns (or upconing). Mass transport modeling studies are usually concerned with the movement within an aquifer system of a solute. These studies have become increasingly important with the current interest on ground water pollution. Heat transport models are usually focused on developing geothermal energy resources, and deformation studies are employed to analyze the effects of ground water removal on land subsidence. This chapter describes the governing equations for flow and solute transport modeling and contains information on both physical and numerical models. Based on a review of published literature, examples of the types and applications of models are also included.

GOVERNING EQUATIONS

The need for ground water models stems from the fact that the equations governing the behavior of the four processes listed earlier are all complex, second-order partial differential equations which are not amenable to direct analytical solution. Ground water models try to circumvent these difficulties by either: (1) simulating the behavior of the aquifer system on a small scale; or (2) using simplifying assumptions or numerical approximations to the governing equations. The basic equations to be addressed herein include the Laplace and Poisson's equations and the equations for transient flow, unconfined aquifers, and solute transport.

Laplace Equation

If a steady-state flow of ground water through an infinitesimal volume is assumed, then by combining Darcy's law and the equation of continuity the following equation can be developed:

$$\frac{\delta}{\delta x}(-K\frac{\delta h}{\delta x})+\frac{\delta}{\delta y}(-K\frac{\delta h}{\delta y})+\frac{\delta}{\delta z}(-K\frac{\delta h}{\delta z})=0 \quad (6.1)$$

where:

$K=K(x,y,z)$ = hydraulic conductivity of aquifer at point (x,y,z)
$h=h(x,y,z)$ = hydraulic head at point (x,y,z)

If K is assumed to be constant throughout the aquifer and independent of direction, that is, the aquifer is assumed to be homogeneous and isotropic, then equation 6.1 becomes:

$$\frac{\delta^2 h}{\delta x^2}+\frac{\delta^2 h}{\delta y^2}+\frac{\delta^2 h}{\delta z^2}=0 \quad (6.2)$$

Equation 6.2 is the well known Laplace equation. The Laplace equation, as applied to hydraulic head in an aquifer, can be used to describe steady-state flow through an isotropic, homogeneous aquifer with no sources of recharge or sinks of removal.

Poisson's Equation

If sources of recharge (such as direct precipitation) or sinks of removal (such as extraction wells) need to be addressed, the Laplace equation must be modified. By considering a volumetric water balance at steady-state conditions, that is, outflow equals inflow, for an infinitesimal volume of aquifer the following equation can be identified:

$$\frac{\delta^2 h}{\delta x^2}+\frac{\delta^2 h}{\delta y^2}=-\frac{R(x,y)}{T} \quad (6.3)$$

where:

$R(x,y)$ = volume of water added per unit time per unit aquifer area
T = transmissivity of the aquifer = hydraulic conductivity times the saturated thickness

Equation 6.3 is known as Poisson's equation, and it can be used to predict two-dimensional, steady-state flow in an isotropic, homogeneous aquifer with possible sources and/or sinks included.

Transient Flow Equation

Equations 6.2 and 6.3 both govern steady-state flow conditions. The more common case encountered is transient flow conditions where the position of the water table (or piezometric surface) is changing with time. By setting the volume outflow rate in an infinitesimal aquifer equal to the volume inflow rate plus the rate of release of water from storage, the following equation is developed:

$$\frac{\delta^2 h}{\delta x^2} + \frac{\delta^2 h}{\delta y^2} = \frac{S}{T}\frac{\delta h}{\delta t} - \frac{R(x,y)}{T} \qquad (6.4)$$

where:

t = time

S = storage coefficient of aquifer = volume of water released per unit area of aquifer per unit drop in head

Equation 6.4 is a two-dimensional equation for addressing transient flow in an isotropic, homogeneous aquifer.

Unconfined Aquifers

Equations 6.2, 6.3 and 6.4 are most often applied to confined aquifers. Equations governing the behavior of unconfined aquifers must take into consideration the fact that the saturated thickness is variable. The Dupuit-Forchheimer assumptions of: (1) horizontal flow; and (2) hydraulic gradient equal to the slope of the water table are needed for development of these equations. The equation for two-dimensional, steady-state flow in an unconfined aquifer becomes:

$$\frac{\delta^2 h^2}{\delta x^2} + \frac{\delta^2 h^2}{\delta y^2} = -\frac{2R(x,y)}{K} \qquad (6.5)$$

The equation for two-dimensional, transient flow in an unconfined aquifer is:

$$\frac{K}{2}\left(\frac{\delta^2 h^2}{\delta x^2} + \frac{\delta^2 h^2}{\delta y^2}\right) = s\frac{\delta h}{\delta t} - R(x,y,t) \qquad (6.6)$$

where:

s = specific yield of aquifer = volume of water released per unit area per unit drop in water table.

Solute Transport Equation

The equation governing the movement of a dissolved species in ground water due to convection and dispersion can be developed by utilizing a conservation of mass approach and employing Fick's law of dispersion. The equation in statement form is:

[net rate of change of mass of solute within the element] = [flux of solute out of the element] − [flux of solute into the element] ± [loss or gain of solute mass due to reactions]

The solute transport equation can take on different forms depending on the assumptions made or reactions considered. The one-dimensional form of the equation for a nonreactive, dissolved constituent in a homogeneous, isotropic aquifer under steady-state, uniform flow is:

$$D_{\ell}\frac{\delta^2 C}{\delta \ell^2} - V_{\vartheta}\frac{\delta C}{\delta \ell} = \frac{\delta C}{\delta t} \quad (6.7)$$

where:

- ℓ = curvilinear coordinate direction taken along the flowline
- $D\ell$ = coefficient of hydrodynamic dispersion in the flow path direction
- $V\ell$ = average linear ground water velocity
- C = solute concentration

The two-dimensional equation for a nonreactive, dissolved chemical species in flowing ground water can be written as follows:[1]

$$\frac{\delta(Cb)}{\delta t} = \frac{\delta}{\delta x_i}(bD_{ij}\frac{\delta C}{\delta x_j}) - \frac{\delta}{\delta x_i}(bCV_i) - \frac{C'W}{\epsilon} \quad (6.8)$$

where:

- C = the concentration of the dissolved chemical species—(mass or M divided by L^3)
- D_{ij} = the coefficient of hydrodynamic dispersion (a second-order tensor)—(L^2/t)
- b = the saturated thickness of the aquifer—(L)
- C' = the concentration of the dissolved chemical in a source or sink fluid—(M/L^3)
- V_i = the seepage velocity in the direction of x_i—(L/t)
- W = the volume flux per unit area—(L/t)

The first term on the right side of Equation 6.8 represents the change in concentration due to hydrodynamic dispersion. The second term describes the effects of convective transport, while the third term represents a fluid source or sink.[1]

TYPES OF MODELS

As shown in Table 6.1, ground water models can be divided into two broad groups: physical models and numerical models. There are six classes of models within these two groups and each is described below.

Physical Models

The earliest attempts at modeling ground water were of the physical type. Physical models can be divided into two classes: scale models and analogs.

Scale Models

Scale models are actual physical replicas of an aquifer that have been scaled down for study in the laboratory. The most common scale models are the soil column (one-dimensional model), the Hele-Shaw apparatus (two-dimensional model), and the sand tank (three-dimensional model). In these models, media is placed in such a way that it parallels the soil structure of the aquifer of concern. The models are then subjected to certain stresses such as water removal or injection, or contaminated recharge. The response of the models is obtained through direct measurements. The behavior of the prototype aquifer to real life stresses can then be predicted by using scale relationships.

Table 6.1. Types of Ground Water Models.

1. Physical
   a. Scale Models
      - soil column
      - Hele-Shaw
      - sand tank
   b. Analog Models
      - electric
      - thermal
      - mechanics
2. Numerical Models
   a. Analytical Models
   b. Stochastic Models
   c. Computer Models

Because of size restrictions, the scale factors will not be the same in all directions or for all soil parameters. For example, the equation for unsteady flow in a confined aquifer (written in polar coordinates) is:

$$\frac{\delta^2 h}{\delta r^2} + \left(\frac{1}{r}\right)\frac{\delta h}{\delta r} + \frac{1}{r^2}\left(\frac{\delta^2 h}{\delta \theta^2}\right) = \frac{S}{T}\frac{\delta h}{\delta t} \tag{6.9}$$

where:

h = hydraulic head—(L)
r = radius from well—(L)
θ = angle from a given plane
S = storage coefficient of the aquifer—(dimensionless)
T = transmissivity of the aquifer—(L^2/t)

The same equation applied to a scale model is:

$$\frac{\delta^2 h_m}{\delta r_m{}^2} + \left(\frac{1}{r_m}\right)\left(\frac{\delta h_m}{\delta r_m}\right) + \frac{1}{r_m{}^2}\left(\frac{\delta^2 h_m}{\delta \theta_m{}^2}\right) = \frac{S_m}{T_m}\left(\frac{\delta h_m}{\delta t_m}\right) \tag{6.10}$$

If U is defined as the ratio of the model parameter divided by the actual aquifer parameter, five different scale relationships can be identified:

$$U_r = \frac{r_m}{r},\ U_z = \frac{h_m}{h},\ U_t = \frac{t_m}{t},\ U_s = \frac{S_m}{S},\ U_T = \frac{T_m}{T} \tag{6.11}$$

Through substitution and algebraic manipulation, the equation governing the behavior of the model can be expressed in terms of its own measurable parameters and the various scale ratios. The equation is:

$$\frac{\delta^2 h_m}{r_m{}^2} + \left(\frac{1}{r_m}\right)\left(\frac{\delta h_m}{\delta r_m}\right) + \frac{1}{r_m{}^2}\left(\frac{\delta^2 h_m}{\delta \theta_m{}^2}\right) = \frac{S_m}{T_m}\left(\frac{\delta h_m}{\delta t_m}\right)\left(\frac{U_T \cdot U_t}{U_s \cdot U_r{}^2}\right) \tag{6.12}$$

The above relationship holds only if

$$\left(\frac{U_T \cdot U_t}{U_s \cdot U_r{}^2}\right) = 1 \tag{6.13}$$

Therefore, in the use of a physical scale model three of the scale ratios can be manipulated with the other ratio being dictated by Equation 6.13.[2] The behavior of the model to stresses can then be measured and predictions can be made of the behavior of actual aquifers through use of the scale factors.

Three examples of scale models will be cited: Heutmaker, Peterson, and Wheatcraft;[3] Michael, Gelhar, and Wilson;[4] and Williams.[5] First, Michael, Gelhar, and Wilson[4] used a vertical Hele-Shaw model to determine expected patterns of fresh and salt ground water flow for several potential ground water development schemes for Long Island, New York.

Heutmaker, Peterson, and Wheatcraft[3] used a laboratory sand-packed model to study the mechanics of buoyant plume movement and the entrainment of salt water by the plume. Simulated waste effluent was injected into a density-stratified aquifer system under dynamic ground water conditions and the effects on plume mechanics of varying several injection parameters, such as injection depth and rate, type of injection source, strength of the ambient flow field, and density of ambient receiving water, were observed. In every experiment conducted during this study, a buoyant plume of injected effluent, which was clearly distinct from the resident aquifer liquids, formed and rose vertically into the lower portion of the freshwater lens where it was subjected to the freshwater flow field and migrated downgradient with the fresh water, still keeping its identity as a distinct plume of injected effluent. Three of the injection parameters exerted significant control on the movement of the injection plumes, including the depth and rate of effluent injection, and the strength of the freshwater flow field.

Williams[5] also used a two-dimensional Hele-Shaw model for studying wastewater injection into a fresh-saline ground water system. Three different field conditions were simulated: injection under static conditions, injection into an ambient flow field without a fresh-saline water interface, and injection into an ambient flow field with a fresh-saline water inferface. Both single- and double-well injections were studied for the latter case of injection in the presence of an interface. The density of the saline water was 1.026 g/cu cm for the majority of the tests while that of the injected fluid was 1.005 g/cu cm. The test results indicated that when injection is below the fresh-saline water interface, the injected fluid, under the action of the buoyant forces, rises through a relatively narrow, vertical channel to the interface and then spreads laterally along that surface. The distance increases with the distance below the interface at which injection takes place. Consequently, injection at or above the interface appeared to be preferable to injection below the interface from an economic as well as a pollution standpoint.

Analog Models

Analog models are based on the fact that a direct analogy can be made between ground water flow and some easily measurable phenomena from a different field of study. One type of analog is the electric analog. In this type of model, the properties of an aquifer (permeability, storage coefficient, etc.) are simulated by various electronic components (resistors, capacitors, etc.), and the voltage across these components is analogous to the potential (or head) of water in the aquifer. Three of the more common electric analogs are identified as follows:

Table 6.2. Analogies to Ground Water Flow (Representative Units are Shown in Brackets).[8]

VARIABLE	GROUND WATER	ELECTRICITY	HEAT
Potential	Head, h [cm]	Voltage, V [Volts]	Temperature, T [C]
Quantity transported	Volume discharge rate [cm^3 s^{-1}]	Electrical charge [Coulomb]	Heat [calorie]
Physical property of medium	Hydraulic conductivity, K [cm s^{-1}]	Electrical conductivity, σ [mhos m^{-1}]	Thermal conductivity, K [cal cm^{-1} s^{-1} °C^{-1}]
Relation between potential and flow field	Darcy's law $\mathbf{q} = -K\,\mathbf{grad}\,h$ where $\mathbf{q}$ is specific dicharge [cm s^{-1}]	Ohm's law $\mathbf{i} = -\sigma\,\mathbf{grad}\,V$ where $\mathbf{i}$ is electrical current [Amperes]	Fourier's law $\mathbf{q} = -K\,\mathbf{grad}\,T$ where $\mathbf{q}$ is heat flow [cal cm^{-2} s^{-1}]
Storage quantity	Specific storage, S_s [cm^{-1}]	Capacitance, C [microfarad]	Heat capacity, C_v [cal cm^{-3} °C^{-1}]

(1) Resistor-Capacitance Network Analog
The resistor-capacitance (R-C) network dissipates electrical energy in somewhat the same way a porous medium consumes ground water energy to let water travel through its voids.

(2) Conductive-Liquid/Conductive-Solid Analog
These analogs utilize the difference in conductance of different materials to create a voltage potential. This voltage potential is analogous to the velocity potential or head in ground water flow.

(3) Electrical Resistance Analog
This is a specialized case of the resistance-capacitance network analog. In this model variable resistors (rheostats) are used to simulate the fillable/drainable porosity of a soil. Hence these analogs will be most useful in studying time-varying recharge phenomena.

Listed in Table 6.2 are the relationships between aquifer parameters and their electrical analogs.

Shamberger and Domenico[6] described an electrical analog model for the ground water system in the Las Vegas Valley, Nevada. The model was used to measure aquifer response to pumping by measuring appropriate voltages and currents. The analog model was considered to be a useful tool in analyzing ground water management options.

Another example of an analog is the thermal analog. It is known that the flow of heat in a uniform body and steady-state ground water flow both obey the Laplace equation as follows:

$$\nabla^2 Q = 0 \text{ or } \nabla^2 h = 0 \tag{6.14}$$

where:

∇ = Laplacean Operator ($\nabla = \delta/\delta_x + \delta/\delta_y + \delta/\delta_z$)

By adding a heat source or sink to a given material and measuring temperatures an analogy to ground water flow can be developed. Analogous parameters for thermal analogs are also listed in Table 6.2.

Another model analog for ground water flow can be constructed with a stretched thin rubber membrane. Small slopes of the membrane surface can be expressed in polar coordinates as follows:[7]

$$\frac{d^2z}{dr^2} + \frac{1}{r}\frac{dz}{dr} = \frac{-W_m}{T_m} \tag{6.15}$$

where:

dz = deflection at a radial distance dr from a central deflecting point
W_m = weight of membrane/area
T_m = uniform membrane tension

If the right side of equation 6.15 can be set equal to zero, by using lightweight (small W_m) material or turning the model vertical, it becomes analogous to the steady-state ground water flow equation in polar coordinates as follows:

$$\frac{\delta^2 h}{\delta r^2} + \frac{1}{r}\frac{\delta h}{\delta r} = 0 \tag{6.16}$$

Therefore, measurements of the deflections of the membrane from protrusions become analogous to drawdowns from a well. Analogous parameters for mechanical analogs are listed in Table 6.3.

With the advent of high-speed computational capabilities, the use of physical models for predicting ground water flows has decreased. Physical models are also limited by space, time, cost, and accuracy deficiencies. However, scale models are currently being used in laboratory studies of the transport and fate of selected ground water contaminants. Analog models are not applicable to the movement of contaminants in ground water.

Numerical Models

Numerical models are currently the most popular types of ground water models. Numerical models can be grouped into three classes: (1) analytical models; (2) stochastic models; and (3) computer models.

Table 6.3. Structural Mechanics Analogy to Ground Water Flow.[8]

VARIABLE	GROUND WATER	MECHANICS
Unknown variable	Head, *h*	Displacement, $\delta = \begin{pmatrix} u \\ v \end{pmatrix}$
First derivative quantity	**grad** $h = \begin{pmatrix} \frac{\partial h}{\partial x} \\ \frac{\partial h}{\partial y} \end{pmatrix}$	Strain, $\epsilon = \begin{pmatrix} \frac{\partial u}{\partial x} \\ \frac{\partial v}{\partial y} \\ \frac{\partial u}{\partial y} + \frac{\partial v}{\partial x} \end{pmatrix}$
Conjugate variable	$\mathbf{q} = \begin{pmatrix} q_x \\ q_y \end{pmatrix}$	Stress, $\sigma = \begin{pmatrix} \sigma_{xx} \\ \sigma_{yy} \\ \tau_{xy} \end{pmatrix}$
Physical property medium	Hydraulic conductivity, $[K] = \begin{pmatrix} K & 0 \\ 0 & K \end{pmatrix}$	Elastic moduli, $[C] = \begin{pmatrix} \frac{E}{1-v^2} & \frac{v}{1-v^2} & 0 \\ \frac{v}{1-v^2} & \frac{E}{1-v^2} & 0 \\ 0 & 0 & G \end{pmatrix}$ where E is Young's modulus, v is Poisson's ratio, and G is shear modulus, $G = \frac{E}{2(1+v)}$
Assembled finite element matrix equation	$[G]\{h\} = \{B\}$ where $[G]$ is conductance matrix and $\{B\}$ is recharge matrix	$[S]\{\delta\} = \{F\}$ where $[S]$ is stiffness matrix and $\{F\}$ is load matrix

Analytical Models

Analytical models are usually developed by considering highly idealized conditions or using significant simplifying assumptions to obtain a solution to the governing equations. The most famous examples of analytical solutions are for the steady-state flow of ground water under either confined or unconfined conditions using the Dupuit-Forchheimer assumptions. The Dupuit-Forchheimer assumptions are: (1) purely horizontal flow; and (2) flow uniformly distributed with depth. Utilizing these assumptions and employing Darcy's law gives rise to two relationships describing the flow of ground water.

$$Q=\frac{2\pi KD(h_2-h_1)}{\ln(r_2/r_1)} \qquad \text{(6.17a, confined)}$$

or

$$Q=\frac{\pi K(h_2{}^2-h_1{}^2)}{\ln(r_2/r_1)} \qquad \text{(6.17b, unconfined)}$$

where:

Q = volumetric flow rate—(L^3/t)
h = hydraulic head—(L)
K = hydraulic conductivity—(L/T)
D = depth of aquifer—(L)
r = radius from well—(L)

Equations 6.17a and 6.17b are overly simplistic for all but the most general studies. Conversely, many other analytical solutions have been developed for extremely idealized or specific situations.

Several techniques have been developed for analyzing ground water flow problems. The use of the "separation of variables" technique for solving the governing second-order partial differential equation has been widespread. Freeze and Cherry[9] used the technique to solve the equation for steady-state flow through a confined aquifer. Van Der Kamp[10] used the technique to develop an equation describing periodic flow and wave propagation in ground water. An early application of the technique was described by Toth[11] relative to the problem of subsurface flow to parallel drains. Kirkham and Powers[12] have developed a general solution using the separation of variables technique; they have also outlined a general procedure for applying their solution to a variety of ground water flow problems. Cohen and Miller[13] present a comprehensive listing of analytical solutions to a wide variety of ground water situations and concerns.

The use of analytical solutions to delineate flow paths has been applied extensively to the problem of subsurface flow to parallel drains. For example, Jury[14] has used the solution generated by Kirkham and a numerical integration technique to develop contaminant travel times for parallel drain situations. Reilly[15] has used a dimensionless solution to develop a technique for determining contaminant travel times from the top of an aquifer to a pumping well. McLin[16] has expanded the work of Jury[14] by using a dimensionless analytical solution to verify and calibrate a numerical model. The model is then used to make multiple iterations for different geologic situations. McLin[16] also used a numerical integration scheme to calculate pollutant breakthrough curves.

Bear[17] discussed several analytical solutions to relatively simple, one-dimensional solute transport problems. However, even simple solutions tend

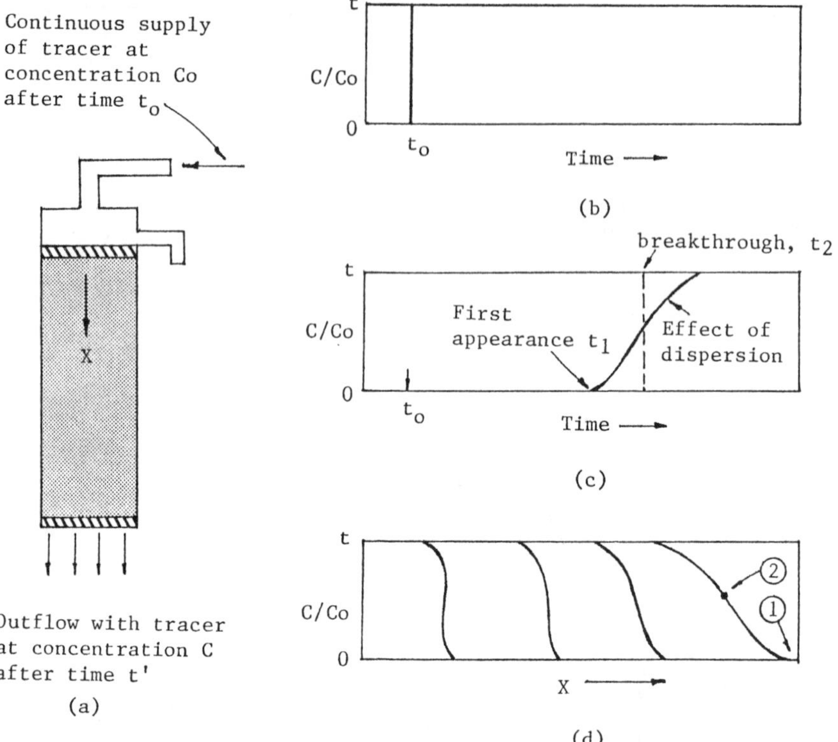

Figure 6.1. Longitudinal dispersion of a tracer passing through a column of porous medium: (a) column with steady flow and continuous supply of tracer after time t_o; (b) step-function-type tracer input relation; (c) relative tracer concentration in outflow from column (dashed line indicates plug flow condition and solid line illustrates effect of mechanical dispersion and molecular diffusion); and (d) concentration profile in the column at various times.

to get overwhelmed with advanced mathematics. As an example, consider the one-dimensional flow of a solute through the soil column shown in Figure 6.1a. The boundary conditions represented by the step-function input are described mathematically as:

$$
\begin{aligned}
&C(1,0) = 0 && 1 \geq 0 \\
&C(0,t) = C_0 && t \geq 0 \\
&C(\infty,t) = 0 && t \geq 0
\end{aligned}
$$

For these boundary conditions the solution to Equation 6.7 for a saturated homogeneous porous medium is:

$$\frac{C}{C_o} = \frac{1}{2}\left[\text{erfc}\frac{(1-\bar{v}t)}{2\sqrt{D_\ell t}} + \exp\left(\frac{\bar{v}_\ell}{D_\ell}\right)\text{erfc}\left(\frac{1+\bar{v}t}{2\sqrt{D_\ell t}}\right)\right] \tag{6.18}$$

where:

erfc represents the complementary error function; l is the distance along the flow path; and $\bar{v}$ is the average linear water velocity.

For conditions in which the dispersivity ($D\ell$) of the porous medium is large or when l or t is large, the second term on the right-hand side of equation 6.18 is negligible. Equation 6.18 can be used to compute the shapes of the breakthrough curves and concentration profiles illustrated in Figure 6.1c and 6.1d.

Analytical models represent an attractive alternative to both physical and numerical models in terms of decreased complexity and input data requirements. However, analytical models are often only feasible when based on significant simplifying assumptions, and these assumptions may not allow the model to accurately reflect the conditions of interest. Additionally, even some of the simplest analytical models tend to involve complex mathematics.

Stochastic Models

Stochastic modeling involves using statistical methods applied to large amounts of aquifer data to generate empirical relationships between the various properties of the aquifer and its behavior. The objective of a stochastic model is that, given a specific input, the model will generate as an output the "expected value" of a certain characteristic with a specified variability. Stochastic models are often used in making management decisions relative to regional aquifer systems.

Three examples of stochastic models will be cited: Rao, Rao, and Kashyap;[18] Sagar and Clifton;[19] and Smith and Freeze.[20] Ground water levels in an aquifer as affected by external causes such as precipitation, outflows from the aquifer, and pumping were modeled by Rao, Rao, and Kashyap.[18] Both multi-input/single-output and multivariate models have been developed and tested. It was found that these models can be used to evaluate the time lag between precipitation and changes in ground water levels, the effect of the stream flows on water levels in adjacent wells, and the ground water resources in a region.

Smith and Freeze[20] conducted a stochastic analysis of one-dimensional, steady-state ground water flow through a bounded domain by using Monte Carlo simulation techniques. The flow domain was divided into a finite set of discrete blocks. Hydraulic conductivity values in neighboring blocks were autocorrelated by assuming that the spatial variations in conductivity could be represented by a first-order, nearest neighbor, stochastic process model. An integral scale was defined to characterize the average distance over which conductivity values in the block system are autocorrelated. This model leads to a realistic representation of the spatial variations in hydraulic conductivity in a discrete block medium.

Stochastic ground water flow modeling by means of a second-order, uncertainty analysis technique is discussed by Sagar and Clifton.[19] This technique is based on a Taylor series expansion of the state variables of interest (hydraulic heads and Darcian velocities) about the expected values of the model parameters. The method has been incorporated into the computer code PORSTAT, which solves the two-dimensional, stochastic ground water flow equation coupled with the deterministic heat transfer and mass transport equation using integrated finite differences coupled to a direct-equation solver. Uncertain variables that can be considered in the application of PORSTAT are: (1) hydraulic conductivities (x- and y-direction); (2) specific storage; (3) boundary conditions; and (4) initial conditions. The output from PORSTAT consists of the expected values, variances, and covariances of hydraulic heads and Darcian velocities. The application of PORSTAT to a sample problem is also presented by Sagar and Clifton.[19]

In addition, Bakr, et al.[21] used statistical methods to examine spatial variability in subsurface flows. Bathals, Rao and Spooner[22] have developed a statistical procedure for analyzing regional aquifer systems. Carlsson and Carlstedt[23] used statistical analysis of pump-test data to calculate average values of transmissivity and permeability. Other stochastic modeling examples are described by Delleur, et al.;[24] Gelhar;[25] Newman and Yakowitz;[26] Ross, Koplik and Crawford;[27] Sagar;[28] and Yakowitz.[29]

The use of stochastic methods to analyze ground water systems has been fairly limited to analyses of regional problems or situations with significant amounts of extant input data. Stochastic models are often limited in usage by their large input data requirements.

Computer Models

The most popular approach for modeling ground water behavior is through the use of numerical techniques involving computer-based solutions. Numerical modeling techniques incorporate analytical models that are so large they require the use of computers capable of multiple iterations to converge on a solution. Computer models can be grouped into two broad classifications based on their spatial approach to the aquifer system: (1) lumped models; and (2) distributed models.

Lumped models attempt to predict the behavior of the aquifer as a whole unit. The approach used is to estimate the total change in a given parameter as the difference between total input and output. For example, a simple water balance equation for a stream-connected phreatic aquifer system can be represented by:[30]

$$n\frac{dh}{dt} = q_n + q_a - q_o - q_p \pm q_\ell \qquad (6.19)$$

where:

h = average thickness of the ground water zone
t = time
n = average effective porosity
q_n = natural recharge rate per unit surface area
q_a = artificial recharge rate per unit surface area
q_o = natural aquifer outflow rate per unit surface area
q_p = aquifer pumping rate per unit surface area
q_ℓ = river leakage rate per unit surface area

The corresponding mass balance equation for the stream-connected phreatic aquifer system would be:[30]

$$n\frac{d(h_c)}{dt} = q_n c_n + q_a c_a - q_o c - q_p c \pm q_\ell c + nhr \qquad (6.20)$$

where:

r = volumetric source-sink term that accounts for contaminant additions or degradation within the flow zone
c = average aquifer concentration
c_n = concentration in natural recharge
c_a = concentration in artificial recharge

Hence, an aquifer-wide accounting procedure for the parameters listed in equations 6.19 and 6.20 will provide the data necessary for a predictive model.

Gelhar and Wilson[31] have described a generalized lumped parameter ground water model developed based on simultaneous water and solute balances for a phreatic aquifer. The basic behavior of the model is characterized by two response times, one associated with the hydraulics and the other with the solute. The model has been applied to simulate the impact of highway deicing salts on ground water quality in eastern Massachusetts. The results, obtained by digital computer simulation, were found to be in reasonable agreement with observed trends over a 15-year period. The effects of various highway deicing alternatives have been simulated and the dependence on the aquifer parameters has been demonstrated.

The second classification of computer models are the distributed models. Distributed models use numerical techniques to describe the behavior of an aquifer at selected points (nodes). Several different numerical techniques exist for solving the relevant partial differential equations. However, the three most popular techniques for ground water studies are finite difference methods, finite element methods, and the method of characteristics. The theory behind the methods and examples of their specific applications are described as follows.

Finite Difference Methods Probably the simplest and most popular numerical technique for solving the relevant partial differential equations governing ground water behavior involves the use of finite differences. The technique is based on approximating the partial differential terms by their truncated Taylor's series expansion. Consider for example, the distribution

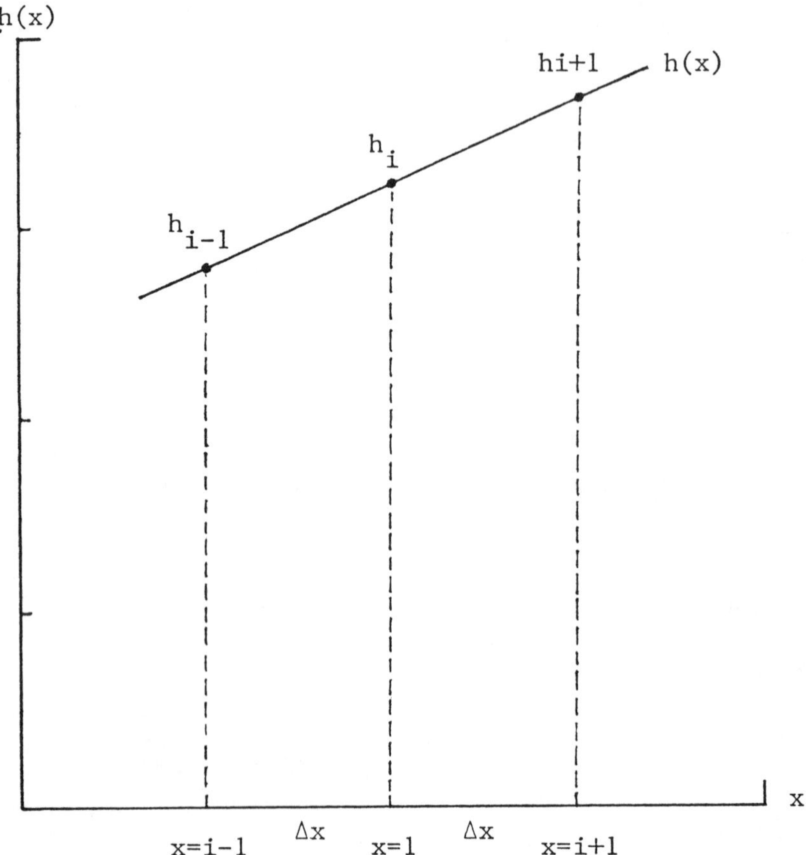

Figure 6.2. Distribution of head in one dimension.

of head in one dimension (Figure 6.2). The finite difference approximation for the rate of change of head at some point i could be written as:

$$\frac{\delta h}{\delta x} \simeq \frac{h(i) - h(i+1)}{\Delta x} \tag{6.21}$$

This is called the first forward approximation in that it involes the i + 1 term. Similarly, the first backward approximation would be written as:

$$\frac{\delta h}{\delta x} \simeq \frac{h(i-1) - h(i)}{\Delta x} \tag{6.22}$$

The central difference approximation is independent of the head at node i, and can be shown as:

$$\frac{\delta h}{\delta x} \simeq \frac{h(i+1) - h(i-1)}{2\Delta x} \tag{6.23}$$

Similarly, the approximations for second-order or higher derivatives can be derived. For example, the second difference at node i could be written:

$$\frac{\delta^2 h}{\delta x^2} \simeq \frac{h(i-1) - 2h(i) + h(i+1)}{(\Delta x)^2} \tag{6.24}$$

The equation for transient flow would be:

$$\frac{\delta^2 h}{\delta x^2} + \frac{\delta^2 h}{\delta y^2} = \frac{S}{T}\frac{\delta h}{\delta t} \tag{6.25}$$

Using the forward difference approximation to both the spatial and temporal derivatives yields the following:

$$\frac{h^k_{i-1,j} - 2h^k_{i,j} + h^k_{i+1,j}}{(\Delta x)^2} + \frac{h^k_{i,j-1} - 2h^k_{i,j} + h^k_{i,j+1}}{(\Delta y)^2} = \frac{S}{T}\left(\frac{h^{k+1}_{i,j} - h^k_{i,j}}{\Delta t}\right) \tag{6.26}$$

where:

i = denotes the position in x direction
j = denotes the position in y direction
k = elapsed time

In equation 6.26 all values of h are known at the k time step and the value of $h_{i,j}$ at the k + 1 time step can be solved for directly. This is called an explicit scheme.

If equation 6.25 is written with the backward difference approximation for the time derivative the following is obtained:

$$\frac{h^{k+1}_{i-1,j} - 2h^{k+1}_{i,j} + h^{k+1}_{i+1,j}}{(\Delta x)^2} + \frac{h^{k+1}_{i,j-1} - 2h^{k+1}_{i,j} + h^{k+1}_{i,j+1}}{(\Delta y)^2} = \frac{S}{T}\left(\frac{h^{k+1}_{i,j} - h^k_{i,j}}{\Delta t}\right) \tag{6.27}$$

Equation 6.27 has five unknowns; however, it can be written for each node in the grid resulting in a set of simultaneous equations. These equations can be solved to give a new value of h at each node for the k+1 time increment. This is called an implicit scheme.

Explicit schemes are simple to formulate, but have severe restrictions on the grid spacing and time increments. Implicit methods are more complicated, but more versatile. They require greater computer storage capacity, but use less running time than explicit methods. Implicit models are superior in that they also permit the use of larger time values.

Table 6.4 lists 17 examples of models using either the finite difference or finite element technique for the solution of the ground water flow model. Six of the 17 involve the finite difference approach (Cosner;[32] Grove;[33] Guvanason and Volker;[34] Khaleel and Redell;[35] Konikow;[36] and Prickett and Lonnquist[37]). Cosner[32] simulated the Lower Cretaceous aquifer in southeastern Virginia by a finite difference digital model. The area modeled is in the vicinity of Franklin, Virginia where the aquifer is about 180 m thick. The aquifer is estimated to be about 650 m thick at a distance of about 55 km to the east. The aquifer consists of alternating permeable and semipermeable sand, gravel, silt and claybeds, and it is overlain by a semipermeable confining bed. Transmissivity is about 19,000 cu ft/day/ft (1,899 cu m/day/m) at Franklin, decreasing to the west and increasing to the east. A prepumping water-level surface was developed by modeling. The use of a modeling factor (T/K′), transmissivity over vertical permeability of the confining bed, is described. Simulation runs to 1941, 1970, and 1972 were in good agreement with historical data. Predictive runs show that if pumpage continues to increase, dewatering will occur at Franklin; however, if pumpage does not increase, additional drawdown will be small.

Grove[33] developed a transient, three-dimensional subsurface waste disposal model to aid the design and testing of such systems. The model uses a finite difference solution to the pressure, energy, and mass transport equations.

A method for evaluating the parameters in a two-dimensional ground water flow equation was presented by Guvanason and Volker.[34] The method employs finite difference approximations and the assumption of a limit on the local variability of transmissivity or hydraulic conductivity. The approach entails a systematic minimization of the difference between recorded and calculated water levels using the Newton-Raphson technique along a path in such a way that the calibrated regions are isolated from the rest. Illustrations of the application of this scheme include those from an idealistic rectangular water table aquifer and a real water table aquifer system.

Table 6.4. Finite Element and Finite Difference Techniques for the Solution of the Ground Water Flow Model.

Author(s) (Year)	Comments
Brunch (1973)[38]	Finite element, weighted residual process for solving unsteady ground water flows in an unconfined aquifer either into or out of a surface reservoir
Cabrero and Marino (1976)[39]	Finite element model of the movement and distribution of a conservative substance in a stream-aquifer system
Cheng (1975)[40]	Computer subroutine package for solving the convective-dispersion material balance equation using quadratic isoparametric finite elements
Cosner (1975)[32]	Finite difference model for the Lower Cretaceous aquifer in southeastern Virginia
Duguid and Reeves (1976)[41]	Finite element, two-dimensional, transient model incorporating advective transport, hydrodynamic dispersion, chemical adsorption, and radioactive decay
Grove (1976)[33]	Transient, three-dimensional subsurface waste disposal model based on the finite difference solution to the pressure, energy, and mass transport equations
Grove (1977)[42]	Finite element model for the transport and reaction of chemical solutes in porous media
Gupta, et al. (1984)[43]	Finite element, three-dimensional model for analyzing flow through large, multilayered ground water systems
Guvanason and Volker (1978)[34]	Finite difference approximations for identification of distributed parameters in ground water basins
Khaleel and Redell (1977)[35]	Finite difference model for the movement of a conservative, noninteracting tracer in a nonhomogeneous, anisotropic porous medium
Konikow (1977)[36]	Finite difference solute transport model for chloride movement in the alluvial aquifer at Rocky Mountain Arsenal in Colorado

Table 6.4, continued

Authors(s) (Year)	Comments
Nishi, Bruch, and Lewis (1975)[44]	Finite element method for hydrodynamic dispersion in seepage from a triangular channel
Pickens and Lennox (1976)[45]	Finite element method for simulating the two-dimensional, transient movement of conservative or nonconservative wastes in a steady-state saturated ground water flow system
Prickett and Lonnquist (1971)[37]	Finite difference approaches for one-, two-, and three-dimensional, nonsteady flows of ground water in heterogeneous aquifers
Seckel (1978)[46]	Transient, three-dimensional ground water flow model using the finite element technique
Skrivan (1977)[47]	Finite element-based flow model for an alluvial aquifer in Imperial County, California
Yeh and Huff (1983)[48]	Implementation and demonstration of the Finite Element model of Water flow through Aquifers (FEWA)

A three-dimensional model describing the two-phase (air-water) fluid flow equations in an integrated saturated-unsaturated porous medium was developed by Khaleel and Redell.[35] Also, a three-dimensional convective-dispersion equation describing the movement of a conservative, noninteracting tracer in a nonhomogeneous, anisotropic porous medium was developed. Finite difference forms of these two equations were solved using an implicit scheme to solve for water or air pressures, an explicit scheme to solve for water and air saturations, and the method of characteristics with a numerical tensor transformation to solve the convective-dispersion equations. A typical two-dimensional drainage problem in agriculture was solved in a nonhomogeneous, integrated saturated-unsaturated medium using the total simulator of fluid flow and convective-dispersion equations. A variety of outputs, such as an equipotential map or a solute concentration map, were obtained at selected time steps. A field problem describing the migration of septic tank wastes around the perimeter of a lake was also considered and solved using the total simulator.

Konikow[36] used a solute transport model to predict the movement of dissolved chemicals in ground water at the Rocky Mountain Arsenal, near Denver, Colorado. The model couples a finite difference solution to the ground water flow equation, with the method-of-characteristics solution to the solute transport equation. From 1943 to 1956 liquid industrial wastes containing high chloride concentrations were disposed into unlined ponds at the Arsenal. Wastes seeped out of the unlined disposal ponds and spread for many square miles in the underlying shallow alluvial aquifer. Since 1956 disposal has been into an asphalt-lined reservoir, which contributed to a decline in ground water contamination by 1972. The simulation model quantitatively integrated the effects of the major factors that controlled changes in chloride concentrations and accurately reproduced the 30-year history of chloride ground water contamination.

Prickett and Lonnquist[37] described several ground water resource evaluation models within which a finite difference approach is used to formulate the equations of ground water flow. Generalized computer program listings are given that can simulate one-, two-, and three-dimensional, nonsteady flow of ground water in heterogeneous aquifers under water table, nonleaky, and leaky artesian conditions. Programming techniques involving time varying pumpage from wells, natural or artificial recharge rates, the relationships of water exchange between surface waters and the ground water reservoir, the process of ground water evapotranspiration, and the mechanism of converting from artesian to water table conditions are also included.

Finite Element Methods Whereas the finite difference techniques approximate the partial differential equations by a differential approach, the finite element method approximates the equations by the integral approach. The finite element technique involves solving a differential equation for ground water behavior by means of a variational calculus. Consider the equation for two-dimensional, nonsteady ground water flow in a nonhomogeneous aquifer:

$$\frac{\delta}{\delta x}(K_x b \frac{\delta h}{\delta x}) + \frac{\delta}{\delta y}(K_y b \frac{\delta h}{\delta y}) + Q_s = S\frac{\delta h}{\delta t} \tag{6.28}$$

The solution to this equation is equivalent to finding a solution for h that minimizes the variational function as follows:

$$F = \iint [\frac{Kx}{2}(\frac{\delta h}{\delta x})^2 + \frac{Ky}{2}(\frac{\delta h}{\delta y})^2 + (S\frac{\delta h}{\delta t} - Q_s)h]dxdy \tag{6.29}$$

To obtain a numerical solution for equation 6.29, the aquifer is subdivided into "finite elements." The parameters Kx, Ky, S and Q_s are kept constant for a given element, but they may vary from element to element. The differential $\delta F/\delta h$ is then evaluated for each node and equated to zero. The resulting system of simultaneous equations can then be readily solved by computer.[7]

Because of the complexity of the mathematical description, a finite element method model would not be as easy to manipulate and apply to a generalized problem. As such, most finite element models have been developed for specific problems.

Eleven of the 17 examples of models listed in Table 6.4 use the finite element technique for the solution of the ground water flow model. Three examples relate to ground water flow considerations (Brunch;[38] Nishi, Brunch, and Lewis;[44] and Skrivan).[47] Brunch[38] used a finite element, weighted residual process to solve a nonlinear, partial differential equation describing unsteady ground water flows in an unconfined aquifer either into or out of a surface reservoir. Nishi, Brunch, and Lewis[44] employed the finite element method using rectangular elements to solve for hydrodynamic dispersion in seepage from a triangular channel, and for the influence of electro-osmosis on flow in an aquifer with a recharge and discharge well. Finally, Skrivan[47] developed a flow model using finite element techniques for an alluvial aquifer in the Ocotillo-Coyote Wells basin in Imperial County, California. The model was calibrated with historical data from 1925 to 1975. Evaluations of the impacts of various pumping regimes on ground water levels in the study area were then made through 1995. It was found that continued pumping at a high rate after 1995 may cause saline water to flow toward the potable ground water in and around Ocotillo.

Comprehensive finite element models for ground water flows and hydraulic conditions have been developed by Yeh and Huff[48] and Gupta, et al.[43] Yeh and Huff[48] developed the Finite Element model of Water flow through Aquifers (FEWA). Point as well as distributed sources/sinks are included to represent recharges/pumpings and rainfall infiltrations. All sources/sinks can be transient or steady-state. Either completely confined or completely unconfined aquifers, or partially confined and partially unconfined aquifers can be dealt with effectively. Discretization of a compound region with very irregular curved boundaries is made easy by including both quadrilateral and triangular elements in the formulation. Large field problems can be solved efficiently by including a pointwise iterative solution strategy as an optional alternative to the direct elimination solution method for the matrix equation approximating the partial differential equation of ground water flow. FEWA also includes transient flow through confining leaky aquifers lying above and/or below the aquifer of interest.

Gupta, et al.[43] developed the FE3DGW (Finite Element Three-Dimensional Ground Water) code for analyzing flow through large, multilayered ground water systems. The code has the capability to model noncontinuous and continuous layering, time-dependent and constant sources/sinks, and transient as well as steady-state flow. The code offers a wide choice in specifying boundary conditions like prescribed heads, nodal injection or withdrawal, constant or spatially varying infiltration rates, and elemental sources/sinks. Initial conditions can be prescribed as vertically hydrostatic or variable hydraulic head. The heterogeneity in aquifer permeability and porosity can be described by geologic unit or explicitly for given elements.

Finite element models for the physical transport of conservative pollutants in ground water systems have been developed by Cabrero and Marino[39] and Cheng.[40] Transient, two-dimensional solutions have been developed by Cabrero and Marino[39] for describing the movement and distribution of conservative substances in stream-aquifer systems. Cheng[40] presents a computer subroutine package for solving the convective-dispersion material balance equation using quadratic isoparametric finite elements. This modeling approach has been applied to the problem of sea water encroachment in coastal aquifers, brine movement through causeways, and dispersion of pollutants in ground water.

Pickens and Lennox[45] simulated the two-dimensional, transient movement of conservative or nonconservative wastes in a steady-state ground water flow system by using the finite element method based on a Galerkin technique. Examples involving the movement of nonconservative contaminants described by distribution coefficients and examples with variable input concentration are given. The model can also be applied to environmental problems related to ground water contamination from waste disposal sites.

The partial differential equation that describes the transport and reaction of chemical solutes in porous media was solved by Grove[42] using the Galerkin finite element technique. Comparisons of the finite element methods to the finite difference methods, and to analytical results, indicated that a high degree of accuracy may be obtained using the finite element method. The technique was applied to a field problem involving an aquifer contaminated with chloride, tritium, and strontium-90.

Duguid and Reeves[41] developed a two-dimensional, transient model for flow of a dissolved constituent through porous media. Mechanisms for advective transport, hydrodynamic dispersion, chemical adsorption, and radioactive decay were included in the mathematical formulation. Implementations of quadrilateral finite elements, bilinear spatial interpolation, and Gaussian elimination were used in the numerical formulation. The material transport model is compatible with the moisture transport model for predicting advective Darcy velocities for porous media which may be partly unsaturated. In addition to a description of the mathematical formulation, the results of two computer simulations were described. One simulation was a comparison with a well known analytical treatment and was intended as a partial validation. The other simulation, a seepage-pond problem, was a more realistic demonstration of the capabilities of the model.

A transient, three-dimensional ground water flow model was developed by Seckel[46] from an already existing two-dimensional version. The finite element technique was used for the solution of the boundary value problem that governs flow in saturated-unsaturated porous media. The model was used to study water table fluctuations on a spray irrigation site in St. Charles, Maryland. The model considers precipitation, evaporation, transpiration, and the application of spray effluent in evaluating the hydrologic response of the irrigation fields. To provide efficient and effective wastewater treatment, the model could facilitate the design, operation, and maintenance of a spray irrigation facility.

Method of Characteristics The finite difference and finite element techniques have been applied to both ground water flow and solute transport studies. The method of characteristics is a technique used specifically for solute transport problems, especially those situations where convective transport dominates.

The approach is not to solve the transport equation directly, but rather to solve an equivalent system of ordinary differential equations. The ordinary differential equations are obtained by rewriting the transport equation using the fluid particles as the point of reference. That is, instead of observing how the concentration changes with time at a fixed position in space, changing concentrations associated with fluid movement are noted. Therefore, the velocity distribution represents necessary information. In two dimensions, the end result is three equations for x-velocity, y-velocity and concentration, the solutions of which are called the characteristic curves, hence the name, methods of characteristics.[49]

This method is accomplished numerically by introducing a set of moving points (or reference particles) that can be traced within the stationary coordinates of a finite difference grid. Points are placed in each finite difference block and then allowed to move a distance proportional to the length of the time increment and the velocity at that point (see Figure 6.3). The moving points effectively simulate convective transport because the concentrations at each node vary as different points having different concentrations enter and leave the area of that block. Once the convective effect is determined, the remaining parts of the transport equation are solved using finite difference approximations and matrix methods.[49] Such a procedure is used in the two-dimensional solute transport model developed by Konikow and Bredehoeft.[1]

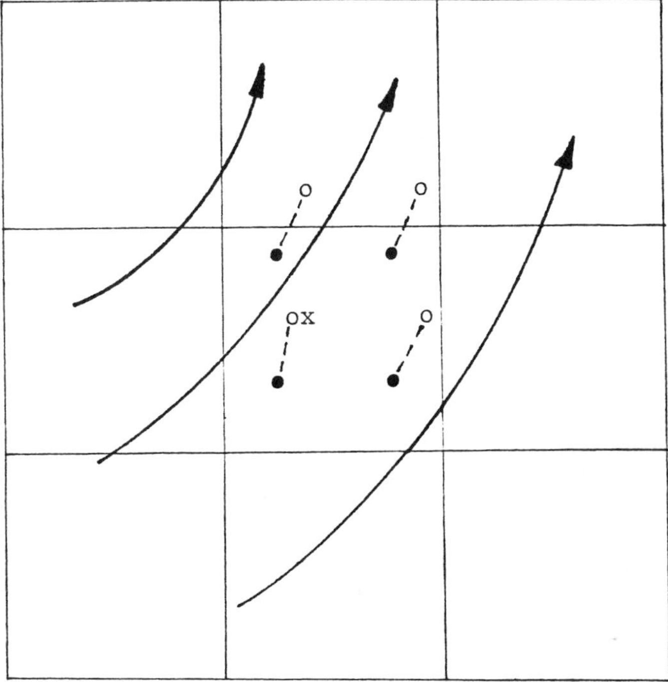

Figure 6.3. Finite difference grid showing reference particles.

Numerical modeling techniques are not a cure-all and can meet with considerable difficulties. Both the finite difference and finite element techniques can be plagued by the problems of numerical dispersion and oscillations when applied to solute transport problems. Numerical dispersion tends to smear or flatten out the effects of solute transport while numerical oscillation shows an inconsistent pattern in the results (Figure 6.4). This is especially true in cases where convection dominates the transport process. Techniques have been devised to minimize the effects of these numerical difficulties. Upstream weighting can reduce oscillations and the use of small grid spacings can reduce the effects of numerical dispersion.

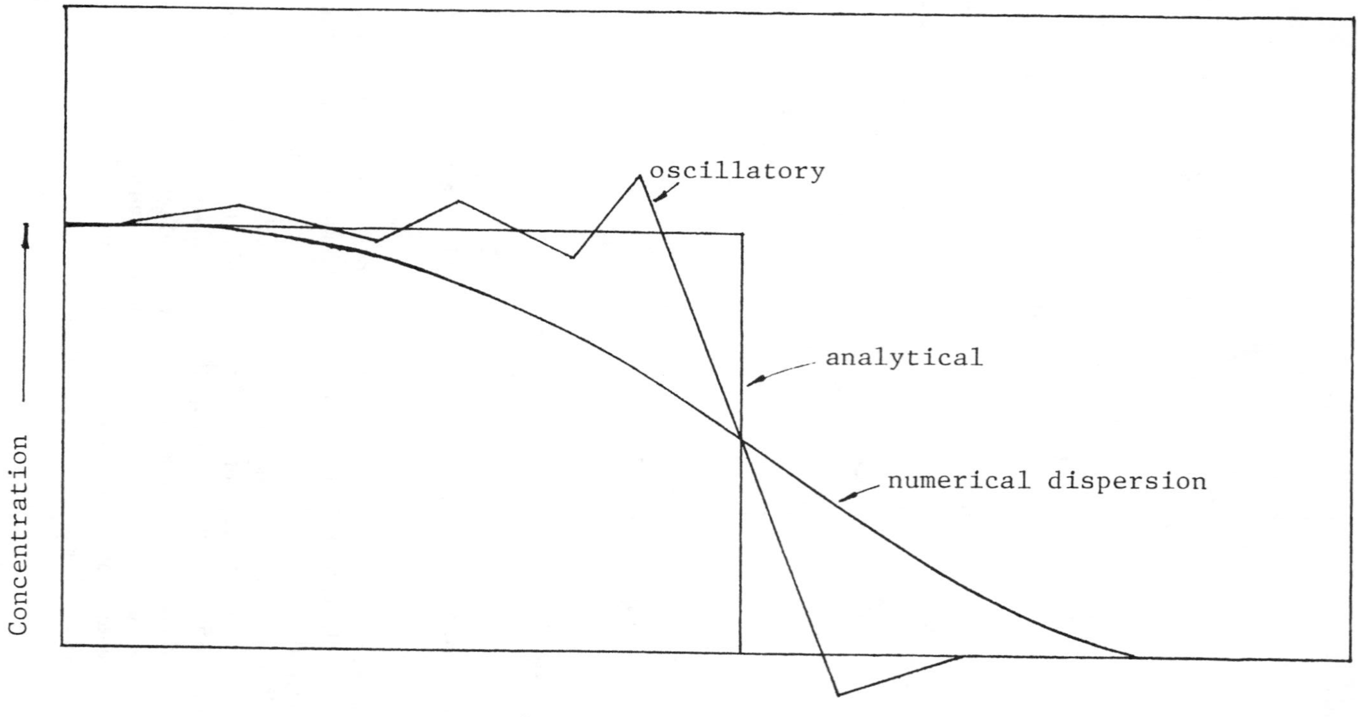

Figure 6.4. Typical numerical solutions for problem having convection only.

EXAMPLES OF MODELS

Numerous flow and/or solute transport models for ground water have been developed within the last several years. These models can be considered in accordance with the following categories: (1) determination of aquifer characteristics; (2) natural and artificial recharge; (3) surface and ground water relationships; (4) conjunctive use; (5) ground water management; (6) pollutant transport; (7) salt water intrusion and salinity; (8) pollutant source evaluation; and (9) miscellaneous, including construction dewatering, mineral dissolution, and freshwater storage.

Several overview studies of the status of model development and usage have been published: Anderson;[50] Appel and Bredehoeft;[51] Bachmat, et al.;[52] and Mercer and Faust.[53,54] A book on ground water modeling has been published by Wang and Anderson.[8] Anderson discusses the use of models to simulate the movement of contaminants through ground water flow systems. Mercer and Faust[53,54] discuss various types of models and their applications.

Appel and Bredehoeft[51] identified the following types of problems for which models have been or are being developed: ground water flow in saturated or partially unsaturated material, land subsidence resulting from ground water extraction, flow in coupled ground water-stream systems, coupling of rainfall-runoff basin models with soil moisture accounting aquifer flow models, interaction of economic and hydrologic considerations, transport of contaminants in an aquifer, and effects of proposed development schemes for geothermal systems. The status of the modeling activity for various models is reported by Appel and Bredehoeft[51] as being in a developmental, verification, operational, or continued improvement phase.

Bachmat, et al.[52] assessed the status of 250 models used internationally as tools for ground water-related water resources management. The 250 models are divided into the following categories: prediction, management, identification, and data management. Problem issues relative to model usage include accessibility to users, communications between managers and technical personnel relative to model limitations and assumptions, and inadequacies of input data.

Aquifer Characterization

Models can be used for the development of information on aquifer characteristics such as hydraulic conductivities, transmissivities, and storage coefficients. Table 6.5 contains summary comments on 12 references useful for aquifer characterization. Hydraulic conductivity is the focus of two references (Bakr, et al.;[21] and Gelhar[25]); transmissivity and storativity is addressed in five references (Corapcioglu and Brutsaert;[55] Navarro;[56] Newman and Yakowitz;[26] Rushton;[57] and Tang and Pinder[58]); evapotranspiration and aquifer outflow is highlighted in one reference (Daniels[59]); and hydrodynamic dispersion coefficients are described in two references (Murty and Scott;[60] and Umari, Willis, and Liu[61]). Models for evaluating pumping test results are described by the Texas Water Development Board[62] and Warner and Yow.[63]

Table 6.5. Models for Determining Aquifer Characteristics.

Author(s) (Year)	Comments
Bakr, et al. (1978)[21]	Stochastic analysis of spatial variability of hydraulic conductivity
Corapcioglu and Brutsaert (1977)[55]	Viscoelastic model using a minimal number of bulk parameters such as transmissivity, the compressibilities of primary and secondary consolidation, and the viscosity of the secondary consolidation
Daniels (1976)[59]	Use of stream flow records for estimating ground water evapotranspiration
Gelhar (1977)[25]	Analysis of the effects of hydraulic conductivity variations on ground water flows
Murty and Scott (1977)[60]	Determination of hydrodynamic dispersion coefficients in aquifers
Navarro (1977)[56]	Modified optimization method for estimating transmissivity and storativity parameters
Newman and Yakowitz (1979)[26]	Statistical approach for estimating spatially varying aquifer transmissivities
Rushton (1978)[57]	Model for estimating transmissivity and the storage coefficient from pumping tests
Tang and Pinder (1979)[58]	Deterministic solution for transmissivity using a finite difference approach
Texas Water Development Board (1973)[62]	Model for evaluating the drawdown from pumping one or more wells
Umari, Willis, and Liu (1979)[61]	Optimization approach for identifying aquifer dispersivities with two-dimensional, transient ground water contaminant transport
Warner and Yow (1979)[63]	Use of programmable hand calculators for pumping or injection well calculations

Bakr, et al.[21] suggested that the complex variation of hydraulic conductivity in natural aquifer materials is represented in a continuum sense as a spatial stochastic process which is characterized by a covariance function. Assuming statistical homogeneity, the theory of spectral analysis was used to solve perturbed forms of the stochastic differential equation describing flow through porous media with randomly varying hydraulic conductivity. From analyses of unidirectional mean flows which are perturbed by one- and three-dimensional variations of the logarithm of the hydraulic conductivity, local relationships between the head variance and the log conductivity variance were obtained. Gelhar[25] also noted that the variation of hydraulic conductivity can be characterized by spatial covariance functions and spectral density functions in the wave-number domain. The effects of more realistic spatial structure were demonstrated through the theoretical analysis of some elementary aquifer flow problems.

A viscoelastic aquifer model was applied by Corapcioglu and Brutsaert[55] to analyze and predict the piezometric drawdown and the compaction observed in a major confined aquifer system in the San Joaquin Valley in California. A field approach was used based on a minimal number of general parameters. The parameters, which are the transmissivity, the compressibilities of primary and secondary consolidation, and the viscosity of the secondary consolidation, were derived by calibration of the model by means of trial and error simulation of two years of records of drawdown and compaction. It was found that in the absence of drawdown records, the identification also may be accomplished with compaction records in addition to a standard pumping test.

Navarro[56] described a modified optimization method for estimating transmissivity and storativity parameters for the fast calibration of mathematical models of transient aquifers. Four field examples of the use of the method were given and discussed in terms of relative accuracy of results. Rushton[57] presented a numerical model for analyzing pumping test results to estimate the transmissivity and the storage coefficient from well-water levels measured during both the withdrawal and recovery phases. The model was applied to a practical test and it yielded satisfactory results.

A statistically based approach to the problem of estimating spatially varying aquifer transmissivities on the basis of steady-state water level data was presented by Newman and Yakowitz.[26] The method involves solving a family of generalized nonlinear regression problems and then selecting one particular solution from this family by means of a comparative analysis of residuals. A linearized error analysis of the solution was included. This analysis allowed one to estimate the covariance of the transmissivity estimates as well as the square error of the estimates of hydraulic head. In addition to the explicitly statistical orientation of the method, it has an additional feature of permitting the user to incorporate a priori information about the transmissivities. This information may be based on actual field data such as pumping tests, or on statistical data accumulated from similar aquifers elsewhere.

Finally, relative to transmissivity, Tang and Pinder[58] deterministically solved the inverse problem for transmissivity by using a finite difference approach. Steady-state hydraulic heads calculated from a preliminary experiment were used as input data. It was shown that if the data on the hydraulic head are known at a sufficiently large number of points, there is no difficulty in setting up a finite difference grid and solving for transmissivity with an acceptable error.

Daniels[59] described a method for determining outflow from an aquifer subjected to constant evapotranspiration (ET). The unaffected aquifer outflow (streamflow) recession after recharge becomes a straight line on a semilogarithmic graph after sufficient time has elapsed. The slope of this recession is, in logarithm to base 10 form, $-Tf/0933a^2S$ where T and S are the transmissivity and storage coefficient of the aquifer, a is the distance from the stream to the ground water divide, and t is the time since recharge occurred (sudden case) or began (constant case). A set of dimensionless type curves was developed for the case of constant ET (leakage) which diverges from and lies below the unaffected case. Streamflow records for an example basin, Indian Creek near Troy, Alabama were analyzed by using the method.

The hydrodynamic dispersion coefficients in aquifers can be determined from observed values of solute concentrations. For a two-dimensional aquifer in which the concentrations of a solute are known, an algorithm was developed to determine the values of longitudinal and transverse dispersivities.[60] Umari, Willis, and Liu[61] used an optimization approach involving the method of quasilinearization to identify unknown aquifer dispersivities in two-dimensional, transient ground water contaminant transport.

The Texas Water Development Board[62] presented a methodology for evaluating the drawdown produced from pumping one or more wells in a well field located in either a confined or unconfined homogeneous aquifer. This drawdown can be evaluated at various locations in the aquifer for various elapsed times since pumping has started. The well drawdown model, called IMAGEW-I, utilizes as input the engineering properites of the aquifer pumping/recharge rates for selected time intervals, and the coordinate locations for all wells (both operating wells and observation wells) in the well field. The output from the well drawdown model is a table of drawdown levels at every observation well for various elapsed times since pumping began.

Finally, small programmable calculators have recently become available that will allow many routine pumping or injection well calculations to be made rapidly, accurately, and inexpensively.[63] Programs and example calculations have been described for the exponential integral and multiple well-multiple rate pumping or injection well equations using water well and oil-field units.

Natural and Artificial Recharge

In order to account for natural and man-induced hydraulic changes in ground water systems, models for both natural and artificial recharge may be needed. For example, Krishnamurthi[64] presented a model to determine natural ground water recharge by using transient soil moisture data measured as a function of vertical position. The model is based upon the finite difference form of the Richards equation. The assumption is made that moisture content varies in the linear range of pressure—moisture and conductivity—moisture relations to obtain a solution to a particular field problem. The parameters of the model, estimated by a linear model of the moisture data, characterize the hydraulic properties of soil and their spatial variability. Application of the model to the High Plains of Colorado provided monthly recharge rates that correlated adequately with the time series date of water table elevations, and events of precipitation and ephemeral stream flow.

Table 6.6 contains summary comments on ten references describing models for artificial recharge. These models can be used for system design and evaluation relative to ground water hydraulics, or for examining the potential impacts of recharge water on the quality of the ground water resources. Models for square or circular recharge basins are presented in one reference (Bianchi and Haskell[65]); for a ditch in one reference (Gureghian[70]); and for recharge wells in one reference (Chowdhury and Shakeja[67]). Models for artificial recharge involving treated or untreated municipal or industrial wastewater are described in five references (Bouwer;[66] Dracup and Kaylus;[68] Willis and Dracup;[72] Willis;[73] and Wilson, Rasmussen, and O'Donnell[74]). Durbin and Morgan[69] and Swain[71] describe case studies of models for ground water basins with significant artificial recharge. Durbin and Morgan[69] developed a well-response model for current and future artificial recharge in the Bunker Hill basin in San Bernardino, California. Swain[71] updated an analog model and developed flow and solute transport models for predicting artificial recharge effects on ground water in the Palm Springs area of California. Although not listed in Table 6.6, the needs and requirements for artificial recharge in the United States, including some information on modeling, has been recently summarized.[75]

Table 6.6. Models for Artificial Recharge.

Author(s) (Year)	Comments
Bianchi and Haskell (1968)[65]	Model for a ground water mound beneath a square or circular recharge basin
Bouwer (1970)[66]	Analog model for the hydraulic conductivity of an aquifer under rectangular recharge basins
Chowdhury and Shakeja (1978)[67]	Model for well recharge in a leaky artesian aquifer overlain by an unconfined aquifer
Dracup and Kaylus (1974)[68]	Model to optimize the assimilative waste capacity of unconfined aquifers
Durbin and Morgan (1978)[69]	Well response model for natural and artificial recharge of the Bunker Hill ground water basin in San Bernardino, California
Gureghian (1978)[70]	Model from seepage from a ditch with vertical sides
Swain (1978)[71]	Model for water level and water quality effects of artificial recharge in the upper Coachella Valley in California
Willis and Dracup (1973)[72]	Model for optimizing wastewater recharge via well injection or basin spreading
Willis (1976)[73]	Models for predicting the potential impacts of wastewater recharge via spreading basins
Wilson, Rasmussen, and O'Donnell (1976)[74]	Model for pit recharge influences on ground water levels and quality in alluvial basins

The rising and falling hydrograph of a ground water mound beneath the center of a square or rectangular recharge basin is of considerable operational importance. Bianchi and Haskell,[65] using the Dupuit-Forchheimer assumptions, theoretically described this hydrograph and then tested for agreement between the theory and actual field observations taken at the center of two experimental recharge ponds. The study also weighed considerations of scale, measurement, and geologic reality as to their significance in the reliability of an idealized mathematical model for the rise and fall of a ground water mound.

Solutions of Boussinesq's equation for ground water seepage from a ditch with vertical sides extending in depth to a horizontal impermeable floor were obtained numerically by Gureghian[76] using the finite difference and finite element methods for the case when the seepage rate from the ditch into the soil is constant with time. Both solutions agreed satisfactorily with experimental results from a Hele-Shaw physical model. The finite difference method also was used to obtain a numerical solution for the reverse situation when water seeps out of the soil into the ditch at a constant rate.

Chowdhury and Shakeja[67] developed theoretical solutions for predicting the rate of recharge through a well in a leaky artesian aquifer overlain by an unconfined aquifer of limited areal extent. The solutions can be used to compute the well recharge rate on the basis of aquifer characteristics and the steady infiltration rate of the soil.

Treated or untreated municipal or industrial wastewater has been frequently used as recharge water in artificial recharge projects.[75] For example, Bouwer[66] described the design concept of a recharge system near Phoenix, Arizona. The system consists of six parallel, rectangular recharge basins with wells located midway between the basins and on both sides. Activated sludge effluent is dosed intermittently to the basins. Using an analog technique, the horizontal and vertical hydraulic conductivity of the underlying aquifer has been determined from the recharge rate and the water level response in two observation wells.

Mathematical models have been used to optimize the assimilative waste capacity of unconfined aquifers.[68,72] Dracup and Kaylus[68] assumed that the aquifer is used conjunctively with surface water supply sources. The wastewaters are introduced into the ground water aquifer system by either well injection or by basin spreading. The model has been applied to a ground water basin using data from the Whittier Narrows and Santee Water Reclamation projects in California. In the general model described by Willis and Dracup,[72] three treatment processes are available to reduce constituent concentrations present in the wastewaters: (1) dilution; (2) surface treatment of each constituent; and (3) the assimilative capacity of the unsaturated and saturated zones of the aquifer system. The total cost for supplying the dilution water and the cost for surface treatment of each constituent is minimized by the model.

Willis[73] presented a mathematical model for assessing the potential impact of the artificial recharge of municipal wastewater effluents on the subsurface environment. Wastewaters are assumed to be recharged to a shallow unconfined ground water aquifer in surface spreading basins. A Galerkin finite element simulation model predicts the time and spatial variations of conservative and nonconservative constituent concentrations within the soil profile. The model incorporates time dependent nonequilibrium adsorption and chemical and/or biochemical reactions. The model can be used to estimate (1) the kinetic, dispersion, and adsorption parameters characterizing quality changes within the soil system, and (2) the initial distribution of a contaminant within the sorbed or solution phases at the inception of the spreading cycle.

The effects of pit recharge of industrial blowdown effluent on local ground water levels and quality in the Tucson, Arizona area were evaluated by Wilson, Rasmussen, and O'Donnell.[74] Fourteen pit trials ranging from 1 day to 185 days duration were implemented during a two-year period. The total volume of effluent recharged was 157 ac-ft. Pit recharge on a limited basis did not appear to offset the general decline in local ground water levels. However, recharge deteriorated the quality of ground water in shallow (50 m) wells but not in a deeper (100 m) well. A finite element model was used to simulate aquifer response during a pit recharge test; however, the results were limited by the injudicious selection of elements and boundary configurations.

Surface and Ground Water Relationships

Aquifers and surface streams and reservoirs are often interconnected with interdependent hydraulic characteristics. Models have been developed and used to demonstrate and manage these systems. For example, Bathals, Rao, and Spooner[22] described a generalized linear model used to analyze ground water flow in a stream-well-aquifer system. In the model, historical records of pumping rates, ground water levels and stream stages, and estimated storage coefficients are used as inputs. The utility of this procedure for analysis of regional aquifer systems was demonstrated by using both field and hypothetical data.

Durbin, Kapple, and Freckleton[77] developed a system of interacting hydrologic models for the Salinas Valley ground water basin in central coastal California. These models include the small stream model, river model, two-dimensional ground water model, and three-dimensional ground water model. The small stream model simulates ground water recharge from small streams that are tributary to the Salinas River; the river model simulates ground water recharge from the surface water discharge in the Salinas River; and the two-dimensional and three-dimensional ground water models simulate hydraulic head in the ground water basin.

A model having both surface water and ground water phases and their interaction has been developed to simulate the hydrology of a basin.[78] The model has been applied to the Little Arkansas River basin of south-central Kansas. Input to the model includes precipitation, climatic conditions, boundary and initial conditions, and basin constants. Provision has been made for modeling withdrawals from both ground water and surface water for consumptive use. The computer output includes streamflow hydrographs at selected points, ground water levels, and ground water recharge and discharge from the basin. The model is designed to be general in nature so that it will be applicable to different basins with similar geographic and geological conditions. Its structure is such that the basin is divided vertically into four layers and a stream network. The four vertical layers are, in order from top to bottom: a surface layer, an upper soil zone, a lower soil zone, and a ground water layer (aquifer). Provision is also made to subdivide each layer horizontally into blocks, or subbasins, that are determined by considering topography, soil types, and climatic data.

A three-dimensional, finite difference model was developed by Kraeger-Rovey[79] for simulating steady and unsteady, saturated and unsaturated flow in a stream-aquifer system. The effects of stream flow on ground water movement are treated by applying appropriate boundary conditions. Contributions of ground water to river flow are quantified by including seepage rates in the computation of river discharge. The model produced results that match observed data for a study area which consisted of a 40-mile reach of the Arkansas Valley of southeastern Colorado. Computed estimates of river discharge at each end of the study area and water table elevations throughout the region agreed reasonably well with observed data. An analysis of the sensitivity of the model results to variations in several input parameters was included.

Conjunctive Use

Conjunctive use of surface and ground water for meeting water supply needs is frequently practiced. Both technical and institutional issues need to be addressed in planning a conjunctive use program.[80] Several models for optimizing conjunctive use have been developed; for example, Morel-Seytoux[81] developed a specific hydrologic model of a stream-alluvial aquifer system for the purpose of designing rules and regulations which maximize the beneficial uses of the waters of a state within the law. The model is particularly suited when decisions on pumping rates are to be reviewed on a frequent regular basis.

Linear quadratic control models were utilized to analyze problems of conjunctive use of surface and ground water in two California settings: (1) Yolo County, where a finite element hydrologic model was developed to provide physical data for an economic model; and (2) Kern County, where ground water overdrafting has been a long standing issue.[82] In addition, a multicell ground water simulation model was constructed by Haimes[83] and applied in the Dayton, Ohio area. An example problem was formulated and successfully solved where the conjunctive use of ground water, streams and a surface reservoir was considered.

A generalized planning model has been developed by Labodie, et al.[84]

This model uses a network approach which is capable of modeling systems with a large-scale and complex water storage, transport, and distribution morphology. The model, called Program CONSIM, is also capable of considering stream-aquifer interaction, including the effects of pumping and artificial recharge on a monthly or weekly basis. Complex institutional factors such as formal water rights and informal exchange agreements can also be considered. The model as designed can be used for locating and sizing new facilities, impacts of transfer of water rights, incorporating new demands for energy development, instream uses, transbasin diversions, and conjunctive use of surface and ground water.

Ground Water Management

Ground water management can encompass a wide range of issues, and many types of models may be pertinent. Table 6.7 contains summary comments on 16 references related to models for ground water management. Included are models for ground water usage and allocation (Colarullo, Heidari, and Maddock;[85] Cummings and McFarland;[86] Gorelick;[87] and Heidari[88]); determining aquifer response to pumping schemes (Herrmann;[89] and Planert[90]); determining the effects of development activities (Kaufman;[91] Pimental;[92] and Wilson and Gerhart[93]); and determining management policy for regional aquifer systems (Maddox;[94] Nichols;[95] Robertson and Mallory;[96] Robson;[97] and Simundich[98]). Ruppert and Clausen[99] address interbasin transfer as a response to aquifer depletion, and Schwarz[100] discusses the management of aquifers as storage systems.

Table 6.7. Models for Ground Water Management.

Author(s) (Year)	Comments
Colarullo, Heidari, and Maddock (1984)[85]	Hydraulic management model for optimal strategy for allocating limited freshwater supplies and containing wastes in a hypothetical aquifer
Cummings and McFarland (1974)[86]	Management model for decision rules for ground water usage for irrigation and salinity control
Gorelick (1983)[87]	Review of hydraulic or policy evaluation models and water allocation models
Heidari (1982)[88]	Model which applies linear systems theory and linear programming in ground water management in south-central Kansas

Table 6.7, continued

Author(s) (Year)	Comments
Herrmann (1976)[89]	Model for predicting aquifer response to increased withdrawals in the lower North Platte Valley in Wyoming
Kaufman (1978)[91]	Model for determining pollution source contributions to ground water quality in the Las Vegas Valley of Nevada
Maddox (1975)[94]	Models for the analysis of hydrologic data and water management policy in the Quincy Basin in the Columbia River basin
Nichols (1977)[95]	Simulation model of the Englishtown aquifer in a 750-square-mile area of New Jersey
Pimental (1977)[92]	Survey of models for predicting the effects of geothermal power development in the Imperial Valley in southern California, and for designing optimal monitoring systems for pollution surveillance
Planert (1976)[90]	Model for simulating the effects of two pumping plans on the outwash sand and gravel aquifer near Columbus, Indiana
Robertson and Mallory (1977)[96]	Regional model for the Floridan aquifer in an 875-square-mile area north of Tampa Bay, Florida
Robson (1974)[97]	Water quality model for the shallow alluvial aquifer near Barstow, California
Ruppert and Clausen (1973)[99]	Economic model for studying interbasin water transfer or population migration as responses to ground water depletion
Schwarz (1976)[100]	Linear models for managing aquifers as storage systems
Simundich (1971)[98]	Model for total dissolved solids in an aquifer in the Raymond Basin of California
Wilson and Gerhart (1979)[93]	Two-dimensional ground water flow model for predicting changes in the potentiometric surface of the Floridan aquifer in response to phosphate mining in west-central Florida

A ground water hydraulic management model was used by Colarullo, Heidari, and Maddock[85] to identify the optimal strategy for allocating limited freshwater supplies and containing wastes in a hypothetical aquifer affected by brine contamination from surface disposal ponds. The present cost of pumping from a network of potential supply and interception wells is minimized over a five-year planning period, subject to a set of hydraulic, institutional, and legal constraints. Cummings and McFarland[86] presented a management model that may be useful in analyzing decision rules for the conjunctive management of ground water reserves for use in irrigation and salinity control.

In a review of distributed parameter management models, Gorelick[87] indicated that they can be divided into two general categories: hydraulics or policy evaluation, and water allocation. Ground water hydraulic management models enable the determination of optimal locations and pumping rates of numerous wells under a variety of restrictions placed upon local drawdown, hydraulic gradients, and water production targets. Ground water policy evaluation and allocation models can be used to study the influence upon regional ground water use of institutional policies such as taxes and quotas. Furthermore, fairly complex ground water-surface water allocation problems can be handled using system decomposition and multilevel optimization. Classified separately by Gorelick[87] are methods for ground water quality management aimed at optimal waste disposal in the subsurface. This classification is composed of steady-state and transient management models that determine disposal patterns in such a way that water quality is protected at supply locations.

A ground water management model based on linear systems theory and linear programming is described by Heidari.[88] The model maximizes the total amount of ground water that can be pumped from a system subject to the physical capability of the system and institutional constraints. This model has been applied to the Pawnee Valley area of south-central Kansas. The results of this application support extant studies about future ground water resources of the valley.

Numerous models for determining aquifer responses to pumping schemes have been developed. Two examples will be cited. Herrmann[89] employed a time-dependent model to predict aquifer response to increased withdrawals in the lower North Platte Valley in Goshen County, Wyoming. Safe yields were determined for the alluvial aquifer under varying discharge and recharge conditions. Planert[90] used a digital model to simulate the effects of two pumping plans on the outwash sand and gravel aquifer in a well field near Columbus, Indiana. The results of the modeling effort can be used to manage the existing well field in addition to planning for well field expansion.

The effects of land development activities on ground water quality were studied for the Las Vegas Valley in Nevada.[91] The current and potential sources of contamination, along with the extant ground water quality, were evaluated in order to develop management alternatives and minimize adverse impacts. Pimental[92] conducted a survey of models potentially applicable for geothermal energy development in the Imperial Valley in southern California. Preliminary work is described for a computer-modeling study to determine the effects of such activity in one region of the valley upon water quality in domestic artesian water wells in an area downstream from the geothermal wells. Various models are identified for estimating contaminant concentration and transport dynamics, and for designing optimal monitoring systems for pollution surveillance of those domestic water wells. Finally in terms of the effects of development activities, Wilson and Gerhart[93] used a two-dimensional ground water flow model to predict changes in the potentiometric surface of the Floridan aquifer resulting from ground water development for proposed and existing phosphate mines during 1976–2000. The modeled area covered over 15,000 sq km in west-central Florida.

Models can also be used for determining management policy for regional aquifer systems. For example, Maddox[94] reported on a model for analyzing water management policy for the Quincy Basin, the northernmost of three ground water basins within the Columbia River basin in the northwestern United States. Nichols[95] utilized a simulation model to evaluate aquifer and confining layer coefficients and to test alternative concepts of flow system hydrodynamics for the Englishtown aquifer in New Jersey. The model was useful in developing management strategies for reversing the decline of water levels in the aquifer.

A regional ground water model of the Floridan aquifer was developed and calibrated for an 875-square-mile portion of the rapidly developing area north of Tampa Bay in Florida.[96] The calibrated model can be used to evaluate management policies relative to ground water withdrawals from the Floridan aquifer. A water quality model for the shallow alluvial aquifer near Barstow, California was developed by Robson,[97] and it provided an excellent means of evaluating the cause and effect relations associated with ground water pollution. Finally, a model useful for evaluating total dissolved solids control options was formulated for the saturated zone of an aquifer in the Raymond Basin in California.[98]

Ruppert and Clausen[99] developed an economic model for comparing two management alternatives for combating ground water mining in an area dependent upon a constant supply of ground water. The management alternatives included development of more water through alternative sources of supply such as improved natural recharge, importation and artificial recharge, and wastewater reuse; or it can incur reductions in income and in population because of migration. The implications of the model for local ground water management districts in Kansas are discussed. Finally, Schwarz[100] developed two models for managing ground water aquifers as storage systems. The models were based on standard and extended linear programming routines.

Pollutant Transport

Pollutant transport is an obvious concern relative to ground water quality management and the development of ground water protection programs. Table 6.8 contains summary comments on 13 references describing models for addressing pollutant transport in the subsurface environment. Nelson[101] summarized the state of the art for unsaturated flow and transport modeling. This summary is based on an in-depth evaluation of available unsaturated flow and transport models, and the results of a symposium to elucidate the present strengths and weaknesses in unsaturated flow and transport analysis. The remaining 12 references in Table 6.8 can be considered in terms of models addressing hydrodynamic processes only (Friedrichs, Cole, and Arnett;[102] Hunt;[103] Konikow and Bredehoeft;[1] Redell and Sunada;[104] Ross, Koplik, and Crawford;[27] Ross;[105] Tyagi and Todd;[106] and Van der Veer;[107] and models addressing hydrodynamic, abiotic, and possibly even biotic processes together (Bredehoeft and Pinder;[108] Gershon and Nir;[109] Hamilton;[110] and Tagaments[111]).

Table 6.8. Models for Pollutant Transport in the Subsurface Environment.

Author(s) (Year)	Comments
Bredehoeft and Pinder (1973)[108]	Application of a physical-chemical model at Brunswick, Georgia to illustrate contaminant movement
Friedrichs, Cole, and Arnett (1977)[102]	Description of the Hanford Pathline Calculational Program to predict transport by advection only
Gershon and Nir (1969)[109]	Effects of boundary conditions on one-dimensional system models accounting for hydrodynamic dispersion, diffusion, radioactive decay, and simple chemical interactions
Hamilton (1982)[110]	Use of three different mass transport models to analyze the conditions at one field site
Hunt (1978)[103]	Use of a model to determine the time requirements for equilibrium to be reached assuming a continuous point source or source of finite size
Konikow and Bredehoeft (1978)[1]	General solute transport model which can be applied to a wide range of flow conditions and source characteristics by accounting for convective transport, hydrodynamic dispersion, and mixing

Table 6.8, continued

Author(s) (Year)	Comments
Nelson (1982)[101]	Summary of the state of the art for unsaturated flow and solute transport modeling
Redell and Sunada (1969)[104]	Model of the convective-dispersive process of pollutant movement in ground water systems as used in a saltwater interface problem
Ross, Koplik, and Crawford (1978)[27]	Model for assessing the impact of fault movement and formation of breccia pipes on the risks from radioactive waste disposal
Ross (1984)[105]	Conceptual model for the movement of water in deep unsaturated zones of acid areas with negligible recharge
Tagaments (1973)[111]	One-dimensional dispersion equation which includes nonlinear isotherms for solute adsorption on soils
Tyagi and Todd (1971)[106]	Study of dispersion of pollutants in saturated porous media as related to the inverse Schmidt number and the molecular peclet number
Van der Veer (1978)[107]	Convective-dispersion equation for two-dimensional ground water flow problems involving a semipervious boundary

The Hanford Pathline Calculational Program (HPCP) is a numerical model for predicting the movement of fluid particles within the Hanford, Washington aquifer system, or within similar ground water systems.[102] The HPCP is a simple transport model wherein only advective changes are considered. Application of the HPCP to test cases for which semianalytical results are obtainable showed that with reasonable time steps and the grid spacing requirements the HPCP gives good agreement with the semianalytical solutions.

Hunt[103] developed analytical solutions for the effects of instantaneous, continuous, and steady-state sources of pollution in uniform ground water flow. Then the solutions were used to determine how long a continuous source must be in place before steady-state conditions are approached, to determine the effect of a finite aquifer depth upon solutions for an infinite aquifer depth, to calculate maximum concentrations for instantaneous sources, and to determine the time required for solutions for a point source and a source of finite size to approach each other.

The solute transport model developed by Konikow and Bredehoeft[1] has been described earlier. The model is both general and flexible in that it can be applied to a wide range of problem types. It is applicable to one- or two-dimensional problems involving steady-state or transient flow. The model computes changes in concentration over time caused by the processes of convective transport, hydrodynamic dispersion, and mixing (or dilution) from fluid sources. The model assumes that the solute is nonreactive and that gradients of fluid density, viscosity, and temperature do not affect the velocity distribution. However, the aquifer may be heterogeneous and (or) anisotropic. The model couples the ground water flow equation with the solute transport equation. The computer program uses an alternating-direction implicit procedure to solve a finite difference approximation to the ground water flow equation, and it uses the method of characteristics to solve the solute transport equation. The latter uses a particle tracking procedure to represent convective transport and a two-step explicit procedure to solve a finite difference equation that describes the effects of hydrodynamic dispersion, fluid sources and sinks, and divergence of velocity.

A computer simulation of the convective-dispersive process of the movement of pollutants in ground water systems was developed and tested for several conditions by Redell and Sunada.[104] An implicit numerical technique was used to solve the flow equation and the method of characteristics used to solve the convective-dispersion equation. The model was tested by applying it to a longitudinal dispersion problem and a longitudinal and lateral dispersion problem. The model was successfully used for a saltwater inferface problem.

Ross, Koplik, and Crawford[27] considered the influence of random geologic events on the transport of pollutants in the subsurface environment. The prediction of such transport requires the solution of a partial differential equation whose coefficients are random processes. The authors developed such a solution and used the method to assess the impact of fault movement and formation of breccia pipes on the risks from radioactive waste disposal. In a related study of deep unsaturated zones with negligible recharge, Ross[105] developed a conceptual model for describing physical forces affecting pollutant transport. Interest in deep unsaturated zones of arid areas has been increased due to hazardous waste disposal needs.

Tyagi and Todd[106] derived an analytical solution for predicting the dispersion of pollutants in homogeneous, isotropic and saturated porous media. In order to define the dispersion phenomenon, the coefficients of longitudinal and transverse dispersion were determined, and relationships for the two coefficients under specific boundary conditions were presented. Since the variables characterizing the properties of the fluid, porous medium, and tracer influence the dispersion of a pollutant, two dimensionless parameters, namely, the inverse Schmidt number and the molecular Peclet number, were obtained. Using these two parameters, experimental data on longitudinal and transverse dispersion were compiled and analyzed over a wide range of the flow velocity, grain size, and molecular diffusion of tracer.

A convective-dispersive equation for addressing the transport of pollutants in ground water systems has been solved by Van der Veer[107] by using an isoparametric, quadrilateral finite element method. Mixed boundary conditions and irregular distribution of nodes can be handled easily by the finite element technique. The calculations by Van der Veer[107] show that the numerical model can be used as an effective predictor of the distribution of contaminants.

In addition to the advection and dispersion hydrodynamic processes influencing pollutant transport in the subsurface environment, abiotic and biotic processes may also be involved. Abiotic processes may include adsorption, ion exchange, chemical reaction, and chemical complexation; biotic processes include bacterial conversion and biodegradation. A physical-chemical mass transport model has been developed by Bredehoeft and Pinder;[108] this model has been successfully used for analyzing a ground water contamination problem at Brunswick, Georgia. Hamilton[110] described a study wherein three different mass transport models were used to analyze existing ground water quality at one site, and to simulate conditions under various management alternatives.

Adsorption is one abiotic process that can lead to the retention of potential ground water pollutants in the subsurface environment. Tagaments[111] has developed a one-dimensional dispersion equation which incorporates a nonlinear isotherm and is solved by a finite difference scheme. The nonlinear isotherm was chosen because the adsorption of many solutes on activated charcoal and soils has been found to exhibit nonlinear rather than linear isotherms. Finally, Gershon and Nir[109] studied the effects of initial and boundary conditions on the distribution of a tracer inflow through porous media. The effects of hydrodynamic dispersion, diffusion, radioactive decay, and chemical interactions were included in the model development.

Saltwater Intrusion and Salinity

Saltwater intrusion into coastal aquifers is a concern in those areas where ground water pumpage is excessive. This type of intrusion can also occur in overpumped noncoastal freshwater aquifers underlain by saline zones. The total dissolved solids in ground water can also be increased by irrigation practices and saline seeps. Table 6.9 provides summary comments on 12 references describing models for saltwater intrusion and salinity. Although not

Table 6.9. Models for Saltwater Intrusion and Salinity.

Author(s) (Year)	Comments
Christensen (1978)[112]	Model for simulating the injection of fresh water through multiple wells into a thin confined or leaky saline aquifer with a head gradient
Kashef (1968)[113]	Model for the movement of fresh and salt waters in coastal aquifers in which fresh water is withdrawn by wells
Kashef (1975)[114]	Use of the method of hydraulic forces to locate the saline water-freshwater interface in artesian aquifers, and exploration of the effectiveness of potential saltwater retardation strategies
Konikow and Bredehoeft (1974)[115]	Model for predicting changes in dissolved solids in ground water as a result of irrigation practices
Labodie, Khan, and Helway (1976)[116]	Application of the Accelerated Salt Transport Method to the lower San Luis Rey River basin; Method includes flow and TDS model as well as an optimizing model
Land (1975)[117]	Digital model for saltwater intrusion and aquifer response in southeast Palm Beach County, Florida
Meyer (1973)[118]	Modeling of freshwater aquifers with salt-water bottoms, with particular reference to the island of Oahu in Hawaii
Pagenkopf (1978)[119]	Equilibrium model for various metals in saline seep waters
Pinder (1975)[120]	Model of saltwater intrusion of the Biscayne aquifer near Miami, Florida
Rofail (1977)[121]	Model for saltwater intrusion where the hydrological boundaries can have any shape and the impervious bed can be considered according to any given function of time
Segol, Pinder, and Gray (1975)[122]	Computer program for numerical simulation of saltwater intrusion in coastal aquifers
Williams and Liu (1975)[123]	Electric analog model to simulate the response of coastal aquifers to tides; the model has been applied in an area of the island of Oahu in Hawaii

included in Table 6.9, Atkinson, et al.,[124] have described current and potential saltwater intrusion problems in the contiguous United States. The references in Table 6.9 can be considered in terms of those with models addressing saltwater intrusion in coastal aquifers (Kashef;[113] Kashef;[114] Land;[117] Pinder;[120] Segol, Pinder, and Gray;[122] and Williams and Liu[123]); saline water intrusion in noncoastal aquifers due to overpumping (Labodie, Khan, and Helway;[116] Meyer;[118] and Rofail[121]); total dissolved solids build-up due to irrigation (Konikow and Bredehoeft[115]) or saline seeps (Pagenkopf[119]); or changes in quality due to freshwater injection into saline aquifers (Christensen[112]).

Kashef[113] developed a mathematical model of the movement of fresh and salt waters in coastal aquifers in which fresh water is withdrawn by wells. The location of the fresh-saltwater interface in artesian and unconfined aquifers is determined on the basis of analysis of hydraulic forces in the body of fresh water. The three-dimensional flow systems are developed by analogy to two-dimensional, finite-length systems. Saltwater intrusion was studied combining the effects of natural flow and multiple well pumpage. Kashef[114] subsequently did additional work on the retardation of saltwater intrusion in coastal aquifers.

Land[117] developed a digital model for saltwater intrusion in Palm Beach County, Florida with the model accounting for lowering of water levels by additional drainage and an increase in ground water withdrawals. Pinder[120] solved the set of nonlinear partial differential equations which describe the movement of the saltwater front in a coastal aquifer by the Galerkin-finite element method. Pressure and velocities are obtained simultaneously in order to guarantee continuity of velocities between elements. A layered aquifer is modeled either with a functional representation of permeability or by a constant value of permeability over each element. The Galerkin-finite element method was subsequently used to simulate the two-dimensional movement of the saltwater front in the Cutler area of the Biscayne aquifer near Miami, Florida. The computer program developed to solve the two-dimensional equations of ground water flow and mass transport using the Galerkin-finite element method is presented by Segol, Pinder, and Gray.[122]

An electric analog model to describe the response of coastal aquifers to tides was presented by Williams and Liu.[123] The model results were compared to field observations at a shallow, coral-limestone aquifer at Ewa Beach, Oahu, Hawaii.

A salinity management study has been conducted by Labodie, Khan, and Helway[116] on the lower San Luis Rey River basin in order to project the impacts of applying the Accelerated Salt TRANsport (ASTRAN) method over a 20-year planning period. The ASTRAN method is an operational approach to salinity control which encourages the application of ground water downstream from where it is pumped. A management algorithm was developed for implementing the ASTRAN method in an optimal manner. The algorithm is composed of (1) a model simulating ground water flow and TDS concentration, and (2) an optimizing model. General conclusions from operating the management algorithm for several salt balance levels are for the same required amount of imported water, the ASTRAN method can achieve a 50% reduction in salinity degradation rate in comparison with the normal practice in the basin of applying water where it is pumped.

Meyer[118] developed some mathematical methods for modeling of a ground water aquifer in which fresh water floats upon and mixes with salt water. Skillful management of the transition zone between fresh and salt water is essential to maximize the amount of fresh water which can be developed and to control encroachment of saline water into wells. Rofail[121] considered the problem of stratified ground water flow (salt water intrusion) in the general case where the hydrological boundaries can have any shape and the impervious bed can be considered according to any given function of time. The fluid density was assumed to be constant throughout each layer. The effect at the boundaries was considered, whether the boundaries were a river or a continuation of the aquifer with different parameters. The systems of equations developed by Rofail[121] were taken in alternative matrix forms. The three-time level implicit scheme was used, and the multi-sweep method was applied for the computation.

Konikow and Bredehoeft[115] developed a digital computer model to predict changes in dissolved solids concentrations in response to spatially and temporally varying hydrologic stresses. The model was applied to the Arkansas River valley in southeastern Colorado where salinity increase in ground water and surface water are primarily due to irrigation practices. The model simulated flow as well as changes in water quality for the river and the aquifer. Dissolved solid concentrations calculated by the model were within 10% of the observed values for both the aquifer and the river approximately 80% of the time.

Saline seep waters are often observed to contain concentrations of trace metals that are higher than those commonly encountered in nonpolluted surface waters.[119] These waters also contain extremely high concentrations of sodium, magnesium, and sulfate. An equilibrium model has been developed by Pagenkopf[119] that predicts the concentrations of aluminum, cadmium, cobalt, copper, nickel, lead, and zinc in the saline seep waters. Ionic activity coefficients are calculated from major component analyses data. Best agreement between calculated and observed concentrations of trace metals was obtained for aluminum, copper, and lead. Agreement for cadmium and nickel was good provided the pH values are above eight. The solutions appear to be super saturated with respect to cobalt carbonate. Agreement for zinc was good for some samples but poor for others.

Concentrated land development in Florida's coastal regions often leads to saltwater intrusion in coastal aquifers. It has been proposed that excess freshwater could be artificially recharged to these coastal aquifers in order to drive out the intruded salt water by the formation of a freshwater lens. The purpose of a study by Christensen[112] was to develop methods for analyzing this process. A numerical model based on the finite element method was developed by which the boundaries of a growing freshwater lens can be traced in two horizontal dimensions. This model simulates the injection of fresh water through multiple wells into a thin confined or leaky saline aquifer in which there may be a preexisting head gradient. The use of the model has been demonstrated by application to the peninsula of Pinellas County, Florida.

Pollutant Source Evaluation

Models have been developed for evaluating specific ground water pollution sources, and Table 6.10 has summary comments on 16 such models. The models can be considered in terms of those related to waste injection (Heidari and Cartwright;[125] Henry, et al.;[126] Larson, et al.;[127] Peterson and Lau;[128] Thomas;[129] U.S. Geological Survey;[130] Waller, Turk, and Dingman;[131] and Wheatcraft[132]); radioactive waste disposal (Alexander;[133] Arnett, et al.;[134] Grove;[135] and Gureghian, Beskid, and Marmer[76]); spray irrigation for wastewater disposal (McDonald and Fleck;[136] and Ostrowski[137]); sanitary landfills (Fungaroli[138]); and fly ash disposal (Theis and Marley[139]).

Heidari and Cartwright[125] described the results of a preliminary theoretical study of the fate of liquid industrial wastes injected into deep geologic formations in Illinois. An industrial well was used as a model and the geology of the area was idealized into a 15-layer homogeneous and anisotropic mathematical model. The finite element method was tested and proved to be an effective mathematical tool in the solution of the equation of flow. The flow and pressure buildup showed that the rocks are capable of receiving greater volumes of waste than are now being injected without endangering the integrity of the aquifer or the confining layer.

Henry, et al.[126] studied the use of deep saline aquifers as possible reservoirs for the disposal of liquid wastes. The investigation consisted of measurements on a hydraulic model of the Floridan aquifer in Florida and theoretical analysis of the system based on solution of the governing differential equations. Distributions of temperatures, salinity, and flow patterns as determined from the laboratory model and mathematical analyses were similar to field data, thus indicating that such analyses can be effective in predicting conditions in the field.

Table 6.10. Models for Some Pollutant Categories and Waste Disposal Practices.

Author(s) (Year)	Comments
Alexander, et al. (1984)[133]	Release rates of radionuclides and chemical reaction products from engineered barrier systems
Arnett, et al. (1976)[134]	Application of ground water modeling to the subsurface transport of radioactive wastes at the Hanford facility near Richland, Washington
Fungaroli (1976)[138]	Model for quantity and time variation of landfill leachates
Grove (1970)[135]	Model for radioactive ionic concentrations in ground water systems as a function of time and position
Gureghian, Beskid, and Marmer (1978)[76]	Two-dimensional mathematical model for ground water flow and solute transport from low-level radioactive waste tailings ponds
Heidari and Cartwright (1974)[125]	Model for predicting fate of liquid industrial wastes injected via wells into deep geologic formations
Henry, et al. (1972)[126]	Laboratory and mathematical models for evaluating disposal of liquid wastes into deep saline aquifers
Larson, et al. (1977)[127]	Model for wastewater treatment plant effluent injection into a lava aquifer system in Hawaii
McDonald and Fleck (1978)[136]	Model for ground water impacts of wastewater storage and spray irrigation of crops in Muskegon County, Michigan
Ostrowski (1976)[137]	Model for ground water impacts from spray irrigation systems for wastewater treatment
Peterson and Lau (1974)[128]	Use of sand-tank and Hele-Shaw models in study of wastewater injection in Hawaii
Theis and Marley (1975)[139]	Model for predicting ground water contamination by heavy metals from the land disposal of fly ash
Thomas (1973)[129]	Model for migration patterns of industrial waste products injected into disposal wells
U.S. Geological Survey (1976)[130]	Three-dimensional model for calculating effects of liquid waste disposal in deep saline aquifers

Table 6.10, continued

Author(s) (Year)	Comments
Waller, Turk, and Dingman (1978)[131]	Laboratory and mathematical models for a deep well waste disposal facility in western New York
Wheatcraft (1977)[132]	Model for wastewater treatment plant effluent injected into an Hawaiian aquifer

Larson, et al.[127] studied the potential impact of the injection of a wastewater treatment plant effluent into the saline-water part of a lava aquifer system. The plant was located near Kahului, Maui, Hawaii. To determine the distribution of injected fluid, a computer model was constructed to enable consideration of (1) an aquifer system containing fresh and saline water, (2) heterogeneous aquifer characteristics, (3) nonuniform boundary conditions, and (4) concurrent withdrawal of saline water near the injection site. A sharp interface was assumed to exist between fresh water and saline water in the aquifer, and a numerical model was employed to simulate ground water flow in the freshwater zone. A simulated injection rate of 6.2 cfs resulted in a steady-state wastewater plume extending 1,200 ft from the injection site and displacing the interface between fresh water and saline water 500 ft seaward. Three-dimensional analysis indicated that if fluid was injected into the saline water zone and if vertical hydraulic conductivity were less than horizontal conductivity the injected plume of wastewater would be contracted in size at the top of the freshwater zone and expanded at the bottom relative to isotropic conditions. Peterson and Lau[128] conducted a laboratory study to ascertain the mechanics of treated effluent injection in Hawaii. Both a sand-tank model and a vertical Hele-Shaw model were used. Calibration of the results of numerical modeling with the results of sand-tank model studies was reported by Wheatcraft.[132]

Thomas[129] focused on the same injection well concerns by noting that where injection of massive quantities of waste materials occurs over long time periods, it is important to predict the migration patterns of the waste solutions, the reservoir pressures created, and the extent to which adsorption strips the waste solution of toxic materials. Desire for such information is predicated primarily on a concern for the effects that waste solution injection may have on both surface and subsurface freshwater supplies. Furthermore, it is also possible that waste solutions will migrate into adjacent hydrocarbon-bearing reservoirs and interfere with oil and gas production. It is possible to employ an immiscible reservoir simulator and, with appropriate artifices, handle a miscible displacement process in a reasonable way. The development and application of such a model to the aqueous waste injection problem is discussed by Thomas.[129]

A transient, three-dimensional subsurface waste-disposal model has been developed by the U.S. Geological Survey[130] to provide a methodology to design and test waste disposal systems. The model is a finite difference solution to the pressure, energy, and mass-transport equations. Equation parameters such as viscosity and density are allowed to be functions of the equations' dependent variables. Multiple user options allow the choice of X, Y, and Z artesian or R and Z radial coordinates, various finite difference methods, iterative and direct matrix solution techniques, restart options, and various provisions for output display.

Finally, relative to injection wells, the feasibility and long-range impacts of deep well waste disposal in western New York was evaluated by the U.S. Geological Survey in their study of a Lackawanna, New York disposal well.[131] The well was drilled for the proposed disposal of acidic, iron-rich waste pickle liquor from a steel mill. The hydraulic, seismic, and geochemical aspects of deep well disposal were explored. Hydraulic reactions under field pressure and temperatures were simulated by mathematical models. Chemical reactions were predicted by laboratory models. Permeability, transmissivity, and fracturing potential for the sandstone and dolomite sequences at the injection levels were developed. Deep well disposal in western New York was found feasible only for short periods because of excessive head buildup.

Disposal of radioactive wastes to surface disposal sites at the Hanford Atomic Reservation has resulted in some radioactive and chemical contamination of the underlying ground water flow systems.[134] In support of the Hanford ground water management effort a preliminary conceptual model of the ground water system has been formulated as a step in understanding possible contaminant movement. The mathematical modeling capabilities needed for better understanding and prediction of such contaminant movement are discussed by Arnett, et al.,[134] and the state of the art in ground water modeling as related to Hanford is summarized.

Issues of concern in the subsurface transport of radioactive material include retention/removal via ion exchange or adsorption. Grove[135] presented an equation based on instantaneous ion exchange and on a linear adsorption isotherm to predict radioactive ionic concentrations in ground water systems as a function of time and position. This equation accounts for radioactive decay, ion exchange, and longitudinal hydraulic dispersion. Data necessary for solution include the ground water velocity, ion-exchange distribution coefficient, exchange ratio, and the dispersion coefficient.

The discharge of low-level radioactive waste into tailings ponds is a potential source of ground water contamination. The estimation of the radionuclides from tailings retention systems depends on reasonably accurate estimates of the movement of both water and solute. A two-dimensional mathematical model having predictive capability for ground water flow and solute transport has been developed.[76] The flow equation has been solved under transient conditions. The simultaneous solution of both equations is achieved through the finite element technique using isoparametric elements, based on the Galerkin solution; the weighting functions used in the solution of the mass transport equation have a nonsymmetric form. The predictive capability of the model is demonstrated using an idealized case based on analyses of field data obtained from the sites of operating uranium mills.

Engineered barrier systems can be used to minimize radioactivity transport from geologic repositories containing radioactive waste materials. Alexander, et al.,[133] described a conceptual model for deriving a repository source term based on probabilistic estimates of: (1) the penetration time for metal barriers, and (2) radionuclide release rates for individually simulated waste packages after penetration has occurred. The conceptual model assumes that release rates are explicitly related to such time-dependent processes as mass transfer, dissolution and precipitation, radionuclide decay, and variations in the geochemical environment. The conceptual model takes into account the reduction in the rates of waste-form dissolution and metal corrosion due to a buildup of chemical reaction products. Cumulative releases from engineered barrier systems are calculated by summing the releases from a probabilistically generated population of individual waste packages.

McDonald and Fleck[136] and Ostrowski[137] developed models for the ground water impacts of wastewater spray irrigation systems. Ostrowski[137] addressed two-dimensional infiltration, evapotranspiration and moisture distribution in an effluent spray irrigation field by using the equation governing saturated-unsaturated flow in porous media. A finite element formulation of the flow problem was obtained by using the Galerkin variational method. The area investigated was the oldest operating spray irrigation field of the west treatment system of the community of St. Charles, Maryland.

A digital model was developed by McDonald and Fleck[136] to study the impact on ground water conditions of the Muskegon County, Michigan wastewater disposal system. At the disposal site, wastewater is stored on two 850-acre lagoons and then spray-irrigated on crop land. About 70 miles of drainage tile, which underlies the irrigated land, has caused the water table to be lowered substantially. The decline in water levels has been partially offset by irrigation and leakage from the lagoons; at some places the water table is higher than it was prior to construction. Predictive simulations by the model were used to study the effects of varying tile drainage, amount of irrigation water applied, lagoon leakage, and natural ground water recharge.

Fungaroli[138] reported on an investigation designed to provide a means for predicting the movement of sanitary landfill pollutants in subsurface regions, to develop criteria for the evaluation of landfill site suitability, and to appraise remedial measures for reducing undesirable contaminant movement. Both a laboratory sanitary landfill and a field sanitary landfill were used. In addition, a model for predicting the quantity and time variation of leachate was also presented along with the significant parameters which control leachate generation.

Finally, the land disposal of fly ash creates the potential for ground water pollution via the migration of heavy metals from the fly ash. Theis and Marley[139] gathered data from both the laboratory and the field for purposes of assessing the impact of heavy metal leachates from fly ash on local water quality, investigated the mechanism of heavy metal attenuation in different soil environments, and evaluated a ground water quality model for heavy metals to be used for managerial and predictive purposes.

Miscellaneous Models

Ground water models have also been developed for addressing the effects of construction dewatering, mineral dissolution, and freshwater storage. For example, Marie[140] developed a model for studying the effects on ground water levels at the Indiana Dunes National Lakeshore caused by construction dewatering. The model analysis indicated that the planned dewatering would cause a drawdown of about 1.2 m under the western-most pond of the lakeshore, and that this drawdown would cause the pond to go almost dry—less than 0.15 m of water remaining in about 1% of the pond—under average conditions during the 18-month dewatering period.

A general differential equation, describing the dissolution of minerals and the transport of their products in porous media, was derived and simplified by Mercado[141] for the case of steady-state, nonequilibrium distribution of solutes in confined carbonate aquifers. The numerical solution of the equation, which represents the hydrogeochemical model of a given aquifer, was based on the transformation of aquifer space coordinates into time coordinates of "tracing points" moving along ground water streamlines, followed by an inverse transformation which yields theoretical water quality maps. The applicability of the model for integrating hydrologic and geochemical data in aquifers was demonstrated for the limestone aquifer of Roswell, New Mexico.

Faulkner[142] suggested that flow-net techniques may be the most practical and economical way to analyze the flow regime of a cavernous carbonate aquifer with well developed circulation. Flow-net analysis was applied successfully to the Silver Springs ground water drainage basin in north-central peninsular Florida by likening the springs to a continuously discharging well whose cone of depression is the entire drainage basin.

The storage of fresh water in horizontal, confined saline aquifers is technically feasible. Experimental results with physical aquifer models involving single- and multiple-injection/production wells were used by Kimbler, Kazman, and Whitehead[143] to verify computer programs for: prediction of the recovery efficiency (the ratio of retrieved-to-injected fresh water); frontal position of the freshwater bubble; position of the saline water—freshwater interface resulting from gravitational laydown, dispersion, and mixing; and calculation of the dispersivity. For salaquifers with preexisting ground water movement, an additional program must be used to design a system of injection/withdrawal bounding wells that will neutralize the movement and create a zone of stagnation for the storage project. A model for addressing the use of bounding wells to control fluid movement in underground water storage projects was developed by Langhettee.[144]

PROBLEMS WITH COMPUTER MODELS

Because of their increased popularity and wide availability, it is important to note the limitations of ground water models. Table 6.11 contains a listing of some example problems and limitations. The first limitation of computer models is that they can have significant data requirements. Models can require a variety of input data and this data may be required for several years. Additionally, some available data may not be useful. For example, values for hydraulic conductivities may be available from a series of pump tests. However, when this information is used in the model, the results may not accurately predict the behavior of the aquifer. These values must be changed until the model does accurately reflect the aquifer conditions. Hence, even though the values obtained may be correct, they do not reflect the overall hydraulic conductivity of the aquifer and as such, are of little value in modeling studies.

The second limitation associated with computer models is their required boundary conditions. Because each model has a finite size associated with it, certain constraints or conditions must be applied at the outermost nodes. These are most often constant head or constant flow conditions. These mandated numerical conditions may not be truly reflective of the actual field conditions at the points represented by the nodes.

Computer models can be very precise in their predictions, but these predictions are not always accurate. The accuracy of the output is highly dependent on the accuracy of the input. Additionally, some models may exhibit difficulty in handling areas of dynamic flow such as occur very near wells. Finally, numerical problems, as discussed previously, can also plague models.

Table 6.11. Problems with Computer Models.[8]

1. Data Requirements
   - head over time
   - concentration over time
   - hydraulic conductivity
   - aquifer dispersivities
2. Boundary Conditions
   - constant head
   - constant flow
3. Accuracy
   - dependent on input data
   - dynamic flow
   - numerical problems
   - sensitivity analysis
4. Complicated
   - FEM
5. Expensive
   - large computer
6. Time-consuming
   - may require extensive efforts
7. Not truly predictive
8. Must deal with uncertainty
9. Misused

Sensitivity analyses may be desirable in aquifer studies. For example, Aquado, Sitar, and Remson[145] presented a method for determining in which areas detailed knowledge of aquifer characteristics and conditions is most critical to the success of a management plan. These questions are answered by using sensitivity analysis to determine how variations in parameters and input data affect the optimal solution of a linear programming management model. The model uses finite element or finite difference approximations of the ground water equations as constraints. The optimal locations and discharge rates of wells have been determined for dewatering a rectangular area to a specified level while minimizing the steady-state total pumping rate and maintaining hydraulic heads in the dewatered area at or below the specified value. The area is in a small aquifer having constant head boundaries. Sensitivity analysis has shown that the optimal, steady-state solution is most sensitive to hydraulic conductivity at and near the aquifer boundaries parallel to the length of the dewatered area. Thus, field exploration and testing should be concentrated on determination of hydraulic conductivity in those areas.

Another problem associated with some computer models is that they can be quite complicated from a mathematical perspective. This is especially true of the finite element models, and also the comprehensive models that are written for three dimensions. For these models, it is imperative that their availability be accompanied by extensive documentation as to the workings of the model. In addition, computer modeling studies can be expensive. This is due to the typically larger computer storage and time requirements of most models.

Computer modeling can also be a time consuming venture. This is especially true if insufficient data is available. In fact, even with sufficient data available, models can sometimes require months to be calibrated. Calibration is essential. That is, all models must be calibrated so that they accurately reflect the conditions of the aquifer being modeled. In this sense, models are not truly predictive tools. This is especially true of the solute transport models. In cases involving solute transport, the model must be calibrated until it accurately reflects the documented movement of a given solute. Once this is achieved, only then can the model be used to make future predictions as to movement of the solute plume. Hence, the models require contamination to exist before predictions can be made about the movement of contaminants.

Uncertainty relative to model assumptions and useability should be recognized. For example, McLaughlin[146] addressed the role of uncertainty in the Rockwell Hanford Operations ground water model development and application program at the Hanford Site in Washington. Methods of applying statistical and probability theory in quantifying the propagation of uncertainty from field measurements to model predictions are discussed. It is shown that measures of model accuracy or uncertainty provided by a statistical analysis can be useful in guiding model development and sampling network design.

The final limitation or problem associated with computer models is that they have, in the past, been misused. Misuse in this sense means that models have been applied to cases where they were not appropriate. More important, misuse of models has come in the form of blind faith in the results of model applications. The output from any modeling study must be interpreted prior to its being used. It is imperative that the output be accurate and realistic. The interpretation of accuracy and reality can only be made by professionals trained in ground water hydrology and quality management.

SELECTED REFERENCES

1. Konikow, L. F. and Bredehoeft, J. D., "Computer Model of Two-Dimensional Solute Transport and Dispersion in Ground Water," Book 1, Chapter 2, 1978, Techniques of Water-Resources Investigations of the United States Geological Survey, Washington, DC.
2. DeWiest, R. J. M., "Steady State Flow," *Geohydrology,* 1st Ed., 1965, John Wiley, New York, New York, pp. 204–238.
3. Heutmaker, D. L., Peterson, F. L. and Wheatcraft, S. W., "A Laboratory Study of Waste Injection into a Ghyben-Herzberg Groundwater System Under Dynamic Conditions," Report No. TR-107, Mar. 1977, Hawaii University, Honolulu, Water Resources Research Center, Office of Water Research and Technology, Washington, DC.
4. Michael, A., Gelhar, L. W. and Wilson, J. L., "Hele-Shaw Model of Long Island Aquifer System," *Journal of the Hydraulics Division, ASCE,* Vol. 98, No. HY9, Sept. 1972, pp. 1701–1714.
5. Williams, J. A., "Well Injection Into a Two-Phase Flow Field: A Hele-Shaw Model Investigation," Report No. TR-108, May 1977, Hawaii University, Honolulu, Water Resources Research Center, Office of Water Research and Technology, Washington, DC.
6. Shamberger, H. A. and Domenico, P. A., "Application of Analog Techniques to Water Management," presented at A.S.C.E. Fourteenth Hydraulics Division Conference, held Aug. 25, 1965, University of Arizona, Tucson, Arizona, pp. 1–22.
7. Todd, D. K., "Groundwater Modeling Techniques," *Groundwater Hydrology,* 2nd Ed., 1980, John Wiley, New York, New York, p. 401.
8. Wang, H. F. and Anderson, M. P., *Introduction to Groundwater Modeling—Finite Difference and Finite Element Methods,* 1982, W. H. Freeman, San Francisco, California.
9. Freeze, R. A. and Cherry, J. A., "Appendix III: Example of an Analytical Solution to a Boundary-Value Problem," *Groundwater,* 1st Ed., 1979, Prentice-Hall, Englewood Cliffs, New Jersey, 1979, pp. 534–535.
10. Van Der Kamp, G. S., "Theory for Periodic Flow of Groundwater," *Periodic Flow of Groundwater,* 1973, Institute of Earth Sciences, Free University, Amsterdam, The Netherlands.
11. Toth, J. A., "Groundwater Motion in Small Drainage Basins," *Journal of Geophysical Research,* Vol. 167, No. 11, 1962, pp. 4375–4382.
12. Kirkham, D. and Powers, W. L., "Physical Artifices for Solving Flow Problems," *Advanced Soil Physics,* 1st Ed., 1972, John Wiley and Sons, New York, New York, pp. 85–130.
13. Cohen, R. M. and Miller, W. J., "Use of Analytical Models for Evaluating Corrective Actions at Hazardous Waste Disposal Facilities," *Proceedings of the*

Third National Symposium on Aquifer Restoration and Ground Water Monitoring, 1983, National Water Well Association, Worthington, Ohio, pp. 85–97.

14. Jury, W. A., "Solute Travel-Time Estimates for Tile-Drained Fields: I. Theory," *Proceedings of the Soil Science Society of America,* Vol. 39, No. 6, Nov.–Dec. 1975, pp. 1020–1024.
15. Reilly, T. E., "Convective Contaminant Transport to Pumping Well," *Journal of the Hydraulics Division, ASCE,* Vol. 104, No. HY12, Dec. 1978, pp. 1565–1575.
16. McLin, S. G., "Lumped Parameter Hydrosalinity Model," Ph.D. Dissertation, 1980, New Mexico Mining and Technology Institute, Socorro, New Mexico.
17. Bear, J., "Groundwater Quality Problem (Hydrodynamic Dispersion)," *Hydraulics of Groundwater,* 1st Ed., 1979, McGraw-Hill Book Company, New York, New York, pp. 263–275.
18. Rao, A. R., Rao, R. G. and Kashyap, R. L., "Stochastic Models for Ground Water Levels," Report No. TR-67, Aug. 1975, National Technical Information Service, U.S. Department of Commerce, Springfield, Virginia.
19. Sagar, B. and Clifton, P. M., "Stochastic Groundwater Flow Modeling Using the Second-Order Method," RHO-BW-SA-364P, Mar. 1984, Rockwell Hanford Operations, Rockwell International, Richland, Washington.
20. Smith, L. and Freeze, R. A., "Stochastic Analysis of Steady State Ground Water flow in a Bounded Domain 1. One-Dimensional Simulations," *Water Resources Research,* Vol. 15, No. 3, June 1979, pp. 521–528.
21. Bakr, A. A., et al., "Stochastic Analysis of Spatial Variability in Subsurface Flows, 1, Comparison of One- and Three-Dimensional Flows," *Water Resources Research,* Vol. 14, No. 2, Apr. 1978, pp. 263–271.
22. Bathals, C. T., Rao, A. R. and Spooner, J. A., "Systematic Development of Methodologies in Planning Urban Water Resources for Medium Size Communities: Application of Linear Systems Analysis to Ground Water Evaluation Studies," Report No. PWRRC-TU-91, Feb. 1977, National Technical Information Service, U.S. Department of Commerce, Springfield, Virginia.
23. Carlsson, G. and Carlstedt, A., "Estimation of Transmissivity and Permeability in Swedish Bedrock," *Nordic Hydrology,* Vol. 8, No. 2, 1977, pp. 103–116.
24. Delleur, J. W., et al., "Systematic Development of Methodologies in Planning Urban Water Resources for Medium Size Communities Phase I," Report No. Tr-74, Apr. 1976, National Technical Information Service, U.S. Department of Commerce, Springfield, Virginia.
25. Gelhar, L. W., "Effects of Hydraulic Conductivity Variations on Ground Water Flows," *Proceedings, Second International IAHR Symposium on Stochastic Hydraulics,* Water Resources Publications, 1977, pp. 409–431.
26. Newman, S. P. and Yakowitz, S., "A Statistical Approach to the Inverse Problem of Aquifer Hydrology, 1. Theory," *Water Resources Research,* Vol. 15, No. 4, Aug. 1979, pp. 845–860.

27. Ross, B., Koplik, D. M. and Crawford, B. S., "Statistical Approach to Modeling Transport of Pollutants in Groundwater," W-7405-ENG-48, 1978, Analytic Sciences Corporation, Reading, Massachusetts.
28. Sagar, B., "Solution of Linearized Boussinesq Equation with Stochastic Boundaries and Recharge," *Water Resources Research,* Vol. 15, No. 3, June 1979, pp. 618–624.
29. Yakowitz, "Model-Free Statistical Methods for Water Table Production," *Water Resources Research,* Vol. 12, No. 5, Oct. 1976, pp. 836–844.
30. McLin, S. G. and Gelhar, L. W., "A Field Comparison Between the USBR-EPA Hydrosalinity and Generalized Lumped Parameter Models," *Proceedings of the Canberra Symposium,* IAHS-AISH Pub. No. 128, 1979, pp. 339–348.
31. Gelhar, L. W. and Wilson, J. L., "Ground Water Quality Modeling," *Ground Water,* Vol. 12, No. 6, Nov.–Dec. 1974, pp. 399–408.
32. Cosner, O. J., "Predictive Computer Model of the Lower Cretaceous Aquifer, Franklin Area, Southeastern Virginia," Report No. USGS/WRI-51-74; USGS/WRD-75/041, Apr. 1975, National Technical Information Service, U.S. Department of Commerce, Springfield, Virginia.
33. Grove, D. B., "A Model for Calculating Effects of Liquid Waste Disposal in Deep Saline Aquifers," Report No. USGS/WRI-76-61, June 1976, National Technical Information Service, U.S. Department of Commerce, Springfield, Virginia.
34. Guvanason, V. and Volker, R. E., "Identification of Distributed Parameters in Ground Water Basins," *Journal of Hydrology,* Vol. 36, No. 3/4, Feb. 1978, pp. 279–293.
35. Khaleel, R. and Redell, D. L., "Simulation of Pollutant Movement in Groundwater Aquifers," OWRT-A-030-TEX(1), May 1977, Water Resources Institute, Texas A&M University, College Station, Texas.
36. Konikow, L. F., "Modeling Chloride Movement in the Alluvial Aquifer at the Rocky Mountain Arsenal, Colorado," Geological Survey Water-Supply Paper 2044, 1977, United States Department of the Interior, Washington, DC.
37. Prickett, T. A. and Lonnquist, C. G., "Selected Digital Computer Techniques for Groundwater Resource Evaluation," Bulletin 55, 1971, Illinois State Water Survey, Urbana, Illinois.
38. Brunch, J. C., Jr., "Nonlinear Equation of Unsteady Ground Water Flow," *Journal of the Hydraulics Division, ASCE,* Vol. 99, No. 3, Mar. 1973, pp. 395–403.
39. Cabrero, G. and Marino, M. A., "A Finite Element Model of Contaminant Movement in Ground Water," *Water Resources Bulletin,* Vol. 12, No. 2, Apr. 1976, pp. 317–335.
40. Cheng, R. T., "Finite Element Modeling of Flow Through Porous Media," OWRT-C-4026 (9006) (4), 1975, Department of Civil Engineering, State University of New York at Buffalo, Buffalo, New York.

41. Duguid, J. O. and Reeves, M., "Material Transport Through Porous Media: A Finite-Element Galerkin Model," ESD Publication 733, Mar. 1976, Environmental Sciences Division, Oak Ridge National Laboratory, Oak Ridge, Tennessee.
42. Grove, D. B., "The Use of Galerkin Finite-Element Methods to Solve Mass-Transport Equations," Report No. USGS/WRI-77-49; USGS/WRD/WRI-78/011, Oct. 1977, U.S. Geological Survey, Denver, Colorado.
43. Gupta, S. K., et al., "Finite-Element Three-Dimensional Ground-Water (FE3DGW) Flow Model: Formulation, Computer Source Listings, and User's Manual," BMI-ONWI-548, Oct. 1984, Office of Nuclear Waste Isolation, Battelle Project Management Division, Columbus, Ohio.
44. Nishi, T. S., Brunch, J.C., Jr. and Lewis, R. W., "Finite Element Solutions to Two Groundwater Flow Problems, One Including Dispersion and the Other the Influence of Electro-Osmosis," UCAL-WRC-W-428, 1975, Water Resources Center, University of California-Davis, Davis, California.
45. Pickens, J. F. and Lennox, W. C., "Numerical Simulation of Waste Movement in Steady Ground Water Flow Systems," *Water Resources Research,* Vol. 12, No. 2, Apr. 1976, pp. 171–184.
46. Seckel, K. W., "Feasibility Study for Development of a Transient Three-Dimensional Ground Water Flow Model Utilizing the Finite Element Method," Technical Report No. 51, 1978, Maryland University, College Park, Maryland.
47. Skrivan, J. A., "Digital-Model Evaluation of the Groundwater Resources in the Ocotillo-Coyote Wells Basin, Imperial County, California," Report No. USGS/WRD/WRI-78/012, USGS/WRI-77-30, Nov. 1977, Water Resources Division, U.S. Geological Survey, Menlo Park, California.
48. Yeh, G. T. and Huff, D. D., "FEWA: A Finite Element Model of Water Flow Through Aquifers," ORNL-5976, Nov. 1983, Oak Ridge National Laboratory, Oak Ridge, Tennessee.
49. Geo Trans, Inc., "Notes for Ground Water Models Workshop," Holcomb Research Institute Modeling Workshop, 1982, Butler University, Indianapolis, Indiana.
50. Anderson, M. P., "Using Models to Simulate the Movement of Contaminants through Ground Water Flow Systems," *CRC Critical Reviews, in Environmental Control,* Vol. 9, Issue 2, Nov. 1979, pp. 97–156.
51. Appel, C. A. and Bredehoeft, J. D., "Status of Ground Water Modeling in the U.S. Geological Survey," Circular No. 737, 1976, U.S. Geological Survey, Washington, DC.
52. Bachmat, Y., et al., "Utilization of Numerical Groundwater Models for Water Resources Management," EPA-600/8-78/012, June 1978, U.S. Environmental Protection Agency, Ada, Oklahoma.
53. Mercer, J. W. and Faust, C. R., "Ground Water Modeling: An Overview," *Ground Water,* Vol. 18, No. 2, Mar.–Apr. 1980a, pp. 108–115.

54. Mercer, J. W. and Faust, C. R., "Ground Water Modeling: Mathematical Models," *Ground Water,* Vol. 18, No. 3, May–June 1980b, pp. 212–227.
55. Corapcioglu, N. Y. and Brutsaert, W., "Viscoelastic Aquifer Model Applied to Subsidence Due to Pumping," *Water Resources Research,* No. 3, June 1977, pp. 597–604.
56. Navarro, A., "A Modified Optimization Method of Estimating Aquifer Parameters," *Water Resources Research,* Vol. 13, No. 6, Dec. 1977, pp. 935–939.
57. Rushton, K. R., "Estimating Transmissivity and Storage Coefficient from Abstraction Well Data," *Ground Water,* Vol. 16, No. 2, Mar.–Apr. 1978, pp. 81–85.
58. Tang, D. H. and Pinder, G. F., "A Direct Solution of the Inversed Problem in Groundwater Flow," *Advances in Water Resources,* Vol. 2, June 1979, pp. 97–99.
59. Daniels, J. F., "Estimating Ground Water Evapotranspiration from Stream Flow Records," *Water Resources Research,* Vol. 12, No. 3, June 1976, pp. 360–364.
60. Murty, V. V. and Scott, V. H., "Determination of Transport Model Parameters in Groundwater Aquifers," *Water Resources Research,* Vol. 13, No. 6, Dec. 1977, pp. 941–947.
61. Umari, A., Willis, R. and Liu, P. L., "Identification of Aquifer Dispersivities in Two-Dimensional Transient Ground Water Contamination Transport: An Optimization Approach," *Water Resources Research,* Vol. 15, No. 4, Aug. 1979, pp. 815–831.
62. Texas Water Development Board, "Well Field Drawdown Model, IMAGEW-1: Program Documentation and User's Manual," Report No. TDWR/US-S7302, Feb. 1973, Austin, Texas.
63. Warner, D. L. and Yow, M. G., "Programmable Hand Calculator Programs for Pumping and Injection Wells: I—Constant or Variable Pumping (Injection) Rate, Single or Multiple Fully Penetrating Wells," *Ground Water,* Vol. 17, No. 6, Nov.–Dec. 1979, pp. 532–537.
64. Krishnamurthi, N., "Mathematical Modeling of Natural Groundwater Recharge," *Water Resources Research,* V. 13, No. 4, 1977, pp. 720–724.
65. Bianchi, W. C. and Haskell, E. E., Jr., "Field Observation Compared with Dupuit-Forchheimer Theory for Mound Heights Under a Recharge Basin," *Water Resources Research,* Vol. 5, No. 1, 1968, pp. 1049–1057.
66. Bouwer, H., "Groundwater Recharge Design for Renovating Wastewater," *Journal of the Sanitary Engineering Division, ASCE,* Vol. 96, Feb. 1970, pp. 59–74.
67. Chowdhury, P. K. and Shakeja, S. K., "Drainage by Recharge Wells in a Leaky Aquifer," *Journal of Hydrology,* Vol. 36, No. 1/2, Jan. 1978, pp. 87–93.

68. Dracup, J. A. and Kaylus, W. J., "Simulation of the Diffusion of Dissolved Salts in Aquifers," Report No. UCAL-WRC-W-340, Feb. 1974, Water Resources Center, University of California at Los Angeles, Los Angeles, California.
69. Durbin, T. J. and Morgan, C. O., "Well-Response Model of the Confined Area, Bunker Hill Ground Water Basin, San Bernardino County, California," Report No. USGS/WRD/WRI-78/079; USGS/WRI-77/129, July 1978, National Technical Information Service, U.S. Department of Commerce, Springfield, Virginia.
70. Gureghian, A. B., "Solutions of Boussinesq's Equation for Seepage Flow," *Water Resources Research,* Vol. 14, No. 2, Apr. 1978, pp. 231–236.
71. Swain, L. A., "Predicted Water-Level and Water-Quality Effects of Artificial Recharge in the Upper Coachella Valley, California, Using a Finite-Element Digital Model," Report No. USGS/WRD/WRI-78/081; USGS/WRI-77-29, Apr. 1978, National Technical Information Service, U.S. Department of Commerce, Springfield, Virginia.
72. Willis, R. and Dracup, J. A., "Optimization of the Assimilative Waste Capacity of the Unsaturated and Saturated Zones of an Unconfined Aquifer System," Report No. UCLA-NEG-7394, Dec. 1973, School of Engineering and Applied Science, California University, Los Angeles, California.
73. Willis, R., "Optimal Management of the Subsurface Environment: Parameter Identification," Report No. OWRT-A-052-NY(1), May 1976, National Technical Information Service, U.S. Department of Commerce, Springfield, Virginia.
74. Wilson, L. G., Rasmussen, W. O. and O'Donnell, D. F., "Feasibility of Modelling the Influences of Pit Recharge on Ground Water Levels and Quality in Alluvial Basins," Report No. OWRT-A-056-ARIZ(2), July 1976, National Technical Information Service, U.S. Department of Commerce, Springfield, Virginia.
75. O'Hare, M. P., et al., *Artificial Recharge of Ground Water—Status and Potential in the Contiguous United States,* 1986, Lewis Publishers, Inc., Chelsea, Michigan.
76. Gureghian, A. B., Beskid, N. J. and Marmer, G. J., "Predictive Capabilities of a Two-Dimensional Model in the Ground Water Transport of Radionuclides," Report No. 18, 1978, National Technical Information Service, U.S. Department of Commerce, Springfield, Virginia.
77. Durbin, T. J., Kapple, G. W. and Freckleton, J. R., "Two-Dimensional and Three-Dimensional Digital Flow Models for the Salinas Valley Ground Water Basin, California," Report No. G.S. WRI-78-113, Nov. 1978, U.S. Geological Survey, Menlo Park, California.
78. Knapp, R. M., "Development and Field Testing of a Basin Hydrology Simulator," *Water Resources Research,* Vol. 11, No. 6, Dec. 1975, pp. 879–888.

79. Kraeger-Rovey, C. E., "Numerical Model of Flow in a Stream-Aquifer System," Hydrology Paper No. 74, Aug. 1975, Colorado State University, Fort Collins, Colorado.
80. McArthur, R. E. and Brammer, D. B., "Institutional Framework for Conjunctive Use of Surface and Ground Water in Tupelo-Lee County, Mississippi: Phase 2," OWRT-A-137-MS(1), Apr. 1983, Office of Research and Technology, U.S. Department of the Interior, Washington, DC.
81. Morel-Seytoux, H. J., "Simple Case of Conjunctive Surface-Ground-Water Management," *Ground Water,* Vol. 13, No. 6, Nov.–Dec. 1975, pp. 506–515.
82. Gardner, B. D., Howitt, R. E. and Moore, C. V., "An Empirical Assessment of Conjunctive Use and Water Pricing Policy," OWRT-B-187-CAL(5), Aug. 1980, Office of Water Research and Technology, U.S. Department of the Interior, Washington, DC.
83. Haimes, Y. Y., "Hierarchical Management of Ground and Surface Water Systems via the Multicell Approach," Report No. OWRT-B-062-OHIO (1), July 1975, National Technical Information Service, U.S. Department of Commerce, Springfield, Virginia.
84. Labodie, J. W., Phamwon, S. and Lazaro, R. C., "River Basin Network Mode for Conjunctive Use of Surface and Groundwater: Program CONSIM," OWRT-B-201-COLO(5), June 1983, Office of Water Research and Technology, U.S. Department of the Interior, Washington, DC.
85. Colarullo, S. J., Heidari, M. and Maddock, T., III, "Identification of an Optimal Groundwater Management Strategy in a Contaminated Aquifer," *Water Resources Bulletin,* Vol. 20, No. 5, Oct. 1984, pp. 747–760.
86. Cummings, R. G. and McFarland, J. W., "Groundwater Management and Salinity Control," *Water Resources Research,* Vol. 10, No. 5, Oct. 1974, pp. 909–915.
87. Gorelick, S. M., "A Review of Distributed Parameter Ground Water Management Modeling Methods," *Water Resources Research,* Vol. 19, No. 2, Apr. 1983, pp. 305–319.
88. Heidari, M., "Application of Linear System's Theory and Linear Programming to Ground Water Management in Kansas," *Water Resources Bulletin,* Vol. 18, No. 6, Dec. 1982, pp. 1003–1012.
89. Herrmann, R., "Shallow Aquifers Relative to Surface Water, Lower North Platte River Valley, Wyoming," *Water Resources Bulletin,* Vol. 12, No. 2, Apr. 1976, pp. 371–380.
90. Planert, M., "Digital-Model Analysis to Predict Water Levels in a Well Field Near Columbus, Indiana," Report No. USGS/WRD/WRI-76/049; USGS/WRI76-63, May 1976, National Technical Information Service, U.S. Department of Commerce, Springfield, Virginia.

91. Kaufmann, R. F., "Land and Water Use Effects on Ground Water Quality in Las Vegas Valley," EPA/600/2-78/179, Aug. 1978, U.S. Environmental Protection Agency, Environmental Research Laboratory, Ada, Oklahoma.
92. Pimental, K. D., "Survey of Models to Predict the Effect of Geothermal Power Development on Domestic Water Supplies and to Design Pollution Monitoring Networks," Report No. CONF-770854-1, Aug. 1977, National Technical Information Service, U.S. Department of Commerce, Springfield, Virginia.
93. Wilson, W. E. and Gerhart, J. M., "Simulated Change in Potentiometric Levels Resulting from Groundwater Development for Phosphate Mines, West-Central Florida," *Journal of Hydrology,* Vol. 43, No. 1/4, Oct. 1979, pp. 491–515.
94. Maddox, G. E., "Use of Digital Models to Manage Ground Water," *Proceedings of a Symposium Conducted by the Irrigation and Drainage Division of the American Society of Civil Engineers,* Logan, Utah, 1975, pp. 568–579.
95. Nichols, W. D., "Digital Computer Simulation Model of the Englishtown Aquifer in the Northern Coastal Plain of New Jersey," Water Resources Investigation 77-73, 1977, Water Resources Division, U.S. Geological Survey, Trenton, New Jersey.
96. Robertson, A. F. and Mallory, M. J., "A Digital Model of the Floridan Aquifer, North of Tampa, Florida," Water Resources Investigation 77-64, Oct. 1977, Water Resources Division, U.S. Geological Survey, Tallahassee, Florida.
97. Robson, S. G., "Feasibility of Digital Water-Quality Modeling Illustrated by Application at Barstow, California," USGS-WRI-46-73, 1974, Water Resources Division, U.S. Geological Survey, Menlo Park, California.
98. Simundich, T. M., "A Groundwater Quality Model: A Hybrid Computer Simulation," Master's Thesis, 1971, Department of Computer Science, University of California, Los Angeles, California.
99. Ruppert, R. W. and Clausen, G. S., "Interbasin Transfer or Migration: An Economic Analysis of Two Responses to Ground Water Depletion," OWRRA-039-KAN(1), 1973, Kansas Water Resources Research Institute, Kansas State University, Manhattan, Kansas.
100. Schwarz, J., "Linear Models for Groundwater Management," *Journal of Hydrology,* Vol. 28, No. 2–4, Feb. 1976.
101. Nelson, R. W., "Summary of the Unsaturated Hydrologic Flow and Transport Project Conducted for NRC," PNL-SA-10700, Aug. 1982, Batelle Pacific Northwest Laboratories, Richland, Washington.
102. Friedrichs, D. R., Cole, C. R. and Arnett, R. C., "Hanford Pathline Calculational Program: Theory, Error Analysis and Applications," Report No. 18, June 1977, National Technical Information Service, U.S. Department of Commerce, Springfield, Virginia.

103. Hunt, B., "Dispersive Sources in Uniform Ground Water Flow," *Journal of the Hydraulics Division, ASCE,* Vol. 104, No. NY1, Jan. 1978, pp. 75–85.

104. Redell, D. L. and Sunada, D. K., "Computer Simulation of Waste Transport in Ground Water Aquifers: Ground Water," OWRR-A-001-COLO, 1969, Natural Resources Center, Colorado State University, Fort Collins, Colorado.

105. Ross, B., "Conceptual Model of Deep Unsaturated Zones with Negligible Recharge," *Water Resources Research,* Vol. 20, No. 11, Nov. 1984, pp. 1627–1629.

106. Tyagi, A. K. and Todd, D. K., "Dispersion of Pollutants in Saturated Porous Media," Report No. OWRR-B-020-ARIZ, 1971, Arizona University, Department of Hydrology and Water Resources, Tucson, Arizona.

107. Van der Veer, P., "Exact Solutions for Two-Dimensional Ground Water Flow Problems Involving a Semi-Pervious Boundary," *Journal of Hydrology,* Vol. 37, No. 1/2, Apr. 1978, pp. 159–168.

108. Bredehoeft, J. D. and Pinder, G. F., "Mass Transport in Flowing Groundwater," *Water Resources Research,* Vol. 9, No. 1, Feb. 1973, pp. 194–210.

109. Gershon, N. D. and Nir, A., "Effects of Boundary Conditions of Models on Tracer Distribution in Flow Through Porous Mediums," *Water Resources Research,* Vol. 5, No. 4, Aug. 1969, pp. 830–839.

110. Hamilton, D. A., "Groundwater Management Strategy for Michigan. Groundwater Modeling: Selection, Testing and Use. Volume 1," MI/DNR/GW-82-09, Sept. 1982, Michigan Department of Natural Resources, Lansing, Michigan.

111. Tagaments, T., "Longitudinal Dispersion with Non-Linear Adsorption in Porous Media," Report No. TR-52 (B), 1973, Office of Water Research and Technology, Washington, DC.

112. Christensen, B. A., "Mathematical Methods for Analysis of Fresh Water Lenses in the Coastal Zones of the Florida Aquifer," Report No. WRRC-PUB-44, Dec. 1978, Florida University, Gainesville, Water Resources Research Center, Office of Water Research and Technology, Washington, DC.

113. Kashef, A. I., "Diffusion and Dispersion in Porous Media-Salt Water Mounds in Coastal Aquifers," Report No. 11, 1968, North Carolina Water Resources Research Institute, North Carolina State University, Raleigh, North Carolina.

114. Kashef, A. I., "Management of Retardation of Salt Water Intrusion in Coastal Aquifers," PB-244 721/7SL, 1975, National Technical Information Service, U.S. Department of Commerce, Springfield, Virginia.

115. Konikow, L. F. and Bredehoeft, J. D., "Modeling Flow and Chemical Quality Changes in an Irrigated Stream-Aquifer System," *Water Resources Research,* Vol. 10, No. 3, June 1974, pp. 546–562.

116. Labodie, J. W., Khan, I. A. and Helway, O. J., "Salinity Management Strategies for the Lower San Luis Rey River Basin," Report No. OWRT-C-G301(5229)(1), Sept. 1976, National Technical Information Service, U.S. Department of Commerce, Springfield, Virginia.

117. Land, L. F., "Digital Model for Determining Aquifer Response in Southeast Palm Beach County, Florida," 1975, U.S. Geological Survey, Miami, Florida.
118. Meyer, C. F., "Mathematical Modeling of Fresh-Water Aquifers Having Salt-Water Bottoms," 1973, General Electric Tempo, Santa Barbara, California.
119. Pagenkopf, G. K., "Predicted Trace Metal Concentrations in Saline Seep Waters," Report No. MUJWRRC-99, Nov. 1978, National Technical Information Service, U.S. Department of Commerce, Springfield, Virginia.
120. Pinder, G. F., "Numerical Simulation of Salt-Water Intrusion in Coastal Aquifers. Part 1," Report No. WRP-75-2, 1975, National Technical Information Service, U.S. Department of Commerce, Springfield, Virginia.
121. Rofail, N., "A Mathematical Model of Stratified Groundwater Flow," *Hydrological Sciences Bulletin,* Vol. 22, No. 4, Dec. 1977, pp. 503–512.
122. Segol, G., Pinder, G. F. and Gray, W. G., "Numerical Simulation of Saltwater Intrusion in Coastal Aquifers, Part 2," Report No. WRP-75-3, 1975, National Technical Information Service, U.S. Department of Commerce, Springfield, Virginia.
123. Williams, J. A. and Liu, T., "Response to Tides of Coastal Aquifers: Analog Simulation vs. Field Observations," Report No. TR-86, June 1975, National Technical Information Service, U.S. Department of Commerce, Springfield, Virginia.
124. Atkinson, S. F., et al., *Salt Water Intrusion: Status and Potential in the Contiguous United States,* 1986, Lewis Publishers, Inc., Chelsea, Michigan.
125. Heidari, M. and Cartwright, K., "Analysis of Liquid-Waste Injection Wells in Illinois by Mathematical Models," Report No. WRC-RR-77; UILU-WRC-74-0077, Feb. 1974, Illinois University, Urbana, Illinois.
126. Henry, H. R., et al., "Exploration of Multiphase Fluid Flow in a Saline Aquifer System Affected by Geothermal Heating," Report No. BER-150-118, July 1972, Alabama University, Bureau of Engineering Research, Huntsville, Alabama.
127. Larson, S. P., et al., "Simulation of Wastewater Injection into a Coastal Aquifer System Near Kahului, Maui, Hawaii," *ASCE Proceedings of Hydraulics in the Coastal Zone,* College Station, Texas, 1977, pp. 107–116.
128. Peterson, F. L. and Lau, L. S., "Subsurface Waste Disposal by Injection in Hawaii: A Conceptual Formulation and Physical Modeling Plan," Report No. TM-41, Nov. 1974, Hawaii University, Honolulu, Water Resources Research Center, Office of Water Research and Technology, Washington, DC.
129. Thomas, G. W., "Mathematical Model for Predicting Migration Patterns of Aqueous Waste Resolutions," Report PB-232 885/4WP, Aug. 1973, National Technical Information Service, U.S. Department of Commerce, Springfield, Virginia.

130. U.S. Geological Survey, "A Model for Calculating Effects of Liquid Waste Disposal in Deep Saline Aquifers, Part I—Development, Part II—Documentation," Report No. USGS/WRI-76-61, 1976, Reston, Virginia.
131. Waller, R. M., Turk, J. T. and Dingman, R. J., "Potential Effects of Deep-Well Waste Disposal in Western New York," USGS Report 1053, 1978, U.S. Geological Survey, Washington, DC.
132. Wheatcraft, S. W., "Numerical Modelling of Liquid Waste Injection into Porous Media Saturated with Density-Stratified Fluid: A Progress Report," Report No. TKLM-55, Dec. 1977, Hawaii University, Honolulu, Water Resources Research Center, Office of Water Research and Technology, Washington, DC.
133. Alexander, D. H., et al., "Conceptual Model for Deriving the Repository Source Term," PNL-SA-12940, Nov. 1984, Pacific Northwest Laboratories, Richland, Washington.
134. Arnett, R. C., et al., "Conceptual and Mathematical Modeling of the Hanford Groundwater Flow Regime," 1976, Energy Research and Development Administration, Atlantic Richfield Manford Co., Richland, Washington.
135. Grove, D. B., "A Method to Describe the Flow of Radioactive Ions in Ground Water," Sandia Laboratories Contract Report SC-CR-70-6139, Dec. 1970, U.S. Geological Survey, Denver, Colorado.
136. McDonald, M. G. and Fleck, W. B., "Model Analysis of the Impact on Groundwater Conditions of the Muskegón County Wastewater Disposal System, Michigan," Open-file Report 78-99, Jan. 1978, Water Resources Division, U.S. Geological Survey, Lansing, Michigan.
137. Ostrowski, J. T., "Behavior of Ground Water Subject to Irrigation of Effluent, A Case Study," PB-252 330/4WP, 1976, National Technical Information Service, U.S. Department of Commerce, Springfield, Virginia.
138. Fungaroli, A. A., "Pollution of Subsurface Water by Sanitary Landfills, Volume 1," Publication No. SW-12 rg, 1976, U.S. Environmental Protection Agency, Solid Waste Management Office, Cincinnati, Ohio.
139. Theis, T. L. and Marley, J. J., "Contamination of Groundwater by Heavy Metals from the Land Disposal of Fly Ash," Report No. 18, June–Sept. 1975, National Technical Information Service, U.S. Department of Commerce, Springfield, Virginia.
140. Marie, J. R., "Model Analysis of Effects on Water Levels at Indiana Dunes National Lakeshore Caused by Construction Dewatering," Report No. USGS/WRD/WRI-76/048; USGS/WRI-76-82, July 1976, National Technical Information Service, U.S. Department of Commerce, Springfield, Virginia.
141. Mercado, A., "The Kinetics of Mineral Dissolution in Carbonate Aquifers as a Tool for Hydrological Investigations, Part II, Hydrogeochemical Models," *Journal of Hydrology,* Vol. 35, No. 3/4, Nov. 1977, pp. 365–384.
142. Faulkner, G. C., "Flow Analysis of Karst Systems with Well Developed Underground Circulation," *Proceedings of the U.S. Yugoslavian Symposium,* U.S. Geological Survey, Tallahassee, Florida, June 1976.

143. Kimbler, O. K., Kazman, R. G. and Whitehead, R. R., "Cyclic Storage of Fresh Water in Saline Aquifers," Report No. Bull-10, Oct. 1975, Louisiana Water Resources Research Institute, Louisiana State University, Baton Rouge, Louisiana.

144. Langhettee, E. J., "The Use of Bounding Wells to Control Flux in Underground Water Storage Projects," OWRT-A-027-LA(4), Aug. 1974, Louisiana Water Resources Research Institute, Louisiana State University, Baton Rouge, Louisiana.

145. Aquado, E., Sitar, N. and Remson, I., "Sensitivity Analysis in Aquifer Studies," *Water Resources Research,* Vol. 13, No. 4, Aug. 1977, pp. 733–737.

146. McLaughlin, D.C., "Hanford Groundwater Modeling: Statistical Methods for Evaluating Uncertainty and Assessing Sampling Effectiveness," Report No. RMA-8310, Jan. 1979, National Technical Information Service, U.S. Department of Commerce, Springfield, Virginia.

CHAPTER 7

Pollutant Source Prioritization

There are multiple sources of ground water pollution, including waste dumps, sanitary and chemical landfills, septic tanks systems, municipal and industrial wastewater ponds, and oil and gas field activities. With increasing emphasis on ground water quality protection, there is a growing need to systematically evaluate and prioritize man-made pollution sources. Site selection is continuing for activities which could become future pollution sources if appropriate natural control measures are not utilized and man-made control measures implemented. Ground water pollution potential should be an important consideration in various permit processes. Additionally, there is an expanding need for the planned monitoring of existing and potential ground water pollution sources. Since the costs for extensive monitoring can be prohibitive, attention must be given to those areas, or "hot spots," with greater hydrogeological vulnerability to ground water pollution. One technique for ground water pollution source evaluation and prioritization, site selection, permitting, and monitoring planning involves the use of empirical assessment methodologies.

Empirical assessment methodologies, in the context of ground water quality management, refer to approaches which lead to the development of numerical indices or classifications of the ground water pollution potential of man's activities. This potential can be based on pollutant characteristics, ground water system vulnerability, and/or a combination of both issues. Interpretation of the ground water pollution potential based on the resultant index or classification must be based on professional judgment. These methodologies have traditionally dealt with site selection and evaluation for sanitary or chemical landfills and with evaluation of liquid waste pits, ponds, and lagoons (surface impoundments).

GENERAL COMMENTS

Some pollutant source prioritization methodologies are approaching two decades in age; however, the majority have been developed and applied during the late 1970s and 1980s. Table 7.1 lists nine example methodologies to be described along with their primary area of application. Table 7.2 summarizes the features of empirical assessment methodologies listed in Table 7.1. The methodologies typically focus on a numerical index, with larger numbers generally used to denote greater ground water pollution potential; however, some methodologies encourage the classification or ranking of pollution potential without extensive usage of numerical indicators.

Table 7.1. Applications of Nine Empirical Assessment Methodologies.

Methodology	Examples of Application
Surface Impoundment Assessment	Evaluation of extant liquid impoundments (pits, ponds, and lagoons).
Landfill Site Rating	Evaluation of extant or new sanitary landfill sites.
Waste-Soil-Site Interaction Matrix	Evaluation of new industrial solid or liquid waste disposal sites.
Site Rating System	Chemical landfill site selection or evaluation.
Hazard Ranking System	Ranking of hazardous waste sites for remedial actions.
Site Rating Methodology	Ranking of hazardous waste sites for remedial actions.
Brine Disposal Methodology	Evaluation of extant or planned practices for brine disposal from oil and gas field activities.
Pesticide Index	Ranking of pesticides based on their ground water pollution potential.
DRASTIC	Evaluation of the potential for ground water pollution at a specific site given its hydrogeological setting.

Table 7.2. Summary Features of Empirical Assessment Methodologies.

Numerical indices or classifications of ground water pollution potential
Multiple factors and relative importance weighting
Measurement techniques for factors and scaling (scoring) of importance weights
Indices based on summation of factor scores or products of scores
Need for careful interpretation with professional judgment

Methodologies typically contain several factors for evaluation, with the number and type, and importance weighting varying from methodology to methodology. Methodologies also include descriptions of measurement techniques for the factors, with information provided on the scaling of the factors (points). Final integration of information may involve summation of factor scores or their multiplication followed by summation.[1]

Empirical assessment methodologies can be utilized for relative evaluations and not absolute considerations of ground water pollution. Considerable professional judgment is needed in the interpretation of results. However, they do represent approaches which can be used, based on minimal data input, to provide a structured procedure for source evaluation, site selection, and monitoring planning.

SURFACE IMPOUNDMENT ASSESSMENT

The surface impoundment assessment (SIA) method is based on work by LeGrand.[2] The method was developed for evaluating wastewater ponds,[3] and it yields a sum index with numerical values ranging from 1 to 29. The index is based on four factors: the unsaturated zone, the availability of ground water (saturated zone), ground water quality, and the hazard potential of the waste material. Numerical values for the unsaturated zone range from 0 to 9, for the availability of ground water from 0 to 6, for ground water quality from 0 to 5, and for hazard potential of waste from 1 to 9.

The unsaturated zone rating is based on considering earth material characteristics as well as zone thickness. Table 7.3 provides the basis for the evaluation, with the categories of earth materials based on permeability and secondarily upon sorption character. In rating a particular locality where hydrologically dissimilar layers exist, the waste is more likely to move through the more permeable zones and avoid the impermeable zones. In such cases, the earth material should be rated as the more permeable of the two or more layers which might exist. The availability of ground water factor considers the ability of the aquifer to transmit ground water, thus it is dependent upon aquifer permeability and saturated thickness. Table 7.4 provides information on the types of earth material and thicknesses for various ratings. The letters accompanying the rating matrices in Tables 7.3 and 7.4 are for the purpose of identifying the origin of the rating and documenting the process.

The ground water quality factor is based upon criteria associated with the Underground Injection Control program of the U.S. Environmental Protection Agency. Table 7.5 contains information on the rating.[3] If ground water has high total dissolved solids (TDS) the rating is lower since potential ground water uses which would be curtailed would be limited. If the ground

Table 7.3. Rating of the Unsaturated Zone in the SIA Method.[3]

Earth Material Category		I	II	III	IV	V	VI
Unconsolidated rock		gravel, medium to coarse sand	fine to very fine sand	sand with <15% clay, silt	sand with >15% but <50% clay	clay with <50% sand	clay
Consolidated rock		cavernous or fractured limestone, evaporites, basalt lava fault zones	fractured igneous and metamorphic (except lava) sandstone (poorly cemented)	sandstone (moderately cemented) fractured shale	sandstone (well cemented)	siltstone	unfractured shale, igneous and metamorphic rocks
Representative permeability							
in gpd/ft^2		>200	2–200	0.2–2	<0.2	<0.02	<0.002
in cm/sec		$>10^{-2}$	10^{-4}–10^{-2}	10^{-5}–10^{-4}	$<10^{-5}$	$<10^{-6}$	$<10^{-7}$
Rating Matrix							
Thickness of the un-saturated zone (in meters)	>30	9A	6B	4C	2D	0E	0F
	>10≤30	9B	7B	5C	3D	1E	0G
	>3≤10	9C	8B	6C	4D	2E	0H
	>1≤ 3	9D	9F	7C	5D	3E	1F
	>0≤ 1	9E	9G	9H	9I	9J	9K

Table 7.4. Rating Ground Water Availability in the SIA Method.[3]

Earth Material Category	I	II	III
Unconsolidated rock	gravel or sand	sand with 50% clay sand	clay with 50% sand
Consolidated rock	cavernous or fractured rock, poorly cemented sandstone, fault zones	moderately to well cemented sandstone, fractured shale	siltstone unfractured shale and other impervious rock
Representative permeability			
in gpd/ft^2	>2	0.02–2	<0.02
in cm/sec	$>10^{-4}$	10^{-6}–10^{-4}	10^{-6}
Rating Matrix			
Thickness of saturated zone (meters) ≤30	6A	4C	2E
3–30	5A	3C	1E
≤ 3	3A	1C	OE

Table 7.5. Rating Ground Water Quality in the SIA Method.[3]

Rating	Quality
5	≤500 mg/l TDS or a current drinking water source
4	>500 – ≤1,000 mg/l TDS
3	>1,000 – ≤3,000 mg/l TDS
2	>3,000 – ≤10,000 mg/l TDS
1	>10,000 mg/l TDS
0	No ground water present

water is serving as a drinking water supply, the rating is 5 regardless of the TDS concentration. The waste hazard potential factor is associated with the potential for causing harm to human health. Examples of hazard potential ratings of waste materials classified by source are in Table 7.6. The ratings consider waste characteristics such as toxicity, mobility, persistence, volume, and concentration. Table 7.6 includes a range of ratings for several sources, with the concept being that in cases where there is considerable pretreatment, the lowest rating may be used. The waste hazard potential rating based on wastes classified by type can also be used.[3]

Table 7.6. Examples of Contaminant Hazard Potential Ratings of Waste Classified by Source in the SIA Method.[3]

SIC	Number	Description of Waste Source	Hazard Potential Initial Rating
02		Agricultural Production—Livestock	
	021	Livestock, except dairy, poultry and animal specialties	3 (5 for feedlots)
	024	Dairy farms	4
	025	Poultry and eggs	4
13		Oil and Gas Extraction	
	131	Crude petroleum and natural gas	7
	132	Natural gas liquids	7
	1381	Drilling oil and gas wells	6
20		Food and Kindred Products	
	201	Meat products	3
	202	Dairy products	2
	203	Canned and preserved fruits and vegetables	4
	204	Grain mill products	2
28		Chemicals and Allied Products	
	2812	Alkalies and chlorine	7–9
	2816	Inorganic pigments	3–8
	2819	Industrial inorganic chemicals, not elsewhere classified	3–9
29		Petroleum Refining and Related Industries	
	291	Petroleum refining	8
	295	Paving and roofing materials	7
	299	Miscellaneous products of petroleum and coal	7

Summation of the ratings for each of the four factors in the SIA method yields an overall evaluation for the source. An additional consideration is the degree of confidence of the investigator as well as data availability for the specific site. An overall evaluation of the final rating is suggested, with the ratings being either A, B, or C. A rating of A denotes high confidence and is given when the data used has been site-specific. Ratings of B and C denote moderate and low confidence, respectively, and are given when data has been obtained from a generalized source, or extrapolated from adjacent sites.

The SIA method has been used in a national survey of pits, ponds, and lagoons by the U.S. Environmental Protection Agency. The results provide

Table 7.7. Example of Landfill Site Rating.[2]

		Key Hydrologic Factors									
Step 1	**Point Value**	**0**	**1**	**2***	**3**	**4**	**5**	**6**	**7**	**8**	**9**
Determine distance on ground between contamination source and water supply	Distance in feet	30	50	75	100	150	200	300	500	1,000	2,500 or more
		*Where water table lies in permeable consolidated rocks (II in Step 4), no more than 2 followed by ·) points should be allotted on distance scale.									
Record Point Value											
Step 2	**Point Value**	**0**	**1**	**2***	**3**	**4**	**5**	**6**	**7**	**8**	**9**
Estimate the depth to water table	Depth in feet of water table below base of contamination source more than 5% of the year.	0	2	4	7	15	25	50	75	100	200 or more
		*Where water table lies in permeable or moderately permeable consolidated rocks (II in Step 4), no more than 2 (followed by ·) points should be allotted, regardless of greater depth to water table.									
Record Point Value											
Step 3	**Point Value**	**0**	**1**	**2**	**3**	**4**	**5**				
Estimate water-table gradient from contamination site	Water-table gradient and flow direction (related, in part, to land slope)	gradient greater than 2 percent toward water supply and is the anticipated direction of flow	gradient greater than 2 percent toward water supply but not the anticipated direction of flow	gradient less than 2 percent toward water supply and is the anticipated direction of flow	gradient less than 2 percent toward water supply but not the anticipated direction of flow	gradient almost flat	gradient away from all water supplies that are closer than 2,500 feet				
Record Point Value											

Table 7.7, continued

Step 4 — **Estimate permeability-sorption for the site of the contamination source**

		Clean Coarse Gravel		Clean Coarse Sand		Clean Fine Sand		Sand with a little Clay		Layers of Sand and Clay		Clayey Sand		Mixture of Sand and Clay		Sandy Clay		Clay	
		I	II	I	II	I	II	I	II	I	II	I	II	I	II	I	II	I	II
(1)	100 +	OA	OA	OA	OA	2A	2A	4A	4A	5A	5A	6A	6A	7A	7A	8A	8A	9A	9A
Thickness in feet of unconsolidated material over bedrock	100	OB	OJ	OB	OJ	2B	2D	4B	3D	5B	4H	5D	4K	7B	5K	8B	6K	9B	6M
	90	OB	OJ	OB	OJ	2B	1E	4B	3D	5B	4H	5D	4K	6B	5K	7C	5L	8C	6M
	80	OC	OK	OC	OK	2B	1E	4B	3D	5B	4H	5D	4K	6B	4M	7C	5L	8C	5M
	70	OC	OL	OC	OL	2B	IF	4C	3E	5C	4J	5E	4L	6C	4M	7D	4P	8D	5M
	60	OD	OL	OD	OL	2C	IF	4C	2E	5C	3G	5E	3J	6C	4N	7D	4P	8D	5N
	50	OD	OM	OE	OM	2C	1G	4C	2E	4D	3G	5F	3J	6D	3J	7E	4Q	8E	4R
	40	OE	OM	OE	OM	1B	OS	4D	2F	4E	3H	5F	3K	6E	3K	6G	4Q	7F	4S
	30	OF	ON	OF	ON	1C	OT	3B	2F	4F	3H	5G	2G	6F	2J	6G	3L	7G	3M
	20	OG	OP	OG	OP	ID	OU	3C	1H	4G	2G	5H	2H	5H	2K	6H	2L	7H	3N
	10	OH	OQ	OH	OQ	OQ	OV	2D	1J	3F	1J	4C	1J	5J	1K	6J	1L	6L	2M
(2)	0	5Z	OZ	5Z	OZ	5Z	OZ	5Z	OZ	5Z	OZ	5Z	OZ	5Z	OZ	5Z	OZ	5Z	OZ

Record Point Value

Point Value is determined from Matrix. For single type of unconsolidated material over bedrock, point value is determined by its thickness alone. For combination of unconsolidated materials, point value must be interpolated.

I—over shale or other poorly permeable, consolidated rock

II—over permeable or moderately permeable, consolidated rocks (some basalts, highly fractured igneous and metamorphic rocks, and cavernous carbonate rocks—also fault zones).

(1)—suffix A means because of depth, bedrock is not to be considered, for example, a coastal plain situation

(2)—suffix Z means bedrock is at surface, i.e., there is no soil

Step 5	Total Point Value	0–5	6–7	8–13	14–20	21–25	26—32
Add all Point Values determined in Steps 1 through 4 above. Record Total Point Value	Description of Site in Relative Hydrogeologic Terms only. (Without regard to type of contaminant.)	VERY POOR to POOR because one or more key factors must have values of less than 2.		FAIR If no separate value is less than 2	GOOD to VERY GOOD If all separate values are 3 or greater	If all separate values are 3 or greater	EXCELLENT If all separate values are 3 or greater

Special Site Identifier

Step 6	A	B	C
Sensitivity of Aquifer (choose appropriate category)	A permeable, extensive aquifer capable of easy contamination.	Aquifer of moderate permeability not likely to be contaminated over a large area from a single contamination source.	Limited aquifer of low permeability, or slight contamination potential from a source.
Step 7	**A**	**B**	**C**
Degree of confidence in accuracy of rating values (choose appropriate category)	Confidence in estimates of ratings for the parameters is high, and estimated ratings are considered to be fairly accurate.	Confidence in estimates of ratings for the parameters is fair.	Confidence in estimates of ratings for the parameters is low, and estimated ratings are not considered to be accurate.

Table 7.7, continued

Step 8 Miscellaneous Identifiers (add if appropriate)	A. Alluvial valley—a common hydrogeologic setting—especially important because of the general high permeability and prevalence of down-gradient water supplies B. Designates property boundary when ground distance from a contamination site is to boundary rather than to a water supply C. Special conditions require that a comment or explanation to be added to the evaluation D. Cone of pumping depression near a contamination source, which may cause contaminated ground water to be diverted toward the pumped well E. Distance recorded is that from a water supply to the estimated closest edge of an existing plume rather than to the original source of contamination F. Indicates the contamination source is located on a ground water discharge area, such as a flood plain, and would likely cause minimal ground water contamination M. Mounding of the water table beneath a contamination site—common beneath waste sites where there is liquid input or reduced infiltration capacity P. Percolation may not be adequate—the permeability-sorption digit suggests the degree to which percolation may be a problem, a digit of 7 or more being a special warning of poor percolation Q. Designates a "recharge or transmission" part of an extensive aquifer that is sensitive to contamination—may be suggested by a low rating on the permeability-sorption scale and A or B rating for Step 6 S. Indicates that the most likely water supply to be contaminated is a surface stream, rather than a well or spring

Completion of Numerical Rating

Step 9

Completion of site numerical rating

The total point value determined in Step 5 is recorded and then followed in sequence by the individual point values for the four key hydrogeologic factors: distance, depth to water table, water-table gradient, and permeability-sorption. This is followed, in turn, by the special site identifier suffixes: aquifer sensitivity, degree of confidence, and miscellaneous identifiers. An example of a site rating with brief explanations and interpretations is shown below.

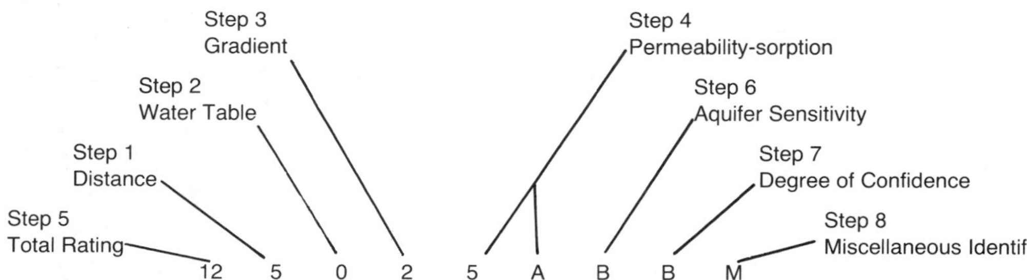

Explanation of sequence of digits and letters

12— Total point value as shown in Step 5
5— The first digit is rating for ground distance—Step 1
0— The second digit is rating for depth to water table—Step 2
2— The third digit is rating for water-table gradient—Step 3
5— The fourth digit is rating for permeability-sorption—Step 4
A— Represents a closely defined position (5A) in permeability-sorption scale—Step 4
B— Represents sensitivity of an aquifer to be contaminated—Step 6
B— Represents degree of confidence or reliability of overall rating—Step 7
M—Indicates special conditions (mounding of water table in this case)—Step 8

a perspective on the nature and potential magnitude of the ground water pollution potential of this source category. In addition, this methodology has been used in a study of several central Oklahoma areas containing septic tank systems.[4]

LANDFILL SITE RATING

The landfill site rating method (called the LeGrand-Brown method) is based upon the experience gained by many individuals to objectively establish the more-favorable and less-favorable site conditions for prevention of adverse ground water impacts. Four key hydrogeological factors or variables are used. The following four factors are considered to represent the simplest and most easily determined and effective factors for a wide variety of applications: (1) distance from a contamination source to the nearest well or point of water use, (2) depth to the water table, (3) gradient to the water table, and (4) permeability and attenuation capacity of the subsurface materials through which the comtaminant is likely to pass. An example of the LeGrand-Brown methodology is shown in Table 7.7.[2] The methodology requires the sequential usage of the nine listed steps. The methodology can be used in the comparative evaluation of potential landfill sites or for prioritizing the ground water pollution concerns for existing landfills in a geographical area.

WASTE-SOIL-SITE INTERACTION MATRIX

The Waste-Soil-Site Interaction Matrix was developed for assessing industrial solid or liquid waste disposal on land.[5] The method involves summation of the products of various waste-soil-site considerations, with the resultant numerical values ranging from 45 to 4,830. The methodology includes ten factors related to the waste, and seven factors associated with the site of potential waste application. Table 7.8 contains a description of the waste factors and their numerical scores, and Table 7.9 lists the soil-site factors with their associated weights. The waste factors listed in Table 7.8 are related to the potential effects of the industrial waste in terms of toxicity and disease transmission; its behavioral characteristics relative to chemical and biological persistence, mobility, viscosity, solubility, and acidity/basicity; and its volume and concentration. Ranges of the numerical scores for each of these waste factors are listed in Table 7.8; detailed information on the data needs and/or testing required for each factors are contained elsewhere.[5] Soil-related factors listed in Table 7.9 include permeability and sorption characteristics. Site hydrology is addressed in terms of the depth to the water table,

Table 7.8. Waste Factors in Waste-Soil-Site Interaction Matrix.[5]

Group	Factor
Effects	Human Toxicity (Ht)—ability of a substance to produce injury once it reaches a susceptible site in or on the body. Based on severity of effect all substances grouped into those with no toxicity, slight toxicity, moderate toxicity, and severe toxicity. The Ht values range from 0 (no toxicity) to 10 (maximum toxicity).
	Ground Water Toxicity (Gt)—related to minimum concentration of waste substance in ground water which would cause damage or injury to humans, animals, or plants. The Gt value is a function of the lowest concentration which would cause damage or injury to any portion of the ecosystem; the Gt values range from 0 (non-toxic) to 10 (very toxic).
	Disease Transmission Potential (Dp)—considers mode of disease contraction, pathogen life state, and ability of the pathogen to survive. Disease contraction includes direct contact, infection through open wounds, and infection by vectors (usually insects). Pathogen life state includes pathogenic microorganisms with more than one life state (virus and fungi), one life state (vegetative pathogens), and those which cannot survive outside their host. The ability of the pathogen to survive includes survival in air, water, and soil environments. The Dp values range from 0 (no effect) to 10 (maximum effect).
Behavioral (Behavioral Performance)	Chemical Persistence (Cp)—related to the chemical stability of toxic components in the waste. Consideration is given to the concentration of toxic components after one-day and six-day contact with soil from potential disposal site. The Cp values range from 1 (very unstable toxic component) to 5 (very stable toxic component).
	Biological Persistence (Bp)—related to the biodegradability of the waste as determined by biochemical oxygen demand (BOD) and theoretical oxygen demand (TOD). The Bp values range from 1 (very biodegradable) to 4 (non-biodegradable).
	Sorption (So)—related to the mobility of the waste in the soil environment. Consideration is given to initial concentration of toxic component(s) in waste and well as one-day following mixing with soil from potential disposal site. The So values range from 1 (very strong sorption) to 10 (no sorption).

Table 7.8, continued

Behavioral (Behavioral Properties)	Viscosity (Vi)—related to the flow of the waste toward the water table. Consideration is given to the waste viscosity measured at the average maximum temperature of the site during its proposed months of use. The Vi values range from 1 (very viscous) to 5 (viscosity of water).
	Solubility (Sy)—along with sorption, solubility relates to the mobility of the waste in the soil environment. Waste solubility is measured in pure water at 25°C and pH of 7. The Sy values range from 1 (low solubility) to 5 (very soluble). In case the waste is miscible with water, Sy is equal to 5.
	Acidity/Basicity (Ab)—considers the influence of acidic or basic wastes on the solubility of various metals. Acidic wastes tend to solubilize metals whereas basic wastes tend to immobilize metals through precipitation. The Ab values range from 0 (no effect) to 5 (maximum effect).
Capacity Rate	Waste Application Rate (Ar)—related to the volumetric application rate of the waste at the site, the sorption characteristics of the site (NS to be discussed in Table 7.9), and the concentration of toxic component(s) in the waste. The Ar values range from 1 (low volumetric application rate of a low concentration waste to a site having high sorptive properties) to 10 (high volumetric application rate of a high concentration waste to a site having low sorptive properties).

the hydraulic gradient of the ground water, and infiltration. The distance from the disposal site to the nearest point of water use and the thickness of the porous layer are two additional site factors listed. Again, ranges of the numerical scores for each of these soil-site factors are given in Table 7.9; detailed information on the data needs and/or testing required for each factor are contained elsewhere.[5]

Table 7.10 represents an example interaction matrix resulting from the usage of this methodology, with the total summation of the products being 990. Ten classes used for interpretation are as follows: Class 1 (45–100 points), Class 2 (100–200), Class 3 (200–300), Class 4 (300–400), Class 5 (400–500), Class 6 (500–750), Class 7 (750–1,000), Class 8 (1,000–1,500), Class 9 (1,500–2,500), and Class 10 (greater than 2,500). Classes 1 through 5 are considered acceptable, and classes 6 through 10 are considered unacceptable. This methodology can be used for the comparative evalu-

Table 7.9. Soil-Site Factors in Waste-Soil-Site Interaction Matrix.[5]

Group	Factor
Soil	Permeability (NP)—relates to permeability of site materials. Clay is considered to have poor permeability, fine sand moderate permeability, and coarse sand and gravel good permeability. The NP values range from 2.5 (low permeability) to 10 (maximum permeability).
	Sorption (NS)—relates to sorption characteristics of site materials. The NS values range from 1 (high sorption) to 10 (low sorption).
Hydrology	Water Table (NWT)—considers the fluctuating boundary free water level and its depth. The zone of aeration occurs above the water table and is important to oxidative degradation and sorption. The NWT values range from 1 (deep water table) to 10 (water table near surface).
	Gradient (NG)—relates to the effect of the hydraulic gradient on both the direction and rate of flow of ground water. The NG values range from 1 (gradient away from the disposal site in a desirable direction) to 10 (gradient toward point of water use).
	Infiltration (NI)—relates to the tendency of water to enter the surface of a waste disposal site. Involves consideration of the maximum rate at which a soil can absorb precipitation or water additions. A site with a large amount of infiltration will have greater ground water pollution potential. The NI values range from 1 (minimum infiltration) to 10 (maximum infiltration).
Site	Distance (ND)—relates to the distance from the disposal site to the nearest point of water use. The greater the distance the less chance of contamination because waste dilution, sorption, and degradation increase with distance. The ND values range from 1 (long distance from disposal site to use site) to 10 (disposal site close to use site).
	Thickness of Porous Layer (NT)—refers to porous layer at the disposal site. The NT values range from 1 (about 100 ft or more of depth) to 10 (about 10 ft of depth).

ation of potential waste disposal sites. It can also be used for prioritizing the ground water pollution concerns for existing waste disposal sites in a geographical area. It has been used in a study of several central Oklahoma areas containing septic tank systems.[4]

Table 7.10. Example of Waste-Soil-Site Interaction Matrix.[5]

Waste \ Soil	Soil	Soil Group: Permeability NP (2½–10)	Soil Group: Sorption NS (1–10)	Hydrology Group: Water Table WT (1–10)	Hydrology Group: Gradient NG (1–10)	Hydrology Group: Infiltration NI (1–10)	Site Group: Distance ND (1–10)	Site Group: Thickness of Porous Layer NT (1–10)	Total
Waste	P* \ P*	**5**	**4**	**5**	**2**	**6**	**7**	**1**	**30**
		Effects Group							
Human Toxicity Ht (0–10)	8	40	32	40	16	48	56	8	240
Groundwater Toxicity Gt (0–10)	5	25	20	25	10	30	35	5	150
Disease Transmission Potential DP (0–10)	0								
		Behavioral Group							
Chemical Persistence Cp (1–5)	3	15	12	15	6	18	21	3	90

Biological Persistence Bp (1–4)	4	20	16	20	8	24	28	4	120
Sorption So (1–10)	5	25	20	25	10	30	35	5	150
Viscosity Vi (1–5)	2	10	8	10	4	12	14	2	60
Solubility Sy (1–5)	1	5	4	5	2	6	7	1	30
Acidity/ Basicity Ab (0–5)	1	5	4	5	2	6	7	1	30
		Capacity—Rate Group							
Waste Application Rate Ar (1–10)	4	20	16	20	8	24	28	4	125
Total	33	165	132	165	66	198	231	33	990

*P = Point score

SITE RATING SYSTEM

A methodology for chemical landfill site selection/evaluation was developed by Hagerty, Pavoni, and Heer, Jr.[6] The methodology also includes the rating of waste materials, with the factors being human toxicity, ground water toxicity, disease transmission potential, biological persistence, and waste mobility. The site rating system portion of the overall methodology considers the influence of ten factors described in Table 7.11. Infiltration potential, bottom leakage potential, and ground water velocity are factors affecting waste transmission and range in point scores from 0 to 20. Filtering

Table 7.11. Factors in Site Rating System.[6]

Group	Priority[a]	Factor
Soil	1	Infiltration Potential (Ip)—relates to the potential for water to enter a waste deposit. The parameter is a function of the infiltration rate of the cover soil, and the field capacity and depth of the cover soil. The Ip values range from 0.02 (most desirable) to 20 (least desirable) points.
	1	Bottom Leakage Potential (Lp)—relates to the water passing through the soil layer beneath the waste site and into the ground water system. The parameter is a function of the bottom soil permeability and thickness of the bottom soil layer. The Lp values range from 0.02 (most desirable) to 20 (least desirable) points.
	2	Filtering Capacity (Fc)—relates to the ability of bottom soils to remove solid particles traveling downward in a fluid suspension. The parameter is dependent upon the pore spaces between individual soil grains and is inversely proportional to the average grain size (particle diameter) in the soil stratum. The Fc values range from about 0 (good filtering capacity) to 16 (poor filtering capacity.
	2	Adsorptive Capacity (Ac)—relates to the ability of bottom soils to remove organic and inorganic minerals from suspension and solution in a migrating fluid by adsorption. The parameter is dependent on the organic content and cation exchange capacity of the bottom soil. The Ac values range from 0 (good adsorptive capacity) to 16 (poor adsorptive capacity).
Ground Water	3	Organic Content (Oc)—relates to the role of the organic content of ground water as a substrate for reproduction of pathogenic organisms. The parameter is dependent on the biochemical oxygen demand of the ground water. The Oc values range from 0 (mini-

Table 7.11, continued

Group	Priority[a]	Factor
		mal potential for pathogenic microorganism reproduction) to 10 (maximal potential for pathogenic microorganism reproduction).
	3	Buffering Capacity (Bc)—relates to the ability of the ground water system to neutralize any acidic or basic characteristics of entering waste materials. The parameter is related to the ground water pH, acidity, and alkalinity. The Bc values range from 0 (strong buffer) to 10 (weak buffer).
	4	Potential Travel Distance (Td)—relates to the potential for dispersal of pollutants through a ground water system. The parameter is related to the distance of travel from a point directly beneath the waste site through the ground water and surface water systems to the sea. The greater the distance the greater the potential number of water uses. The Td values range from 0 (less than 500 ft) to 5 (more than 50 miles).
	1	Ground Water Velocity (Gv)—relates to the time of waste transmission through the ground water system. This parameter is a function of the coefficient of permeability and the hydraulic gradient. The Gv values range from 0 (minimal velocity and waste transmission) to 20 (maximal velocity and waste transmission).
Air	4	Prevailing Wind Direction (Wp)—relates to the transport of any airborne toxics or pathogens from the waste site in relation to the distribution of the human population around the site. The parameter considers the prevailing wind direction in four quadrants along with the population and population nodes in the quadrants. The area of influence encompasses a 42-km radius around the site. The Wp values vary from 0 (wind direction primarily away from population centers) to 5 (wind direction primarily toward population centers).
	4	Population Factor (Pf)—relates to the human population around the site that could be adversely affected by escaping hazardous materials. The parameter is related to the total population within a 42-km radius of the site. The Pf values range from 0 (no population exposed) to 7 (10,000,000 people exposed).

[a]Priority 1 denotes those factors which would immediately affect waste transmission.
Priority 2 denotes those factors which would affect waste transmission after contact with water.
Priority 3 denotes those factors representing the present conditions of receiving ground water.
Priority 4 denotes those factors outside the immediate disposal site.

capacity and adsorptive capacity affect waste transmission after contact with water, and they are assigned points ranging up to 16. The organic content and buffering capacity of ground water are associated with present conditions, and they are assigned points ranging from 0 to 10. Potential travel distance, prevailing wind direction, and population are factors outside the immediate disposal site and related to potential human exposure. These three factors are assigned numerical scores from 0 to 7. Composite scores for a given site can range up to 129. Detailed information on the data needs, testing required, and/or the necessary calculations for evaluating each factor are described elsewhere.[6] The methodology can be used for comparing potential waste disposal sites as well as evaluating the environmental pollution potential of existing sites.

HAZARD RANKING SYSTEM

The U.S. Environmental Protection Agency has developed the Hazard Ranking System (HRS) as a method for ranking facilities for remedial action according to risks to health and the environment.[7] It can be used in the prioritization of sites for inclusion in the Superfund program. It can also be used for selecting a new site for waste disposal. The HRS is a scoring system designed to address the full range of problems resulting from releases of hazardous substances, including surface water, air, fire and explosion, and direct contact, in addition to ground water contamination. The HRS applies a structured value analysis approach. Three migration routes of exposure, ground water, surface water, and air are evaluated, and their scores are combined to derive a score representing the relative risk posed by the facility. Two additional routes of exposure, (1) fire and explosion and (2) direct contact, are measures of the need for emergency response.

Application of the HRS results in three scores for a hazardous waste facility or potential Superfund site. One score, S_M, reflects the potential for harm to humans or the environment as a result of migration of a hazardous substance away from the facility by routes involving ground water, surface water, or air. The S_M is a composite of separate scores for each of the three routes. Another score, S_{FE}, reflects the potential for harm from materials that can explode or cause fires. The third, S_{DC}, reflects the potential for harm as a result of direct human contact with hazardous materials at the facility (i.e., no migration need be involved). The score for each hazard mode (migration, fire and explosion, and direct contact) or route is obtained by considering a set of factors that characterize the hazardous potential of the facility or site. A comprehensive listing of factors for all of the hazard modes is given in Table 7.12.[7] Each factor is assigned a numerical value

Table 7.12. Comprehensive List of Rating Factors in Hazard Ranking System.[7]

Hazard Mode	Factor Category	Factors: Ground Water Route	Factors: Surface Water Route	Factors: Air Route
Migration	Containment	Containment	Containment	
	Route Characteristics	Depth to aquifer of concern	Site slope and intervening terrain	
		Net precipitation	1 year 24-hour rainfall	
		Permeability of unsaturated zone	Distance to nearest surface water	
			Flood potential	
	Waste Characteristics	Physical State	Physical State	Volatility and Physical State
		Persistence	Persistence	Reactivity
		Toxicity	Toxicity	Incompatibility
				Toxicity
	Hazardous Waste Quantity	Total waste quantity	Total waste quantity	Total waste quantity
	Targets	Ground water use	Surface water use	Distance to nearest human population
		Distance to nearest well down gradient	Distance to a sensitive environment	Population within 1-mile radius
		Population served by ground water drawn with 3-mile radius	Population served by surface water drawn within 3-mile radius	Distance to sensitive environment
				Land use

Table 7.12, continued

Hazard Mode	Factor Category	Factors: Ground Water Route	Factors: Surface Water Route	Factors: Air Route
Fire and Explosion[a]	Containment	Containment		
	Waste Characteristics	Direct evidence of ignitibility or explosivity		
		Ignitibility		
		Reactivity		
		Incompatibility		
	Hazardous Waste Quantity	Total waste quantity		
	Targets	Distance to nearest human population		
		Distance to nearest building		
		Distance to sensitive environment		
		Land use		
		Population within 2-mile radius		
		Number of buildings within 2-mile radius		

Table 7.12, continued

		Factors		
Hazard Mode	**Factor Category**	**Ground Water Route**	**Surface Water Route**	**Air Route**
Direct Contact[a]	Accessibility	Accessibility		
	Containment	Containment		
	Waste Characteristics	Toxicity		
	Targets	Population within a 1-mile radius		
		Distance to a critical habitat		

[a]Same factors apply to each of the three routes.

(generally on a scale of 0 to 3, 5, or 8) according to prescribed guidelines. This value is then multiplied by a weighting factor to yield the factor score. The factors scores are then combined by following established guidelines: scores within a factor category are additive, but the factor category scores are multiplicative.

In computing S_{FE} or S_{DC}, or an individual migration route score, the product of its factor category scores is divided by the maximum value the product can have and the resulting ratio is multiplied by 100, thus normalizing scores to a 100-point scale. Computation of S_M is slightly more complex since S_M is a composite of the scores for the three possible routes: ground water (S_{gw}), surface water (S_{sw}), and air (S_a). The S_M is obtained from the equation:

$$S_M = \frac{1}{1.73}\sqrt{S_{gw}^2 + S_{sw}^2 + S_a^2}$$

The factor 1/1.73 arises from the vector addition of the three route scores after the individual scores are normalized to a common denominator. This means of combining them gives added weight to routes with higher scores.

The HRS does not result in quantitative estimates of the probability of harm from a facility or site or the magnitude of the harm that could result. Rather, it is a device for rank ordering facilities or sites in terms of the potential hazard they present. Risks are generally considered to be a function of the probability of an event occurring and the magnitude or severity should the event occur. Applying this approach to hazardous substance facilities or sites, the probability and magnitude of a release are generally functions of the following areas:

(1) Manner in which the hazardous material is contained
(2) Route by which its release would occur
(3) Characteristics of the harmful substance
(4) Amount of hazardous substance
(5) Likely targets

These areas have been examined and representative factors were chosen to address each in the HRS.[7]

SITE RATING METHODOLOGY

The Site Rating Methodology (SRM) has also been used for the prioritization of Superfund sites in terms of remedial actions. The SRM has three elements: (1) a system for rating the general hazard potential of a site, (2) a system for modifying the general rating based on site-specific problems, and (3) a system for interpreting the ratings in meaningful terms. The first element is called the Rating Factor System, the second is called the Additional Points System, and the third is called the Scoring System.[8] The Rating Factor System is used for the initial rating of a waste disposal site based on a set of 31 generally applicable rating factors. As shown in Table 7.13, the 31 factors have been divided into four categories based on their focus. The four categories are: (1) receptors, (2) pathways, (3) waste characteristics, and (4) waste management practices. For each of the factors, a four-level rating scale has been developed which provides factor-specific levels ranging from 0 (indicating no potential hazard) to 3 (indicating a high potential hazard). The rating scales are also listed in Table 7.13. These scales have been defined so that the rating factors can typically be evaluated on the basis of readily available information from published materials, public and private records, contacts with knowledgeable parties, or site visits.

The rating factors do not all assess the same magnitude of potential environmental impact. Consequently, a numerical value called a multiplier has been assigned to each factor in accordance with the relative magnitude of impact that it does assess. Multipliers were originally defined based on the judgment of individuals from several technical disciplines and were revised based on the results of field testing. These values are multiplied by the appropriate factor ratings to result in factor scores for each of the rating factors.

The Additional Points System addresses special features of a facility's location, design, or operation that cannot be handled satisfactorily by rating factors alone.[8] These features might present hazards that are unusually serious, unique to the site, or not assessable by rating scales. For example, an extremely high population density near a site should be considered even more hazardous than the rating factor for "population within 1,000 ft" indicates. Power lines running through sites that contain explosive or flammable wastes, though not typical of waste disposal sites in general, should be considered a potential hazard. Finally, the function of the nearest offsite building might indicate a serious threat of human exposure exists, even though types of functions cannot be quantitatively evaluated by rating scales the way distance can be. In such cases, raters can assign a greater hazard potential score to a site than it might otherwise receive by using the Additional Points System. Guidance on suggested point assignments is in Table 7.14.[8] In order to maintain the objectivity of the rating methodology while allowing the assignment of additional points, the following limits have been placed on the number of additional points that can be assigned in each rating factor category: receptors, 50 points; pathways, 25 points; waste characteristics, 20 points; and waste management practices, 30 points.

The Scoring System is based on all rating factors and additional points

Table 7.13. Rating Factors and Scales for the Site Rating Methodology.[8]

Rating Factors	Rating Scale Levels 0	1	2	3
Receptors				
Population within 1,000 feet	0	1 to 25	26 to 100	Greater than 100
Distance to nearest drinking water well	Greater than 3 miles	1 to 3 miles	3,001 feet to 1 mile	0 to 3,000 feet
Land use/zoning	Completely remote (zoning not applicable)	Agricultural	Commercial or industrial	Residential
Critical environments	Not a critical environment	Pristine natural areas	Wetlands, flood plains, and preserved areas	Major habitat of an endangered or threatened species
Pathways				
Evidence of contamination	No contamination	Indirect evidence	Positive proof from direct observation	Positive proof from laboratory analyses
Level of contamination	No contamination	Low levels, trace levels, or unknown levels	Moderate levels or levels that cannot be sensed during a site visit but which can be confirmed by a laboratory analysis	High levels or levels that can be sensed easily by investigators during a site visit
Type of contamination	No contamination	Soil contamination only	Biota contamination	Air, water, or food stuff contamination
Distance to nearest surface water	Greater than 5 miles	1 to 5 miles	1,001 feet to 1 mile	0 to 1,000 feet
Depth to ground water	Greater than 100 feet	51 to 100 feet	21 to 50 feet	0 to 20 feet

Net precipitation	Less than −10 inches	−10 to +5 inches	+5 to +20 inches	Greater than +20 inches
Soil permeability	Greater than 50% clay	30% to 50% clay	15% to 30% clay	0 to 15% clay
Bedrock permeability	Impermeable	Relatively impermeable	Relatively permeable	Very permeable
Depth to bedrock	Greater than 60 feet	31 to 60 feet	11 to 30 feet	0 to 10 feet
		Waste Characteristics		
Toxicity	Sax's level 0 or NFPA's level 0	Sax's level 1 or NFPA's level 1	Sax's level 2 or NFPA's level 2	Sax's level 3 or NFPA's levels 3 or 4
Radioactivity	At or below background levels	1 to 3 times background levels	3 to 5 times background levels	Over 5 times background levels
Persistence	Easily biodegradable compounds	Straight chain hydrocarbons	Substituted and other ring compounds	Metals, polycyclic compounds, and halogenated hydrocarbons
Ignitability	Flash point greater than 200°F or NFPA's level 0	Flash point of 140°F, to 200°F, NFPA's level 1	Flash point of 80°F, to 140°F, or NFPA's level 2	Flash point less than 80°F, or NFPA's levels 3 or 4
Reactivity	NFPA's level 0	NFPA's level 1	NFPA's level 2	NFPA's levels 3 or 4
Corrosiveness	pH of 6 to 9	pH of 5 to 6 or 9 to 10	pH of 3 to 5 or 10 to 12	pH of 1 to 3 or 12 to 14
Solubility	Insoluble	Slightly soluble	Soluble	Very soluble
Volatility	Vapor pressure less than 0.1 mm Hg	Vapor pressure of 0.1 to 25 mm Hg	Vapor pressure of 25 to 78 mm Hg	Vapor pressure greater than 78 mm Hg
Physical state	Solid	Sludge	Liquid	Gas

Table 7.13, continued

Rating Factors	Rating Scale Levels			
	0	1	2	3
	Waste Management Practices			
Site security	Secure fence with lock	Security guard but no fence	Remote location or breachable fence	No barriers
Hazardous waste quantity	0 to 250 tons	251 to 1,000 tons	1,001 to 2,000 tons	Greater than 2,000 tons
Total waste quantity	0 to 10 acre feet	11 to 100 acre feet	101 to 250 acre feet	Greater than 250 acre feet
Waste incompatibility	No incompatible wastes are present	Present, but does not pose a hazard	Present and may pose a future hazard	Present and posing an immediate hazard
Use of liners	Clay or other liner resistant to organic compounds	Synthetic or concrete liner	Asphalt—base liner	No liner used
Use of leachate collection systems	Adequate collection and treatment	Inadequate collection or treatment	Inadequate collection and treatment	No collection or treatment
Use of gas collection systems	Adequate collection and treatment	Collection and controlled flaring	Venting or inadequate treatment	No collection or treatment
Use and condition of containers	Containers are used and appear to be in good condition	Containers are used but a few are leaking	Containers are used but many are leaking	No containers are used

Table 7.14. Guidance for Additional Points System in the Site Rating Methodology.[8]

Example Situation	Suggested Point Allotment
Receptors (50 Points Maximum)	
Use of site by nearby residents, especially children (For example, a site may be remote and/or fenced, but may still be used frequently by children as a play area or by adults with recreational vehicles.)	0 to 4 points if used sparingly by adults, 4 to 10 points if used regularly by adults, 10 to 20 points if used regularly by children
Type of building nearby (For example, a school vs a warehouse.)	0 to 6 points for public use buildings (e.g., shopping centers), 5 to 15 points for schools and hospitals
Presence of major surface water supplies, aquifers, or aquifer recharge areas near the site	0 to 30 points depending on the proximity of the drinking water supply and the extent to which it is used
Type of adjacent land use (For example, dairy farms, meat packing plants, orchards, and municipal water treatment plants would cause extreme concern	0 to 10 points for recreational uses, 10 to 30 points for food or water-related uses
Presence of economically important natural resources (e.g., shellfish beds, agricultural lands)	0 to 20 points depending on the number of people affected
Presence of major transportation routes	0 to 2 points for railways, 2 to 6 points for roads, and 4 to 10 points for foot paths or bicycle trails
Residential population over 100 people within 1,000 feet	1 point per 25 people up to 10 points

Table 7.14, continued

Example Situation	Suggested Point Allotment
Pathways (25 Points Maximum)	
Erosion and runoff, susceptibility to washout from a flood, and slope instability	0 to 4 points if a potential problem, 4 to 8 points if a moderate problem, 8 to 12 points if a severe problem
Seismic activity	0 to 10 points depending on the most likely adverse effects
Waste Characteristics (20 Points Maximum)	
Substances that are carcinogenic, teratogenic, or mutagenic	4 points per substance
Any high-level radioactive wastes	5 points if in minute quantities, 15 points if in significant quantities
Substances with a high bioaccumulation potential	2 points per substance
Substances that are infectious	0 to 5 points for wastes containing known transmittable pathogens depending on the potency of the wastes

Table 7.14, continued

Example Situation	Suggested Point Allotment
Waste Management Practices (30 Points Maximum)	
No training or safety measures for personnel (active sites)	0 to 4 points depending on the number of people and their responsibilities
Open burning (active sites)	0 to 10 points depending on regularity of the practice and the type of waste burned
Site abandonment	0 to 5 points depending on the reasons for the abandonment
No waste mapping or records	0 to 8 points depending on the presence of hazardous or incompatible wastes
Heat sources or power lines near areas having explosives or flammable wastes	0 to 8 points depending on the proximity and potential for ignition
Less than 18 inches of cover over inactive landfills	0 to 4 points for no apparent problems; 4 to 8 points for blowing trash; 6 to 12 points for hazardous vapors
Less than 6 inches of daily cover on active landfills	0 to 2 points for no apparent problems; 2 to 4 points for blowing trash; 3 to 6 points for hazardous vapors
Total quantity of waste over 250 acre-feet	1 point per 10 acre-feet up to 15 points
Quantity of hazardous waste over 2,000 tons (2,370 cubic yards)	1 point per 4,000 tons (4,750 cubic yards) up to 25 points

that are used to rate a site. Each subscore is based on those rating factors and additional points in that factor category which are used to rate a site. All of these scores are normalized so that they are on a scale of 0 to 100. Associated with every hazard potential score is a percentage of missing and assumed data. These percentages highlight scores that are based on large amounts of missing data and, in a general way, measure the reliability of the scores.[8]

Once a site has been rated using the SRM, the scores must be interpreted. Assessing the meaning of a score can be approached by either relative or absolute interpretation.[8] Relative interpretations are made by means of rankings. When a number of sites all require attention, rankings can be used to determine a preferential ordering. For example, when there are too many facilities to be addressed with available resources, rankings can be used to help determine the order of sites for collection of additional background information; surveys of sites; complete investigations of sites; implementation of remedial actions; and preparation of enforcement cases. Sites can be ranked in several ways including by overall scores, subscores, combinations of scores, and percentages of missing data. Overall scores are likely to be the most useful basis for ranking waste sites because all rating factors are included in the score. The subscores, however, may also provide valuable rankings. A scale developed for absolute interpretations can be used as a convenient point of reference for deciding "how bad is bad" at a site, and as a tool for deciding how urgent it is to respond to the problems at a site.[8] For example, Figure 7.1 is a flexible scale for overall hazard potential scores that can be used as a guide for defining absolute levels of hazard. Based on this use of hazard potential scores, a rater might decide to pursue in-depth investigations of sites classified high and very high before making initial surveys of sites classified very low and low.

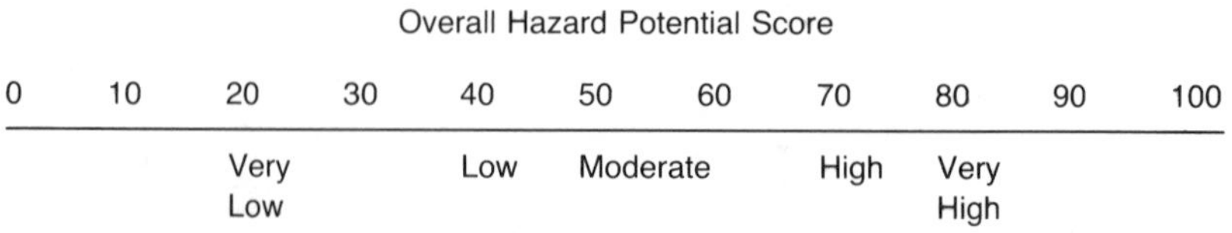

Overall Hazard Potential Score

0	10	20	30	40	50	60	70	80	90	100
		Very Low		Low	Moderate		High	Very High		

Figure 7.1. Guidelines for interpreting the level of potential hazard based on the Site Rating Methodology.[8]

BRINE DISPOSAL METHODOLOGY

A methodology for assessing the potential for ground water contamination from oil and gas field brine disposal was developed and applied in Michigan.[9] Numerical ratings were assigned to either individual oil and gas fields, or townships, depending upon the location in the state. The concept of the methodology is shown in Figure 7.2.[9] The numerical ratings were such that a higher number would indicate a greater potential hazard to ground water resources. Factors considered in assigning ratings were the method of brine disposal, volume disposed, subsurface geology, oil and gas well density, and proximity to water wells (domestic and municipal). Ratings range from 5 to a possible maximum of 59.

The method of brine disposal was the first factor considered. Separate ratings are assigned for disposal in pits and injection wells. Disposal in pits would present the greatest ground water contamination hazard and was consequently assigned an initial rating of 10. This rating is increased according to the total recorded amount of brine discharged to pits over the life of the field. In a few cases where the total amount of brine discharged to pits over the life of the field was small (approximately 1,000 barrels or less), the rating was marked with an asterisk to indicate a lower ground water contamination potential than such a rating would normally indicate. Disposal wells were given progressively lower initial ratings according to the type of well and equipment used. These ratings were also increased according to the total amount of brine injected. Fields that have used annular wells, in which brine is injected between outer and inner casings to relatively shallow disposal formations, were assigned an initial rating of 4. Fields using wells in which brine was injected through the center casing were assigned an initial rating of 3. Better protection against brine leakage through the casing, and consequently a lower rating, is provided by the use of tubing and packers within the casing. Wells using tubing alone were assigned an initial rating of 1. If brine is discharged directly to an aquifer or injected under pressure, the initial rating adjusted for total volume is multiplied by 2. If brine is discharged under pressure directly to an aquifer, the rating is multiplied by 4.

Vertical isolation and shale thicknesses were determined from oil and gas well logs. Lower ratings were assigned for greater thicknesses of shale and greater vertical isolation. Isolation distances for pits were calculated from the vertical isolation between the pit and aquifer. Vertical isolation for injection wells is the distance between the base of the aquifer and the top of the disposal zone.

Oil and gas well density was expressed as number of wells per square mile determined from surface area of the field and total number of wells, including abandoned wells. Fields with higher densities were assigned higher

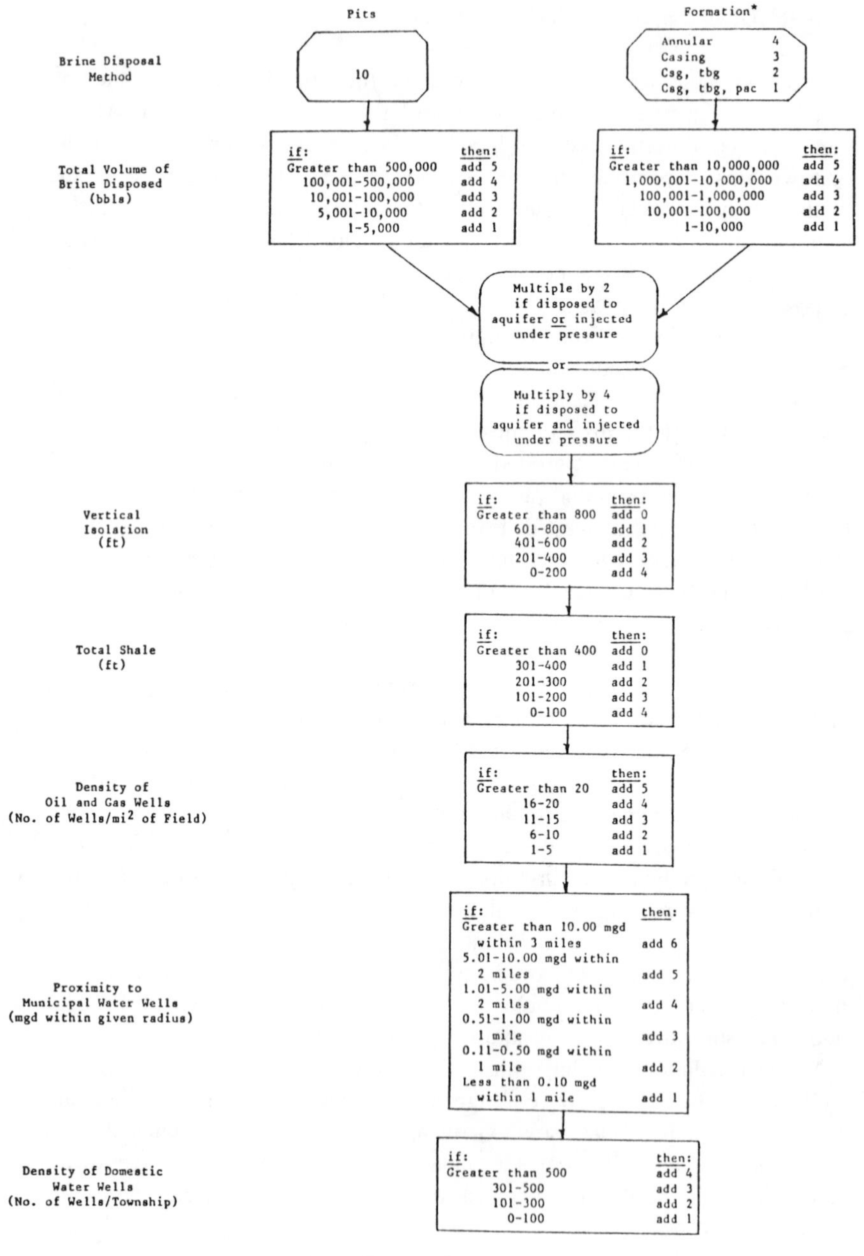

Figure 7.2: Diagram for determining ground water contamination potential factor.

ratings due to the greater possibility of upward migration of brine through well bores and accidental spills.

Proximity to municipal water wells was assessed on the basis of distance between an oil field and municipal wells and production of the municipal wells. Higher ratings were assigned to oil fields near municipal fields with relatively high production. Assuming a larger amount of drawdown from high production municipal wells, a larger area was considered in assigning ratings. Proximity to domestic wells was estimated from well density figures expressed in number of wells per township. These figures were determined from well records kept by the Michigan Department of Natural Resources.

Application of this brine disposal methodology to specific geographical areas can aid in identifying potential "hot spots" in terms of ground water pollution. This information can then be used in developing an areawide ground water monitoring network.

PESTICIDE INDEX

Pesticide contamination of ground water has recently become an issue of concern, particularly in agricultural areas. A large number of physical, chemical, and biological processes have been found to influence pesticide behavior and transport and fate in the subsurface environment. Pesticide characteristics which may have an influence include aqueous solubility, melting point, vapor pressure, Henry's constant, octanol-water partition coefficient, sorption coefficient, and degradation half-life.[10] Detailed evaluation of the ground water pollution potential of a pesticide at a site would require the following site-specific information: climatological data including daily records of rainfall, evapotranspiration, temperature, and net radiation; irrigation, crop, and pesticide management practices; soil profile characteristics including depth to ground water, total porosity, volumetric soil-water content at field capacity and permanent wilting point, soil bulk density, soil organic carbon content, and ground water net recharge rate; and crop parameters such as rooting depth and rooting density.

Rao, Hornsby, and Jessup[10] have suggested a simple scheme for ranking the relative potentials of different pesticides to intrude into ground water. This ranking scheme does not require the detailed pesticide and site characteristics information which would be necessary in a complete mathematical model of the system. The scheme addresses pesticide transport through the crop root zone and the intermediate vadose zone. The following equations and definitions are used in the ranking scheme:[10]

$$AF = M_2/M_o = \exp(-B)$$

where:

AF = attenuation factor between 0 and 1 = index for pesticide mass emission from the vadose zone

M_2 = amount of pesticide entering ground water

M_o = amount of pesticide applied at soil surface

and

$$B = \frac{0.693\ tr}{t^{1/2}}$$

where:

t_r = time required for pesticide to travel through the root zone and intermediate vadose zone

$t^{1/2}$ = degradation half-life of the pesticide

and

$$t_r = \frac{(L)(RF)(FC)}{q}$$

where:

L = distance from the soil surface to ground water

RF = retardation factor

FC = volumetric soil-water content at field capacity

q = net recharge rate

and

$$RF = 1 + \frac{(BD)(OC)(Koc)}{FC} + \frac{(AC)(Kh)}{FC}$$

where:

BD = soil bulk density

OC = soil organic carbon content

Koc = sorption coefficient of pesticide on soil

AC = air-filled porosity of soil

Kh = Henry's constant for pesticide

Rao, Hornsby, and Jessup[10] have suggested that the attenuation factor (AF) index can be used by regulatory agencies in the preliminary evaluation of pesticides to be monitored in geographical areas with ground water susceptible to pesticide pollution. Usage of this index is based on the following simplifying assumptions:

(1) Vadose zone properties are independent of depth
(2) An average ground water recharge rate can be computed given local rainfall, irrigation, and evapotranspiration data
(3) A Koc value can be estimated for each pesticide, based on the assumption that hydrophobic interactions are dominant
(4) An average $t^{1/2}$ value can be estimated for each pesticide

DRASTIC INDEX

A numerical rating scheme, called DRASTIC, has been developed for evaluating the potential for ground water pollution at a specific site given its hydrogeological setting.[11] The rating scheme is based on seven factors chosen by a large number of ground water scientists from throughout the United States. Information on these factors is presumed to exist for all locations in the United States. In addition, these scientists also established relative importance weights and a point rating scale for each factor. The acronym DRASTIC is derived from the seven factors in the rating scheme:

D = depth to ground water
R = recharge rate (net)
A = aquifer media
S = soil media
T = topography (slope)
I = impact of the vadose zone
C = conductivity (hydraulic) of the aquifer

Determination of the DRASTIC index involves multiplying each factor weight by its point rating and summing the total. The higher sum values represent greater potential for ground water pollution. At a given site being evaluated, each factor is rated on a scale of 1 to 10 indicating the relative pollution potential of the given factor at that site. Once all factors have been assigned a rate, each rate is multiplied by the assigned weight, and resultant numbers are summed as follows:

$$D_rD_w + R_rR_w + A_rA_w + S_rS_w + T_rT_w + I_rI_w + C_rC_W = \text{Pollution Potential}$$

where:

r = rating for the site
w = importance weight for the parameter

Table 7.15 displays the rating and weight for the depth to ground water factor.[11] Two importance weights are shown, one for general usage, and one for evaluation of the ground water pollution potential of agricultural chemicals. Table 7.16 contains the rating and weight information for the net recharge factor.[11] Figure 7.3 contains the rating and weight information for the aquifer media factor, while Table 7.17 does similarly for the soil media factor.[11] Table 7.18 contains the pertinent information for the topography factor, and Figure 7.4 does similarly for the impact of the vadose zone factor.[11] Finally, Table 7.19 summarizes the pertinent information for the hydraulic conductivity of the aquifer factor.[11]

DRASTIC can be used in several ways in conjunction with pollutant source prioritization. Geographical areas having greater vulnerability to

Table 7.15. Evaluation of Depth to Ground Water Factor in DRASTIC.[11]

Depth to Ground Water (feet)	
Range	Rating
0–5	10
5–15	9
15–30	7
30–50	5
50–75	3
75–100	2
100+	1
Weight: 5	Agricultural Weight: 5

Table 7.16. Evaluation of Net Recharge Factor in DRASTIC.[11]

Net Recharge (inches)	
Range	Rating
0–2	1
2–4	3
4–7	6
7–10	8
10+	9
Weight: 5	Agricultural Weight: 5

Table 7.17. Evaluation of Soil Media Factor in DRASTIC.[11]

Soil Media	
Range	Rating
Thin or absent	10
Gravel	10
Sand	9
Shrinking and/or aggregated clay	7
Sandy loam	6
Loam	5
Silty loam	4
Clay loam	3
Nonshrinking and nonaggregated clay	1
Weight: 2	Agricultural Weight: 5

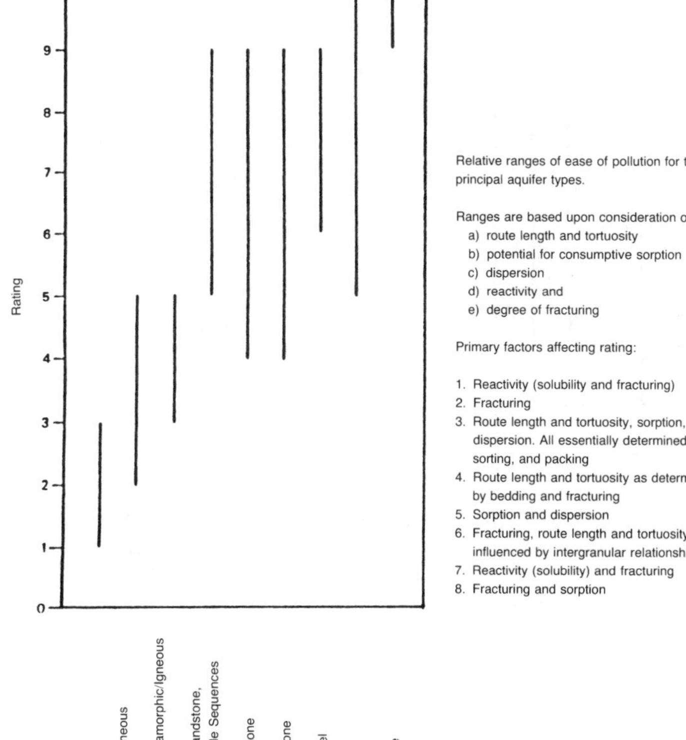
10
9
8
7
6
5
4
3
2
1
0
Rating
Relative ranges of ease of pollution for the principal aquifer types.
Ranges are based upon consideration of:
a) route length and tortuosity
b) potential for consumptive sorption
c) dispersion
d) reactivity and
e) degree of fracturing
Primary factors affecting rating:
1. Reactivity (solubility and fracturing)
2. Fracturing
3. Route length and tortuosity, sorption, dispersion. All essentially determined by grain size, sorting, and packing
4. Route length and tortuosity as determined by bedding and fracturing
5. Sorption and dispersion
6. Fracturing, route length and tortuosity, influenced by intergranular relationships
7. Reactivity (solubility) and fracturing
8. Fracturing and sorption
8. Massive Shale
2. Metamorphic/Igneous
5. Weathered Metamorphic/Igneous
4. Thin Bedded Sandstone, Limestone, Shale Sequences
6. Massive Sandstone
7. Massive Limestone
3. Sand and Gravel
2. Basalt
1. Karst Limestone
Primary Media

Figure 7.3. Evaluation of the aquifer media factor in DRASTIC.[11]

Table 7.18. Evaluation of Topography Factor.[11]

Topography (percent slope)	
Range	**Rating**
0–2	10
2–6	9
6–12	5
12–18	3
18+	1
Weight: 1	Agricultural Weight: 3

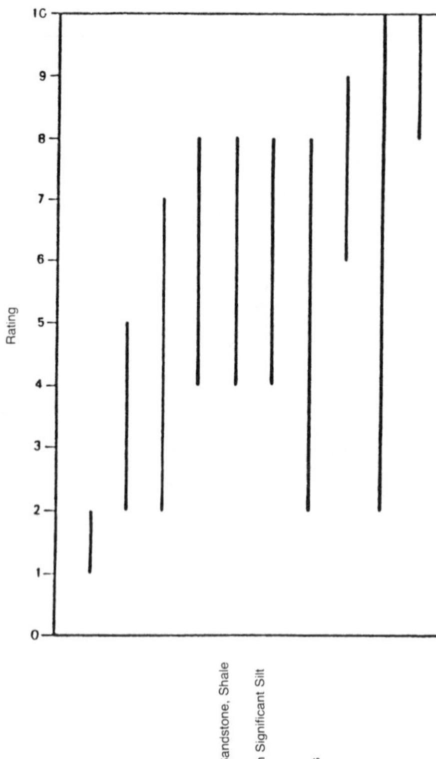

Relative impact of the principal Vadose Zone Media types. Range based upon:

a) Path length and tortuosity
b) potential for dispersion and consequent dilution
c) reactivity (solubility)
d) consumptive sorption
e) fracturing

Primary factors affecting rating:

1. Consumptive sorption and fracturing
2. Fracturing and reactivity
3. Fracturing; path length as influenced by intergranular relationships
4. Fracturing; path length and tortuosity as influenced by bedding planes, sorption, and reactivity
5. Path length and tortuosity as impacted by bedding grain size; sorting and packing; sorption
6. Path length and tortuosity as influenced by grain size, sorting, and packing
7. Reactivity and fracturing

Figure 7.4. Evaluation of the impact of vadose zone media factor in DRASTIC.[11]

Table 7.19. Evaluation of Hydraulic Conductivity Factor in DRASTIC.[11]

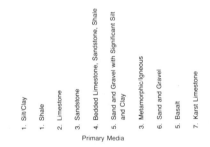

Hydraulic Conductivity (GPD/Ft^2)	
Range	**Rating**
1–100	1
100–300	2
300–700	4
700–1000	6
1000–2000	8
2000+	10
Weight: 3	Agricultural Weight: 2

ground water pollution can be identified as part of the general planning for a ground water monitoring program. DRASTIC can also be used in evaluating the pollution potential of existing waste disposal sites, as well as in the selection process for new sites.

DEVELOPMENT OF EMPIRICAL ASSESSMENT METHODOLOGIES

Several generic steps are associated with the development of numerical indices or classifications of the ground water pollution potential of man's activities. These include factor identification, assignment of importance weights, establishment of scaling functions, and application and field verification. Several techniques can aid in the accomplishment of each of these steps, including paired comparisons, the nominal group process, and standards of practice.

Source Characterization and Factors Indicative of Ground Water Pollution

Existing methodologies have traditionally characterized pollution sources such as sanitary or chemical landfill disposal sites and liquid waste pits, ponds, and lagoons. Existing methodologies generally consider the characteristics of the waste materials, or components thereof, which may be transported to aquifers, and the characteristics of the local surface and subsurface environment relative to the inducement or attenuation of pollutant movement from the source to the aquifer. Some methodologies give consideration to the atmospheric transport of waste materials or components, and the associated human population exposure. For example, Table 7.20 summarizes the waste rating factors included in two empirical assessment methodologies for landfills and one for surface impoundment. Waste rating factors are biological, chemical, or physical parameters which address toxicity, persistence, and mobility. To continue the example, Table 7.21 lists the site rating factors for the three methodologies which take into account soil, ground water, and air parameters.

If a methodology is being developed for a new source type, the pollutional characteristics of the source must be considered along with the factors which could be chosen as indicators of the ground water pollution potential. A common feature of empirical assessment methodologies is that information from the pertinent factors is aggregated into one composite numerical index or classification for interpretation.

Table 7.20. Summary of Waste Rating Factors in Empirical Assessment Methodologies.

Waste Rating Factor	Method[a]
1. Human toxicity	HPH, PNM
2. Ground water toxicity	HPH, PNM
3. Disease transmission potential	HPH, PNM
4. Biological persistence	HPH, PNM
5. Chemical persistence	PNM
6. Sorption properties	PNM
7. Viscosity	PNM
8. Acidity/basicity	PNM
9. Waste application rate	PNM
10. Waste mobility	HPH
11. Waste hazard potential (source/type)	LeG

[a]HPH = Hagerty, Pavoni, and Heer.[6]
LeG = LeGrand[2] and LeGrand and Brown.[12]
PNM = Phillips, Nathwani and Mooij.[5]

Table 7.21. Summary of Site Rating Factors in Empirical Assessment Methodologies.

Site Rating Factors	Method[a]
A. Soil Parameters	
1. Soil permeability	HPH, LeG, PNM
2. Filtering capacity (sorption)	HPH, PNM, LeG
3. Adsorptive capacity	HPH
4. Depth to ground water table (unsaturated soil thickness)	HPH, PNM, LeG
5. Infiltration	PNM
6. Infiltration capacity (field capacity)	HPH
B. Ground Water Parameters	
7. Ground water gradient	LeG, PNM
8. Ground water velocity	HPH
9. Distance from source to point of use	LeG, PNM
10. Organic content	HPH
11. Buffering capacity	HPH
12. Ground water quality	LeG
13. Travel distance	HPH
C. Air Parameters	
14. Prevailing wind direction	HPH
15. Population factor	HPH

[a]HPH = Hagerty, Pavoni, and Heer.[6]
LeG = LeGrand[2] and LeGrand and Brown.[12]
PNM = Phillips, Nathwani and Mooij.[5]

Assign Relative Importance Weights to Factors

The second step in the development of a methodology is the assignment of relative importance weights to the source-transport factors, or at least the arrangement of them in a rank ordering of importance. Table 7.22 lists some importance weighting or ranking techniques which could be used to achieve this step. These techniques have been used in numerous environmental decisionmaking projects. Brief descriptions of two reference sources from Table 7.22 will be presented as illustrations of the techniques.

Eckenrode[13] developed six methods for collecting the judgments of experts concerning the relative value of sets of criteria and compared their reliability and time efficiency. The methods were ranking, rating, three versions of paired comparisons, and a method of successive comparisons. In each of three situations, six criteria were comparatively evaluated by the judges. The results of these experiments showed that there were no significant differences in the sets of criterion weights derived from collecting the judgment data by any of the methods, but that ranking was by far the most efficient method. A fourth experiment was conducted to develop baseline data on the time required to make comparative judgments vs number of items to be judged, by the ranking method and by the simplest paired comparisons method. Ranking is increasingly more efficient than paired comparisons as the number of items to be judged increases from 6 to 30.

The Nominal Group Process Technique (NGT), an interactive group technique, was applied to the identification and rating of factors important in siting nuclear power plants.[19] The NGT was derived from social-

Table 7.22. Selected Importance Weighting Techniques.

Reference Source	Importance Weighting Technique
Dalkey[14]	Delphi
Dean and Nishry[15]	Paired comparison
Dee, et al.[16]	Ranked pairwise comparison
Eckenrode[13]	Ranking Rating Paired comparison (3 types) Successive comparison
Edwards[17]	Multi-attribute utility measurement
School of CEES and Oklahoma Biological Survey[18]	Ranked pairwise comparison
Voelker[19]	Nominal group process

psychological studies of decision conferences, management science studies of aggregating group judgments, and social work studies of problems surrounding citizen participation in program planning. It has gained wide acceptance in health, social service, education, industry, and government organizations. Basically, it consists of four steps: (1) nominal (silent and independent) generation of ideas in writing by a panel of participants, (2) round-robin listing of ideas generated by participants on a flip chart in a serial discussion, (3) discussion of each recorded idea by the group for the purpose of clarification and evaluations, and (4) independent voting on priority ideas, with group decision determined by mathematical rank-ordering.

Establish Scaling Functions

Several approaches have been used to scale the data associated with factors in empirical assessment methodologies. Development of scaling functions is the third step in the development of an empirical assessment methodology. Examples of techniques for this purpose include the use of: (1) linear scaling based on the range of data, (2) letter or number assignments designating data categories, (3) functional curves, or (4) the paired comparison technique.

Odum, et al.[20] utilized a scaling technique in which the actual measures of the assessment factors for each alternative plan are normalized and expressed as a decimal of the largest measure for that variable. This represents linear scaling based on the maximum change. The approach used by Aller, et al.[11] in DRASTIC is generally dependent on linear scaling.

A letter scaling system is used in the Voorhees and Associates method.[21] This methodology incorporates 80 assessment factors oriented to the types of projects conducted by the U.S. Department of Housing and Urban Development. The scaling system consists of the assignment of a letter grade from A+ to C− for the impacts, with A+ representing a major beneficial impact and C− an undesirable detrimental change. Relative to ground water pollution potential, letter scaling could be used to represent vulnerable or nonvulnerable conditions for the chosen factors.

Functional curves can also be used to accomplish impact scaling for assessment factors. The functional curve is used to relate the objective evaluation of an assessment factor to a subjective judgment regarding its quality or numeral score based on a range from high likelihood of pollution to low likelihood.[16] A paired comparison technique can also be used for assigning scale values to alternatives based on their impact on assessment factors. The paired comparison technique for accomplishing scaling is described in Dean and Nishry.[15]

Other tools for developing scaling functions include use of existing literature and/or standards of practice. As an example, many structures such as pits and lagoons have design specifications and/or guidelines for operation and maintenance. Some of these may have been written specifically to prevent or minimize ground water pollution. These criteria can be used to determine scaling functions; that is, the minimum which would be acceptable. For some waste sites such as injection wells, state regulations provide criteria for developing scaling functions. With specific criteria, established injection wells can be evaluated to determine if they are in compliance with regulations or if they exceed (are better than) what is required.

Application and Field Verification

All of the steps discussed so far lead to the evaluation of the pollution potential of various sites. Field verification can be used to check the accuracy of the empirical assessment methodology, make refinements as needed, and to define the measures necessary to mitigate pollution.

SUMMARY

Ground water can become polluted from many source types, including sanitary and chemical landfills, surface impoundments, septic tank systems, land application of municipal wastewaters and sludges, and oil and gas field activities. Several methodologies have been developed for prioritizing the pollution potential of these source types. To develop empirical assessment methodologies, existing methodologies could be modified or new ones developed. In either case, the basic approach will involve the following five steps: (1) identify pertinent pollutant source and transport factors, (2) assign relative importance weights to the factors, (3) characterize the range of numerical or qualitative data for each of the identified source-transport factors and develop scaling approaches, (4) aggregate the information into a composite score, and (5) apply the methodology and modify as necessary.

SELECTED REFERENCES

1. Canter, L. W., "Methods for Assessment of Ground Water Pollution Potential," *Ground Water Quality,* Ward, C. H., Giger, W., and McCarty, P. L., Editors, 1985, John Wiley and Sons, Inc., New York, New York, pp. 270–306.
2. LeGrand, H. E., "System of Reevaluation of Contamination Potential of Some Waste Disposal Sites," *Journal American Water Works Association,* Vol. 56, August 1964, pp. 959–974.

3. U.S. Environmental Protection Agency, "A Manual for Evaluating Contamination Potential of Surface Impoundments," EPA/570/9-78-003, June 1978, Office of Drinking Water, Washington, DC.
4. Canter, L. W. and Knox, R. C., *Septic Tank System Effects on Ground Water Quality,* 1985, Lewis Publishers, Inc., Chelsea, Michigan.
5. Phillips, C. R., Nathwani, J. D., and Mooij, H., "Development of a Soil-Waste Interaction Matrix for Assessing Land Disposal of Industrial Wastes," *Water Research,* Vol. 11, November 1977, pp. 859–868.
6. Hagerty, D. J., Pavoni, J. L., and Heer, J. E., Jr., *Solid Waste Management,* 1973, Van Nostrand Reinhold, New York, New York, pp. 242–262.
7. Caldwell, S., Barrett, K. W., and Chang, S. S., "Ranking System for Releases of Hazardous Substances," *Proceedings of the National Conference on Management of Uncontrolled Hazardous Waste Sites,* 1981, Hazardous Materials Control Research Institute, Silver Spring, Maryland, pp. 14–20.
8. Kufs, C., et al., "Rating the Hazard Potential of Waste Disposal Facilities," *Proceedings of the National Conference on Management of Uncontrolled Hazardous Waste Sites,* 1980, Hazardous Materials Control Research Institute, Silver Spring, Maryland, pp. 30–41.
9. Western Michigan University, "Hydrogeologic Atlas of Michigan," Plate 33, 1981, Department of Geology, Kalamazoo, Michigan.
10. Rao, P. S., Hornsby, A. G., and Jessup, R. E., "Indices for Ranking the Potential for Pesticide Contamination of Groundwater," *Proceedings of Soil and Crop Science Society of Florida,* Vol. 44, 1985, pp. 1–8.
11. Aller, L., et al., "DRASTIC: A Standard System for Evaluating Ground Water Pollution Potential Using Hydrogeologic Settings," EPA/600/2-85/018, May 1985, U.S. Environmental Protection Agency, Robert S. Kerr Environmental Research Laboratory, Ada, Oklahoma.
12. LeGrand, H. E. and Brown, H. S., "Evaluation of Ground Water Contamination Potential from Waste Disposal Sources," 1977, Office of Water and Hazardous Materials, U.S. Environmental Protection Agency, Washington, DC.
13. Eckenrode, R. T., "Weighting Multiple Criteria," *Management Science,* Vol. 12, No. 3, November 1965, pp. 180–192.
14. Dalkey, N. C., "The Delphi Method: An Experimental Study of Group Opinion," Memorandum RM-5888-PR, June 1969, The Rand Corporation, Santa Monica, California.
15. Dean, B. V. and Nishry, J. J., "Scoring and Profitability Models for Evaluating and Selecting Engineering Products," *Journal Operations Research Society of America,* Vol. 13, No. 4, July–August 1965, pp. 550–569.
16. Dee, N., et al., "Environmental Evaluation System for Water Resources Planning," 1972, Battelle-Columbus Laboratories, Columbus, Ohio.
17. Edwards, W., "How to Use Multi-Attribute Utility Measurement for Social Decision Making," SSRI Research Report 76-3, August 1976, Social Science Research Institute, University of Southern California, Los Angeles, California.
18. School of Civil Engineering and Environmental Science and Oklahoma Biological Survey, "Mid-Arkansas River Basis Study—Effects Assessment of Alternative Navigation Routes from Tulsa, Oklahoma to Vicinity of Wichita, Kansas," June 1974, University of Oklahoma, Norman, Oklahoma.

19. Voelker, A. H., "Power Plant Siting, An Application of the Nominal Group Process Technique," ORNL/NUREG/TM-81, February 1977, Oak Ridge National Laboratory, Oak Ridge, Tennessee.
20. Odum, E. P., et al., "Optimum Pathway Matrix Analysis Approach to the Environmental Decision Making Process—Test Case: Relative Impact of Proposed Highway Alternates," 1971, Institute of Ecology, University of Georgia, Athens, Georgia.
21. Voorhees, A. M. and Associates, "Interim Guide for Environmental Assessment: HUD Field Office Editions," June 1975, Washington, DC.

CHAPTER 8

Ground Water Monitoring Planning

An expanding field of study is encompassed by the catch-all term "ground water monitoring." The technologies, equipment, and expertise available to those interested in gathering information on ground water quantity and quality are manifold. The limits must be imposed by the interested party by deciding upon one to several of many options for a large number of decision factors. Most of these decision factors represent objective decisions, while others tend to be more subjective. A partial listing of some pertinent questions and their contained decision factors, in no particular order, includes:

1. What subsurface phenomenon is to be monitored (i.e., flow, heat or solute transport, or subsidence)?
2. What are the reasons for developing the monitoring program?
3. What will be the objectives of the monitoring program?
4. Will monitoring be needed in the unsaturated zone, saturated zone, or both zones?
5. Are surface geophysical methods appropriate, or will there be a need for subsurface sampling through boreholes (monitoring wells)?
6. If subsurface sampling through boreholes is required, what is the most desirable of the dozen or so available monitoring well drilling methods?
7. What quality parameters of ground water will need to be monitored?
8. What sampling procedures are most appropriate for the needed quality parameters?
9. What analysis procedures are most appropriate for the needed quality parameters?
10. What processes could operate to bias the results of ground water sampling and analysis?
11. Where should monitoring wells be located and at what depth?
12. What is the optimum number and spatial distribution of the monitoring wells?

13. Is there a need to incorporate modeling as part of the monitoring network?
14. In what form should the collected data be stored, transferred, or presented?
15. How much money can be or is allocated for ground water monitoring?

From the above list, it is obvious that developing a ground water monitoring network involves much more than drilling a couple of monitoring wells. It should also be apparent that the actual installation of wells and the conduction of sampling and analysis must be preceded by substantial forethought and study.

As shown in Figure 8.1, there are four basic phenomena that occur in the subsurface: ground water flow, heat transport, subsidence, and solute transport. Ground water flow is the most commonly monitored and best understood phenomenon. In addition, the mechanics of the other three phenomena are directly influenced by ground water flow. With few exceptions, all subsurface monitoring programs will have to include monitoring of the ground water flow patterns. Ground water flow is most commonly assessed by recording the rise or fall of ground water (or piezometric) levels and analytically converting these observations to estimates of flow.

As the name implies, heat transport processes are concerned with the movement or conduction of thermal energy in the saturated subsurface environment. These processes are assessed by monitoring temperature changes in the subsurface. Monitoring subsurface heat transport is important in geothermal energy assessments and waste heat disposal studies. Subsidence refers to the phenomenon wherein land surfaces actually sink due to excessive ground water withdrawals. This mostly irreversible process is highly dependent on ground water movement, but is usually monitored at the land surface by marking elevation changes in a given area. Solute transport monitoring is concerned with the movement of chemical species in the subsurface environment. This type of monitoring is the fastest growing field of study and is primarily related to ground water pollution. Solute transport is influenced by ground water flow and its evaluation most often requires the retrieval and laboratory or field analyses of ground water samples.

Ground water monitoring networks must be carefully planned in order to be cost-effective. This chapter identifies four broad categories of monitoring networks and then delineates a number of steps associated with planning and implementing a network. Specific information is included on planning and conducting a basinwide ambient trend monitoring program and a source monitoring program. Chapter 9 includes an extensive discussion of the variety of technologies, equipment, and methodologies available for actual retrieval and analysis of ground water samples.

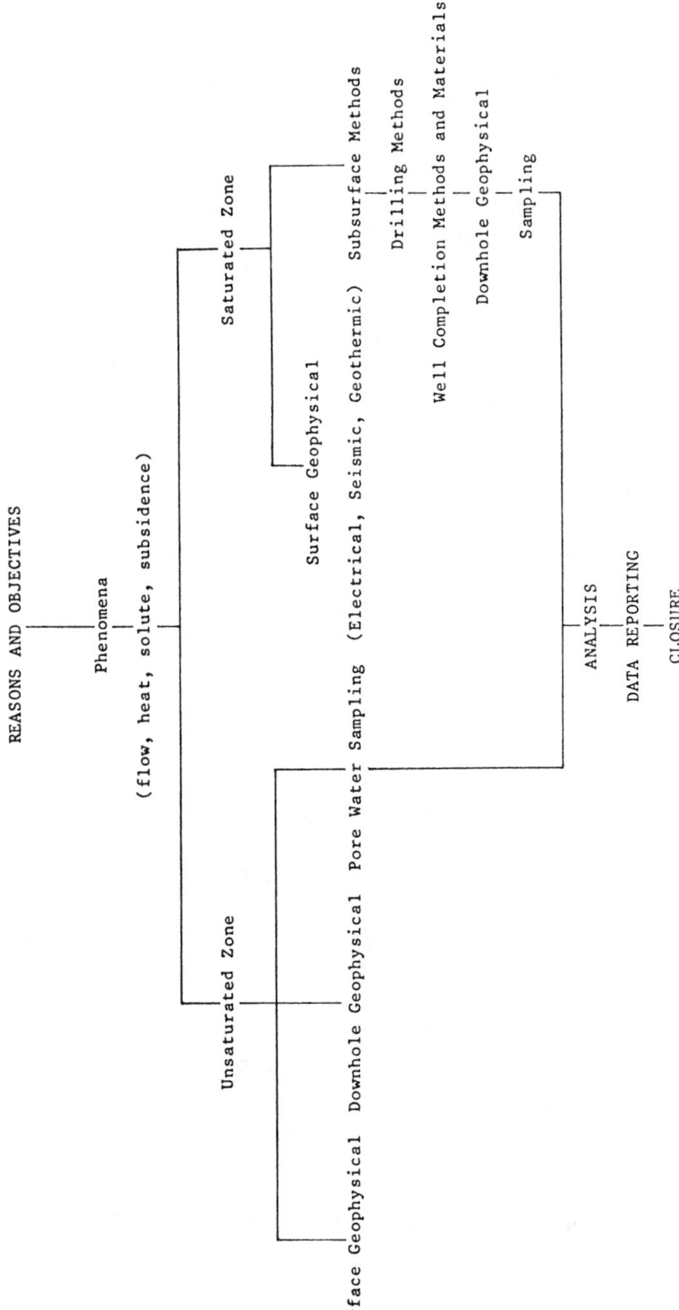

Figure 8.1. Subsurface monitoring considerations.

TYPES OF MONITORING NETWORKS

Todd, et al.[1] have identified four types of ground water monitoring. The first type can be labeled Ambient Trend Monitoring. This type of monitoring is designed to detect temporal and spatial trends in the overall quality of a ground water basin or geographical area. The second type of monitoring is Source Monitoring involving measurement and analysis of potential and actual ground water pollution from identified sources. The last two types are labeled Case Preparation Monitoring, which relates to gathering evidence for enforcement actions; and Research Monitoring, which refers to the development of information on how the subsurface system operates.

In order to reduce the terminology and address monitoring networks on a simpler basis, two broad types of ground water quality monitoring networks can be considered. These two types of networks can be classified according to the area encompassed and their objectives. The first type can be described as Basinwide Ambient Trend Monitoring. As the name implies, this type of monitoring encompasses a large geographical area and concentrates on determining long-term fluctuations in the overall ground water quality of an area. Natural ground water quality levels can also be ascertained. The second type of network can be described as Source Assessment (or Source) Monitoring. This type of monitoring is concerned with assessing the existing or potential impacts on ground water quality from a proposed, active, or abandoned pollutant source. This type of monitoring is localized and concentrates on the changes in ground water quality in the immediate area of the source. Examples of Source Assessment Monitoring totally dominate examples of Basinwide Ambient Trend Monitoring.

The development and implementation of ground water monitoring networks is currently moving away from a "seat of the pants" or "rule of thumb" approach toward more scientific and structured approaches. The main reason for the lag in advancement of ground water monitoring approaches is similar to a never-ending cycle. Simply stated, design of subsurface sampling and analysis networks cannot be accurately accomplished until some subsurface sampling and analysis data become available and these data may be totally useless unless they result from a properly designed system. However, recent advances in technology and knowledge gained from previous experiences have helped to overcome this cycle.

PLANNING GROUND WATER MONITORING NETWORKS

There is a growing amount of information on methodologies for designing ground water monitoring networks. One of the earlier works lists 15 steps for developing and implementing a monitoring network. These steps are:

Step 1—Select area or basin for monitoring
Step 2—Identify pollution sources, causes, and methods of waste disposal
Step 3—Identify potential pollutants
Step 4—Define ground water usage
Step 5—Define hydrogeologic situation
Step 6—Study existing ground water quality
Step 7—Evaluate infiltration potential for wastes at the land surface
Step 8—Evaluate mobility of pollutants from the land surface to the water table
Step 9—Evaluate attenuation of pollutants in the saturated zone
Step 10—Set priorities on sources and causes
Step 11—Evaluate existing monitoring programs
Step 12—Establish alternative monitoring approaches
Step 13—Select and implement the monitoring program
Step 14—Review and interpret monitoring results
Step 15—Summarize and transmit monitoring information

Although the 15 steps seem rather broad and ambitious, the point is made that much work (Steps 1 through 11) must be done prior to actually developing a monitoring network. In other words, it is important to obtain and evaluate existing information. This will aid in identifying the need for and extent of any new monitoring activities.

Naymik[2] outlined a five-step procedure for design and optimization of monitoring networks: (1) preliminary network design and information gathering; (2) initial installation and testing; (3) completion and verification; (4) operation; and (5) project termination. Once again the need for preliminary study was highlighted. This five-step procedure was taken from Pfannkuch and Labno[3] who described several considerations which form the basis on which the rational design of a monitoring system can be achieved. These considerations include: (1) the overall management objectives; (2) the particular monitoring system objectives; and (3) the constraints imposed by the physical and financial conditions. Figure 8.2 indicates how these considerations relate to monitoring system design.[3]

Nacht[4] also emphasized the need for a comprehensive study of existing information and delineation of the goals and objectives of a monitoring network prior to the initiation of physical sampling and analysis. The methodology is presented in the form of a flow chart in Figure 8.3.[4] Lewis[5] reiterated the point that initial study is vital to monitoring network design. His three-phased approach to monitoring network design included: (1) collection and analysis of existing data; (2) test drilling; and (3) compilation of existing information to determine monitoring well design. Finally, LeGrand[6] stated

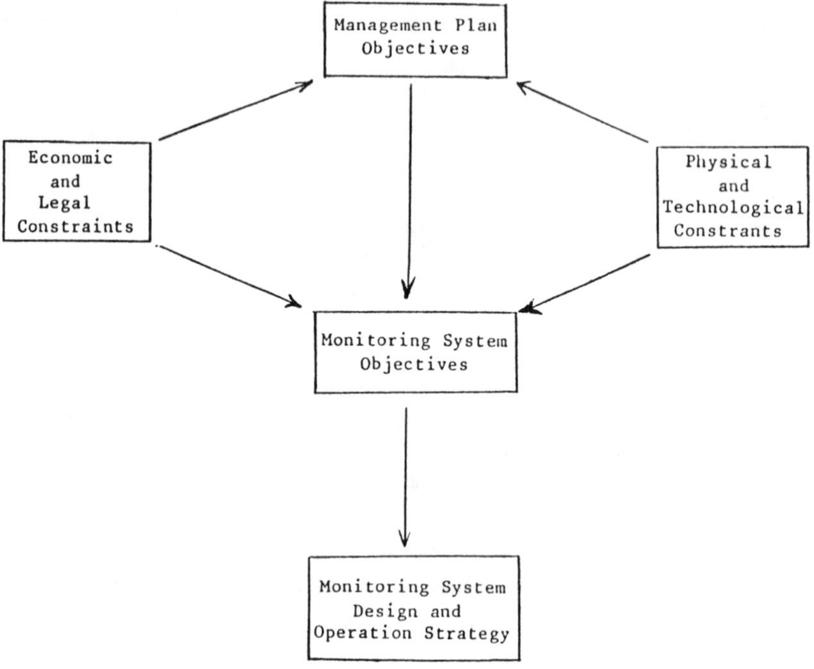

Figure 8.2. Evolution of monitoring system design.[3]

that planning a monitoring program should be based on a preliminary evaluation of the hydrogeologic framework, mobility of particular contaminants, and manmade events causing movement of contaminated water. Once again, the need for preliminary study is stressed.

Reasons for Monitoring

One of the above-listed questions for which an answer is usually provided prior to the development of a monitoring network is, What are the reasons for developing the monitoring program? The most common reasons cited are to meet statutory requirements, or to satisfy legislated mandates. As an example, Roy and Drake[7] point out several water quality programs or laws or potential ground water pollution sources in Massachusetts that now do, or soon will, require ground water monitoring. These include: (1) ground water withdrawal and discharge permitting; (2) landfill operations; (3) the Safe Drinking Water Act (SDWA); (4) the Resource Conservation and Recovery

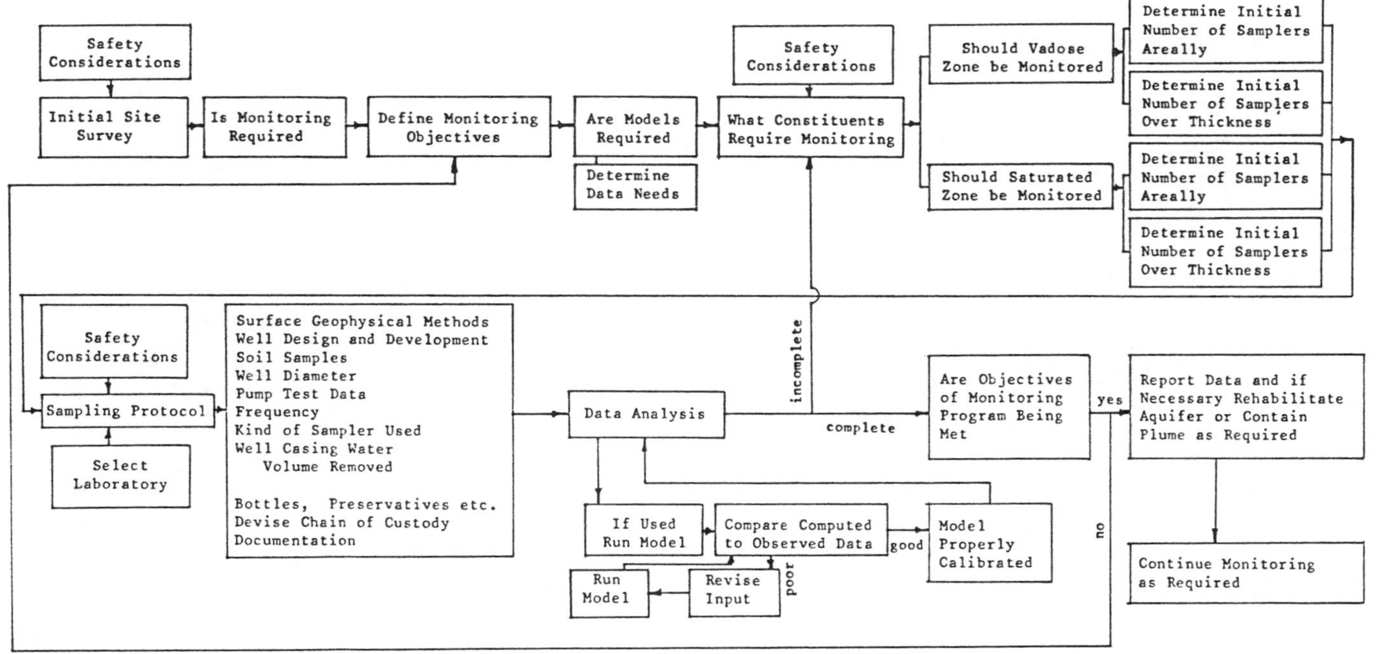

Figure 8.3. Flow chart for a typical monitoring program.[4]

Note: The sequence of events and what is included will vary with the needs and objectives of the monitoring program.

Act (RCRA); (5) the Underground Injection Control Program (UIC); and (6) emergency response efforts. Roy and Drake[7] also identified seven additional reasons, programs, or areas that could require ground water monitoring: (1) monitoring to determine the extent of contamination at hazardous waste sites; (2) monitoring to comply with hazardous waste regulations; (3) wetlands monitoring; (4) aquifer testing for water resources planning; (5) environmental impact review monitoring; (6) regional background water quality monitoring; and (7) research and development monitoring.

Nacht[4] in his discussion on when ground water monitoring should be undertaken, cited nine circumstances, not necessarily based on regulations, that might dictate conduction of ground water monitoring. These include:

1. Location of industrial or commercial facilities handling large hazardous materials or wastes, even if these materials are not listed under RCRA. For example, polychlorinated biphenyls, which are listed under the Toxic Substances Control Act (TSCA).
2. Facilities above heavily used or drinking water aquifers.
3. Geologic materials that are highly fractured or have solution openings where attenuation of pollutants will be minimal and pollutant movement can be rapid and difficult to predict.
4. Facilities located in communities concerned about the effects of industrial activity on the human population.
5. Facilities that have experienced spills or other accidents, or those in which water pollution control devices have a history of failure resulting in a release of pollutants to surface water or ground water. This may be due to equipment malfunction, overloads, or large swings in process wastewater quality.
6. Facilities which must comply with environmental or health regulations or which are undergoing legal proceedings involving pollution, health or similar problems, especially involving solid materials or water.
7. Facilities using or managing compounds, the chemical or health effect properties of which may not be well known.
8. Complaints by neighbors that the local surface water bodies or wells have degraded in quality or are producing foul odors.
9. An observable increase in health problems in the area.

Objectives of Monitoring

As seen from the above, the reasons for developing a monitoring network can be many and varied. It is important that the reasons for developing a monitoring network be identified and kept in perspective for the duration of the monitoring program. Another aspect of monitoring network planning that is closely related to the reasons for monitoring is the objective(s) of the monitoring program. Arguments can be made that the reasons and the objectives are identical. For example, the reasons for monitoring may be to satisfy a legal requirement, hence, one can argue that the objective of the network would be to satisfy the legal requirement. In actuality, the objectives of a monitoring network should be defined much more precisely.

Todd, et al.[1] stated: " . . . the objective of a monitoring program should be to collect, manage, and analyze the data on ground water quality and the sources and causes of ground water pollution, and the other information—geologic, hydrologic, and economic—necessary to enable the EPA and the state(s) involved to fulfill their statutory responsibilities as regards protection of ground water quality. . . ." This type of "objective" statement may be too broad. Voytek,[8] on the other hand, identified the following three objectives of a monitoring well: (1) to provide access to ground water; (2) to determine which pollutants are present and in what concentrations; and (3) to determine the distribution of pollutants both areally and vertically. This type of "objective" statement is possibly too narrow in scope. First, the objectives are related only to monitoring wells—there are many monitoring technologies other than wells. Second, the objectives are related solely to describing physical aspects of the problem. Ideally, a complete objective definition should include statements concerning what technologies are going to be used (for example, monitoring wells); what information the technologies will provide (for example, spatial distribution of a contaminant); and why that information is needed or how it will be used (for example, to design remedial measures).

BASINWIDE AMBIENT TREND MONITORING

A basinwide ambient trend monitoring network is one which is concerned with transient background quality of a given ground water basin or aquifer. Because this type of network is concerned with existing ground water quality, monitoring in the vadose (unsaturated) zone is minimal. The main concerns in planning this type of network are: (1) the number of wells or springs to be included in the network; (2) the spacing and location of the wells and springs; (3) the optimum sampling frequency; (4) the indicator parameters to be used; and (5) data storage, presentation, and interpretation. This section of Chapter 8 will summarize several related studies and delineate several considerations for designing and operating basinwide ambient trend networks.

Examples of State Programs

Most states address their long-term ground water monitoring programs and summarize the results in their biennial water quality inventories as mandated by the Federal Water Pollution Control Act (Section 305 (b) PL 95-217). Many states have already developed impressive sets of data through years of previous monitoring. A few state examples, including both basinwide and source monitoring, are highlighted as follows.

California (California State Water Resources Control Board, 1982).[9]

A statewide monitoring program was initially developed in 1974 with the designation of some 24 "Priority I" ground water basins out of about 500 ground water basins classified statewide. The classification was based on population, average use rate, water quality problems, estimated usable capacity, and the availability of an alternate water source. There are 180 basins classified as Priority II and 300 basins classified as Priority III.

Development of the monitoring networks began with pilot ground water monitoring networks in 4 basins: the 3 small basins lumped together as the Eureka Basin, the Santa Clara Valley, the Kings-Tulare Basin, and the Santa Clara River Valley in Ventura County. The inventory of existing wells and design of the networks were performed by the Department of Water Resources (DWR) in 1977 and monitoring by DWR was started in 1978. Monitoring has continued on an annual basis in these four basins. The results of the first year's sampling showed that concentrations of common minerals were comparable to historical values. High total dissolved solids and sulfates are characteristics of ground water quality in the Santa Clara River Valley, as well as in aquifers above the Corcoran Clay zone of the Tulare Basin. In general, mineral constituents in the other basins were well within acceptable limits.

Ground water monitoring in 1980 and 1981 was also conducted for heavy metals and nutrients, items for which there is little historical data. The reported concentrations for some of these chemical groups were higher than expected. Some equaled or exceeded the Maximum Contaminant Levels (MCL) of the state's Primary Drinking Water Standards. The wells in which MCL's were exceeded were reported to the State Department of Health Services for appropriate health protection action.

At present a structured, step-wise program is used for the development of the monitoring networks for the remaining Priority I ground water basins. For each basin the development consists of the following steps:

Phase I—an inventory of the number and characteristics of the active monitoring networks in the basin,

Phase II—detailed well-by-well inventories for each active well in the networks including well characteristics and monitoring frequency; and

Phase III—network design using existing networks and wells (when possible) and selection of monitoring parameters and frequency.

The basin inventories and the design process are being accomplished under agreements with the USGS.

Following completion of the network design the final two steps in the process are network implementation and network operation. The network implementation phase coordinates the monitoring activities and information retrieval of the agencies and/or individuals involved in the monitoring network. In the network operation phase the Regional Boards, with assistance from DWR, act as the central data collection agency for information from local agencies.

Kansas (Kansas Department of Health and Environment, 1982).[10]

Water quality data are collected from a network of wells to determine the chemical characteristics of ground water in the principal aquifers and evaluate the chemical quality data to determine the adequacy of the network for describing baseline ground water quality, to detect pollution of the principal aquifers in the state, and to determine the significance of the data in respect to state and federal water quality standards imposed by the Safe Drinking Water Act. Data have been obtained at 625 wells in the statewide water quality network or roughly about 6 wells per county. Data determined at these sites include water temperature, specific conductance, pH, and inorganic constituents. Organic constituents also are determined at selected sites. During the period 1976–80, 1,669 chemical analyses were made; 736 samples were tested for metals and 106 samples were examined for total organic carbon, pesticides and radiochemicals. This is a cooperative program between KDHE and the USGS. Each year the 525 water supplies dependent on ground water as a source of supply have a laboratory chemical analysis of a sample taken from the distribution system, and every 3 years the water is tested for presence of heavy metals. This sampling has been done for many years and is an important part of the statewide monitoring network.

The Kansas Geological Survey water investigations, in cooperation with the USGS, have focused on ground water. Present research is directed to assessment of major ground water systems, development of predictive quantitative and qualitative ground water models, geophysical and geochemical investigations, saline systems, acquisition and management of ground water data systems, and problems such as liquid waste disposal, hydrological effects of mining, and economics of ground water and energy.

New Jersey (New Jersey Department of Environmental Protection, 1982).[11]

There are several government agencies engaged in some form of data collection related to ground water in New Jersey. Examples are briefly described below.

USEPA, Region II, Surveillance and Analyses Division—The major purpose of the EPA monitoring program is to characterize ground water contamination associated with hazardous waste sites. Monitoring is undertaken at selected hazardous waste sites located throughout the state. Sites having the greatest priority are sampled regularly and the remaining sites are sampled at intervals determined by their potential hazard. The chemical parameters examined consist of 129 priority pollutants. The information generated by the EPA ground water monitoring program is stored in the STORET computer data base system, where it is readily accessible for tabulation or statistical calculations.

U.S. Geological Survey, Water Resources Division (USGS/WRD), Trenton District Office—The USGS conducts two types of ground water monitoring programs: long-term basic gathering programs and short-term special projects. All programs are conducted as joint, cooperative programs between the USGS and the Division of Water Resources, NJDEP. The 2 types of data gathering programs are designed to monitor and evaluate the quality and quantity aspects of the ground water supplies within the state. Short-term special projects are conducted in response to ground water contamination or supply problems discovered during analysis of data generated by the 2 existing long-term monitoring programs: the Salt Water Encroachment Monitoring Network and the Synoptic Water Level Monitoring Program. The Salt Water Encroachment Monitoring Network consists of approximately 500 wells located in 9 coastal and estuarine counties of the New Jersey Coastal Plain. The Synoptic Water Level Monitoring Program consists of a network of 175 wells distributed throughout the New Jersey Coastal Plain. The program's purpose is to document changes in the piezometric water levels of the major aquifers.

Delaware River Basin Commission (DRBC)—The DRBC monitoring program examines ground water quality and quantity at specific sites within the Delaware River drainage basin. The DRBC collects water level information on approximately 1,800 wells, usually on a monthly basis. Information on the water level, latitude/longitude, and the monitored aquifer is then stored in DRBC files. Ambient ground water information is usually collected only for specific projects.

NJDEP, DWR, Bureau of Potable Water—The Bureau of Potable Water has a threefold ground water monitoring program: when requested, drinking water wells (tap water) are tested for contamination; tests of raw water from new ground water public supplies; and inspection of all water supply company wells and municipal supply wells. The bureau collects approximately 300 to 500 ground water samples per year. Generally, these wells are sampled for pH, hardness, heavy metals and volatile organics.

NJDEP, Division of Waste Management—The purpose of the Division of Waste Management program is to monitor water quality around nonhazardous landfills and to report potential and actual ground water contamination to DEP's enforcement elements. Samples are collected at 4 to 6 landfills each month in response to complaints, to verify the analytical results from a private laboratory submitting a landfill compliance monitoring report, or to test for organic compounds. The division also conducts a permit program which requires all landfills to install wells and conduct compliance monitoring.

NJDEP, Office of Cancer and Toxic Substances Research—One purpose of the office of Cancer and Toxic Substances Research monitoring program is to examine the state's ground water for toxic chemicals. The data generated by this office is being entered into the STORET system.

NJ Department of Transportation, Bureau of Quality Control—The Bureau of Quality Control conducts specific ground water tests at DOT construction sites in compliance with existing construction regulations. Only those parameters likely to be present as a result of construction are monitored.

County Agencies—In addition to the above state and federal ground water monitoring programs, 8 county agencies are planning or have instituted ambient ground water quality monitoring programs, primarily under the auspices of the County Environmental Health Act.

Pennsylvania (Pennsylvania Department of Environmental Resources, 1982).[12]

Ground water quantity investigation began in Pennsylvania in 1925 under a cooperative agreement between the Pennsylvania Geological Survey and the Ground Water Branch of the U.S. Geological Survey. The agreement continued to 1943, by which time, all parts of the state had been covered by reconnaissance studies which were described in 6 area reports and 1 statewide report published by the Pennsylvania Geological Survey. In 1930, a statewide network of observation wells was established to determine long-term trends of ground water levels in rural areas. Selected locations were largely unaffected by local ground water withdrawals. This study has been kept current and in recent years it has been greatly expanded to include nearly every county in the state.

The Ground Water Quality Monitoring Program in Pennsylvania consists principally of public water supply, facility and pollution source monitoring. Public water supply wells and springs are sampled before they are put into service and annually thereafter. Facility monitoring involves the measurement of effluent quantity and quality at permitted facilities equipped with ground water observation points. These points are strategically located around many facilities such as landfills, coal mining areas, industrial waste impoundments, hazardous waste disposal areas, sewage lagoons, spray irrigation sites, etc., with a ground water contamination potential. Pollution monitoring utilizes short-term sampling to gather evidence for enforcement actions of past, existing, or anticipated ground water pollution situations. Data is collected on a regular basis to study ground water quality trends and, as required, for evidence in enforcement actions.

Ambient trend monitoring concerns measurement of ground water quality and deviations, and involves temporal and spatial trends within a ground water basin or area. The existing program consists of sampling a network of public water supply wells and springs and analyzing the samples for a set number of parameters. In the past, information concerning this network of public water supplies was maintained on the Ground-Water Quality Management Information System (G/WQMIS) which utilizes the U.S. Environmental Protection Agency's STORET computer system. However, because of problems with retrieval from G/WQMIS, a new system known as the Ground-Water Quality Information Management System (GW-QIMS) has been developed.

The development and implementation of a regional Ground Water Quality Monitoring program has been initiated under a federally funded 208 grant. The monitoring strategy includes identification and selection of monitoring techniques, ground water basin evaluation and prioritization, location of specific monitoring points in high priority basins, and a cost assessment for implementation purposes. Operation of the program will be facilitated by the development of ground water quality standards and the delineation of ground water basins.

South Dakota (South Dakota Department of Water and Natural Resources, 1980).[13]

There are several sources of ground water quality data in South Dakota. Each of these will be briefly described as follows:

Office of Water Rights—The Office of Water Rights is within the South Dakota Department of Water and Natural Resources (DWNR) and maintains roughly 1200 observation wells within the state. In addition to geologic logs and water level data for each observation well, wells are also sampled every 3 to 5 years for water quality. The Office of Water Rights also maintains files on approximately 1000 irrigation wells. Data on these wells include driller's logs and an initial water quality analysis.

South Dakota Geological Survey (SDGS)—The SDGS is also part of DWNR and has considerable ground water quality data from their numerous county geologic and hydrologic studies, city studies and others. It is estimated the SDGS has water quality data for up to 5000 wells.

U.S. Geological Survey (USGS)—The USGS maintains 193 observation wells in bedrock aquifers within the state (about 3 per county). These have all been sampled at least once and often more for water quality. In addition, the USGS maintains about 60 observation wells in shallow aquifers. These 60 wells, however, are maintained primarily for water level data and may or may not include water quality data. As part of the cooperative county studies with the SDGS, the USGS samples about 20 to 50 private wells for water quality within each county being studied at the time of the study. The older SDGS county reports often include at least some of this data but the more recent SDGS county reports do not.

Office of Drinking Water—The Office of Drinking Water is within the South Dakota Department of Water and Natural Resources and maintains water quality records on the 409 public water supplies within the state. The majority of these public water supplies are from ground water sources. Public water supplies are sampled at least monthly for bacteria (more often for larger supplies) and about every 2 years for common ions (NO_3, Ca, Mg, etc.). In addition, the EPA now requires that trace element analyses be taken from public ground water sources at least once every 3 years (and once a year from surface water sources). The Office of Drinking Water also maintains water quality records on private wells where the owner requests an analysis but these are generally sampled only for bacteria and nitrates.

National Uranium Resource Evaluation Survey (NURE)—As part of the federally sponsored NURE survey, ground water quality samples were collected from about 1200 wells within the state. The samples were primarily taken from private wells in bedrock aquifers (especially the Dakota aquifer). With respect to inorganics, these analyses are remarkably detailed and include analyses for 29 parameters.

Water Resources Institute—A number of studies which include some ground water quality data have been conducted by the Water Resources Institute at South Dakota State University. In addition, the Water Resources Institute has been maintaining water quality analyses from irrigation and stock wells since 1965.

The reviews presented above serve a number of purposes pertinent to this chapter. First, an outline is provided of some of the approaches that have previously been used for monitoring ground water over large areas. Second, reemphasized is the point that significant data, from a variety of sources, may already exist at a given location. Hence, the need for comprehensive preliminary studies is reinforced. Third, if scrutinized carefully, the reviews provide some general network design criteria. For example, the USGS observation wells in South Dakota are spaced to average about three wells per county. These and additional criteria are outlined in the next section.

Design and Operation of Networks

There are three basic aspects to the design and operation of an ambient trend monitoring network. The first critical aspect involves deciding which existing wells and/or springs should be included to form the basic network. Second, a decision needs to be made on the need for and locations of new (to-be-installed) monitoring points. The third aspect of network design involves data handling, storage, and interpretation. The general steps for developing and operating an ambient trend ground water monitoring network can be divided into three phases as depicted in Figure 8.4. Phase I can be termed the preliminary study phase; this phase is dominated by information gathering and interpretation. Phase II is the implementation phase in which new wells are installed and sampling and analysis programs are conducted. Phase III of the process is concerned with presenting and interpreting the data developed as a result of the analyses conducted in the second phase.

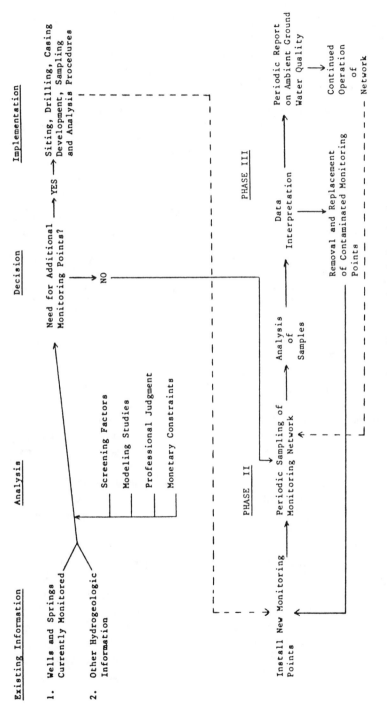

Figure 8.4. Flow chart for ambient trend monitoring planning.

The two major objectives of the preliminary study should be: (1) to assimilate all available information on existing wells and springs; and (2) from this information make a decision on which wells and/or springs should be included as part of the monitoring network. Comments related to additional considerations for Phase I are as follows:

1. It is essential that the preliminary search be comprehensive. As seen in the previous section, ground water information can be collected by several different agencies for a variety of purposes. At a minimum, information should be requested and procured from state water resources agencies, state pollution control agencies, and state and federal geological surveys.
2. Provisions should be made in study budgets and work forces to devote significant time and effort to the collation of gathered information. More specifically, it should be anticipated that existing information will be available in a variety of forms covering, or possibly not covering, many different parameters. Some wells will have water-level data only; others will have water quality analyses. Some wells may have extensive records; others will only be characterized by recent data. Common information, pertinent to the objectives of the ambient trend monitoring network, will need to be sorted, collated, and compiled into a uniform format.
3. Not all useful information will be from monitoring well analyses. Useful information about existing subsurface conditions can also be obtained from soil maps, well logs, and/or geophysical surveys.

After obtaining and collating existing information, decisions need to be made as to what information is useful; and specifically, which wells and/or springs should be utilized in the monitoring network. Clark and Trippler[14] outlined the following screening factors used for the selection of wells and/or springs for the Minnesota ambient ground water quality monitoring program: (1) in conjunction with available data, the wells or springs should provide a statewide (basinwide) overview of ground water quality with respect to naturally occurring constituents and contaminants; (2) wells or springs should be selected to include all principal aquifers, with an emphasis proportional to present use and availability of alternative water supplies; (3) the network as a whole should be integrated with other water resources data networks and projects; and (4) the network should provide data on water quality for studies of regional significance, such as those associated with areas of Karst or intensive irrigation.

Clark and Trippler[14] also described three major elements in their monitoring network: (1) point sampling which refers to nonrepetitive sampling from a single well; (2) point monitoring which involves repetitive sampling from a single well to detect changes in quality; and (3) regional monitoring which involves repeated sampling of many wells to define spatial and temporal trends in water quality. The criteria used by Clark and Trippler[14] for including a well as part of the point sampling network were: (1) points should be systematically distributed with respect to the regional flow system, where known; (2) points should be selected so that analytical data will refine the definition of baseline quality and areal changes in water quality; (3) points should be part of present water resources data bases, if possible; and (4) points should be frequently used, to help assure that samples are representative.

An attempt was made to locate at least one sample every 1,000 square miles (2,590 km^2), or approximately one per county. An attempt was also made to ensure that at least three samples would be taken in each of Minnesota's watersheds.

The wells included for point monitoring had to meet one of the following requirements:[14] (1) points should be selected within cones of depression in major metropolitan area pumping centers; (2) points should be selected from areas where many wells of a single type, such as irrigation wells, are concentrated; (3) points should be selected from hydrogeologically sensitive areas such as the Karst areas of the southeast or the sand-plain areas of west central Minnesota; and (4) points should be selected from areas subjected to induced or artificial recharge from surface water.

Once a well was identified as satisfying the criteria for either point sampling or point monitoring, there were several well data requirements described as essential to the development of a sound data base.[14] These requirements were:[14] (1) adequate information must be available to enable field personnel to positively identify the well; (2) a geologic log of the well borehole must be available to identify the sequence of geologic materials penetrated by the well; (3) well construction information must be provided, including depth drilled, depth of casing, casing material and diameter, length of open hole or type and length of screen, to assure that the well isolates a single aquifer; and (4) reasonable assurance that the well can be sampled must be provided.

If a well is chosen for inclusion in the ambient monitoring program there are several specific points of information regarding each well which must be obtained at or before the time of sampling:[14]

1. STORET site identifier, necessary for entry of chemical data into the federal water quality management system
2. static water level and date of measurement
3. altitude of water-level measuring point (usually estimated from a topographic map)
4. description of the sampling point, including identification of any nearby pollution sources
5. use of the water
6. estimated pumping rate and, if available, annual pumpage
7. unique well number

The probability that existing wells and/or springs completely and satisfactorily meet the needs of an ambient trend monitoring network decrease as the size of the basin or aquifer being studied increases. Although it is highly desirable, from an economic standpoint, to keep the number of new wells to be drilled at a minimum, almost inevitably some new monitoring wells will have to be developed. The need for and siting of these new wells represent two critical decisions to be made in planning a network. Several analytical tools are available to aid in this decision-making process. Numerical ground water models, because of their intensive data requirements, can aid in assessing those areas with insufficient data and can optimize the placement of new wells based on existing data. For example, Moore and McLaughlin[15] discussed a statistical method, borrowed from the mining industry, called "Kriging," which is finding increased popularity in ground water contamination studies for locating monitoring points.

Should it be decided to place new monitoring wells at a given location, planning for well design and construction should take into consideration the possible factors that could interfere with ground water quality monitoring. Walker[16] provides the following recommendations to avoid installation trauma (anomalous initial quality data):

1. To minimize the extent and duration of installation trauma, consider the following to improve installation procedures:
   (a) minimize or eliminate the introduction of water in installing the well; do not use bentonite or other drilling fluids unless absolutely needed;
   (b) if water is needed, attempt to use water with a pH similar to that anticipated of the natural ground water, based on localized knowledge (U.S. Soil Conservation Service, Geological Surveys, etc.);
   (c) exercise extra care in constructing well seals; use multiple seals, especially for deep wells in low-yield media; and
   (d) extend the period of development pumping, especially if the well is developed using surging techniques by which air or other foreign substances may have been introduced into the host medium (soil or rock) and the ground water; however, the pumping rate must be controlled to limit drawdown so that differential pressures on the seals are minimized.
2. To detect the occurrence of installation trauma, review the water quality data carefully and look for:
   (a) trends of increases or decreases in analyte concentrations;
   (b) relationships of trends among analytes, such as between base metals (manganese, iron, copper, etc.) and physico-chemical parameters (Eh, pH, etc.); and between heavy metals (barium, lead, etc.) and organic carbon (TOC); and
   (c) changes in ratios between geochemically related analytes (Mg/Ca, Mn/Fe, Cu/Pb, etc.)
3. To reduce the level of existing trauma, consider:
   (a) slow pumping from the affected well over an extended period, thereby keeping drawdown to a minimum; and
   (b) replace wells that display extreme or slowly dissipating trauma.

Having identified a monitoring network consisting of new and existing wells and/or springs, the primary activities now become sampling and analysis and data reporting. The frequency at which wells must be sampled must strike a balance between the desire for comprehensive data and costs. From the earlier discussion of examples of state programs, it can be concluded that sampling frequency should be different depending on the parameters to be monitored. Some suggested general guidelines are as follows:

1. At least once a year and no more than twice a year, each sampling point should be subjected to routine determinations for the parameters indicative of the naturally occurring constituents of the subsurface such as calcium, sodium, alkalinity, sulfate, chloride, nitrate, phosphate, pH, and total dissolved solids.
2. At a less frequent interval (no greater than once a year), special analyses should be performed on each sampling point for those parameters indicative of pollution such as chemical oxygen demand, total organic carbon, synthetic organics, and heavy metals.

It is important to note that sampling too often is just as misleading as not sampling often enough. In short, large numbers of samples produce large amounts of data which can have a lot of "noise" and be hard to interpret. On the other hand, infrequent sampling may lead to nondetection of an important subsurface water quality stress.

During the operation of the network it is imperative that there exist provisions for removing and/or replacing wells that show contamination if the objective of the network is to provide information on background water quality. Any well that shows signs of contamination, such as high concentrations of synthetic organic chemicals should be removed from the network and its recent data stricken from the record. The well need not be abandoned, as it may prove valuable for inclusion in a source monitoring network.

SOURCE MONITORING

The second type of monitoring network is a source assessment network. This type of network is concerned with determining the existence and magnitude of ground water stresses due to localized sources. This is the most common type of network, with the number of previous applications in the thousands and growing by hundreds every year. Figure 8.5 displays a flow chart for planning a source monitoring network.

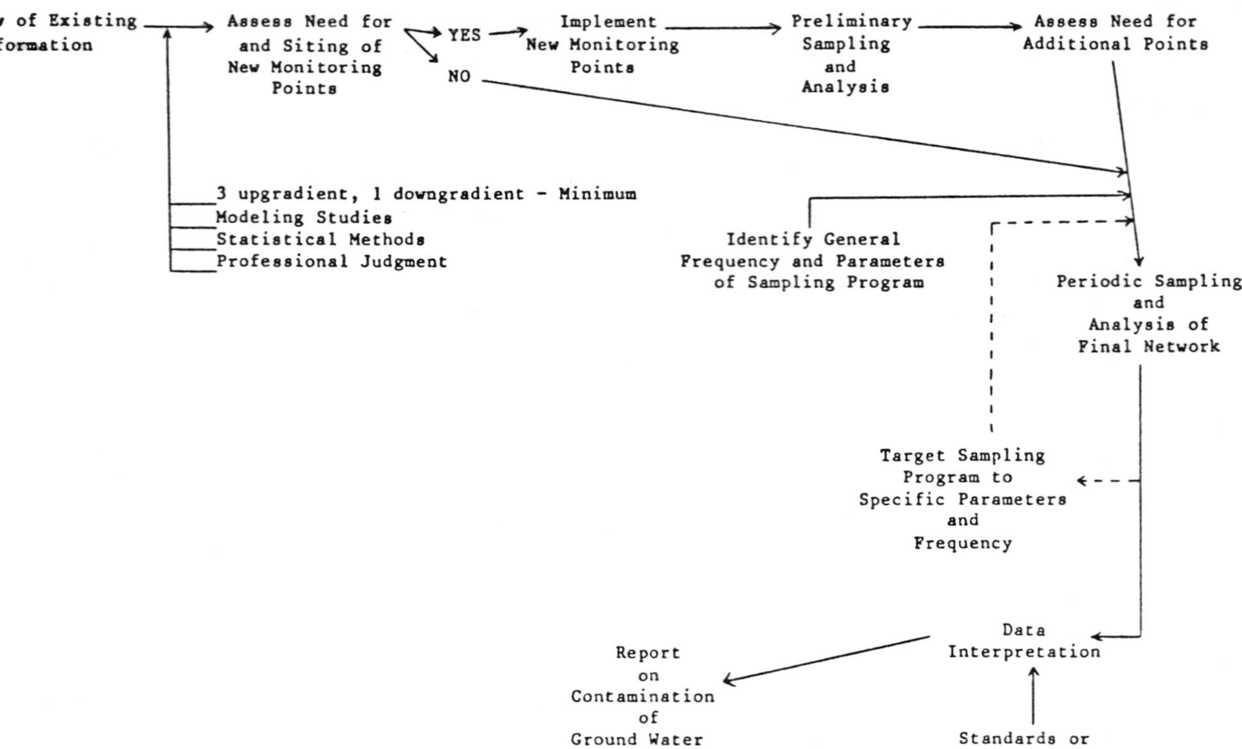

Figure 8.5. Flow chart for development of source monitoring network.

Design of Networks

As discussed previously, the design of any monitoring network should be preceded by a comprehensive preliminary study. In the case of potential and/or existing sources of ground water pollution, information needs to be gathered on both the sources and the subsurface environment. Gillham, et al.[17] have listed the following five stages of preliminary study needed before implementing a source monitoring network:

1. Characterization of the waste:
   (a) determine the physical characteristics of the waste solutions;
   (b) identify the potential contaminant species, and other species that might serve as ground water tracers; and
   (c) evaluate the geochemical characteristics of the contaminant species.

   Much of this information may be available from the producer of the wastes, or specific tests may be required. This is an important first step in that, depending upon the characteristics of the waste, it may serve to limit the extent (and cost) of subsequent field testing.
2. Review of regional hydrology:
   (a) identify major hydrostratigraphic units;
   (b) characterize the respective units, particularly with respect to hydraulic conductivity, type of media (porous, fractured or fractured porous) and the type(s) of heterogeneity; and
   (c) determine the regional ground water flow characteristics (recharge and discharge areas, directions of flow, flow-system scales, and estimate velocities).

   This information can often be obtained from geologic and hydrologic reports, existing well logs, aerial photographs, satellite imagery, soil maps, and reports of other agencies such as local conservation authorities.
3. Preliminary study of local hydrogeology:

   The pertinent questions at this stage of the investigation are similar to those of the regional hydrology review study; however, the information should be obtained at a much greater level of detail. Though information may be obtained from existing reports, in general, detailed field testing will be required. The initial field studies should include geologic test drilling and the installation of piezometers for monitoring hydraulic head, for measuring hydraulic conductivity and for providing an initial survey of the geochemical conditions. Geophysical surveys may also play a useful role at this stage.
4. Characterization of local hydrogeology:

   The field investigations of the previous stage may become an iterative procedure in order to develop stratigraphic interpretations and hydraulic-head distributions that are consistent over the region of detailed interest and within the regional trends.
5. Design of monitoring network:

   Design and installation of the detailed monitoring network should proceed based on the results of stages 1 through 4. Significant considerations should include the vertical and horizontal spacing of intake zones and the length of individual intake zones. These decisions will be based largely on the distribution and extent of individual geologic units. The anticipated extent of the contaminant plume, based on the chemical and geochemical characteristics of the contaminants, is also an important consideration.

To serve as an example of source monitoring network planning, Cook[18] presented a review of some preliminary steps taken at an industrial site in Massachusetts. The steps included: (1) identification of potential chemicals from historic industries; (2) assemblage of all geology and hydrology information with field visits to fill in the gaps in the literature; (3) review of drillers' logs; and (4) conduction of seismic surveys.

Pfannkuch and Labno,[3] in their five-phased procedure for ground water monitoring, described the first phase as being the Preliminary Network Design and Information Gathering Phase. In this phase, there are three areas of information that could influence the design of the monitoring network: (1) infiltration from the pollution source to the subsurface; (2) percolation through the unsaturated zone; and (3) transport and dispersion of disposed substances within the ground water flow system. Clarke, Mutch, and Brother[19] also pointed out the need for assessing pollutant mobility by noting that chemical transport information can be combined with preliminary testing for water quality indicators, GC organic scans and selected heavy metals to design a monitoring program which is cost-effective.

The actual areas of emphasis in a source monitoring network preliminary study can be many and varied. For example, Knox, et al.[20] emphasized the need for a preliminary study prior to the development and implementation of remedial action plans. The possible areas of study and reasons for their inclusion are as follows:[20]

1. Physical/Chemical Characterization of Source—A complete physical/chemical characterization of a waste can be used to predict potential pollutants.
2. Variability of Wastes—The variability of the wastes at a site must be considered in order that a pollutant-specific monitoring program is not implemented for a highly variable waste. A highly variable waste source would include a landfill that accepted both hazardous and nonhazardous constituents.
3. Time Factors—Data on the period a waste source has been in existence can give insight to the magnitude of the problem and aid in siting new wells.
4. Physical/Chemical Characterization of Pollutant(s)—A complete physical/chemical characterization of the known or suspected pollutant(s) will aid in assessing their potential mobilities and appropriate analysis techniques. An example would be the case of hydrocarbons which tend to "float" on top of phreatic aquifers.
5. Information on Transport and Fate of Pollutant(s)—Having characterized the pollutant(s), it is then necessary to obtain information on the transport and fate of these pollutant(s) in the subsurface environment. Information on the attenuating capacity of the given soils for the pollutant(s) is important. The behavior of the pollutant(s) under the different pH environments of the subsurface is important in that some pollutants can precipitate or solubilize under particular conditions. The final outcome should be information on the ability of the particular pollutant(s) to actually migrate

through the soil structure and reach the ground water. Feasible mitigation measures will be a function of this ability.

6. Areal Extent, Depth, Amount of Pollutant(s)—If possible, an estimate of the areal extent, depth, and amount of pollutant(s) in the subsurface environment needs to be obtained. The objective is to obtain an estimate of the magnitude of the problem. Ideally, data from existing monitoring wells, well logs, etc., should be used to estimate the magnitude of the problem. However, this information is often not available. In most cases an "educated guess" of the magnitude of the problem will have to be made by considering broad data such as soil types, topography, climatological features, and the duration of the problem.
7. Geologic Setting and Generalized Soil Profiles—Determination of the types of soils is important for ascertaining the capacity of the pollutant(s) to move through the subsurface. Certain soils will possess higher tendencies to attenuate the pollutant(s) (through adsorption, precipitation, filtration, etc.) than others.
8. Soil Physical/Chemical Characteristics—Once a general soil type has been identified, it is then necessary to characterize this type both physically and chemically. Physical characterization of the soil type will provide information on the ability of the soil to filter the pollutant(s). Chemical characterization will provide information on the ability of the soil to chemically remove a given pollutant through adsorption, precipitation, etc.
9. Depth to Ground Water—The depth to ground water in conjunction with the soil physical/chemical characterization will give insight as to how long it will take a pollutant(s) to actually reach the aquifer if, in fact, it will and provide basic information for determining drilling costs.
10. Ground Water Flow Patterns and Volumes—Although usually only obtainable after implementing monitoring operations, generalized ground water flow patterns and volumes can aid in siting new wells.
11. Aquifer Characteristics—Identification of aquifer characteristics will be essential for any analysis of ground water flow and pollutant transport. This information becomes extremely important if ground water modeling studies are to be initiated.
12. Existing Monitoring Well Locations and Procedures—Identification of existing monitoring well locations and the parameters monitored, and use of this information, can save both study time and costs.

Having reviewed available existing information, the next component of source monitoring network planning is to assess the need for and site new monitoring well locations. For example, Mooreland and Wood[21] point out that water level, water quality, and geologic information might indicate that some of the existing and/or initial monitoring wells are not ideally located to sample ground water most likely to be affected by a source. Hence, the need for new, accurately sited monitoring locations.

Probably the most important aspect of a source monitoring network is the siting and placement of new monitoring wells. The success of the network will depend heavily on whether or not samples representative of the subsurface environment can be obtained, hence it is vital that the finite number of monitoring points be situated such that they produce an accurate description of the subsurface. Obviously, the more monitoring wells placed around a source, the more certain one can be that they will detect on-site contamination and any off-site migration; these are two of the objectives for a source monitoring network outlined by Hagger and Clay.[22] However, there must be a balance between monitoring costs and comprehensiveness. For example, Absalon and Starr[23] point out that regulations under the Resource Conservation and Recovery Act (RCRA) require a minimum of four monitoring wells; one upgradient and three downgradient. Although good in intent, these regulations may be limited. First, it is not always easy to ascertain the hydraulic gradient at a site without first drilling boreholes. Second, as pointed out by Absalon and Starr,[23] the hydraulic gradient at a site can be very transient and actually reverse itself at times. For example, leakage from a landfill can cause localized mounding which masks the true hydraulic gradient if monitoring wells are placed too close to the source. Furthermore, a well intended to be downgradient at a source can be rendered ineffective if located within the cone of depression of a removal well.

Additional considerations related to siting new monitoring wells include the following:

1. Three down, one up—The initial source monitoring network should consist of three wells downgradient and one well upgradient as recommended by RCRA. However, one should attempt to spatially vary the three downgradient wells. In order to increase the "area" of aquifer monitored, the downgradient wells should be placed in a triangular arrangement. Moreover, this triangular arrangement should be skewed downgradient with two wells within the existing plume of contamination and the third one located downgradient from the leading edge of the plume. This arrangement should not only provide data on the spatial variation of ground water levels and contaminant concentrations, but should also allow for observation and/or calculation of plume migration rates by monitoring the time-varying concentrations of progressively downgradient wells.
2. Predicting Hydraulic Gradients—There are several methods for attempting to predict the nature of the subsurface hydraulic gradient given no or minimal existing information. Consideration can be given to regional geography and local topography.
   (a) Consider regional geography—Many states now have maps of major ground water formations including recharge areas. These maps can give insight as to the regional movement of the ground water and should be helpful for local sources.
   (b) Local topography—In many instances, ground water formations, especially shallow unconfined aquifers, follow a subdued version of surface topography. Hence, downgradient wells could be sited in areas topographically downgradient from the source, and vice versa for upgradient wells.
3. Predicting Plume Dimensions—Oftentimes the existence of a contaminant plume is documented by a contaminated well or wells. If additional wells need to be sited within or outside the plume, methods exist for aiding this process, including the use of geophysical techniques and field surveys.
   (a) Geophysical Techniques—Many geophysical techniques are now being utilized

to delineate areas of ground water contamination. These methods can aid in siting wells for monitoring plume movement. Chapter 9 contains information on geophysical techniques.

(b) Field Surveys—Prior to initiating any monitoring activities, a thorough field survey should be conducted. Areas of information from the field survey which could aid in siting monitoring wells may be subtle. Consideration should be given to areas of vegetative stress. Additionally, one should scan areas of seepage (if any) to detect odors and/or discoloration of rocks. All of these phenomena might be indicative of contaminated ground water and could give insight as to the general movement of the plume.

4. Unique Hydrogeology—When dealing with unique hydrogeological conditions, especially limestone or Karst-type aquifers, the general guidelines for monitoring granular type aquifers are best forgotten. Quinlan and Ewers[24] have developed a recommended monitoring strategy for carbonate terrains.
5. Additional Wells—An important consideration relative to a source monitoring network is that planning activities include provisions for additional monitoring wells. Budgets should include monies for installation, development, sampling and analysis for at least twice the number of proposed initial monitoring points. Rarely will four monitoring points be situated accurately and yield sufficient data for a successful source monitoring program.

Finally, depicted in Figures 8.6 through 8.14 is a number of hypothetical hydrogeologic landfill settings. Although only schematic and somewhat oversimplified, these figures serve to illustrate the general principles of leachate flow and could aid in the siting of monitoring wells for specific sources.[25]

Operation of Networks

The operation of a source monitoring network will consist mainly of sampling and analyzing water from the monitoring well. The need for an adequate sampling program is enforced by Seanor and Brannaka[26] who noted that advanced laboratory techniques are very costly and labor consuming. Chapter 9 provides additional information on ground water sampling.

There are three issues associated with the sampling of ground water monitoring wells. First, what are the appropriate sampling procedures or techniques? Second, what is the optimum sampling frequency? Third, what water quality parameters should be measured? Similar to the situation for basinwide ambient trend monitoring, the sampling frequency for a source monitoring network should be such that it is comprehensive without overstudying the site. Nacht[27] describes an analytical method for determining sample frequency based on a hazard rating for the pollutants involved. More general guidelines could be developed based on the fact that source monitoring is concerned with a highly localized area; hence, the sampling frequency should be great enough to detect the movement of the contaminant plume.

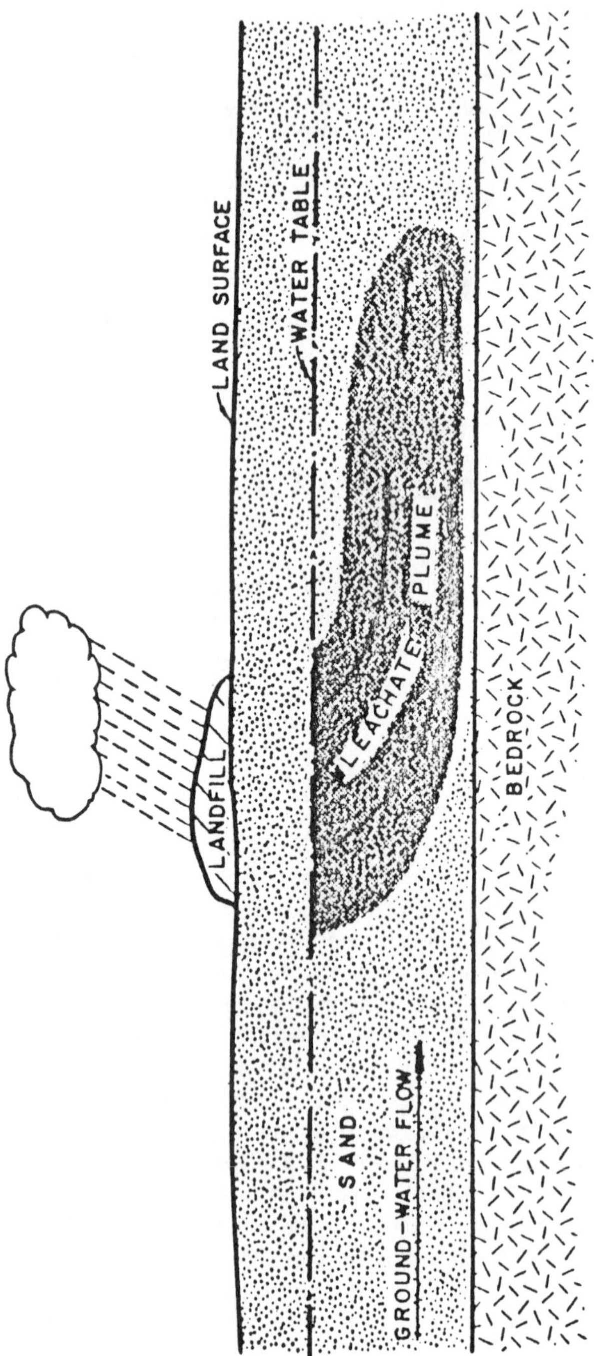

Figure 8.6. Single aquifer with a deep water table.[25]

Note: Leachate percolates downward from the landfill to the underlying aquifer and then moves downgradient as a bulb or plume in the direction of ground water flow. The mass of leachate may: 1) sink to the bottom of the aquifer if of a heavier specific gravity, or 2) float at or near the top of the water-bearing unit if the leachate is predominantly hydrocarbon in nature.

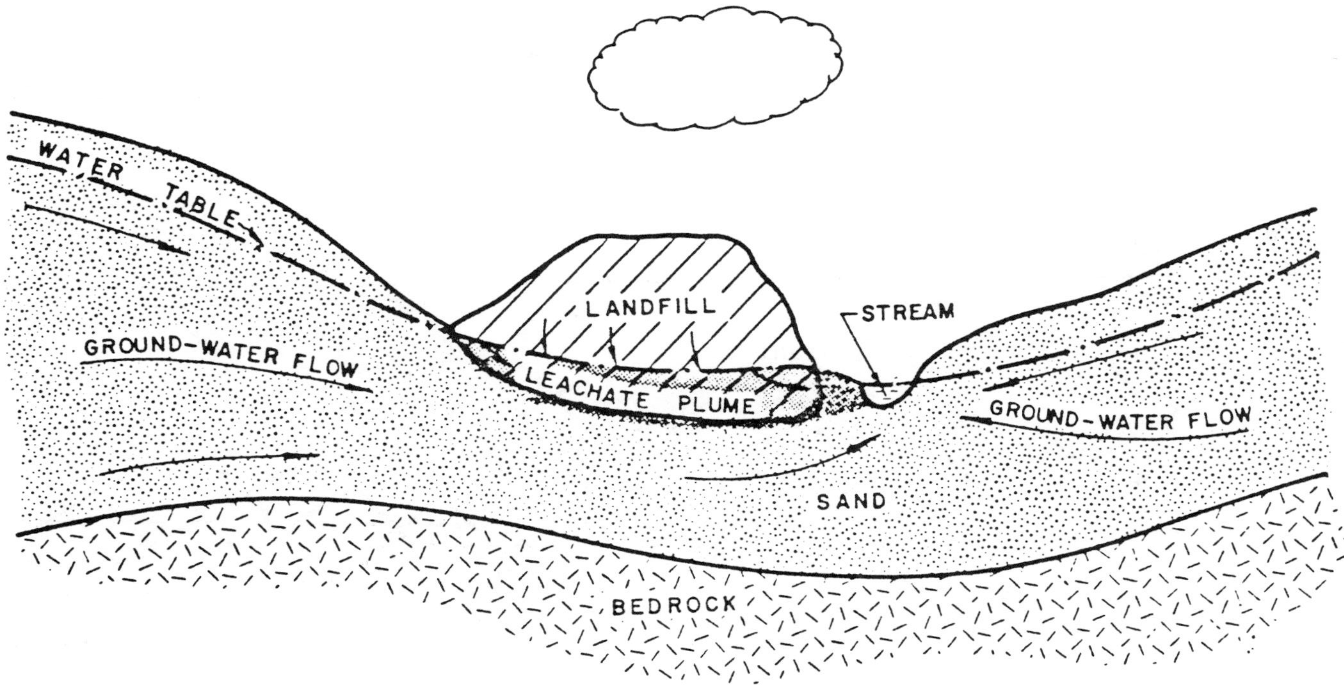

Figure 8.7. Groundwater discharge areas.[25]

Note: Landfills located within the zone of saturation are always in contact with ground water moving from topographically higher recharge areas to a stream discharge point. In such cases, leachate is transported with the ground water to the stream where it becomes diluted by surface water.

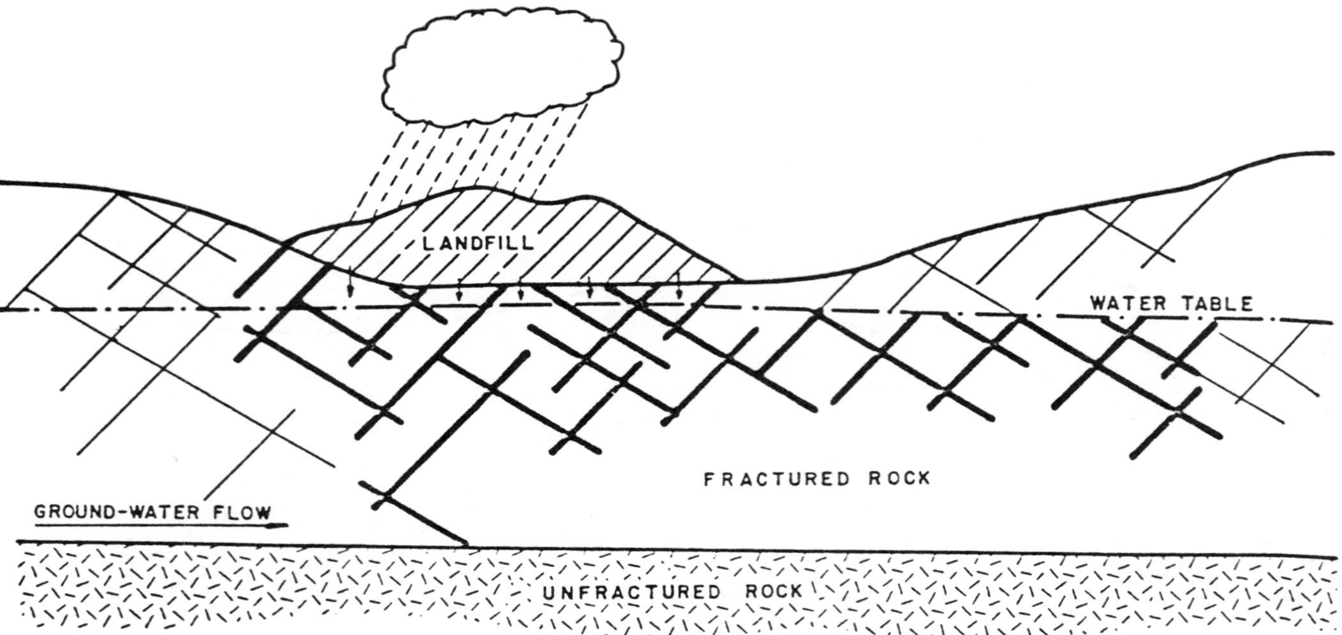

Figure 8.8. Fractured rock surface with a high water table.[25]

Note: Leachate migrates downgradient along interconnected rock fractures to some lower natural discharge area or a pumping well.

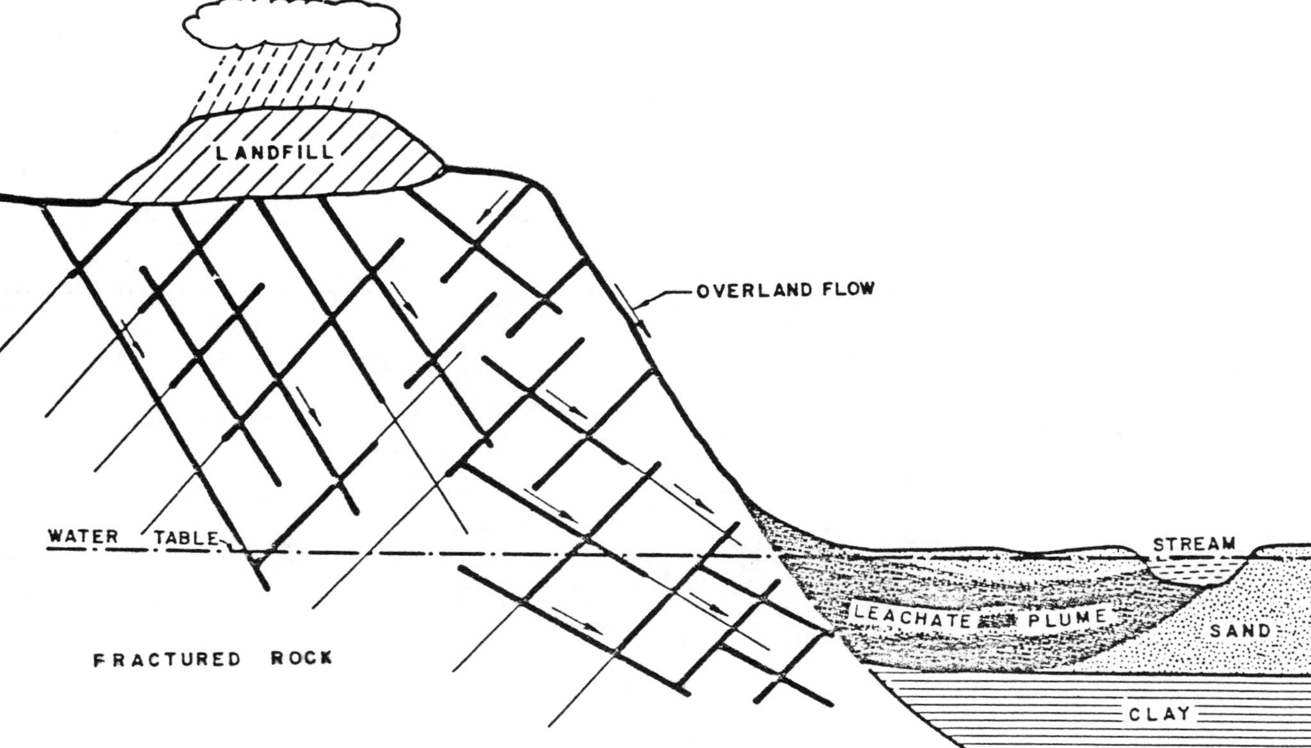

Figure 8.9. Fractured rock surface with a deep water table.[25]

Note: Leachate flows into and through interconnecting fractures and discharges either at the surface or into the subsurface where it moves with the ground water to some more distant discharge point.

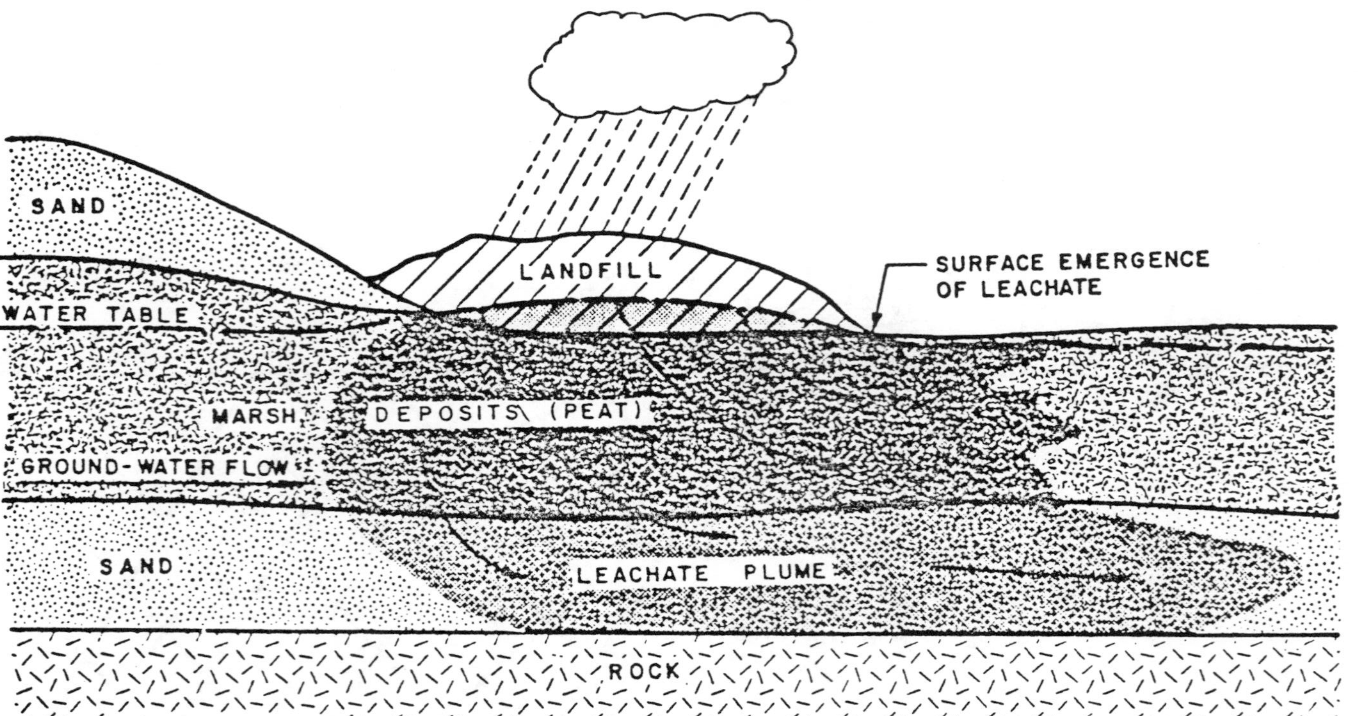

Figure 8.10. Marsh deposit underlain by an aquifer.[25]

Note: The water table is high, and a mound is formed at the base of the landfill. Leachate migrates downward through the marsh material to the aquifer. In many cases, surface emergence of leachate will occur at the toe of the slope. Some contaminants may be attenuated within the marsh deposits. The portion reaching the water table moves through the aquifer with the ground water to some surface discharge point.

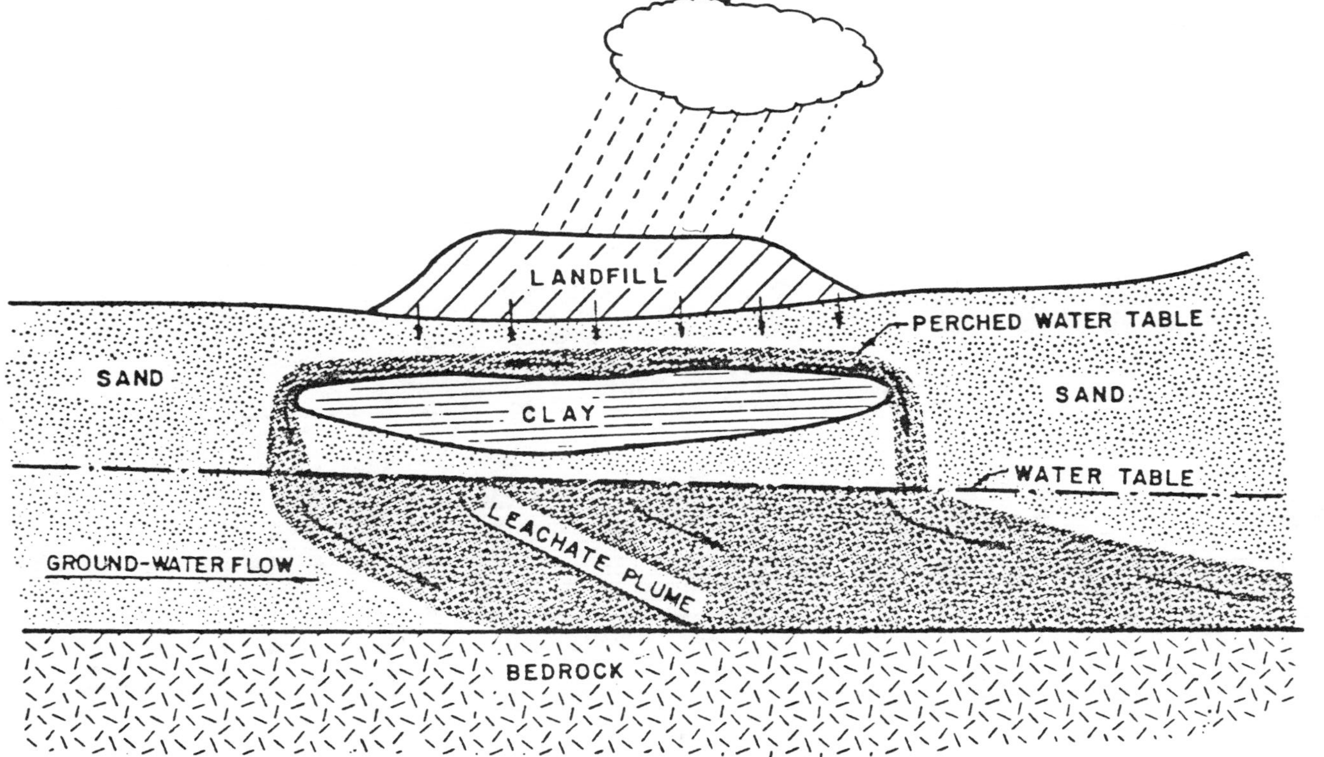

Figure 8.11. Permeable sand layer underlain by a clay layer.[25]

Note: The water table is deep. Leachate percolates downward under the landfill, forming a perched water table before finally reaching the actual water table.

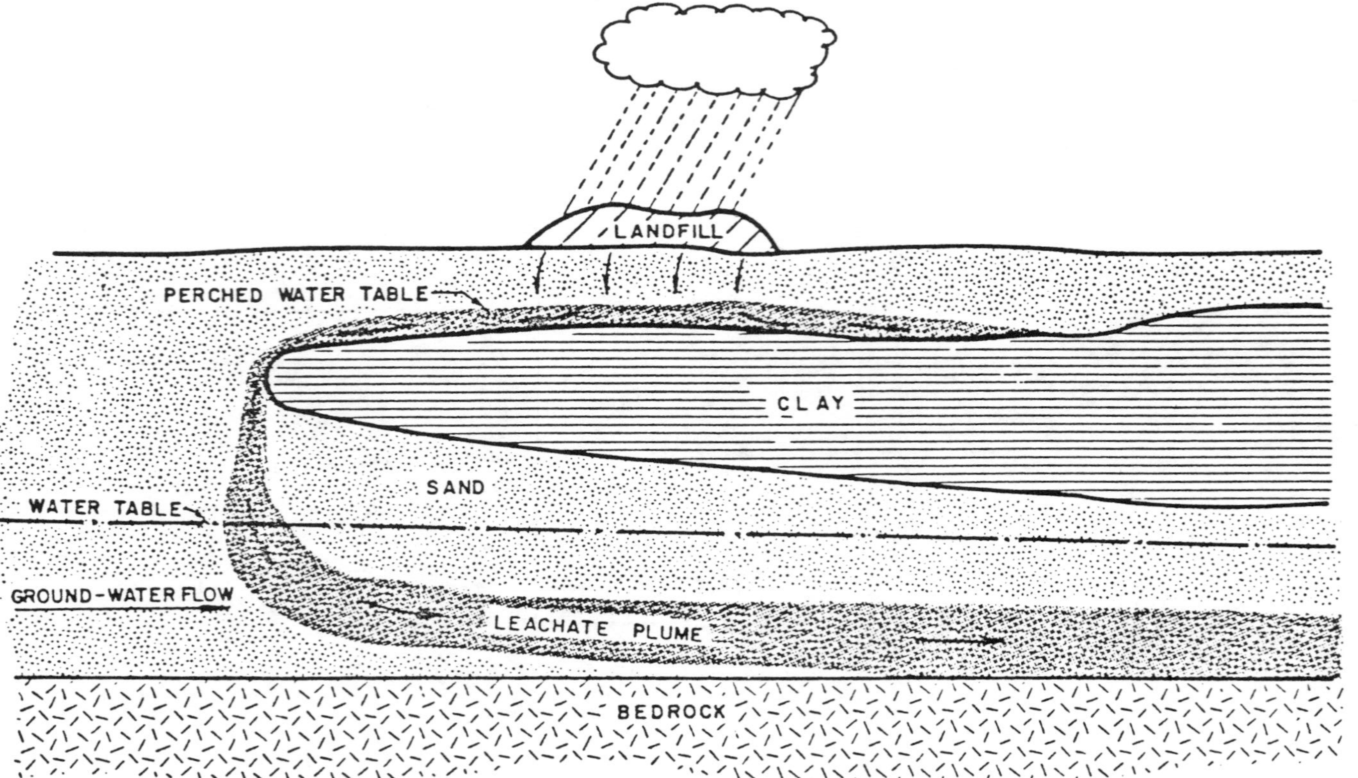

Figure 8.12. Perched water table condition.[25]

Note: Leachate percolates to the perched water table and flows downgradient to the end of the confining layer where it moves downward to the actual water table.

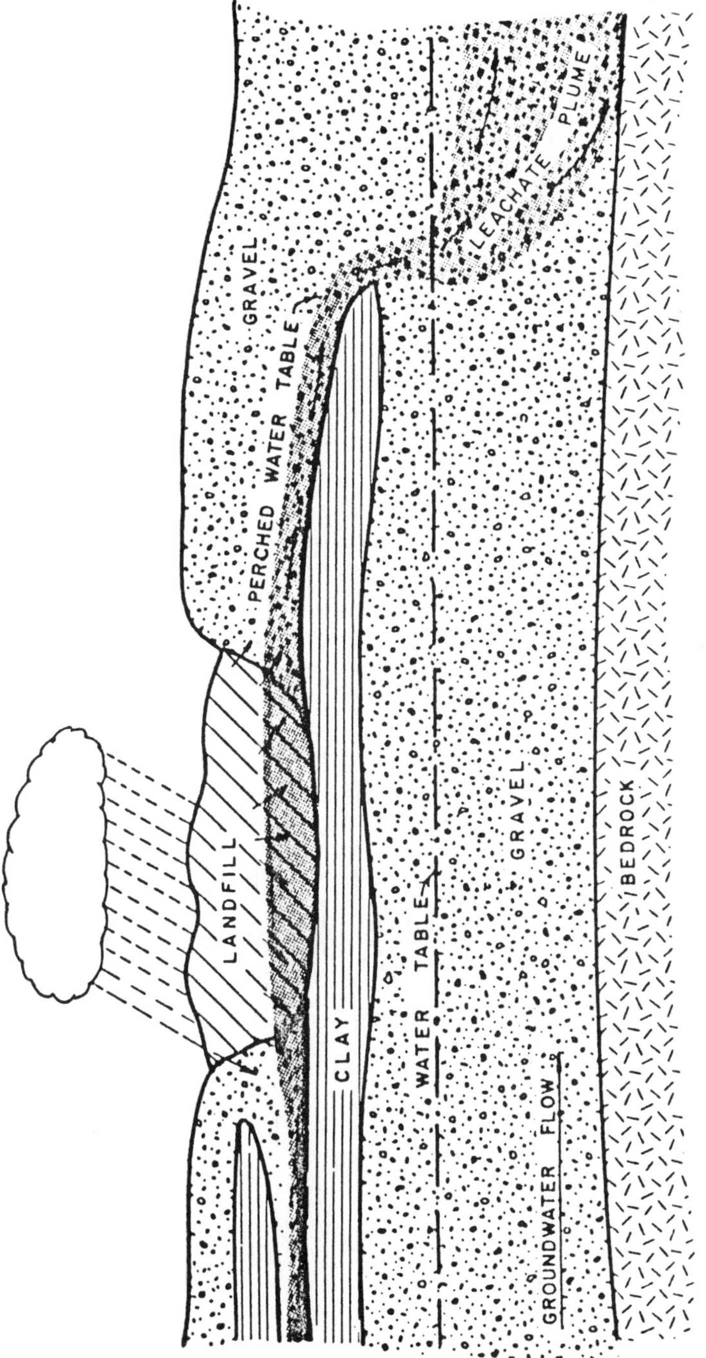

Figure 8.13. Abandoned gravel pit with a clay layer at its base.[25]

Note: A perched water table (leachate) will build up under the landfill and flow laterally through the ground above the clay until it is free to percolate to the main water table.

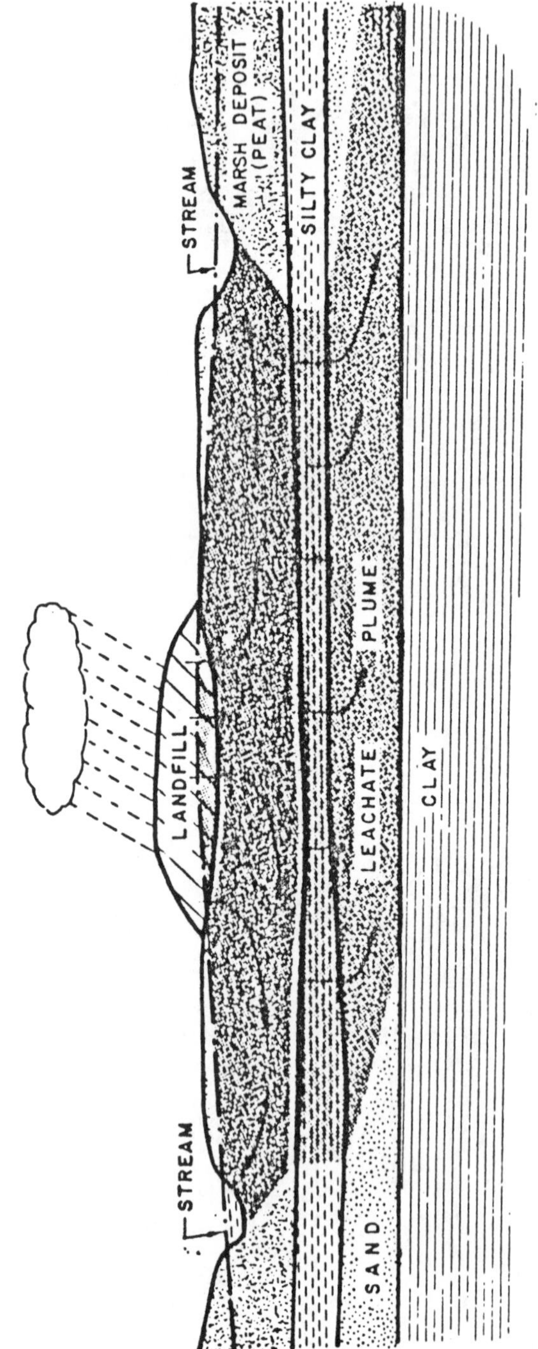

Figure 8.14. Marsh deposits bounded on either side by streams and underlain by a shallow aquifer.[25]

Note: Leachate from the landfill may move horizontally through the marsh materials to the stream, or vertically downward as ground water recharge to the aquifer.

Conversely, only in rare cases does ground water flow at a rate greater than on the order of meters per year. Consequently, it could be recommended that source monitoring wells be sampled at least twice a year and no more than once a month.

The parameters for which ground water samples should be analyzed can be preliminarily determined by the initial study of the waste source. Once contamination is detected, the list of parameters could be targeted to those contaminants which accurately depict plume movement, taking into consideration the different retardation rates for the different contaminants.

The final operational phase of a source monitoring program would be data collection, reporting and interpretation. Many of the potential biases associated with ground water quality data will be discussed in Chapter 9. The one overriding concern with ground water quality data interpretation is whether or not what has been sampled represents a problem. In other words, given data that is accurate, is this data indicative of pollution? What can be used as a basis for comparison? The two most accepted bases for comparison are drinking water and/or ground water quality standards, which may or may not be readily available; and background water quality information, which must be obtained from a nearby, uncontaminated well.

SELECTED REFERENCES

1. Todd, et al. "Monitoring Groundwater Quality: Monitoring Methodology," EPA-600/4-76-026, June 1976, U.S. Environmental Protection Agency, Las Vegas, Nevada.
2. Naymik, T. G. "Modeling as a Tool in Monitoring Well Network Design," *The Second National Symposium on Aquifer Restoration and Ground Water Monitoring*, 1982, National Water Well Association, Worthington, Ohio, pp. 151–155.
3. Pfannkuch, H. O. and Labno, B. A. "Design and Optimization of Ground-Water Monitoring Networks for Pollution Studies," *Ground Water*, Vol. 14, No. 6, Nov.–Dec. 1976, pp. 455–462.
4. Nacht, S. J. "Ground-Water Monitoring System Considerations," *Ground Water Monitoring Review*, 1983a.
5. Lewis, R. W. "Custom Designing of Monitoring Wells for Specific Pollutants and Hydrogeologic Conditions," *The Second National Symposium on Aquifer Restoration and Ground Water Monitoring*, 1982, National Water Well Association, Worthington, Ohio, pp. 187–193.
6. LeGrand, H. "Monitoring Changes in Ground Water Quality," *Water Well Journal*, Vol. 33, No. 6, July 1979.
7. Roy, S. P. and Drake, J. T. "Development of the Massachusetts Ground Water Monitoring Program," *NWWA Symposium*, 1983, National Water Well Association, Worthington, Ohio.

8. Voytek, J. "Application of Downhole Geophysical Methods in Ground Water Monitoring," *The Second National Symposium on Aquifer Restoration and Ground Water Monitoring,* 1983, National Water Well Association, Worthington, Ohio, pp. 276–278.
9. California State Water Resources Control Board, "Water Quality Inventory for Water Years 1980 and 1981," Report No. 82-ITS, July 1982, Sacramento, California.
10. Kansas Department of Health and Environment, "Groundwater Quality Management Plan for the State of Kansas," Jan. 1982, Topeka, Kansas.
11. New Jersey Department of Environmental Protection, "305b Report," 1982, Trenton, New Jersey.
12. Pennsylvania Department of Environmental Resources, "Ground Water Monitoring," 1982, Harrisburg, Pennsylvania.
13. South Dakota Department of Water and Natural Resources, "1979 Water Quality of South Dakota," Sept. 1980, Pierre, South Dakota.
14. Clark, T. P. and Trippler, D. J. "Design and Operation of an Ambient Ground-Water Quality Monitoring Program for Minnesota," *NWWA Symposium,* October 1977, National Water Well Association, Worthington, Ohio.
15. Moore, S. F. and McLaughlin, D. B. "Mapping Contaminated Soil Plumes by Kriging," *Proceedings of the National Conference on Management of Uncontrolled Hazardous Waste Sites,* 1980, Hazardous Materials Control Research Institute, Silver Spring, Maryland, pp. 66–70.
16. Walker, S. E. "Background Ground-Water Quality Monitoring: Well Installation Trauma," *The Third National Symposium on Aquifer Restoration and Ground Water Monitoring,* 1983, National Water Well Association, Worthington, Ohio, pp. 235–246.
17. Gillham, et al. "Groundwater Monitoring and Sample Bias," June 1983, Department of Earth Sciences, University of Waterloo, Toronto, Ontario, Canada.
18. Cook, D. K. "Selection of Monitoring Well Locations in East and North Woburn, Massachusetts," *Proceedings of the National Conference on Management of Uncontrolled Hazardous Waste Sites,* 1981, Hazardous Materials Control Research Institute, Silver Spring, Maryland, pp. 63–69.
19. Clarke, J. H., Mutch, R. D. and Brother, M. R. "Design of Cost-Effective Chemical Monitoring Programs for Land Disposal Facilities," *The Third National Symposium on Aquifer Restoration and Ground Water Monitoring,* 1983, National Water Well Association, Worthington, Ohio, pp. 201–209.
20. Knox, et al. "State-of-the-Art of Aquifer Restoration," June 1984, National Center for Ground Water Research, University of Oklahoma, Oklahoma State University and Rice University, Norman, Oklahoma.
21. Mooreland, J. A. and Wood, W. A. "Appraisal of Ground-Water Quality Near Wastewater-Treatment Facilities, Glacier National Park, Montana," USGS/WRI-82-4, June 1982, U.S. Geological Survey, Washington, DC.
22. Hagger, C. and Clay, P. F. "Hydrogeological Investigation of an Uncontrolled Hazardous Waste Site," *Proceedings of the National Conference on Management of Uncontrolled Hazardous Waste Sites,* 1982, Hazardous Materials Control Research Institute, Silver Spring, Maryland, pp. 45–51.

23. Absalon, J. R. and Starr, R. C. "Practical Aspects of Ground Water Monitoring at Existing Disposal Sites," *Proceedings of the National Conference on Management of Uncontrolled Hazardous Waste Sites,* 1980, Hazardous Materials Control Research Institute, Silver Spring, Maryland, pp. 53–58.
24. Quinlan, J. F. and Ewers, R. O. "Ground Water Flow in Limestone Terrains—Strategy Rationale and Procedure for Reliable, Efficient Monitoring of Ground Water Quality in Karst Areas," *Proceedings of the Fifth National Symposium and Exposition on Aquifer Restoration and Ground Water Monitoring,* 1985, National Water Well Association, Worthington, Ohio, pp. 197–234.
25. Fenn, et al. "Procedures Manual for Ground Water Monitoring at Solid Waste Disposal Facilities," SW-616, 1977, U.S. Environmental Protection Agency, Cincinnati, Ohio.
26. Seanor, A. M. and Brannaka, L. K. "Influence of Sampling Techniques on Organic Water Quality Analyses," *Proceedings of the National Conference on Management of Uncontrolled Hazardous Waste Sites,* 1981, Hazardous Materials Control Research Institute, Silver Spring, Maryland, pp. 143–148.
27. Nacht, S. J. "Monitoring Sampling Protocol Considerations," *Ground Water Monitoring Review,* 1983b.

CHAPTER 9

Ground Water Monitoring and Analysis

This chapter provides an overview of the wide variety of equipment and techniques used for monitoring traditional physical and chemical features of the unsaturated (vadose) and saturated zones of the subsurface environment. Additionally, the latter part of the chapter is devoted to identifying and discussing potential anomalies in ground water quality data and procedures which can serve to reduce the amount of anomalous information generated. Finally, the chapter addresses quality assurance/quality control and statistical analyses of ground water quality data.

MONITORING THE VADOSE ZONE

Although monitoring in the unsaturated zone is the best method of early detection of ground water contamination, it has not been as widely practiced as monitoring of the saturated zone. Although exact definitions vary widely, it is generally accepted that the vadose zone consists of all earth materials up from and including the capillary fringe of the water table to the earth's surface. Everett, et al.[1] describe the vadose as being comprised of three different regions, as depicted in Figure 9.1. These three regions (soil zone, intermediate vadose zone, and capillary fringe) are usually classified according to their geological materials and/or water content.

Comparison of Sampling and Nonsampling Methods

Monitoring in the vadose zone can be undertaken for a number of reasons to determine a variety of characteristics. Morrison[2] classified all vadose zone monitoring equipment according to the characteristics to be monitored. These five groups of characteristics include: (1) soil moisture potential; (2)

VADOSE ZONE

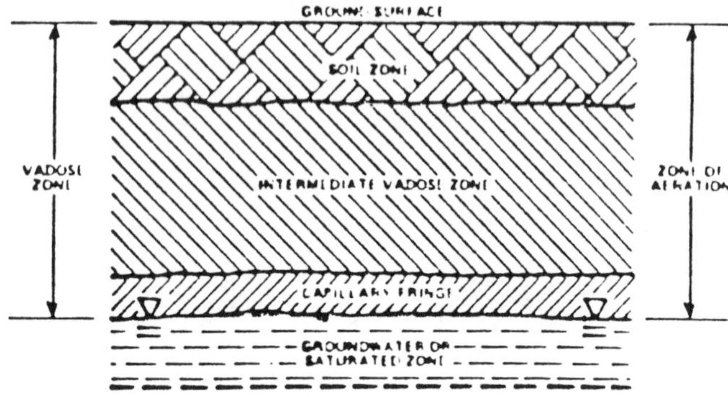

Figure 9.1. Idealized cross section of vadose and ground water zones.[1]

soil moisture content; (3) soil salinity; (4) temperature; and (5) soil pore water. Everett, Wilson, and McMillion[3] tabulated sampling and nonsampling methods and equipment according to the property to be measured and the purpose of monitoring. This tabulation for hazardous waste disposal sites is provided in Tables 9.1, 9.2, and 9.3.[3] Included in the work by Everett, Wilson, and McMillion[3] was a tabulation of the various vadose zone sampling methods and equipment versus fourteen selection criteria. This tabulation appears as Tables 9.4, 9.5, and 9.6.[3]

Suction Cup Lysimeters

In spite of all the information contained in Tables 9.1 through 9.6, the primary soil water quality monitoring device for the vadose zone is the suction cup lysimeter. A lysimeter consists of a porous cup (usually ceramic), casing, a means of applying a vacuum to the cup, and a means of transferring the sample from the buried lysimeter to the earth's surface for analysis. Figure 9.2 shows the components of a vacuum lysimeter, and Figure 9.3 provides information on the sampling and installation of pressure-vacuum lysimeters.

The main objective in using suction cup lysimeters is to pull samples of percolating pore water from the unsaturated zone from below potential sources of contamination. Three limitations of suction cup lysimeters cited by Everett, et al.[4] are: (1) only point samples are provided; (2) the extreme

Table 9.1. Premonitoring of the Vadose Zone at Hazardous Waste Disposal Sites.[3]

Property	Purpose of Monitoring	Approach	Alternative Methods
1. Storage	1. To determine overall storage capacity of the vadose zone	1. Relate storage capacity to depth of water table or depth to confining layer	1. Examine ground water level maps 2. Measure water levels in wells 3. Examine drillers' logs for depth to water table or confining layer 4. Drill test wells
		2. Estimate available porosity	1. Estimate from grain-size data 2. Drill test wells and obtain drill cuttings 3. Neutron moisture logging (available porosity = total porosity-water content, by volume)
	2. To determine regions of potential liquid accumulation (perched ground water)	1. Characterize subsurface stratigraphy	1. Examine drillers' logs 2. Drill test wells and obtain samples for size analyses 3. Natural gamma logging
		2. Locate existing perched ground water zones	1. Examine drillers' logs 2. Drill test wells 3. Neutron moisture logging
2. Transmission of liquid wastes a. Flux	1. To determine infiltration potential	1. Measure infiltration rate in the field	1. Infiltrometers 2. Test plots
	2. To estimate percolation rates in vadose zone	1. Measure unsaturated hydraulic conductivity for use in Darcy's equation	1. Instantaneous rate method 2. Laboratory column studies

Table 9.1, continued

Property	Purpose of Monitoring	Approach	Alternative Methods
		2. Measure or estimate saturated hydraulic conductivity for use in Darcy's equation a. Use core samples or grain-size data	1. Permeameters 2. Estimate from grain-size data using a catalogue of hydraulic properties of soils
		b. Measure saturated hydraulic conductivity in shallow regions	1. Pump in method 2. Air entry permeameter 3. Infiltration gradient 4. Double tube method
		c. Measure saturated hydraulic conductivity in deep regions	1. U.S.B.R. open end casing test 2. U.S.B.R. open hole method 3. Stephens-Neuman method
b. Velocity	1. To estimate the flow rate of liquid pollutants in the vadose zone	1. Estimate from field data on flux	1. Use flux values obtained as above, divide flux values by water content values at field capacity
		2. Tracer studies	1. Field plots, coupled with a depth-wise sequence of suction samplers, using conservative tracer

Table 9.1, continued

Property	Purpose of Monitoring	Approach	Alternative Methods
3. Pollutant mobility	1. To estimate the mobility of potential pollutants in the vadose zone	1. Characterize solids' samples for properties affecting pollutant mobility, cation exchange capacity, clay content, content of hydrous oxides of iron, pH and content of free lime, and surface area	1. Obtain solids' samples (e.g., by drilling test holes) and conduct standard laboratory analyses
		2. Estimate from laboratory or field testing using liquid wastes	1. Batch testing 2. Column studies 3. Field plots

Table 9.2. Sampling Methods for Vadose Zone Monitoring at Hazardous Waste Disposal Sites.[3]

Property	Purpose of Monitoring	Approach	Alternative Methods
1. Chemical and microbial properties of liquid wastes in unsaturated regions of the vadose zone	1. To obtain liquid samples for determination of pollutants in unsaturated regions	1. Obtain soil samples and drill cuttings followed by field or laboratory extraction of pore fluid for analyses of inorganics, other extraction methods for organics and microorganisms	1. Hand auger 2. Power auger 3. Hydraulic squeezer for fluid extraction 4. Field saturated extract methods 5. Laboratory saturated extract methods
		2. Obtain in situ samples of pore fluid	1. Suction cup lysimeters, vacuum-operated type 2. Suction cup lysimeters, vacuum-pressure type 3. Suction cup lysimeters, high pressure-vacuum type 4. Filter candles 5. Hollow fiber filters 6. Membrane filters
2. Chemical and microbial properties of liquid wastes in saturated regions of the vadose zone (perched ground water)	1. To obtain liquid samples for determination of pollutants in saturated regions	1. Obtain an integrated sample	1. Existing cascading wells or installed monitor wells 2. Collection pans and manifolds 3. Tile-drain outflow samplers
		2. Depth-wise samples	1. Profile sampler 2. Multilevel sampler 3. Piezometers

Table 9.3. Nonsampling Methods for Vadose Zone Monitoring at Hazardous Waste Disposal Sites.[3]

Property	Purpose of Monitoring	Approach	Alternative Methods
1. Salinity	1. To determine overall salinity of pore fluids	1. Relate salinity to electrical conductance properties	1. Four-probe method 2. EC probe 3. Salinity sensors
2. Infiltration rate	1. To infer the movement of liquid-borne pollutants based on intake rates at the land surface	1. Measure infiltration in ponds	1. Water budget method 2. Instantaneous rate method 3. Seepage meters
		2. Measure infiltration on land treatment areas	1. Water budget method 2. Infiltrometers 3. Test plots
3. Flux	1. To infer the movement of liquid-borne pollutants based on changes in storage	1. Determine water content changes	1. Gravimetric 2. Neutron moisture logging 3. Gamma ray attenuation 4. Tensiometers 5. Hygrometer/psychrometer 6. Heat dissipation sensor 7. Resistor/capacitor type sensors 8. Remote sensing
	2. To infer the movement of liquid-borne pollutants based on changes in matric potential	1. Measure hydraulic gradient, use Darcy's law	1. Tensiometers 2. Hygrometer/psychrometer 3. Heat dissipation sensors 4. Resistor/capacitor type sensors
		2. Assume unit hydraulic gradient, measure water pressure at a point, and use Darcy's law	1. Tensiometers 2. Hygrometer/psychrometer 3. Heat dissipation sensors 4. Resistor/capacitor type sensors

Table 9.3, continued

Property	Purpose of Monitoring	Approach	Alternative Methods
	3. To directly measure flux	1. Install flow meters	1. Direct flow measurement 2. Heat pulse method
4. Velocity	1. To estimate the velocity of liquid-borne pollutants in the vadose zone	1. Introduce a tracer into liquid waste at land surface and collect depth-wise samples	1. Suction cup lysimeters 2. Vadose zone wells
		2. Estimate from information on infiltration rates	1. Assume steady state and unit hydraulic gradients, divide infiltration value by estimated or measured water content values (volumetric) at field capacity
		3. Estimate from field determined flux values	1. Assume steady state and unit hydraulic gradients, divide flux values by estimated or measured water content values (volumetric) at field capacity

Table 9.4. Categorization of Vadose Zone Monitoring Techniques—Premonitoring Phase.[3]

Property	Type of Site		Applicability		Power Requirements				Depth Limit			Multi Use Capability		Data Collection System			Continuous Sampling	
	New	Old	Lab	Field	AC	DC	ENG	NO	Surf	<30′	>30′	Samp	Meas	Man	Auto	Remote	Yes	No
STORAGE																		
I. Overall storage capacity																		
A. Depth to water table																		
1. water level maps	X	X		X				X	X	X	X	N/A	N/A	X			N/A	N/A
2. sounding in wells	X	X		X		X		X	X	X	X	X	X	X	X	X	X	
3. examine driller's log	X	X	office					X	X	X	X	N/A	N/A	X			N/A	N/A
4. drill test wells	X	X		X			X			X	X	X	X	X				X
B. Available porosity																		
1. grain-size data	X	X	office					X	X	X	X	N/A	N/A	X			N/A	N/A
2. test wells, cuttings	X	X		X			X		X	X	X	X	X	X				X
3. neutron logging	X	X		X	X	X	X		X	X	X	X	X	X	X		X	
II. Perched ground water																		
A. Stratigraphy																		
1. driller's log	X	X	office					X	X	X	X	N/A	N/A	X			N/A	N/A
2. test wells	X	X		X			X		X	X	X	X	X	X				X
3. natural gamma	X	X		X	X	X				X	X	X	X	X	X			X
B. Existing perched ground water																		
1. driller's log	X	X	office					X	X	X	X	N/A	N/A	X			N/A	N/A
2. test wells	X	X		X				X	X	X	X	X	X	X				X
3. neutron logs	X	X		X	X	X			X	X	X	X	X	X	X			X
TRANSMISSION																		
I. Flux																		
A. Infiltration																		
1. infiltrometers	X			X				X	X				X	X	X		N/A	N/A
2. test plots	X	X		X				X	X				X	X	X		N/A	N/A

Table 9.4, continued

Property	Type of Site		Applicability		Power Requirements				Depth Limit			Multi Use Capability		Data Collection System			Continuous Sampling	
	New	Old	Lab	Field	AC	DC	ENG	NO	Surf	<30′	>30′	Samp	Meas	Man	Auto	Remote	Yes	No
B. Percolation rates																		
1. meas unsat K																		
a. inst rate method	X			X				X	X				X	X	X		N/A	N/A
b. lab column	X	X	X					X	X	X	X		X	X			N/A	N/A
2. meas sat K																		
a. permeameters	X	X	X					X	X	X	X		X	X			N/A	N/A
b. est from grain size	X	X	office					X	X	X			X	X			N/A	N/A
c. shallow field methods	X	X		X				X	X				X	X			N/A	N/A
d. deeper field methods	X	X		X			X			X	X	X	X	X				X
C. Velocity																		
1. from field value of flux	X	X	office					X	N/A	N/A	N/A	N/A	N/A	N/A	N/A	N/A	N/A	N/A
2. tracer studies	X	X		X				X	X	X	X	X	X	X				
POLLUTANT MOBILITY																		
(1) chemical properties	X	X	X			X	X	X	X	X		X			N/A	N/A		
(2) batch tests	X	X	X		X		X	X	X	X		X			N/A	N/A		
(3) columns	X	X	X			X	X	X	X	X	X	X			N/A	N/A		
(4) field plots	X	X		X		X		X		X	X	X			N/A	N/A		

Table 9.4, continued

	Sample/Meas	Vol	Reliability and Life Expectancy		Degree of Complexity		Direct Indirect		Media		Effect on Flow Region		Effect of Haz Waste Type on Research	
	Point Line	Bulk	Good	Quest	Simp	Comp	Dir	Ind	Porous	Fract	Yes	No	Yes	No
STORAGE														
I. Overall storage capacity														
A. Depth to water table														
1. water level maps	N/A	N/A	X		X		X		X	X		X		X
2. sounding in wells	X		X		X		X		X	X		X		X
3. examine driller's log	X		X		X		X		X	X		X		X
4. drill test wells	X		X		X		X		X	X		X		X
B. Available porosity														
1. grain-size data	N/A	N/A	X		X			X	X			X		X
2. test wells, cuttings	X		X		X			X	X			X		X
3. neutron logging	X		X			X		X	X	X		X	X	
II. Perched ground water														
A. Stratigraphy														
1. driller's log	X		X		X		X		X	X		X		X
2. test wells	X		X		X		X		X	X		X		X
3. natural gamma	X		X			X		X	X	X		X		X
B. Existing perched ground water														
1. driller's log	X		X		X			X	X	X		X		X
2. test wells	X		X		X		X		X	X		X		X
3. neutron logs	X		X			X		X	X	X		X		X
TRANSMISSION														
I. Flux														
A. Infiltration														
1. infiltrometers	X		X		X		X		X	X	X		X	
2. test plots	X		X		X		X		X	X	X		X	

Table 9.4, continued

	Sample/Meas	Vol	Reliability and Life Expectancy		Degree of Complexity		Direct Indirect		Media		Effect on Flow Region		Effect of Haz Waste Type on Research	
	Point Line	Bulk	Good	Quest	Simp	Comp	Dir	Ind	Porous	Fract	Yes	No	Yes	No
B. Percolation rates														
1. meas unsat K														
a. inst rate method	X		X			X		X	X	X	X		X	
b. lab column	X		X		X			X	X			X	X	
2. meas sat K														
a. permeameters	X		X		X		X		X			X	X	
b. est from grain size	X		X		X			X	X			X		X
c. shallow field methods	X		X			X		X	X			X	X	
d. deeper field methods	X		X			X		X	X	X		X	X	
C. Velocity														
1. from field value of flux	N/A	N/A	N/A	N/A	X		N/A	N/A	X	X	N/A	N/A	X	
2. tracer studies	X		X		X		X		X	X	N/A	N/A	X	
POLLUTANT MOBILITY														
(1) chemical properties	X		X			X	X		X	X		X		X
(2) batch tests	X		X			X	X		X	X		X	X	
(3) columns	X		X		X		X		X			X	X	
(4) field plots		X	X		X		X		X	X		X	X	

Table 9.5. Categorization of Vadose Zone Monitoring Techniques—Sampling Methods.[3]

	Type of Site		Applicability		Power Requirements				Depth Limit			Multi Use Capability		Data Collection System			Continuous Sampling	
Property	New	Old	Lab	Field	AC	DC	ENG	NO	Surf	<30′	>30′	Samp	Meas	Man	Auto	Remote	Yes	No
POLLUTANT MOBILITY																		
I. Chemical/microbial properties w/liquid wastes in unsat region																		
A. Solids samples/ extraction of pore fluid																		
1. obtaining solids samples																		
a. hand auger	X	X		X						X		X	X	N/A	N/A	N/A	N/A	N/A
b. power auger	X	X		X			X				X	X	X	N/A	N/A	N/A	N/A	N/A
2. obtaining solution samples																		
a. hydraulic squeezer	X	X		X				X	N/A	N/A	X	N/A	N/A	N/A	N/A	N/A	N/A	N/A
b. field sat extract	X	X		X				X	N/A	N/A	X	N/A	N/A	N/A	N/A	N/A	N/A	N/A
c. lab sat extract	X	X		X				X	N/A	N/A	X	N/A	N/A	N/A	N/A	N/A	N/A	N/A
B. In situ pore water samples																		
1. lysimeters, vac oper	X			X				X		X		X	X	N/A	N/A	N/A	X	
2. lysimeters, vac pres	X			X	X						X	X	X	N/A	N/A	N/A	X	
3. lysimeters, high-pres vac	X			X	X						X	X	X	N/A	N/A	N/A	X	
4. filter candle	X			X	X					X		X	X	N/A	N/A	N/A	X	
5. hollow fiber filters	X		X					X		X		X		N/A	N/A	N/A	X	
6. membrane filters	X		X					X		X		X		N/A	N/A	N/A	X	

Table 9.5, continued

	Type of Site		Applicability		Power Requirements				Depth Limit			Multi Use Capability		Data Collection System			Continuous Sampling	
Property	New	Old	Lab	Field	AC	DC	ENG	NO	Surf	<30′	>30′	Samp	Meas	Man	Auto	Remote	Yes	No
II. Chemical/microbial properties w/liquid wastes in sat regions																		
A. Integrated samples																		
1. wells	X	X		X	X		X	X			X	X	X	N/A	N/A	N/A	X	
2. collection pans	X	X		X				X		X		X		N/A	N/A	N/A	X	
3. tile outflow	X	X		X				X		X		X		N/A	N/A	N/A	X	
B. Depth-wise samples																		
1. profile samples	X	X		X	X		X			X		X		N/A	N/A	N/A	X	
2. multilevel samples	X	X		X	X		X			X		X	X	N/A	N/A	N/A	X	
3. piezometers	X	X		X				X			X	X	X	N/A	N/A	N/A	X	

Table 9.5, continued

Property	Sample/Meas	Vol	Reliability and Life Expectancy		Degree of Complexity		Direct Indirect		Media		Effect on Flow Region		Effect of Haz Waste Type on Research	
	Point Line	Bulk	Good	Quest	Simp	Comp	Dir	Ind	Porous	Fract	Yes	No	Yes	No
POLLUTANT MOBILITY														
I. Chemical/microbial properties w/liquid wastes in unsat region														
A. Solids samples/ extraction of pore fluid														
1. obtaining solids samples														
a. hand auger	X		X		X		X		X	X	X			X
b. power auger	X		X		X		X		X	X	X			X
2. obtaining solution samples														
a. hydraulic squeezer	X		X		X		X		X			X		X
b. field sat extract	X		X		X		X		X			X		X
c. lab sat extract	X		X		X		X		X			X		X
B. In situ pore water samples														
1. lysimeters, vac oper	X		X		X		X		X	X	X			X
2. lysimeters, vac pres	X		X		X		X		X	X	X			X
3. lysimeters, high-pres vac	X		X		X		X		X	X	X			X
4. filter candle	X		X		X		X		X	X	X			X
5. hollow fiber filters	X			X	X		X		X	X	X			X
6. membrane filters	X			X	X		X		X	X	X			X

Table 9.5, continued

Property	Sample/Meas	Vol	Reliability and Life Expectancy		Degree of Complexity		Direct Indirect		Media		Effect on Flow Region		Effect of Haz Waste Type on Research	
	Point Line	Bulk	Good	Quest	Simp	Comp	Dir	Ind	Porous	Fract	Yes	No	Yes	No
II. Chemical/microbial properties w/liquid wastes in sat regions														
A. Integrated samples														
1. wells		X	X		X		X		X	X	X			X
2. collection pans	X		X		X		X		X	X	X			X
3. tile outflow		X	X		X		X		X	X		X		X
B. Depth-wise samples														
1. profile samples	X		X		X		X		X	X	X			X
2. multilevel samples	X		X		X		X		X	X	X			X
3. piezometers	X		X		X		X		X	X	X			X

Table 9.6. Categorization of Vadose Zone Monitoring Techniques—Nonsampling Methods.[3]

	Type of Site		Applicability		Power Requirements				Depth Limit			Multi Use Capability		Data Collection System			Continuous Sampling	
Property	New	Old	Lab	Field	AC	DC	ENG	NO	Surf	<30′	>30′	Samp	Meas	Man	Auto	Remote	Yes	No
POLLUTANT MOBILITY																		
I. Salinity																		
A. Four-probe method	X	X		X		X			X				X	X			N/A	N/A
B. EC probe	X			X		X				X			X	X	X	X		
C. Salinity sensor	X			X		X				X			X	X	X	X	N/A	N/A
II. Infiltration rates																		
A. Ponds																		
1. water budget	X	X		X				X	X			X	X	X	X	X	N/A	N/A
2. inst rate method	X	X		X				X	X			X	X	X	X	X	N/A	N/A
3. seepage meter	X	X		X				X	X				X	X			N/A	N/A
B. Land treatment areas																		
1. water budget	X	X		X				X	X			X	X	X	X	X	N/A	N/A
2. infiltrometer	X	X		X				X	X				X	X			N/A	N/A
3. test plots	X	X		X				X	X			X	X	X			N/A	N/A
III. Flux																		
A. Gravimeter	X	X		X			X	X			X	X	X	X			N/A	N/A
B. Neutron moist log	X	X		X	X	X					X	X	X	X	X		N/A	N/A
C. Gamma ray att	X	X		X	X	X					X		X	X	X		N/A	N/A
D. Tensiometers																		
1. bourdon tube type	X	X		X				X		X			X	X			N/A	N/A
2. transducer type	X	X		X	X						X		X	X	X	X	N/A	N/A
E. Hygrometer/psych	X	X		X		X					X		X	X	X	X	N/A	N/A
F. Heat diss sensor	X	X		X		X					X		X	X	X	X	N/A	N/A
G. Resistor/capacitor	X	X		X		X					X		X	X	X	X	N/A	N/A
H. Remote sensing	X	X		X		X	X		X				X			X	N/A	N/A
I. Flowmeters	X			X		X				X			X	X	X	X	N/A	N/A

Table 9.6, continued

Property	Sample/Meas	Vol	Reliability and Life Expectancy		Degree of Complexity		Direct Indirect		Media		Effect on Flow Region		Effect of Haz Waste Type on Research	
	Point Line	Bulk	Good	Quest	Simp	Comp	Dir	Ind	Porous	Fract	Yes	No	Yes	No
POLLUTANT MOBILITY														
I. Salinity														
A. Four-probe method		X	X		X			X	X	X		X	X	
B. EC probe	X		X		X			X	X	X		X	X	
C. Salinity sensor	X			X	X			X	X	X		X	X	
II. Infiltration rates														
A. Ponds														
1. water budget		X	X		X			X	X	X		X	X	
2. inst rate method		X	X		X			X	X	X		X	X	
3. seepage meter	X		X		X			X	X	X		X	X	
B. Land treatment areas														
1. water budget		X	X		X			X	X	X		X	X	
2. infiltrometer	X		X		X			X	X	X		X	X	
3. test plots		X	X		X			X	X	X		X	X	
III. Flux														
A. Gravimeter	X		X		X			X	X	X		X		X
B. Neutron moist log	X		X		X			X	X	X		X	X	
C. Gamma ray att		X	X		X			X	X	X		X		X
D. Tensiometers														
1. bourdon tube type	X		X		X			X	X	?		X	X	
2. transducer type	X			X		X		X	X	?		X	X	
E. Hygrometer/psych	X			X		X		X		?		X	X	
F. Heat diss sensor	X			X		X		X	X	?		X	X	
G. Resistor/capacitor	X			X		X		X	X	?		X	X	
H. Remote sensing		X		X		X		X	X	X		X	X	
I. Flowmeters	X			X		X	X	X	X	X		X	X	

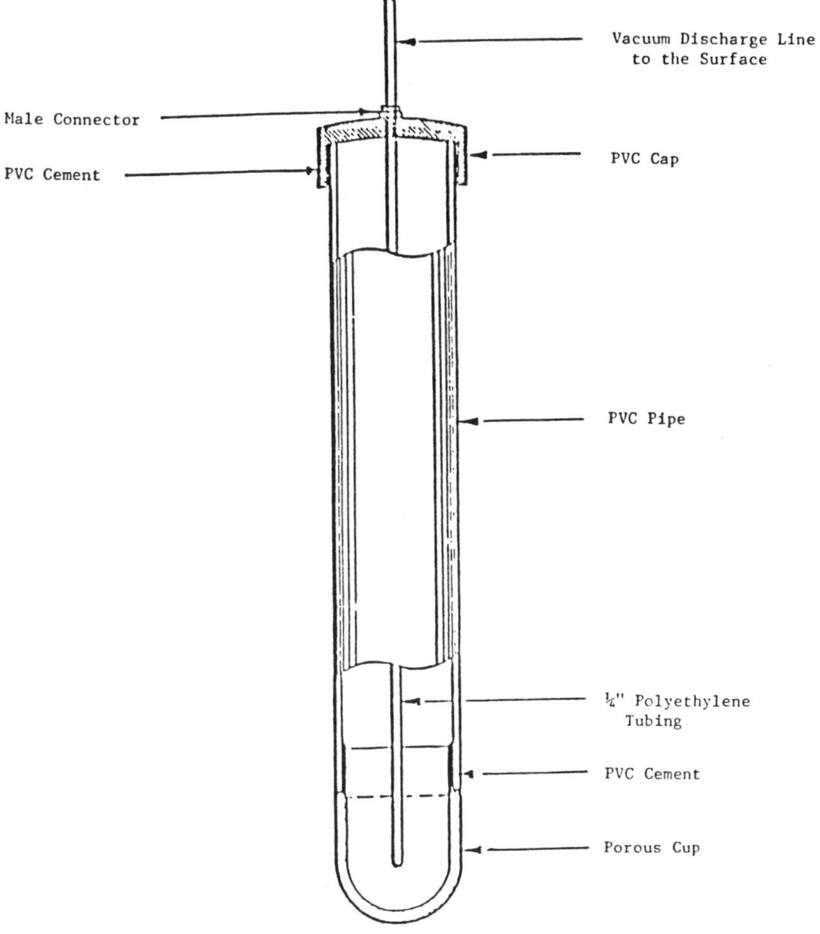

Figure 9.2. Vacuum lysimeter.

difficulty of installation at existing waste disposal facilities; and (3) a tendency to clog. Morrison[2] stated that suction cup lysimeters can also bias results due to the following: (1) although 90% of percolating water flows through larger soil pores, suction cup lysimeters primarily collect water from the slower, smaller pore network; (2) suction disrupts normal drainage and tends to produce samples representative of the area surrounding the sampler rather than a specific depth; and (3) water samples can be contaminated by the material used to construct the unit.

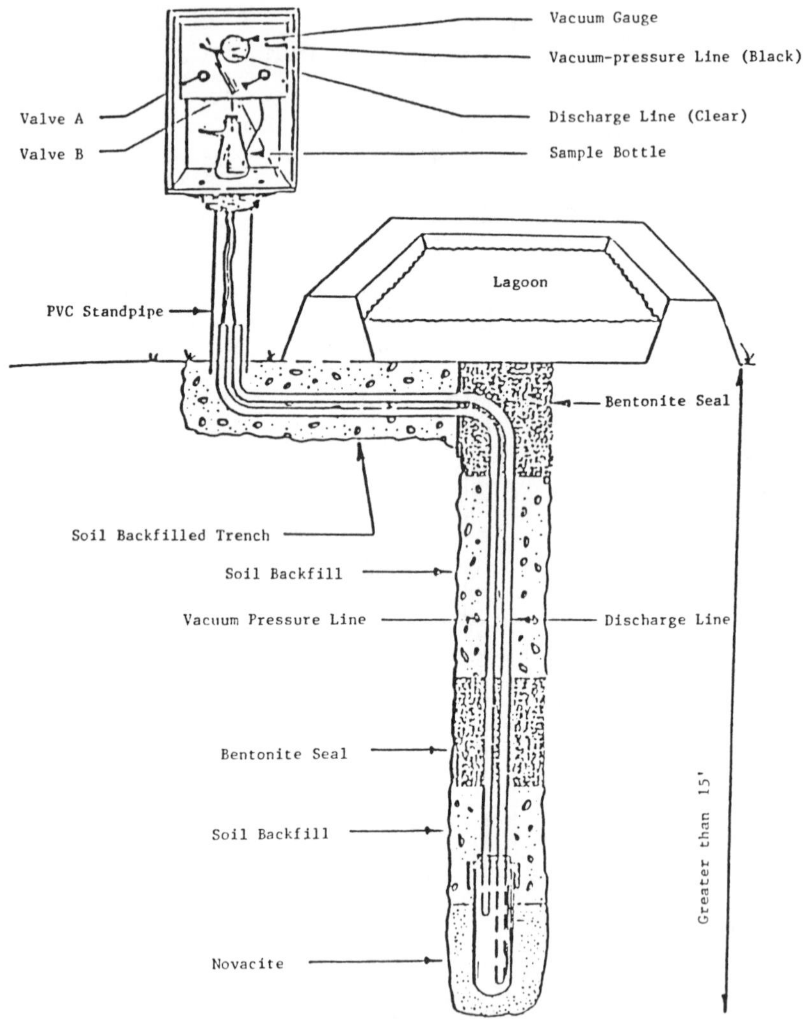

Figure 9.3. Sampling and installation of pressure-vacuum lysimeters.

Everett, et al.[1] have suggested that when constraints limit the use of suction-type sampling devices, the alternatives to consider are: (1) soil sampling; (2) sampling at the water table; and (3) vapor sampling. However, they also point out that soil sampling is a destructive technique that can be applied only once at a given location and that vapor sampling is usually applicable only to sites containing volatile organics.

MONITORING THE SATURATED ZONE (NONSAMPLING APPROACHES)

As depicted in Figure 8.1 in Chapter 8, monitoring in the saturated zone can possibly require the use of several different technologies, and each technology has several types of available equipment. The purpose of this section will be to review surface and downhole geophysical techniques and to point out their advantages and limitations.

Surface Geophysical Techniques

One area of ground water monitoring receiving increased attention is the use of geophysical or geotechnical techniques to characterize the subsurface. Geophysical techniques have traditionally been developed and utilized most extensively in the oil and gas and geothermal energy industries. Recently, existing methods have been modified and new technologies developed that are applicable for studying ground water systems. This is especially so in ground water quality management, where geophysical methods can be used to identify and characterize contaminant plumes, thus minimizing costly subsurface well drilling, sampling and analysis.

The main reason for the increased popularity of geophysical techniques is their relative economics. Physical sampling of the subsurface environment is expensive. Geophysical methods are not an alternative to physical sampling, but are complementary technologies that can aid in optimizing a ground water monitoring network. Noel, Benson, and Glaccum[5] noted that utilization of an integrated approach with traditional direct sampling methods and conventional and contemporary geophysical techniques provides several advantages, including optimum placement of monitoring wells and cost minimization. Walther, et al.[6] suggested that cost-effective assessments of the subsurface require a three-phased approach: (1) preliminary site assessment, using aerial photography, on-site inspections, and other information to approximate hazardous waste site boundaries and contaminant source locations; (2) geophysical surveys to pinpoint sources of contamination and to help define plumes of conductive contaminants; and (3) confirmation of ground water contamination through monitoring well networks designed with the help of the geophysical surveys. DiNitto[7] noted that geotechnical or geophysical techniques can aid in the development of remedial action programs where bedrock aquifers could be contaminated. Finally, Benson, Glaccum, and Beam[8] stated that discrete well sampling programs should be designed with prior knowledge obtained from preliminary surveys using geophysical methods.

Table 9.7. Examples of Surface Geophysical Methods.

A. Electrical
resistivity
self-potential
vertical electrical sounding
B. Electromagnetic
induction
metal detector
magnetometer
ground penetrating radar
C. Seismic
reflection
refraction
D. Remote sensing
aerial photography
soil gas analysis

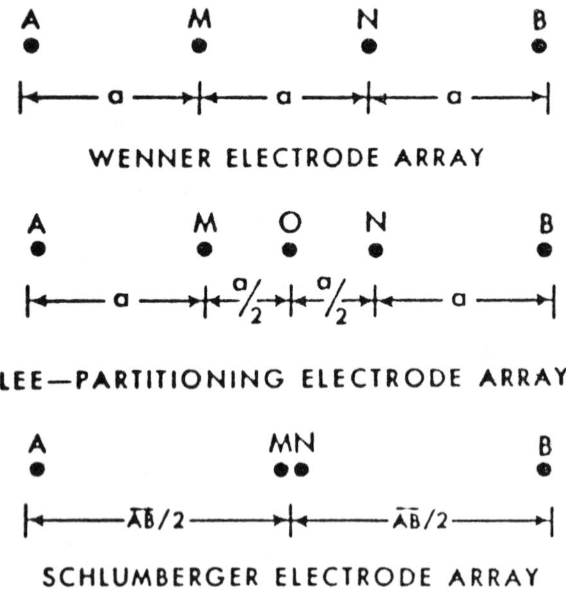

Figure 9.4. Wenner, Lee-Partitioning, and Schlumberger Electrode Arrays.[9]

Note: A and B are current electrodes; M, N and O are potential electrodes; a and AB/2 are electrode spacings.

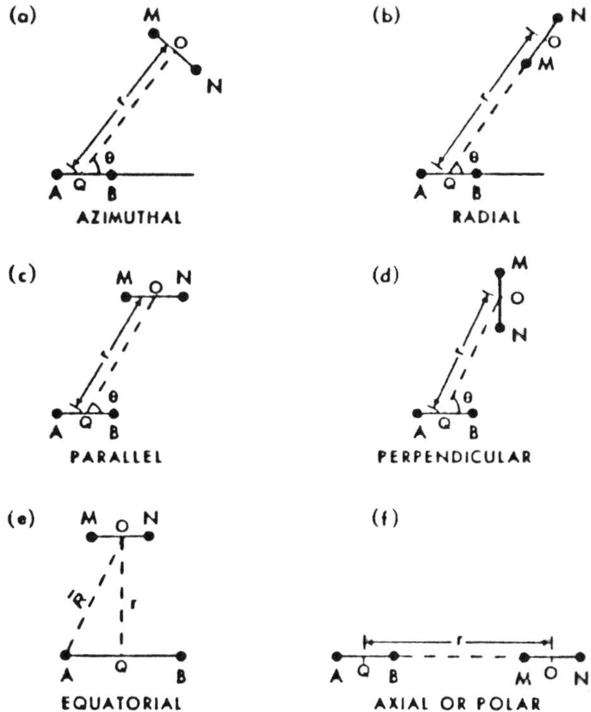

Figure 9.5. Dipole-dipole electrode arrays.[9]

Note: The distance between the current electrodes A and B (current dipole) and the distance between the potential electrodes M and N (measuring dipole) are significantly smaller than the distance, r, between the centers of the two dipoles. The equatorial is more correctly termed as a dipole-dipole array because AB is large.

Surface geophysical methods include many technologies, all of which attempt to assess the nature of the subsurface environment by imposing stresses on the subsurface and measuring the response to the imposed stress from the ground surface. These stresses involve phenomena such as conduction of heat, sound, electricity, magnetism or shock waves. A listing of some common surface geophysical techniques is found in Table 9.7. The general operational principles of some of the more utilized techniques are outlined as follows:

1. Electrical Resistivity Surveys—Resistivity surveys are based on the fact that different geologic materials will show differing resistances to current flow when subjected to an electrical field. Resistivity surveys actually involve placing two sets of elec-

trodes (current electrodes and potential electrodes) in the ground and measuring the response the subsurface yields when subjected to a given electrical field. The three most commonly used electrode arrays are shown in Figure 9.4.[9] Alternative arrays are shown in Figure 9.5.[9] Although an individual survey is depth-specific, complete vertical profiling can be accomplished by varying the spacing of the electrodes. Horizontal profiling of the subsurface is accomplished by maintaining a constant electrode spacing while moving the electrodes horizontally across the area of interest. Gilkeson and Cartwright[9] explained that geologic materials cannot be identified from resistance measurements alone, however, the following generalizations can be made: (1) unsaturated geologic materials have higher resistivity values than the same materials saturated; (2) massive rocks with little pore space have high resistivities; (3) saturated clayey sediments have low resistivities; (4) sand and gravel saturated with ground water of low ionic strength will probably have high resistivities; and (5) geologic materials (including sand and gravel) that are saturated with ground water of high ionic strength may have very low resistivities. Approximate ranges of resistivity for a variety of geologic materials are listed in Figure 9.6.[9]

2. Electromagnetic Methods—The instruments for electromagnetic induction consist of a transmitter coil and a receiver coil. The two coils are spaced at a defined and fixed distance from each other. The electromagnetic field of the transmitter coil induces an alternating current in the soil. This current induces a secondary electromagnetic field. The receiver coil compares the secondary field to the primary field. The secondary field is indicative of the type of subsurface material.[10]

   Magnetometers are instruments which measure variations in the earth's magnetic field due to ferromagnetic objects in the ground. Magnetometers can be used for locating buried drums or abandoned wells or to map certain geologic formations.

   Ground probing (penetrating) radar systems involve a series of antennas which transmit radar signals into the subsurface. A probe is moved over the land surface to re-

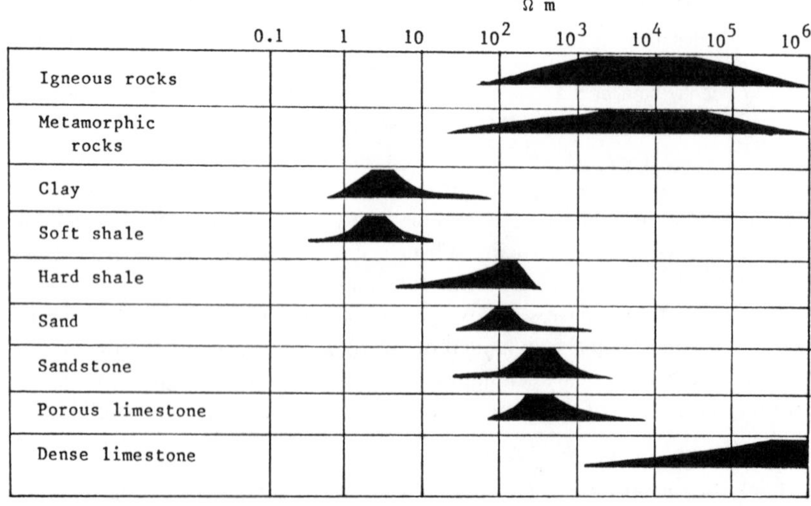

Figure 9.6. Approximate resistivity ranges (ohm-meters) for several geologic materials.[9]

cord the reflected signals with the amount of reflection indicative of the type of geologic formation.

3. Seismic Methods—Seismic reflection and refraction are similar in that both methods assess the subsurface by measuring velocities of compressional waves (P-waves) traveling through the subsurface. These P-waves are generated at the surface by an energy source such as an explosive charge. The waves then travel down into the subsurface and can either be reflected or refracted as shown in Figures 9.7 and 9.8.[11] The velocity of the P-wave is indicative of the specific subsurface conditions.

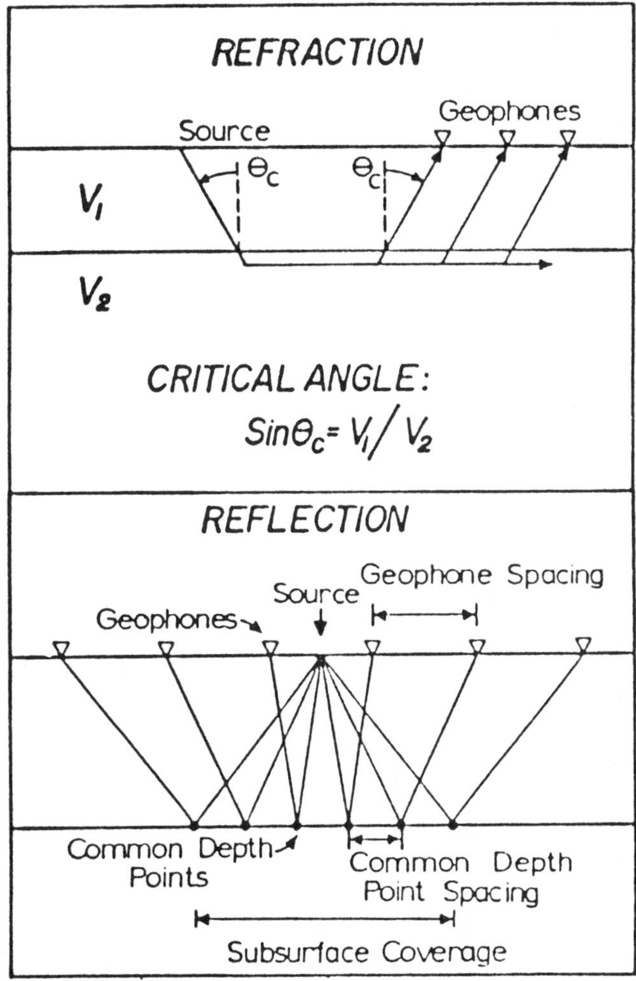

Figure 9.7. Raypath Diagram for seismic refraction and seismic reflection techniques.[11]

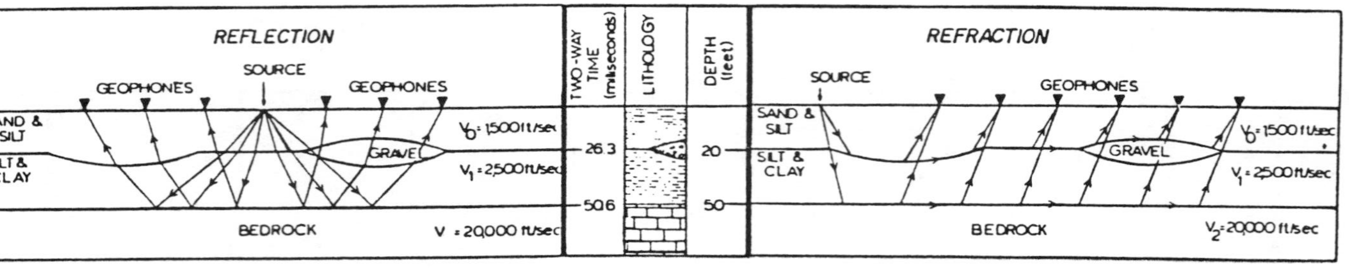

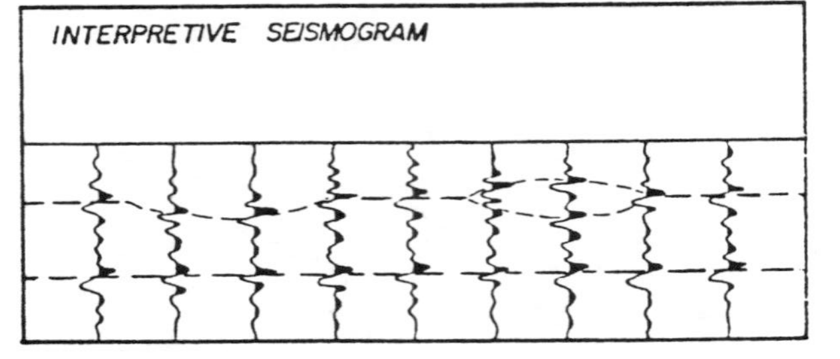

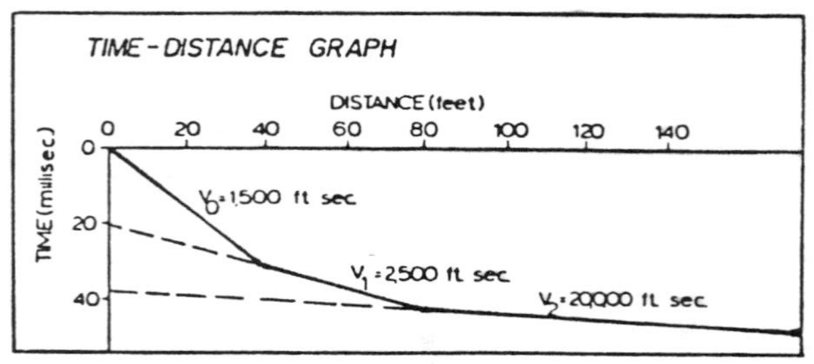

Figure 9.8. Comparison of seismic reflection and seismic refraction techniques using a simple geologic model.[11]

4. Remote Sensing—Although some experts classify all geophysical methods as remote sensing technologies, for this chapter remote sensing methods will be considered to include aerial photography and soil gas analysis. Aerial photographic methods, as the name implies, involves interpretation of aerial photographs taken of a site. If aerial photographs are available for different periods of time, they can be compared to each other to note critical changes in land use—the appearance of vegetative stress over time can be indicative of ground water contamination. Additionally, certain infrared methods can be applied to aerial photographs to map contaminant plumes. The basis for soil gas analysis is that if contaminants exist in the subsurface, especially certain organic and hydrocarbon compounds, they will tend to volatilize into gases. Soil gas analysis involves drilling a hole or driving a sampling tube into the ground and evacuating volumes of soil gas to the surface for the detection of contaminant gases. Figure 9.9 depicts a soil air sampling probe.[12]

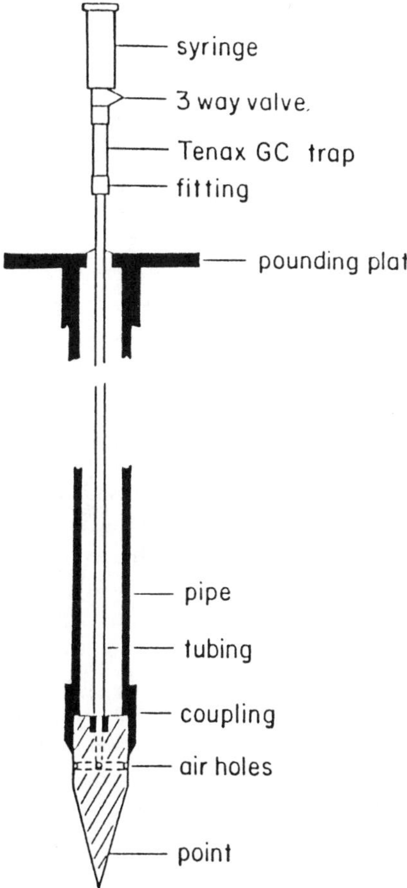

Figure 9.9. Soil air sampling probe.[12]

Several studies have been conducted with surface geophysical methods and several advantages have been identified. For example, Harman and Hitchcock[13] used electrical resistivity in a leachate detection study and found the following advantages over conventional well drilling and sampling: (1) minimal equipment cost; (2) the data produced were immediately available; and (3) with sufficient background information, the interpretation provides a high degree of reliability. Kopsick and Stander[11] outlined a study using high resolution seismic reflection to define layering of unconsolidated strata, cut and fill structures, and bedrock surfaces. The advantages of this method are that it: (1) is applicable to shallow depths; (2) detects the layering of strata (seismic refraction does not); and (3) is nondestructive (no boreholes). Noel, Benson, and Glaccum[5] also noted the advantages of contemporary geophysical techniques, such as electrical resistivity and seismic surveys. The one main advantage is that contemporary techniques operate in a continuous mode, whereas conventional techniques provide data only for discrete points.

One of the more common applications of surface geophysical techniques is for locating buried drums or other objects, especially at hazardous waste disposal facilities. Koerner, et al.[14] discussed a study in which a metal detector, a very low frequency electromagnetic magnetometer, and ground probing radar were found to be of definite value in delineating buried drums. Similarly, Lord, Koerner, and Brugger[15] promoted the use of electromagnetic wave methods as being valuable nondestructive testing procedures for both locating buried containers and detecting seepage plumes. In another study Peters, Schultz, and Duff[16] presented the results of laboratory simulation experiments in which electrical resistivity techniques showed great promise in the detection of leaks in liners.

Geophysical methods can be an aid in ground water monitoring studies, but only under certain conditions and not without some shortcomings or disadvantages. For example, Gilkeson and Cartwright[17] stated that ground water contaminated with trace concentrations of extremely toxic organic constituents, or with petroleum products, will be difficult or impossible to detect by electrical methods. McNeill[18] noted that several factors can affect measured soil resistivities, including soil porosity, soil moisture, clay content and moisture profile. Hitchcock and Harman,[19] in their study of magnetometry and electrical resistivity, noted problems with sand and clay lenses being misinterpreted as contamination, buried pipes causing anomalous readings, and surface obstructions impeding these surveys. Walther, et al.[6] compared electromagnetic induction and complex resistivity for mapping contaminant plumes. Their findings showed that complex resistivity was a slow process which could be affected by radio frequency noise. Additionally, complex resistivity instruments were unable to differentiate between areas of high total dissolved solids and areas of high total organic content.

Noel, Benson and Beam[20] compared resistivity, electromagnetic induction, and ground probing radar methods. They found that although resistivity methods show a larger anomaly than electromagnetic induction methods when a nonconductive (that is, contaminated) region is encountered, this anomaly becomes less apparent in highly resistive geologic materials. Additionally, they noted that resistivity surveys require in-ground probes which limit their utility in covered areas. Resistivity surveys also measure larger volumes of the subsurface than induction methods, hence they are not as good for small plumes. Also, electromagnetic induction methods are now built as mobile units and can operate in a continuous mode. The one limitation of ground probing radar was that the depth of penetration of the radar signals decreased with increasing silt and clay content. In addition, Kopsick and Stander,[11] in their study of seismic reflection and refraction methods, noted the comparative limitations of several geophysical techniques. These limitations are outlined in Table 9.8.[11]

Mehran, Nimmons and Sirota[21] studied the application of organic vapor analysis (OVA) for detecting hydrocarbon leaks. OVA was found to be very good for locating most hydrocarbons, however, water and some low molecular weight hydrocarbons such as methane, ethane, propane, methyl alcohol, and some freons were found not to ionize and can go undetected. The optimal conditions for OVA include: (1) high water table; (2) homogenous, isotropic and coarse-grained material; (3) absence of structural barriers; (4) dry, unsaturated material; and (5) fresh product. Turpin[22] also found OVA to be excellent for identifying "hot spots" of contamination, but limited in ability to qualify and quantify the subsurface contaminants.

More generally, Greenhouse and Slaine[23] outlined some conditions for the practical use of geophysical methods. Leis and Sheedy[24] summarized experience gained through a series of projects designed to locate disposal sites before any field programs involving sampling or monitoring. The methods examined included reviewing aerial photographs, magnetometric/metal detector surveys, and ground penetrating radar surveys. Finally, Horton, et al.[25] discussed how electrical resistivity and ground penetrating radar can be used in a complementary manner since the radar signatures give geological meaning to the resistivity, and the resistivity can be used to estimate radar system capabilities and predict the depth of penetration of the radar signals.

Downhole Geophysical Techniques

Downhole geophysical techniques are similar in concept to surface geophysical techniques, but they differ in their mode of application. The

Table 9.8. Comparison of Parameters and Limitations of Geophysical Techniques.[11]

Technique	Energy	Useful Depth	Strata Resolution	Limitations
Ground penetrating radar (GPR)	Radar signal from ground antenna	3–10 m	Excellent at shallow depths	1. Shallow depth of penetration 2. Radar attenuated by clays and water
Electrical resistivity	Electrical current through ground probe	0–30 m (max 60–90 m)	Decreases with increasing depth	1. Exact depth of investigation unknown 2. Vertical resolution only
Seismic refraction	Compressional waves (hammer, explosive charge)	Unlimited	Excellent for uniformly thick layers	1. Will not detect lower velocity layer overlain by higher velocity layer 2. Lateral resolution poor
Conventional seismic reflection	Compressional wave (explosive charge)	100–10,000 m	Excellent but dependent upon wavelength	1. Minimum depths 2. Not applicable for studies less than 100 m
Refined seismic reflection	Compressional wave (rifle bullet)	2–100 m	Excellent both in vertical and lateral directions and areas of complex structures	1. Frequency content of energy source 2. Resolution decreases with deeper reflections

majority of downhole geophysical techniques involve lowering an instrument down a borehole. While the instrument descends down the hole it imposes some type of stress on the subsurface materials and then measures the response to this stress. This procedure provides a vertical profile of the stress responses. These responses are signatures of the given subsurface material. In essence, this technique produces a description of the lithology of the subsurface at a given point. A wide variety of instruments have been applied to this technique which is given the general name of "logging," that is, the technique gives a vertical "log" of the borehole. The other general category of downhole geophysical methods combines surface and downhole techniques. In these methods, measuring instruments are placed at various

Table 9.9. Downhole Sensing Methods.[26]

Objective	Subsurface Methods
Location of zones of saturation	Electric log Temperature log Neutron log Gamma-gamma log
Physical and chemical characteristics of fluids	Electric log Temperature log Fluid conductivity log Spontaneous potential log Specific ion electrodes Fiber optics D.O., Eh, pH probes
Stratigraphy and porosity	Formation resistivity log Induced polarization log Natural gamma log Spectral gamma log Thermal neutron log Cross borehole radar Cross borehole shear Resistance log Acoustic-transit time log Acoustic-amplitude log Acoustic-wave form log Neutron log Induction log Spontaneous potential log
Flow and direction	Flowmeter Tracer Differential temperature log Water level

Table 9.10. Types of Downhole Equipment.

Acoustical	
	shearwave-crosshole
	3D velocity
	borehole compensated sonic log
	long spaced sonic
	sonic sonde
	shearwave velocity
	velocity
	well seismic tool
Electrical	
	dual laterlog (resistivities)
	dual induction laterlog
	induction spherically focused log
	induction
	induction electric
	induced polarization
	electrical resistivity
	point resistivity
	single point resistivity
	electric log
	resistance
	resistivity
	focused electric sonde
	focused resistivity
	guard logging
	micro-resistivity
	microlog
	microlater log
	micro-spherically focused log
	spontaneous potential
	self potential
	formation conductivity induced
	fluid resistivity
	radar E-M
	electro-magnetic propagation log

depths within the borehole and a stress is imposed at the ground surface near the borehole. The response of the subsurface to this stress is recorded vs the depth of the instruments within the borehole.

Downhole geophysical techniques can be categorized in several ways. In Table 9.9, Adams, Wheatcraft and Hess[26] group downhole methods according to the objectives of the sensing effort. Table 9.10 is adapted from Adams, Wheatcraft and Hess,[26] and is a comprehensive listing of available downhole techniques grouped according to type of instrument. Campbell and

Table 9.10, continued

Nuclear	
	natural gamma
	gamma-gamma density
	dual gamma-gamma density
	gamma ray spectrometer
	natural gamma ray spectrometer
	formation density (gamma)
	density borehole compensated logging
	neutron-neutron porosity
	dual thermal neutron
	neutron
	neutron/borehole compensated logging
	epithermal neutron logging
	compensated neutron log
	sidewall neutron porosity log
	moisture density depth probe
Flow and dimension	
	caliper
	sonic caliper
	flowmeter
	flow measure
Downhole physical and chemical characteristics	
	temperature
	pressure
	pH
	conductivity
	dissolved oxygen
Miscellaneous	
	computer processed data
	drillhole dip and direction indicator
	dipmeter
	downhole camera
	TV camera
	hoist
	hostile environment logging

Lehr[27] have developed a table of suggested logging techniques vs information needed, and this is shown in Table 9.11.

Downhole geophysical methods, like surface geophysical methods, produce data which in its immediate form would mean nothing to an untrained individual. All geophysical methods require expert data interpretation in order to provide accurate information concerning the subsurface environment. Figure 9.10 illustrates this point.[27] This figure shows the results from six different logs of the same borehole in a manmade subsurface environ-

Table 9.11. Suggested Logging Techniques for Ground Water Investigations.[27]

Information Needed on Properties of Rocks, Fluids, Wells or Ground Water System	Conventional Logs Which Might be Utilized
Lithology of aquifers and associated rocks	Electric, sonic, or caliper logs in open holes Radiation logs in open or cased holes
Stratigraphic correlation of aquifers and associated rocks	Electric, sonic, or caliper logs in open holes Radiation logs in open or cased holes
Total porosity or bulk density	Calibrated sonic or gamma-gamma logs in open holes Calibrated neutron logs in open or cased holes
Effective porosity or true resistivity	Calibrated resistivity logs
Clay or shale content	Natural gamma logs
Permeability	No direct measurement by logging May be related to porosity, injectivity, sonic amplitude
Secondary permeability—location of fractures and solution openings	Single point resistivity, or caliper logs, sonic amplitude, borehole television
Specific yield of unconfirmed aquifers	Neutron logs calibrated in percent moisture
Grain size	Possible relationship to formation factor derived from electric logs. Clay content from gamma logs
Location of water level or perched water outside of casing	Electric, fluid resistivity, gamma logs in open hole or inside casing. Neutron or gamma logs outside casing
Moisture content above water table	Neutron logs calibrated in percent moisture
Rate of moisture infiltration	Time interval neutron logs or radioactive tracers Temperature
Direction, velocity and path of ground water flow	Single-well tracer techniques—point dilution and single-well pulse multi-well tracer technique
Dispersion, dilution and movement of waste	Fluid resistivity and temperature log Gamma logs for radioactive wastes samples
Source and movement of water in a well	Injectivity profile flowmeter of tracer during pumping of injection. Differential temperature logs. Time interval neutron of gamma logs

Table 9.11, continued

Information Needed on Properties of Rocks, Fluids, Wells or Ground Water System	Conventional Logs Which Might be Utilized
Chemical and physical characteristics of water—including salinity, temperature, density and viscosity	Calibrated fluid resistivity and temperature logs in hole. Neutron chloride logging outside casing
Determining construction of existing wells, diameter, and position of casing perforations screens	Gamma and caliper logs collar and perforation locator borehole television
Determining optimum length and setting screen	All logs providing data on lithology water-bearing characteristics and correlation and thickness of aquifers
Guide to cementing procedure and determining position of cement	Caliper, temperature, or gamma-gamma logs
Locating corroded casing	Under some conditions caliper, or collar locator
Locating casing leaks or plugged screen	Tracers and flowmeter

ment. Note that the changes in the individual curves correspond fairly well to changes in strata. However, the individual curves sometimes move in completely opposite directions for the same strata.

Historical applications of downhole geophysical techniques have been mainly to determine the physical characteristics of the subsurface environment. However, these methods are now being utilized in ground water quality investigations. Adams, Wheatcraft and Hess[26] have identified four objectives of downhole sensing efforts at hazardous waste sites: (1) to determine the lithology, porosity and structure of the subsurface; (2) to locate zones of saturation; (3) to identify physical and chemical characteristics of subsurface fluids; and (4) to measure ground water flow and velocity.

Voytek[28] presents a discussion of the relative attributes of several downhole geophysical techniques when applied to contaminated aquifers. Similarly, Adams, Wheatcraft and Hess[26] discuss the applicability to hazardous waste site problems of natural gamma logging, fluid conductivity and temperature logging, and a thermal ground water flow meter. Huber[29] provides a discussion of the use of downhole television. Finally, Black, Davis, and Patton[30] discuss the attributes of a piezometric permeability profiler for testing the hydraulic conductivity of small diameter drill holes during drilling.

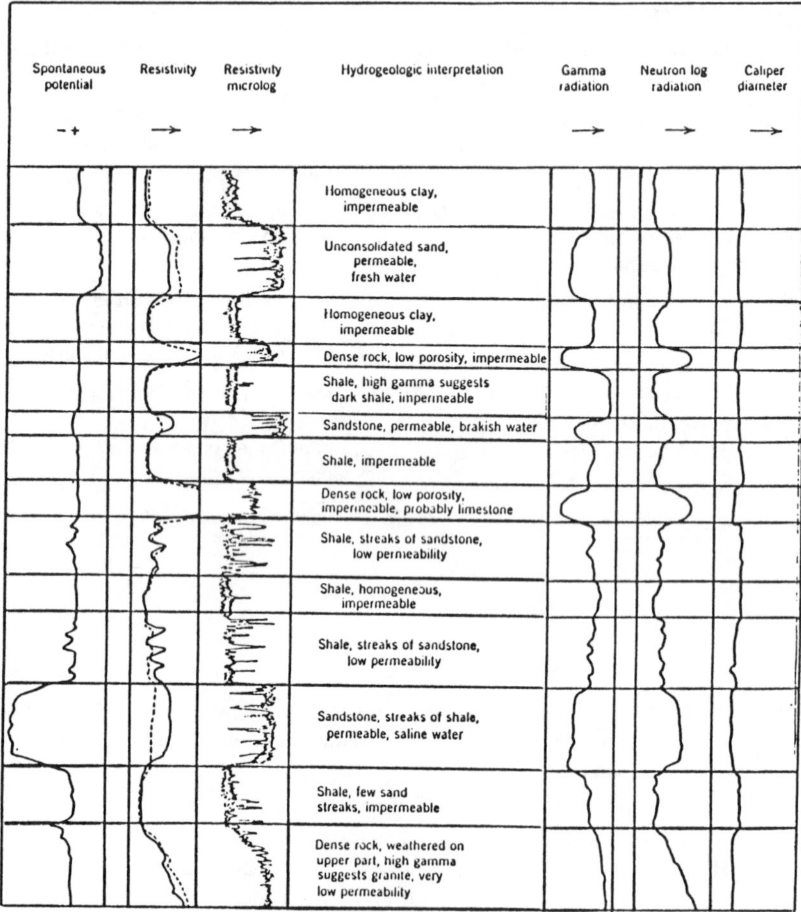

Figure 9.10. A combination of six common logs of a hypothetical test hole showing the hydrogeologic interpretations.[27]

MONITORING WELLS AND SAMPLING

The most frequently used approach in ground water quality monitoring involves the collection and subsequent analysis of samples from wells for preselected water quality constituents. This section will focus on the equipment and procedures available for developing monitoring wells. Because of the wide variety of available equipment, much of the following information addresses only the relative advantages and limitations of the various choices. This type of information should prove valuable to those charged with formulating ground water monitoring programs.

Drilling Methods

Although geophysical techniques can give considerable insight to the subsurface environment, it must be emphasized that these are only complementary technologies. The results from geophysical investigations should be verified by accessing the subsurface environment. This is especially true when monitoring activities are being conducted to determine the existence and/or extent of ground water contamination. The only direct means of sampling ground water is through a borehole that penetrates the aquifer in question. There is a wide variety of methods for drilling boreholes, and some are outlined in Table 9.12. The method selected can have a definite impact on the performance of the borehole when it is used for monitoring, withdrawal, or injection of water.

Scalf, et al.[31] described nine methods of well drilling and discussed the relative advantages and disadvantages of each. This information is tabulated in Table 9.13.[31] Contrary to the large number of methods, Gillham, et al.[32] identified four main types of drilling methods to implant ground water sampling installations: (1) auger drilling; (2) rotary drilling; (3) diamond-bit drilling; and (4) cable-tool drilling. Gillham, et al.[32] also point out that auger drilling poses the least potential for hydrochemical disturbance of samples, but this can cause vertical mixing of different level ground waters. Rotary drilling with fluids poses the greatest threat of chemical alteration of ambient ground water, but can be offset by proper casing placement. Diamond-bit drilling also utilizes drilling water which can cause disturbances, and cable-tool drilling poses problems similar to those of auger drilling.[32]

Selection of Well Casing

Very rarely are boreholes left open as drilled. Most often some type of tubing (casing) is placed in the hole at specified zones. This casing is used to section off various formations and/or to support the walls of the borehole. One of the chief concerns is that the well casing material itself will interfere with representative ground water monitoring. For example, Scalf, et al.[31] pointed out that the cost savings realized by using PVC casing instead of Teflon can be offset by the adsorption of organics on the PVC, or by bleed-

Table 9.12. Water Well Drilling Methods.

Equipment Type	Procedure Description
1. Cable tool	The cable tool drills by lifting and dropping a string of tools suspended on a cable. A bit at the bottom of the tool string strikes the bottom of the hole, crushing and breaking the formation material. Cuttings are removed by baling or recirculaton of a slurry. Casing is usually driven concurrent with drilling operations.
2. Hydraulic rotary	Hydraulic rotary drilling consists of cutting a bore-hole by means of a rotating bit and removing the cuttings by continuous circulation of a drilling fluid as the bit penetrates the formation materials. Drilling fluid is usually a water-based slurry containing some fine material (clay) in suspension.
3. Reverse circulation	Same as (2) except the fluid is forced downhole outside of the casing and returns up the inside of the casing. For (2) the route is down inside the casing and returning up outside the casing.
4. Air rotary drilling	Same as (2) only forced air becomes the drilling fluid.
5. Air-percussion rotary drilling	Rotary technique in which the main source of energy for fracturing rock is obtained from a percussion machine connected directly to the bit.
6. Hollow-rod drilling	Similar to (1) except with more rapid and shorter strokes. Ball and check valve on bit allows water and cuttings to enter and travel up the casing (hollow-rod).
7. Jet drilling	Similar to (6) except water is pumped down inside casing (hollow-rod) and returns up outside the casing.
8. Driven wells	Pipe with a driving plug on the end is driven into the ground by repeated blows from a driving weight.
9. Augered wells	A string of tools is pushed into the subsurface with the rotation of a bit on the end of the string used for cutting. The cuttings are removed by a spiral staircase flight attached to the drill string.

ing of organic constituents in PVC cements. Miller[33] examined the uptake and release of lead, chromium, and trace level organics exposed to synthetic well casings; the results are tabulated in Table 9.14. Additionally, Table 9.15 gives a qualitative comparison of the susceptibilities of acrylonitrile

Table 9.13. Advantages and Disadvantages of Well Drilling Methods.[31]

Method	Advantages	Disadvantages
Mud rotary	(1) Available throughout the U.S. (2) Capable of drilling all formations, hard or soft. (3) Capable of drilling to any depth desired for monitoring. (4) Casing not required during drilling. (5) Formation logging (sampling) is fairly reliable in most formations. (6) Relatively inexpensive.	(1) Drilling fluid mixes with formation fluid and is often difficult to completely remove. (2) Bentonite (if used to minimize fluid loss) will adsorb metals and may interfere with some other parameters, thereby making this drilling method (at least the use of bentonite drilling mud) undesirable where metals are being sampled. (3) Organic/biodegradable additives mixed with the water to minimize fluid loss will interfere with bacterial analyses and organic-related parameters. (4) No information on the position of the water table, and only limited information on water-producing zones is directly available during drilling. Electric logging of rotary drilled wells can substantially add to the accuracy of the driller's log and to water-related information. (5) Circulates contaminants.
Air rotary	(1) No drilling fluid is used, therefore, contamination or dilution of the formation water is not a factor. (2) Air-rotary rigs operate best in hard rock formations. (3) Formation water is blown out of the hole along with the cuttings, therefore, it is possible to readily determine when the first water-bearing zone is encountered. (4) Collection and field analysis (after filtering) of water blown from the hole can provide enough information regarding changes in water quality for some parameters such as chlorides for which only large changes in concentration are significant. (5) Formation sampling ranges from excellent in hard, dry formations to nothing when circulation is lost as in formations such as some limestones or other formations with cavities.	(1) Casing is required to keep the hole open when drilling in soft, caving formations below the water table. This is often a major disadvantage. (2) When more than one water-bearing zone is encountered and where the hydrostatic pressures are different, flow between the zones will occur between the time when the drilling is done and the hole can be properly cased and one zone grouted off.

Table 9.13, continued

Method	Advantages	Disadvantages
	(6) Air rotary rigs are common and readily available throughout most of the U.S.	
Air drilling	(1) Same advantages as with standard air rotary drilling except that soft, caving formations can be drilled. (2) The use of casing minimizes flow into the hole from upper water-bearing layers, therefore multiple layers can be penetrated and sampled for rough field determinations of some water quality parameters.	(1) Air-rotary rigs with casing hammers are not in common use throughout the U.S. and may be difficult to locate in some areas. (2) The cost/hr or /ft is substantially higher than other drilling methods. (3) It is difficult to pull back the casings if it has been driven very deep—say deeper than 50 ft in many formations.
Cable tool	(1) Formation samples can be excellent with driller using a sand-pump bailer. (2) Information regarding water-bearing zones is readily available during the drilling. Even relative permeabilities and rough water quality data from different zones penetrated can be obtained by skilled operators. (3) The cable-tool rig can operate satisfactorily in all formations, but is best suited for caving, large gravel type formations or formations with large cavities above the water table (such as limestones).	(1) Drilling is slow compared with rotary rigs. (2) The necessity of driving the casing along with drilling in unconsolidated formations requires that the casing be pulled back to expose selected water-bearing zones. This process complicates the well completion process and often increases costs. (3) The relatively large diameters required (minimum of 4-in. casing) plus the cost of steel casing result in large costs compared with rotary drilling and plastic casing. (4) It is difficult to place a positive grout seal above the drive shoe of the casing. Therefore, either the drive casing must be totally removed and the seal placed around the outside of an inner casing, or a seal must be placed above the screen but below the drive shoe. Either procedure adds to the cost and time of completion. (5) Cable-tool rigs have largely been replaced by rotary rigs in some parts of the U.S., hence availability may be difficult.

Table 9.13, continued

Method	Advantages	Disadvantages
Special reverse circulation	(1) The formation water is not contaminated by the drilling water. (2) Excellent formation samples can be obtained. (3) When drilling with air, immediate information is available regarding the water-bearing properties of formations penetrated. (4) Caving of the hole in unconsolidated formations is not as great a problem as when drilling with the normal air rotary rig.	(1) Double-wall, reverse-circulation rigs are very rare and expensive to operate. (2) Placing cement grout around the outside of the casing above the screen of the permanent well often is difficult—especially when the screen and casing are placed down through the inner drill pipe before the drill pipe is pulled out.
Solid-stem continuous flight auger	(1) The auger drilling rigs are generally mobile, fast, and inexpensive to operate in unconsolidated formations. (2) No drilling fluid is used, therefore contamination problems are minimized.	(1) Cannot be used in hard rock. (2) Depth limitation varies with equipment and type of soils but approximately 150 ft is maximum. (3) Once the augers have been withdrawn, the degree to which the borehole will remain open is dependent upon the degree of soil consolidation and saturation. Most boreholes will collapse below the water table. (4) Formation samples may not be completely accurate. (5) Depth to the water table may be difficult to determine accurately in deep borings.
Hollow-stem continuous flight auger	(1) The auger drilling rigs are generally mobile, fast, and inexpensive to operate in unconsolidated formations. (2) No drilling fluid is used, therefore contamination problems are minimized. (3) The problem of the hole caving in saturated, unconsolidated material, as when the solid-stem, continuous-flight augur is pulled out of the hole, is overcome by placing the casing and screen down inside the hollow stem before the augers are removed. (4) Natural gamma-ray logging can be done inside the hollow stem which permits defining the nature and thickness of the formations penetrated.	(1) Cannot be used in hard rock. (2) Depth limitation varies with equipment and type of soils but approximately 15 ft is practical. (3) Formation samples may not be completely accurate. (4) Depth to the water table may be difficult to determine accurately in deep borings.

Table 9.13, continued

Method	Advantages	Disadvantages
	(5) A grout seal can be placed around the permanent casing by attaching a cement basket above the screen before setting the assembly inside the hollow stem. Grout is placed in the annulus between the casing and hollow stem and the augers are pulled out. Grout is continuously injected or placed until all augers are removed.	
Bucket auger	(1) No drilling water is required when either drilling above the saturated zone, or below the saturated zone in non-caving formations. (2) After the hole has been drilled, the setting of casing with screen and grouting the outside of the casing to form a seal is relatively easy. (3) Formation sampling is excellent.	(1) The hole diameter is large, hence the annular space is large when small diameter casing is used. This requires careful grouting and backfilling to insure water sample integrity. (2) In caving formations below the water table it is necessary to continuously add water to prevent caving. (3) Use of the bucket auger is restricted to soft formations and depths less than about 50 ft. (4) These rigs are not widely available.
Jetting	(1) Jetting is fast and very inexpensive. (2) Because of the small amount of equipment required, jetting can be accomplished in locations where it would be possible to jet down a well point in the center of a lagoon at a fraction of the cost of using a drill rig. (3) Jetting numerous well points just into a shallow water table is an inexpensive method for determining the water table contours, hence flow direction.	(1) A large amount of foreign water or drilling mud is introduced above and into the formation to be sampled. (2) It is not possible to place a grout seal above the screen to assure depth-discrete sampling. (3) The diameter of the casing is usually limited to two in. therefore, obtaining samples must be either by suction lift, air lift, bailer, or other method applicable to small diameter casings. (4) Jetting is only possible in very soft formations, and the depth limitation is shallow—say 30 ft without special equipment. (5) Large quantities of water are often needed.

Table 9.14. Adsorption and Desorption of Six Organic Compounds on PVC, Polyethylene, and Polypropylene.[33]

SORBATE	PVC		POLYETHYLENE		POLYPROPYLENE	
	Adsorption	Desorption	Adsorption	Desorption	Adsorption	Desorption
Trichlorofluoromethane	None	None	Substantial (75%)	Slight (25%)	Moderate (50%)	Slight (25%)
1,1, 1-Trichloroethane	None	None	Substantial (75%)	Slight (25%)	Moderate (50%)	Slight (25%)
Trichloroethylene	None	None	Substantial (75%)	None	Substantial (75%)	Slight (25%)
1,1, 2-Trichloroethane	None	None	Moderate (50%)	Slight (25%)	Slight (25%)	Slight (25%) (delayed)
Bromoform	None	None	Moderate (50%)	Moderate (50%)	Slight (25%)	Complete (100%) (delayed)
Tetrachloroethylene	Moderate (50%)	Slight (25%)	Complete (100%)	None	Complete (100%)	None

Table 9.15. Resistance of Well Casing Thermoplastics to Common Materials Under Expected Use Conditions.*[34]

	ABS	PVC	SR
Mineral Acids			
hydrochloric (muriatic) acid—30%	+	+	+
sulfuric acid—50%	+	+	+
sulfamic acid—30%	+	+	+
Alkalies			
ammonium hydroxide—30%	+	+	+
calcium hydroxide—30%	+	+	+
sodium hydroxide—30%	+	+	+
Salts			
calcium chloride	+	+	+
potassium chloride	+	+	+
sodium bicarbonate	+	+	+
sodium chloride (salt)	+	+	+
sodium phosphate	+	+	+
sodium sulfite	+	+	+
Oxidizing Agents/Disinfectants			
sodium hypochlorite (bleach solution)—12%	+	+	+
chlorine water	+	+	+
calcium hypochlorite—solution—18%	+	+	+
Organic Acids			
acetic acid—10%	+	+	+
stearic acid	+	+	+
hydroxy acetic acid—10%	+	+	+
Oils and Derived Products			
crude oil—sour	+	+	+
diesel fuel	+	+	+
gasoline	+	+	+
lubricating and thread cutting oils	+	+	–
motor oil	+	+	+
Solvents			
acetone	–	–	–
methyl ethyl ketone	–	–	–
toluene	–	–	–
trichlorothylene	–	–	–
turpentine	–	+	–
xylene	–	–	–
Soaps and Detergents	+	+	+

Table 9.15, continued

	ABS	PVC	SR
Gases			
ammonia	+	+	+
carbon dioxide	+	+	+
hydrogen sulfide	+	+	+
natural gas	+	+	+
oxygen	+	+	+

Key: + denotes resistant
– not resistant

For materials not included in the table, the well casing manufacturer should be consulted. An expanded table on chemical resistance of thermoplastic piping materials is also available from the Plastic Pipe Institute.

*The indicated extent of chemical resistance is for guidance purposes only for conditions of expected usage. As chemical resistance is influenced by stress, temperature and time of contact, this guidance is not necessarily applicable to all conditions.

butadiene styrene (ABS), polyvinyl chloride (PVC), and styrene-rubber (SR) to attack by common materials, including organic compounds.[34] Finally, Nielsen and Yeates[35] have developed Table 9.16 which ranks materials according to their relative inertness. These materials could be used in a variety of ground water sampling equipment in addition to being used as casing.

SAMPLING PROTOCOL

The question of the most appropriate means to prepare or develop a well prior to sampling remains unanswered in a general sense. While it is recognized that representative water quality samples cannot be obtained without first removing the stagnant water in a well, controversy exists over the preferred method for removing this water. Nacht[36] discusses several studies in which significant changes in water quality data (for a number of parameters) were observed due to differing times of pumping or volumes of water purged from a well prior to sampling.

Giddings[37] stated that bore volume purging, intended to improve the performance of a monitoring well, often has exactly the opposite effect. The drawbacks of purging as identified in this study included: (1) it increases turbidity in the sample which can mask the parameters to be analyzed; (2) cascading of water refilling the purged bore can alter the levels of dissolved gases and can cause changes in aquifer water quality; and (3) severe dilution

Table 9.16. Preferred Materials for Use in Ground Water Sampling Devices.[35]

Rigid Materials		Flexible Materials	
Order of Preference	Material	Order of Preference	Material
1	Teflon	1	Teflon
2	Stainless steel 316	2	Polypropylene
3	Stainless steel 304	3	Flexible PVC/Linear polyethylene
4	Polyvinylchloride (PVC)	4	Viton
5	Low-carbon steel	5	Conventional polyethylene
6	Galvanized steel	6	Tygon
7	Carbon steel	7	Silicone/Neoprene

due to water refilling the borehole primarily from deep fracture flow systems may hide the contribution from the shallow flow zone of interest. On the other hand, Unwin and Huis[38] provided results that indicated removal of seven to ten bore volumes by pumping from at or near the water surface is the most reliable way to purge a well.

Keely[39] discussed the attributes of sampling the chemical quality of a well as a function of increasing time of pumpage. This method, called chemical time-series sampling, is promoted based on the following features: (1) it has the potential for locating sources of contaminants; (2) it is able to clearly define minimum purging requirements and sampling radii; and (3) it aids in assessing the need for sampling by methods which create an appreciable local gradient toward the well. The technique also indicates the noncomparability of data collected by grab-sampling wells of different flow rates, pumping durations, and/or construction. In a field application of this technique, Keely and Wolf[40] found that the data inferred the distinct possibility of serious interpretation errors arising from conventional sampling.

SAMPLING EQUIPMENT

There is a number of techniques for retrieving ground water samples from a monitoring well. Scalf, et al.[31] and Nielsen and Yeates[35] both provide descriptions of sampling equipment and their relative advantages/disadvantages. Summary points from both of these studies are highlighted in the following subsections.

Bailers

One of the oldest and simplest methods of sampling monitoring wells involves the use of bailers.[31] Although a bailer may be made of virtually any rigid material, the most common materials in current use are PVC, Teflon or stainless steel.

In most bailers, the top of the bailer is open and the bottom contains a simple "ball and seat" check valve arrangement as shown in Figure 9.11. In order to obtain a sample, the bailer is lowered into the well on a line. This line can be made of almost any material, but for the purpose of detecting trace levels of specific contaminants, a noncontaminating material such as stainless steel or Teflon-coated wire should be used. As the bailer moves down through the water in the well, the check valve remains open, allowing the water to pass through the bailer. At the desired depth, the bailer's descent is stopped. As the bailer is lifted, the weight of the water inside the

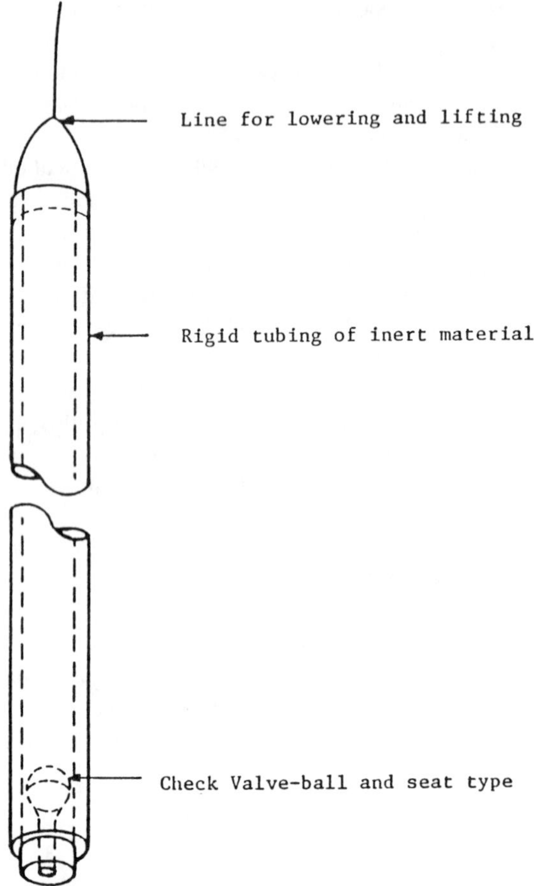

Figure 9.11. Bailer used for ground water sampling.

bailer causes the ball valve to seat thus trapping the sample inside. When the bailer reaches the surface, the sample is decanted into a sample bottle.[35] The relative advantages/disadvantages of bailers are outlined in Table 9.17.

Syringe Devices

Syringe-type samplers have recently been developed in response to the desire for samples not affected by gas pressure changes. Operation of the typical syringe device, as shown in Figure 9.12, involves lowering the device

Table 9.17. Advantages and Disadvantages of Bailers.[31,35]

Advantages	Disadvantages
Bailers can be constructed from a wide variety of materials compatible with the parameters of interest.	The person sampling the well is susceptible to exposure to any contaminants in the sample.
Economical and convenient enough that a separate bailer may be dedicated to each well to minimize cross contamination.	Bailing does not supply a continuous flow of water to the surface.
No external power source required.	It may be difficult to determine the point within the water column that the sample represents.
Low surface to volume ratio reduces outgassing of volatile organics.	Bailer check valves may not operate properly under certain conditions (e.g., high suspended solids content and freezing temperatures).
Bailers can be used to sample water from wells of virtually any depth.	The "swabbing" effect of bailers that fit tightly into a well casing may induce fines from the formation to enter the well, especially if the well has been poorly developed.
Bailers are mechanically very simple, and thus are easily operated and disassembled for cleaning and repair.	Sometimes impractical to evacuate stagnant water in a deep well with a bailer.
Bailers made of flexible material will allow passage through nonplumb wells.	Transfer of water sample from bailer to sample bottle can result in aeration.
Bailers can be made to fit any diameter well, and can be made virtually any length to accommodate any sample volume.	Cross-contamination can be a problem if equipment is not adequately cleaned after each use.
Bailers can provide a "cut" of immiscible contaminants (i.e., petroleum hydrocarbons) from the top of the water column in a well. Transparent bailers are usually utilized for this purpose.	

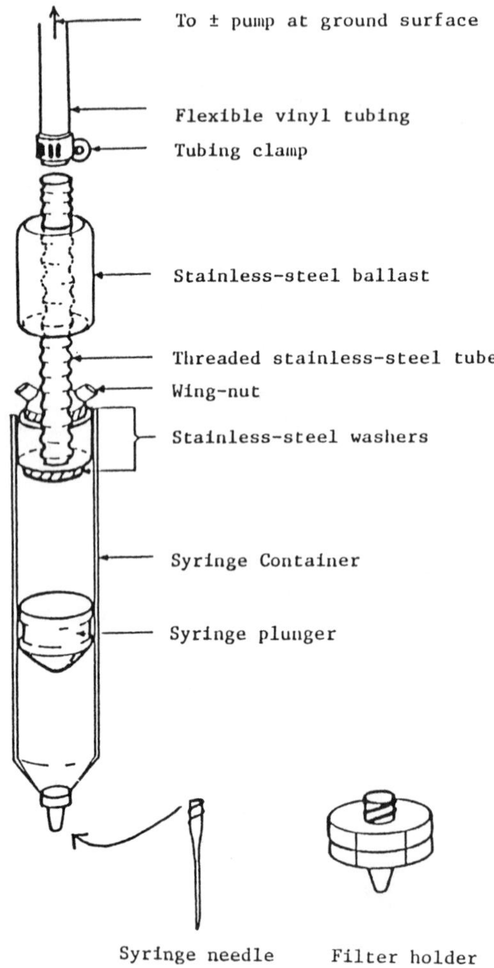

Figure 9.12. Syringe-type sampling device.

to the desired depth within the well. Then, by means of a hand pump or pressurized gas, a negative pressure is applied to the syringe plunger through flexible vinyl tubing running to the ground surface. The negative pressure causes the plunger to rise which then allows water to be drawn into the syringe via the syringe needle. The whole assembly is then retrieved to the surface where the sample can be transferred to a more permanent container. Modifications to the above general procedure involve isolating the sample

Table 9.18. Advantages and Disadvantages of Syringe Devices.[35]

Advantages	Disadvantages
The sample taken with a syringe device does not come in contact with any atmospheric gases and is subjected to a very slight negative pressure, thus neither aeration nor degassing of the sample should occur.	Syringes are inefficient for collecting large samples.
Samples can be collected at discrete intervals and from any depth in a well.	Syringes cannot be used to evacuate a well.
Syringes can be used to sample water from very slowly recharging wells because only a small volume of water is removed from the well.	Syringes are relatively new in this application and may not be as readily available as other sampling devices.
Syringes are or can be made of inert or nearly inert materials.	Sample contamination by components of "homemade" syringe sampling devices is possible unless fabrication materials are carefully selected.
Syringes are not restricted to the limits of suction lift.	The use of syringes is limited to water with a low suspended solids content.
The syringe can be utilized as the sample container, thus removing the possibility of cross-contamination between wells.	Some leakage has been found to occur around the plunger when syringes are used to sample water containing high levels of suspended solids.
Syringes are inexpensive, highly portable, and simple to operate.	
Syringe devices can be used in wells as small as 1½ in. inside diameter.	

from the pressurizing gas and/or preevacuated syringes. The relative advantages/disadvantages of syringe sampling devices are outlined in Table 9.18.

Suction-Lift Pumps

There is a variety of pumps available that can be used when the water table is within suction lift, that is, less than about 20 ft.[31] Centrifugal pumps are the most commonly available, are highly portable, and have pumping rates ranging from 5 to 40 gpm. Most of these require a foot valve on the end of the suction pipe to aid in maintaining a prime. Peristaltic pumps are generally low-volume suction pumps suitable for sampling shallow, small diameter wells. Pumping rates are generally low but can be readily controlled within desirable limits. The advantages/disadvantages of suction-lift pumps are outlined in Table 9.19.

Gas-Drive Devices

The general operation of gas-driven sampling devices as depicted in Figure 9.13 is as followed. Pressurized gas is pumped into the sample container, thus causing the ball check valve to close and effectively pressurize the container. The container is then lowered to a designated depth and the pressure released causing water to flow into the container. After allowing the container to fill, pressurized gas is again pumped into the container causing the water to flow up through the discharge tube to the surface.[35] The advantages/disadvantages of gas-driven sampling devices are outlined in Table 9.20.

Gas-Operated Squeeze (Bladder) Pumps

Gas-operated squeeze (bladder) pumps consist principally of a collapsible membrane inside a long, rigid housing, a compressed gas supply and appropriate control valves.[31] When the pump is submerged, water enters the collapsible membrane through the bottom check valve as shown in Figure 9.14. After the membrane has filled, gas pressure is applied to the annular space between the rigid housing and membrane, forcing the water upward through a sampling tube. When the pressure is released, the top check valve prevents the sample from flowing back down the discharge line, and water from the well again enters the pump through the bottom check valve. The advantages/disadvantages of bladder systems are outlined in Table 9.21.

Table 9.19. Advantages and Disadvantages of Suction-Lift Pumps.[31,35]

Advantages	Disadvantages
The flow rate of most suction-lift pumps is easily controlled.	Sampling is limited to situations in which the potentiometric level is less than 25 ft below the surface.
Suction-lift pumps are highly portable and readily available.	A drop in pressure due to the application of a strong negative pressure (suction) causes degassing of the sample and loss of volatiles.
Most suction-lift pumps are inexpensive in comparison to other sampling devices.	An electric power source is required for peristaltic pumps.
The pump does not contact the sample—only the tubing must be cleaned (peristaltic pump only).	The gasoline motor power source used for most centrifugal pumps provides potential for hydrocarbon contamination of samples.
Suction-lift pumps can be used in wells of any diameter, and can be used in nonplumb wells.	Pumping with centrifugal pumps results in aeration and turbulence.
Hand-operated pumps can operate over a wide range of pumping rates thus allowing for rapid evacuation and low rate sampling.	Centrifugal pumps may have to be primed, providing a possible source of sample contamination.
	Low pumping rates of peristaltic pumps make it difficult to evacuate the well bore in a reasonable amount of time.
	Where the sample comes in contact with the pump mechanism or tubing, the choice of appropriate materials for impellers (centrifugal pump) or flexible pump-head tubing (peristaltic pump) may be restrictive.

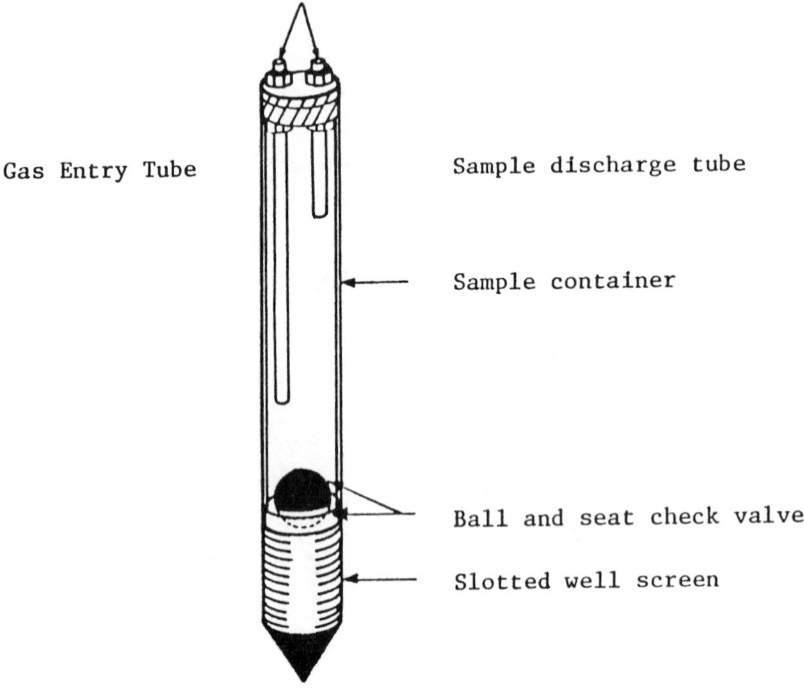

Figure 9.13. Typical gas-driven sampler.

Gas-Driven Piston Pumps

A modification of the pumps previously described is a double-acting, piston-type pump operated by compressed gas.[31] As shown in Figure 9.15, the driving gas enters and exhausts from the gas chambers between the two pistons and the intermediate connector that joins the two pistons. Built-in check valves at each end of the pump allow water to enter the cylinders on the suction stroke and to be expelled to the surface on the pressure stroke. Currently available gas-driven piston pumps are constructed basically of stainless steel, brass, and PVC. Pumping rates vary with the pumping head, but pumping rates of 2.5 to 8 gal/hr have been noted at 100 ft of pumping head. The advantages/disadvantages of gas-driven piston pumps are outlined in Table 9.22.

Table 9.20. Advantages and Disadvantages of Gas-Drive Devices.[35]

Advantages	Disadvantages
Gas-drive devices can be utilized in wells of 1½ in. inside diameter.	If air or oxygen are used as the driving gas oxidation may occur (causing the precipitation of metals), gas-stripping of volatiles may occur, or CO_2 may be driven from the sample (causing a pH shift), consequently air-lift sampling may not be appropriate for many chemically sensitive parameters.
Gas-drive devices are highly portable for most sampling applications and are inexpensive.	
Discrete depth sampling is possible.	
Gas-drive devices can provide delivery of sample at a controlled, nearly continuous rate.	An air compressor or large compressed-air tank must be transported to deep sampling locations, reducing portability.
The use of an inert driving gas (i.e., nitrogen) minimizes sample oxidation and other sample chemical alteration.	Application of excessive air pressure can rupture the gas entry or discharge tubing.
Devices can be installed permanently in boreholes without casing.	Devices installed permanently in boreholes without casing are difficult or impossible to retrieve for repair; proper installation and operation may be difficult to ensure.
Multiple installations can be achieved in a single well or borehole (either temporarily or permanently installed).	
Gas-drive devices can be constructed entirely of inert materials.	
The depth from which samples can be taken with gas-drive devices is limited only by the burst strength of the materials from which the device and tubing are made.	

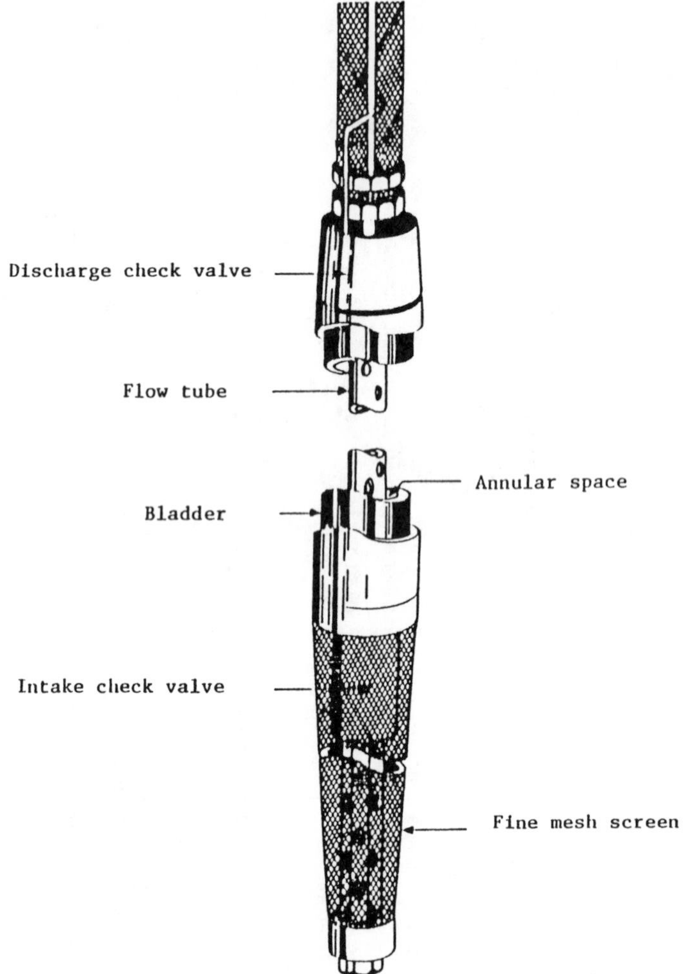

Figure 9.14. Typical bladder pump.

Submersible Electric Pumps

There are currently two different types of submersible electric pumps—the gear-drive and the helical rotor. Basically, both units consist of a motor and driving device that propels water through a discharge line to the surface.

Table 9.21. Advantages and Disadvantages of Gas-Operated Bladder Pumps.[31,35]

Advantages	Disadvantages
Most of these pumps have been designed specifically to sample for low levels of contaminants, so most are or can be made of inert or nearly inert materials.	Deep sampling requires large volumes of gas and longer cycles, thus increasing operating time and expense and reducing portability.
The driving gas does not contact the sample directly, thus problems of sample aeration or gas stripping are minimized.	Check valves in some pump models are subject to failure in water with high suspended solids content.
Bladder pumps are portable, though the accessory equipment may be cumbersome.	Most currently available pump models are expensive, though prices are highly variable.
The relatively high pumping rate (in comparison with other sampling devices) allows well evacuation and large sample volumes to be collected.	Minimum rate of sample discharge of some models may be higher than ideal for the sampling of volatile compounds.
The pumping rate of most of these pumps can be controlled rather easily to allow for both well evacuation at high rates and collection of volatile samples at low rates.	
Most models of these pumps are capable of pumping lifts in excess of 200 ft.	
These pumps are easy to disassemble for cleaning and repair.	
Most models of bladder pumps are designed for use in wells of 2-in. inside diameter; some are available for smaller diameter wells.	
Large diameter bladder pumps (i.e., 3½-in. outside diameter) are available for large diameter monitoring wells.	

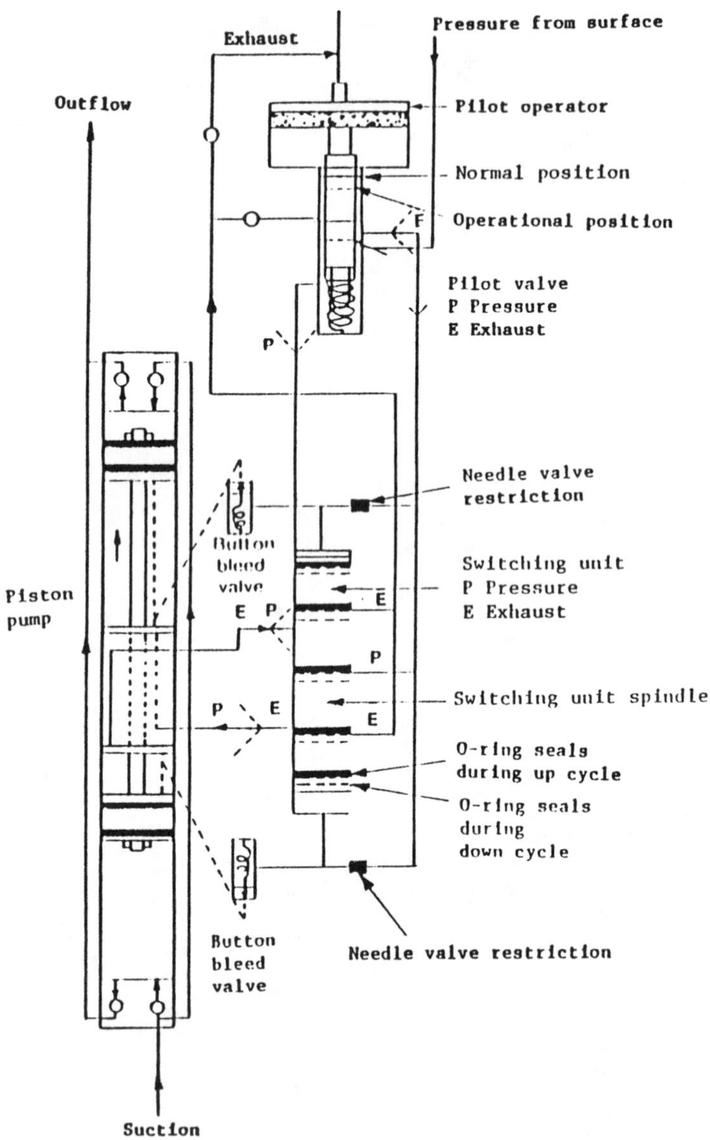

Figure 9.15. Gas-driven, double-acting piston pump.[31]

Table 9.22. Advantages and Disadvantages of Gas-Operated Double-Acting Piston Pumps.[31,35]

Advantages	Disadvantages
Because the sample is isolated from the driving gas, no aeration of the sample occurs.	Piston pumps are relatively expensive in comparison with other sampling devices.
This pump is relatively easy to operate and is easy to disassemble for cleaning and maintenance though some maintenance problems (i.e., with the pump motor or valving mechanism) cannot generally be solved in the field.	Unless the pump intake is filtered, particulate matter may damage the pump's intricate valving mechanism.
	The pump's intricate valving mechanism may cause a series of pressure drops in the sample leading to sample degassing and pH changes.
The pump provides a continuous sample over extended periods of time.	This pump is not highly portable—it must be vehicle-mounted.
Models of this pump are available for wells of 1½-in. inside diameter and for wells of 2-in. inside diameter or greater.	Fixed-length tubing bundles may be inconvenient for shallow, low-yield monitoring wells.
The pump uses compressed gas economically.	The tubing bundles may be difficult to clean in order to avoid cross-contamination.
Pumping lifts of more than 500 ft. can be overcome with this pump.	
Flow rate of the pump can be easily controlled by varying the driving gas pressure on the pump.	
The pump is or can be made of inert or nearly inert materials.	
The moderately high pumping rate at great depths allows for collection of large volumes of sample in a relatively short time.	

Table 9.23. Advantages and Disadvantages of Gear-Drive Electric Submersible Pumps.[35]

Advantages	Disadvantages
These pumps are constructed of inert or nearly inert materials, therefore they are suitable for sampling organics when the optionally available Teflon discharge line is used.	There is no control over flow rates, therefore it is not possible to adjust from a high pumping rate for well purging to a lower rate required for sampling of volatiles.
These pumps are highly portable and totally self-contained, except when auxiliary power sources are employed.	Sampling in wells with high levels of suspended solids may necessitate frequent replacement of gears.
These pumps provide a continuous sample over extended periods of time.	The potential for pressure changes (cavitation) exists at the drive mechanism; however, this has not been adequately evaluated.
Models are available for both 2-in. inside diameter and 3-in. inside diameter (or larger) wells.	
High pumping rates are possible making it feasible to use the pump for both well evacuation and sampling.	
Reasonably high pumping rates can be achieved to depths of 150 ft, and depth range can be extended through the use of an auxiliary power source.	
These pumps are easy to operate, clean and maintain in the field, and replacement parts are inexpensive.	
In comparison to other pumps offering the same performance, these pumps are inexpensive to purchase and operate.	

The main difference in the two types of pumps are the driving devices. The advantages/disadvantages of the two systems are outlined in Tables 9.23 and 9.24, respectively.

ADDITIONAL COMPARISONS AND ISSUES RELATED TO SAMPLING

An exhaustive study of sample retrieval systems has been reported by Gillham, et al.[32] Listed in Table 9.25 are the advantages and disadvantages of various systems as reported by Gillham, et al.[32] Additionally, Table 9.26 shows the suitability of the various sampling systems according to the parameter to be monitored.[32] Table 9.27 shows the characteristics of sampling devices for small diameter wells as outlined by Nielsen and Yeates.[35]

Additional work on sampling equipment has been conducted by McTigue and Kunzel[41] who discussed how cross-contamination, turbidity, and silt intrusion were eliminated in a well by installing gas-displacement samplers. Also, Hatayama[42] discussed two unique ground water sampling techniques: (1) single completion wells with submersible pump sampling; and (2) multiple-cased wells. Johnson[43] outlined the relative advantages and disadvantages of various techniques used for retrieving samples from a hydrocarbon spill site.

In addition to the equipment utilized to retrieve samples, there now exists a wide variety of sampling installation equipment. The trend in recent years has been towards using fewer boreholes by maximizing the amount of information obtained from a single borehole. For example, Rehtlane and Patton[44] discussed the conditions for which multiple-ported piezometers are more economical than multiple standpipe piezometers. Jehn[45] described the use of a submersible pump and packer system to isolate different zones in a multiple-aquifer system. Kerfoot[46] discussed the use of two- and three-dimensional probes for direct ground water flow measurement. The most exhaustive work on sampling installations is by Gillham, et al.[32] Their listing of the advantages and disadvantages of sampling installations is in Table 9.28.[32]

BIAS IN GROUND WATER QUALITY MONITORING

Successful completion of a monitoring well and use of proper sampling equipment will not altogether eliminate sources of bias in water quality sampling. Nacht[36] discussed the following factors which could affect sample integrity: (1) depth the sample is taken in the water column; (2) sample bottle

Table 9.24. Advantages and Disadvantages of the Helical Rotor Electric Submersible Pumps.[35]

Advantages	Disadvantages
Portable and relatively easy to transport in the field to remote locations.	The currently available pump unit is limited to 125 ft of pumping lift.
Well-suited for use in wells of 2-in. inside diameter.	High pumping rates with this pump lead to creation of turbulence, which may cause alteration of sample chemistry.
Relatively high pumping rates are possible with currently available units, thus well evacuation is possible.	Thorough cleaning and repair in the field may be difficult because the pump is moderately difficult to disassemble.
Has been specifically designed for monitoring ground water contamination and therefore is constructed mostly of inert or nearly inert materials.	Water with high suspended solids content can cause operational problems.
	The currently available model is expensive in comparison to other devices offering comparable performance.
	The pump must be cycled on/off approximately every 20 min to avoid overheating of the motor.
	The flow rate cannot be controlled, so the pump may not be suitable for taking samples for analysis of chemically sensitive parameters.

Table 9.25. Advantages and Disadvantages of Sample Collection Methods.[32]

Advantages	Disadvantages
Downhole Collection Devices	
General	
Greater potential to preserve sample integrity than many other methods because water is not driven by pressure differences.	Most devices are unsuitable for flushing, because they provide only discrete, and often very small volumes of water. This problem can be avoided by using another method, which may be disruptive, to flush the installation, prior to using the downhole collection device for sampling.
Bailers	
Inexpensive to purchase or fabricate and economical to operate. This may permit the assignment of one collection device for each installation to be sampled, thereby circumventing problems of cross-contamination.	Usually very time consuming when used for flushing installations, especially when the device has to be lowered to great depths. It can also be very physically demanding on the operator when the device is lowered and raised by hand.
Very simple to operate and require no special skill.	Can cause chemical alterations due to degassing, volatilization or atmospheric invasion when transferring the sample to the storage container.
Easily cleaned, though cleaning of ropes and/or cables may be more difficult.	
Can be made of inert materials.	
Very portable and require no power source.	
Mechanical Depth-Specific Samplers	
Inexpensive to construct.	Some of the materials used can cause contamination (e.g., rubber stoppers).
Very portable and require no power source.	Activating mechanism can be prone to malfunctions.
Stratified sampler is well suited for sampling distinct layers of immiscible fluids.	May be difficult to operate at great depths.
Can be made of inert materials.	Can cause chemical alteration when transferring sample to storage container.
Stratified sampler is easily cleaned.	Difficult to transfer sample to storage container.
	Kemmerer sampler is difficult to clean thoroughly.

Table 9.25, continued

Advantages	Disadvantages
Pneumatic Depth-Specific Samplers	
Can be made of inert materials. Easily portable and require only a small power source (e.g., hand-pump). Solinst sampler and syringe sampler can be flushed downhole with the water to be sampled. Syringe of the syringe sampler can be used as a short-term storage container. Syringe sampler is very inexpensive.	Types that are commercially available are moderately expensive. Westbay sampler is only compatible with the Westbay casing system. Solinst and Westbay samplers are difficult to clean. Materials used in disposable syringes of syringe samplers can contaminate the water. Water sample comes in contact with pressurizing gas in Solinst and Westbay samplers (but not in syringe samplers).
Suction-Lift Methods	
Simple, convenient to operate and easily portable. Inexpensive to purchase and to operate. Easily cleaned. Components can be of inert materials. Depending on the pumping mechanism, these methods can be very efficient for removing standing water from the sampling installations. Provide a continuous and variable flow-rate.	Limited to situations where the water level is less than 7–8 m (23–26 ft) below ground surface. Can cause sample bias as a result of degassing and atmospheric contamination, especially if the sample is taken from an in-line vacuum flask. Can cause contamination if water is allowed to touch pump components.
Positive-Displacement Methods	
General	
Reduced possibility of degassing and volatization because the sample is delivered to ground surface under positive pressure. In some situations, the pressure at ground surface may be substantially less than the natural water pressure in the formation and thus the degassing problem cannot be entirely ignored.	Cost of the commercially available pumps is substantial (roughly \$2,000 to \$5,000). It would therefore not be feasible to dedicate a sampling pump to each sampling point. Can be difficult to clean between sampling sessions. Cleaning of cables and/or delivery tubing is required between sampling points.

Table 9.25, continued

Advantages	Disadvantages
Positive-Displacement Methods, continued	
Sample does not contact the atmosphere.	Commercially available devices are too large for very small-diameter installations such as the bundle piezometers.
Sampling pumps for use in monitoring wells as small as 3.8–5 cm (1.5–2 in.) are commercially available.	
Most of the commercially available devices have a sufficient flow rate for flushing sampling installations.	
Submersible Centrifugal Pumps	
Can pump at large and variable flow rates.	Subject to excessive wear in abrasive or corrosive waters.
Johnson-Keck pumps can fit down wells as small as 5 cm (2 in.).	Conventional submersible pumps cannot be used in installations of diameter less than about 12 cm (4 in.).
Johnson-Keck pump is easily portable.	Potential for contaminating water because of contact with metals and lubricants is larger in conventional pumps.
Conventional pumps are usually much cheaper than the Johnson-Keck pump.	Johnson-Keck pump has intermittent flow (15 min on, 15 min off).
Johnson-Keck pump offers little potential for sample contamination because it is made mostly of stainless steel and Teflon.	
Submersible Piston Pumps	
Gas-drive piston pumps have small power requirements.	Rod pumps require large power source and are permanently mounted.
Gas-drive piston pump of Gillham and Johnson (1981) is inexpensive and can be assigned permanently to sampling point, thereby eliminating problems of cross-contamination.	Difficult to clean.
Double-acting pumps have continuous, adjustable flow rates.	When used as part of an installation, the gas-drive pump of Gillham and Johnson (1981) cannot be retrieved for servicing or repair.
Can be built of inert materials (most commercially available pumps are not, however).	Single-acting pumps have intermittent flow.
Gas-Squeeze Pump	
Can be built of inert materials.	Intermittent but adjustable flow.
Commercially available pumps can fit in installations as small as 5 cm (2 in.).	Require large but portable power source.

Table 9.25, continued

Advantages	Disadvantages
Submersible Piston Pumps, continued	
Can easily be taken apart for cleaning, but can be inconvenient to clean between sampling sessions.	
Easily portable.	
Good potential to preserve sample integrity because the driving-gas does not come in contact with the water sample.	
Gas-Lift Methods	
Simple to construct or are available commercially at relatively low cost.	Can only be used efficiently when roughly ⅓ of the underground portion of the device is submerged.
Can be used in very narrow installations.	Contamination of the sample with the driving gas, atmospheric contamination and degassing are all unavoidable.
Can be easily portable.	Need large power source (gas).
Easily cleaned.	
Gas-Drive Methods	
Can offer good potential for preserving sample integrity, because very little of the gas comes in contact with the water sample, and because the sample is driven by a gradient of positive pressure.	Not very efficient for flushing installations larger than about 2.5 cm (1 in.).
Can be incorporated as part of the sampling installation, thereby removing the possibility of cross-contamination.	Can be difficult to clean between sampling sessions.
The triple-tube sampler is well suited for installations of very narrow diameter; e.g., 0.95 cm (⅜ in.), where the only other possible sampling method is narrow-tube bailers or suction-lift (when applicable).	The driving gas comes in contact with the water, and therefore the beginning and the end of the slug of water obtained at the surface can be contaminated.
Inert materials can be used.	Cannot be retrieved for repair or servicing when used as part of a permanent sampling installation.
	Pump intermittently and at a variable flow rate.
Jet-Pumps	
Can be used at great depths.	Use circulating water which mixes with the pumped water. A large amount of water needs to be pumped before the circulating water has a composition that is close to the water in the installation.
Useful for flushing monitoring installations.	

Table 9.25, continued

Advantages	Disadvantages
Jet-Pumps, continued	
	The water entering the venturi assembly is subjected to a pressure drop (which may be large) and can therefore undergo degassing and/or volatilization.
	The circulating pump at the surface can contaminate the pumped water because of its materials and lubricants.
Destructive Sampling Methods	
Can provide very useful information in reconnaissance surveys and in other specific field situations.	Because no permanent installation is left in the ground, these methods cannot be used for monitoring long-term trends in water quality. In most cases, however, they do not interfere with the implantation of permanent installations.
Most of the techniques are used during the drilling operation and will not interfere with the implantation of a permanent installation.	Can result in large drilling costs.
Coring-extraction methods are the only convenient means of obtaining several parameters related to both the liquid and solid phases (e.g., in very fine-grained formations).	Water contained in cores can be contaminated with drilling fluids and can undergo degassing and volatilization at the ground surface.
Temporary installations can, in some situations, be the most cost-effective way of obtaining preliminary, and/or reconnaissance data.	

style and composition; (3) placement of the sample into the bottle; (4) preservation of the sample prior to analysis; (5) filtering of the sample; and (6) the chain of custody of the sample. The chain of custody is an important consideration for both the ground water sample itself and the results of its analysis (data). Keith, Wilson, and Esposito[47] state that the goals of a monitoring program can only be met if the variability in ground water quality data due to pollution can be distinguished from the variability due to other factors. Additional potential sources of variability include: (1) well construction; (2) pumping schedules; (3) geology-hydrogeology; (4) hydraulic conditions at the sample site; (5) physico-chemical processes in the vadose zone and aquifer; (6) sampling procedure; and (7) analytical laboratory. Barcelona[48] noted the need for cooperation between design, field, and laboratory staffs for cost-effective monitoring. Subjects of mutual concern to these staffs include: (1) proper well construction and development; (2) strict

Table 9.26. Suitability of Sampling Methods.[32]

Inorganic and Radioactive Parameters	Downhole Collection	Downhole Collection: Depth-Specific		Suction-Lift: Suction on:		Positive-Displacement: Centrifugal Pumps		Positive-Displacement: Piston Pumps		Squeeze-Pumps	Gas-Lift	Gas-Drive	Jet-Pumps	Destructive-Sampling	Destructive-Sampling: Temporary Installations	
	Bailers	Mechanical	Pneumatic	liquid phase	gas phase	Conventional	Johnson-Keck	Rod	Gas-Driven					Coring-Extraction	Screened Augers	Multiple Completion Wells
INORGANIC Contamination Parameters																
electrical condition	S	S	S	S	S	S	S	S	S	S	(U)	S	(U)	S	S	(L)
pH	(L)	(L)	S	S	U	S	S	(L)	S	S	U	(L)	L	(L)	S	(L)
redox condition	(L)	(L)	S	(L)	U	(L)	S	L	S	S	U	(L)	L	U	(L)	(U)
Nontoxic Constituents																
chloride	S	S	S	S	S	S	S	S	S	S	S	S	(L)	S	S	S
sulfate	(L)	(L)	S	(L)	(L)	S	S	S	S	S	U	(L)	(U)	(L)	(L)	(L)
sodium	S	S	S	S	S	S	S	S	S	S	S	S	(L)	S	S	(L)
ammonium	(L)	(L)	S	S	U	S	S	S	S	S	U	S	(L)	L	(L)	(L)
calcium, magnesium	S	S	S	S	(L)	S	S	S	S	S	(U)	S	(L)	(L)	(L)	(L)
iron, manganese	(L)	(L)	S	(L)	(U)	S	S	(L)	S	S	U	(L)	(L)	(L)	(L)	(L)
Toxic Constituents																
nitrate	S	S	S	S	S	S	S	S	S	S	(L)	S	(L)	(L)	S	(L)
fluoride	S	S	S	S	S	S	S	S	S	S	(L)	S	(L)	(L)	S	(L)
arsenic	(L)	S	S	(L)	(U)	S	S	(L)	S	S	U	(L)	(U)	(L)	(L)	(L)
selenium	(L)	S	S	(L)	(U)	S	S	(L)	S	S	U	(L)	(U)	(L)	(L)	(L)
barium	(L)	S	S	(L)	(U)	S	S	(L)	S	S	U	(L)	(U)	(L)	(L)	(L)
cadmium	(L)	S	S	(L)	(U)	S	S	(L)	S	S	U	(L)	(U)	(L)	(L)	(L)
chromium	(L)	S	S	(L)	(U)	S	S	(L)	S	S	U	(L)	(U)	(L)	(L)	(L)

lead	(L)	S	S	(L)	(U)	S	S	(L)	S	S	U	(L)	(U)	(L)	(L)	(L)
silver	(L)	S	S	(L)	(U)	S	S	(L)	S	S	U	(L)	(U)	(L)	(L)	(L)
mercury	(L)	S	S	(L)	(U)	S	S	(L)	S	S	U	(L)	(U)	(L)	(L)	(L)
RADIOACTIVE																
radium	S	S	S	S	(L)	S	S	S	S	S	U	S	(L)	(L)	(L)	(L)
gross alpha and beta	(L)	S	S	(L)	U	S	S	(L)	S	S	U	(L)	(L)	(U)	(L)	(L)
BIOLOGICAL																
coliform bacteria	S	(L)	S	S	(L)	L	(L)	U	(L)	S	(U)	(L)	(U)	S	(L)	(L)
ORGANIC																
Drinking-Water Stds.																
endrin	(L)	S	S	S	L	(L)	(L)	U	S	S	U	S	U	S	(L)	(L)
lindane	(L)	S	S	S	L	(L)	(L)	U	S	S	U	S	U	S	(L)	(L)
methoxychlor	(L)	S	S	S	L	(L)	(L)	U	S	S	U	S	U	S	(L)	(L)
toxaphene	(L)	S	S	S	L	(L)	(L)	U	S	S	U	S	U	S	(L)	(L)
2,4-D	(L)	S	S	S	L	(L)	(L)	U	S	S	U	S	U	S	(L)	(L)
2,4,5-TP Silvex	(L)	S	S	S	L	(L)	(L)	U	S	S	U	S	U	S	(L)	(L)
Quality Parameters																
phenols	(L)	(L)	S	(U)	U	(L)	(L)	U	(L)	S	U	(L)	(U)	(U)	(L)	(L)

Key: S: Suitable. L: Limited suitability; can be used for qualitative or approximate information. U: Unsuitable.
(): Procedural modifications can be used to improve suitability to next level.

Table 9.26, continued

Inorganic and Radioactive Parameters	Downhole Collection			Suction-Lift		Positive-Displacement									Destructive-Sampling		
		Depth-Specific		Suction on:		Centrifugal Pumps		Piston Pumps		Squeeze-Pumps					Temporary Installations	
	Bailers	Mechanical	Pneumatic	liquid phase	gas phase	Conventional	Johnson-Keck	Rod	Gas-Driven		Gas-Lift	Gas-Drive	Jet-Pumps	Coring-Extraction	Screened Augers	Multiple Completion Wells
Contamination Parameters																
total organic carbon	(L)	(L)	S	(U)	U	(L)	(L)	U	(L)	S	U	(L)	(U)	(U)	(L)	(L)
total organic halogen	(L)	(L)	S	(U)	U	(L)	(L)	U	(L)	S	U	(L)	(U)	(U)	(L)	(L)
Gasoline Components																
benzene	(L)	(L)	S	(U)	U	(L)	(L)	U	(L)	S	U	(L)	(U)	(U)	(L)	(L)
toluene	(L)	(L)	S	(U)	U	(L)	(L)	U	(L)	S	U	(L)	(U)	(U)	(L)	(L)
xylene	(L)	(L)	S	(U)	U	(L)	(L)	U	(L)	S	U	(L)	(U)	(U)	(L)	(L)
methyl t-butyl ether	(L)	(L)	S	(U)	U	(L)	(L)	U	(L)	S	U	(L)	(U)	(U)	(L)	(L)

Key: S: Suitable.
L: Limited suitability; can be used for qualitative or approximate information.
U: Unsuitable.
(): Procedural modifications can be used to improve suitability to next level.

Table 9.27. Summary of Characteristics of Sampling Devices Available for Small Diameter Monitoring Wells.[35]

Device	Minimum Well Diameter	Approximate Maximum Sampling Depth	Typical Sample Delivery @ Maximum Depth	Flow Controlability	Materials[a] (Sampling Device Only)	Potential for Chemical Alteration	Ease of Operating, Cleaning and Maintenance	Approximate Cost for Complete System[b]
Bailers	½ in.	Unlimited	Variable	Not applicable	Any	Slight-moderate	Easy	<$100–$200
Syringe samplers	1½ in.	Unlimited	0.2 gal	Not applicable	Stainless 316, Teflon or polyethylene/glass	Minimum-slight	Easy	<$100 (50 ml homemade) $1500 (850 ml commercially available)
Suction-lift (vacuum) pumps	½ in.	26 ft	Highly variable	Good	Highly variable	High-moderate	Easy	$100–$550
Gas-drive samplers	1 in.	300 ft	0.2 gpm	Fair	Teflon, PVC, polyethylene	Moderate-high	Easy	$300–$700
Bladder pumps	1½ in.	400 ft	0.5 gpm	Good	Stainless 316, Teflon/Viton, PVC, silicone	Minimum-slight	Easy	$1500–$4000
Gear-drive submersible pumps	2 in.	200 ft	0.5 gpm	Poor	Stainless 304, Teflon, Viton	Minimum-slight	Easy	$1200–$2000
Helical rotor submersible pumps	2 in.	125 ft	0.3 gpm	Poor	Stainless 304, EPDM, Teflon	Slight-moderate	Moderately difficult	$3500
Gas-driven piston pumps	1½ in.	500 ft	0.25 gpm	Good	Stainless 304, Teflon, Delrin	Slight-moderate	Easy to moderately difficult	$3400–$3800

[a]Materials dependent on manufacturer and specification of optional materials.
[b]Costs highly dependent on materials specified for devices and selection of accessory equipment.

Table 9.28. Advantages and Disadvantages of the Main Types of Sampling Installations.[32]

Advantages	Disadvantages
Single-Level Installations	
Suitable in any type of formation.	No information can be obtained regarding vertical distributions of ground water constituents from a single installation (piezometer nests can be used for this purpose but at a much higher cost than multilevel installations).
Simple design.	High cost per sampling point (especially at great depths) as compared to multilevel installations.
The installation itself and, where applicable, the packing and sealing materials can be implanted more easily than in the multilevel installations.	Contaminated water may by-pass installations with short-screened intervals. The problem is also present, but to a lesser extent, in multilevel installations.
No problems of vertical communication between sampling points due to leaky seals.	In most situations, long-screened installations provide concentration and hydraulic-head valves that are spatially averaged over the length of the screen and therefore, may not give accurate measurements of maximum concentrations.
Maximum permissible diameter is limited by size of borehole only. Common sizes vary from 1.2 to 15 cm (½–6 in.) for monitoring installations.	Because of possible dilution in the well, long-screened installations can be used to confirm the presence, but not the absence, of a contaminant.
Most common sizes of installation 5–10 cm (2–4 in.) do not restrict the choice of sample-collection methods.	Long-screened installations can contribute to aquifer contamination by providing a passage from contaminated zones to uncontaminated zones.
Multilevel Installations	
General	
Provide information on vertical distribution of ground water constituents.	The installation itself and, where applicable, the packing and sealing materials are more difficult to place than for single-level installations.
Lower cost per sampling point than single-level installations, especially for large numbers of sampling points of great depths.	The short-screened intervals commonly used in many multilevel installations can miss small zones of contaminated water in heterogeneous materials. The likelihood of this problem can be reduced considerably by using reconnaissance methods (e.g., destructive sampling).
In general, only small volumes of water need to be removed to purge the installation because of the usually small diameters.	

Table 9.28, continued

Advantages	Disadvantages
Multilevel Installations	
General, continued	
	Possibility of vertical communication between sampling points due to leaky seals. Problems of leaky seals can occur in single-level installations, but the consequences are usually less severe than for multilevel installations.
"Flow-through" Wells	
Cheaper than most other multilevel installations.	Basic underlying assumption that water will flow through the well screen without having its course altered cannot be supported for most natural systems.
Simple design and operation.	Suitable only in hydraulically homogeneous formations with no vertical gradient.
Can be installed as easily as single-level installations.	The concentration and hydraulic-head valves are spatially averaged over the length of the screen and, therefore, may not give accurate measurements of maximum concentrations.
The number of sampling points per installation is not limited by the diameter of the borehole.	Can be used to confirm the presence, but not the absence, of a contaminant because of possible dilution in the well.
If the flow-through assumption is valid, there is no need to flush the installation and samples that are representative of a given depth can be obtained with a simple depth-specific sampler.	Can contribute to aquifer contamination by providing a passage from contaminated zones to uncontaminated zones.
Multiple-Piezometers	
Simple design and operation.	Difficulty of installation may increase with number of piezometers per borehole.
Less expensive than piezometer nests of corresponding number of sampling points, and less expensive than most other multilevel installations.	Number of sampling points (i.e., piezometers) per borehole is restricted by the diameter of the individual piezometers and by that of the borehole.
	Bundle piezometers have a higher risk of hydraulic connections between sampling points where seals are required and are therefore most suitable in cohesionless formations that will collapse around the tips.

Table 9.28, continued

Advantages	Disadvantages
Multiple-Piezometers, continued	
	The choice of sample collection methods is usually severely restricted by the commonly small piezometer diameters.
	The small diameters of the individual tubes provide a small storage volume and thus they are generally of limited use in fine-grained materials.
Gas-Drive Installations	
Potential for cross-contamination is small because the sample-collection method is an integral part of the installation.	Difficult to install properly.
Operation is relatively easy and safer than most other installations for hazardous contaminants.	Moderately expensive per sampling point.
Little flushing is necessary because there is little mixing between incoming water from the formation and stagnant water.	Number of sampling points is limited by diameter of borehole. Commonly 3–4 sampling points for 15 cm (6 in.) borehole.
	The device cannot be retrieved for servicing or repairs. If malfunctions occur, the sampling point is lost.
	Hydraulic head measurements are of questionable value because of the presence of check-valves.
	The choice of sample collection method is restricted to gas-drive (and suction-lift for shallow water tables).
Multiple-Port Casings	
Except for the Westbay system, the cost per sampling point is relatively small.	The Westbay system is very expensive, but becomes economically viable for large numbers of sampling points at great depths.
Seals between screens can be provided by permanent packers or with traditional backfilled seals.	Except for the Westbay system, the number of sampling points per installation is limited by diameter of the borehole.
	Assembly and implantation can be difficult.
	The operation of the Westbay system can be difficult and time consuming and requires special operator skills.
	The Westbay system is prone to mechanical difficulties because of the downhole complexity of the system.

adherence to sampling protocols; (3) subsurface reactivity and plume dynamics; and (4) difficulties involved in the application of surface water analytical methodologies to ground water samples.

Keith, Frank, and Mossman[49] emphasized that incorrect laboratory results can lead to incorrect interpretations. This same study outlined a five-point plan for assuring laboratory accuracy. The plan consisted of: (1) developing criteria by which a laboratory can be evaluated in terms of assessing its potential for producing accurate data; (2) establishing a dialogue with the laboratory personnel in which certain questions are asked and certain data are provided when samples are submitted; (3) developing an understanding of what laboratory certification means, and does not mean, in terms of the accuracy of analysis; (4) developing methods of assessing the accuracy of the analytical results; and (5) establishing a contract with the laboratory which provides a financial incentive to produce accurate results and protects the client from paying full cost of documentably unacceptable results. The basic point to be noted is the need for a Quality Assurance/Quality Control (QA/QC) program. Information on QA/QC is presented in the next section.

To serve as an example of the importance of laboratory procedures, Ahern and Grabowski[50] cited five projects in which inaccurate water quality data were produced because of organic chemical interferences with standard analytical laboratory results. The inaccuracies were detected by a lack of correlation between suspected water quality data and other related water quality data, and by inexplicable spatial and temporal trends in the data. In addition, Dunbar et al.[51] reported the following anomalies and their corresponding causes: (1) elevated pH values due to poor well construction procedures; (2) the presence of tetrahydrofuran caused by the use of glue at well casing joints; (3) the temporary occurrence of trihalomethanes produced during the breakdown of synthetic drilling fluid, and (4) cross-contamination of ground water samples due to incomplete equipment decontamination procedures.

Keith and Wilson[52] discuss how some variables in sample collection and analysis may influence the concentration levels of the constituents being monitored. The sample collection variables that can cause chemical alterations of the true quality of a ground water sample include whether a sample is filtered or not, filter type and pore size, pumping mechanism, discharge pipe size, point in time after initiation of pumping that sample is collected, and sample storage and preservation. To illustrate problems with sample analysis, Keith and Wilson[52] developed Figure 9.16. The graphs represent the results of analyses of identical ground water samples by three different laboratories. The fact that the individual lines diverge shows that different labs can produce radically different results. Perhaps the most interesting feature of Figure 9.16 is that the varying data were generated for water quality

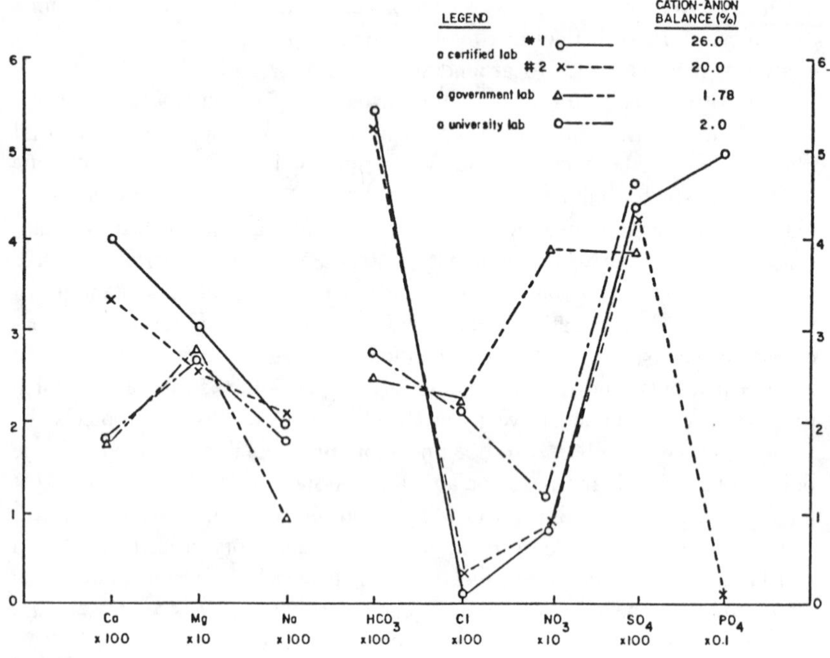

Figure 9.16. Variability in analytical results among three laboratories of one sample.[52]

parameters that can be considered as traditional and which have been routinely analyzed for years.

In summary, Tables 9.29 and 9.30 highlight the in-depth study of Gillham, et al.[32] into sources of bias in ground water quality monitoring. Table 9.29 contains a listing of biases vs quality parameter, and Table 9.30 lists some recommended sampling procedures for various quality parameters.

QUALITY ASSURANCE/QUALITY CONTROL

Quality assurance (QA) refers to a system for ensuring that all information, data, and resulting decisions compiled under a specific ground water monitoring task are technically sound, statistically valid, and properly documented. Quality control (QC) is the mechanism through which quality assurance achieves its goals. Quality-control programs define the frequency and methods of checks, audits, and reviews necessary to identify problems

Table 9.29. Potential Sources of Bias in Ground Water Quality Monitoring.[32]

Inorganic and Radioactive Parameters	Direct	Indirect
Contamination Parameters		
electrical cond.	temperature, suspended particles, precipitation, adsorption/exchange	degassing, changes in Eh, pH
pH }		}
redox condition }	CO_2 degassing, precipitation of carbonates and oxides	O_2 invasion, temperature, microbial activity
Nontoxic Constituents		
chloride	contact with skin	—
sulfate	precipitation of gypsum, reduction	—
sodium	adsorption/exchange, leaching from glass, contact with skin	—
ammonium	adsorption/exchange, volatilization at low pH	—
calcium }		CO_2 degassing
magnesium }	adsorption/exchange	—
iron, manganese	adsorption/exchange, precipitation	O_2 invasion, CO_2 degassing
Toxic Constituents		
nitrate	denitrification, nitrification	—
fluoride	precipitation	—
arsenic }		}
selenium }		}
barium }		}
cadmium }	adsorption/exchange/leaching from sampling equipment, precipitation	}
chromium }		}
lead }		}
silver }		CO_2 degassing, O_2 invasion
mercury }		}
RADIOACTIVE		
radium }		
gross alpha and beta }	adsorption/exchange, precipitation, degassing	

[a] Cross-contamination is a problem for all parameters, but is indicated here as a source of bias only if it is especially likely to occur.

Table 9.29, continued

Biological and Organic Parameters	Direct	Indirect
BIOLOGICAL		
coliform bacteria	cross-contamination[a]	O_2 invasion, pH, precipitation of other compounds
ORGANIC		
Drinking-Water Stds.		
endrin	sorption, cross-contamination,[a] biodegradation during storage	O_2 invasion, pH, precipitation of other compounds
lindane	sorption, cross-contamination,[a] biodegradation during storage	O_2 invasion, pH, precipitation of other compounds
methoxychlor	sorption, cross-contamination,[a] biodegradation during storage	O_2 invasion, pH, precipitation of other compounds
toxaphene	sorption, cross-contamination,[a] biodegradation during storage	O_2 invasion, pH, precipitation of other compounds
2, 4-D	sorption, cross-contamination,[a] biodegradation during storage	O_2 invasion, pH, precipitation of other compounds
2, 4, 5-TP Silvex	sorption, cross-contamination,[a] biodegradation during storage	O_2 invasion, pH, precipitation of other compounds
Quality Parameters		
phenols	sorption, cross-contamination,[a] leaching, biodegradation during storage	O_2 invasion
Contamination Parameters		
total organic carbon	sorption, cross-contamination,[a] leaching, volatilization, biodegradation during storage	O_2 invasion
total organic halogen	sorption, cross-contamination,[a] leaching, volatilization, biodegradation during storage	O_2 invasion
Gasoline Components		
benzene	sorption, cross-contamination,[a] leaching, volatilization, biodegradation during storage	O_2 invasion
toluene	sorption, cross-contamination,[a] leaching, volatilization, biodegradation during storage	O_2 invasion
xylene	diffusion through plastics; sorption, cross-contamination,[a] leaching, volatilization, biodegradation during storage	O_2 invasion
methyl t-butyl ether	diffusion through plastics; sorption, cross-contamination,[a] leaching, volatilization, biodegradation during storage	O_2 invasion

[a] Cross-contamination is a problem for all parameters, but is indicated here as a source of bias only if it is especially likely to occur.

Table 9.30. Summary of Recommended Sampling Procedures.[32]

Inorganic and Radioactive Parameters	Possible Sources of Bias	Possible Materials that Can Cause Bias	Possible Sampling Methods that Can Cause Bias	Recommended Preservatives and/or Treatment
INORGANIC				
Contamination Parameters				
electrical cond.	temperature, suspended particles	—	jet-pump	4°C, or field measurement
pH	CO_2 degassing, O_2 invasion	—	gas-lift, jet-pump	field measurement
redox condition	O_2 invasion, microbial activity	—	jet-pump	
Nontoxic Constituents				
chloride	skin contact	—		none
sodium	skin contact, ads./exch./leaching from glass	glass	jet-pump	4°C
sulfate	gypsum precip., microbial activity	—		field filter, HNO_3 (pH2)
ammonium		—		field filter, 4°C, H_2SO_4 (pH2)
calcium, magnesium	O_2 invasion, pH, ads./exch.	—	gas-lift, jet-pump	field filter, HNO_3 (pH2)
iron, manganese	changes in Eh, pH before filtering	soft or galvanized steel		
Toxic Constituents				
nitrate	microbial activity	—		4°C
fluoride	precip.	—	jet-pump	none
arsenic				
selenium				
barium		all metals		
cadmium	changes in Eh, pH before filtering,		gas-lift, jet-pump,	
chromium	ads./exch. on sample container		conventional submersible-pump	field filter, HNO_3 (pH2)
lead		PVC, PE, PP, all metals		
silver		all metals		
mercury				
RADIOACTIVE				
radium	changes in Eh, pH before filtering,	—	gas-lift, suction-lift	—
gross alpha and beta	degassing	—		—

Table 9.30, continued

Biological and Organic Parameters	Most Likely Sources of Bias	Possible Materials that Can Cause Bias	Possible Sampling Methods that Can Cause Bias	Recommended Preservatives and/or Treatment
BIOLOGICAL				
coliform bacteria	cross-contamination			
ORGANIC				
Drinking-Water Stds.			gas-lift, jet-pump	4°
endrin				
lindane				
methoxychlor				
toxaphene				
2, 4-D	sorption, oxidation biodegradation			
2, 4, 5-TP Silvex				
Quality Parameters		All, except: • Teflon • stainless-steel • glass		
phenols				4°C, H_2SO_4 (pH2)
Contamination Parameters				
total organic carbon	volatilization, sorption, leaching			4°C, H_2SO_4 or HCl (pH2)
total organic halogen			suction-lift, rod-pump, gas-lift, jet-pump, coring-extraction	4°C
Gasoline Components				
benzene	volatilization, sorption, oxidation			
toluene				
xylene	volatilization, sorption, diffusion			4°C
methyl t-butyl ether	volatilization, sorption, oxidation			

and dictate corrective action, thus verifying product quality. Table 9.31 includes definitions of several appropriate terms.[53]

The soundness of an organization's QA/QC program has a direct bearing on the integrity of its sampling and laboratory work. Laboratories should have a rigorous quality control program to guarantee that the measurements that they perform generate precise, accurate, and reliable data. In some states, certain departments and agencies of the state government, which are concerned with monitoring ground water quality, check on the quality control programs of commercial analytical laboratories and issue certifications

Table 9.31. Glossary of Terms for Quality Assurance/Quality Control

AUDIT: A systematic check to determine the quality of operation of some function or activity. Audits may be of two basic types: (1) performance audits in which quantitative data are independently obtained for comparison with routinely obtained data in a measurement system, or (2) system audits of a qualitative nature that consist of an on-site review of a laboratory's quality assurance system and physical facilities for sampling, calibration, and measurement.

DATA QUALITY: The totality of features and characteristics of data that bears on its ability to satisfy a given purpose. The characteristics of major importance are accuracy, precision, completeness, representativeness, and comparability. These characteristics are defined as follows:

- *Accuracy*—the degree of agreement of a measurement (or an average of measurements of the same thing), X, with an accepted reference or true value, T, usually expressed as the difference between the two values, $X-T$, or the difference as a percentage of the reference or true value, $100\ (X-T)/T$, and sometimes expressed as a ratio, X/T. Accuracy is a measure of the bias in a system.
- *Precision*—a measure of mutual agreement among individual measurements of the same property, usually under prescribed similar conditions. Precision is best expressed in terms of the standard deviation. Various measures of precision exist depending upon the "prescribed similar conditions."
- *Completeness*—a measure of the amount of valid data obtained from a measurement system compared to the amount that was expected to be obtained under correct normal conditions.
- *Representativeness*—expresses the degree to which data accurately and precisely represent a characteristic of a population, parameter variations at a sampling point, a process condition, or an environmental condition.
- *Comparability*—expresses the confidence with which one data set can be compared to another.

DATA VALIDATION: A systematic process for reviewing a body of data against a set of criteria to provide assurance that the data are adequate for the intended use. Data validation consists of data editing, screening, checking, auditing, verification, certification, and review.

Table 9.31, continued

ENVIRONMENTALLY RELATED MEASUREMENTS: A term used to describe essentially all field and laboratory investigations that generate data involving (1) the measurement of chemical, physical, or biological parameters in the environment; (2) the determination of the presence or absence of criteria or priority pollutants in waste streams; (3) assessment of health and ecological effect studies; (4) conduction of clinical and epidemiological investigations; (5) performance of engineering and process evaluations; (6) study of laboratory simulation of environmental events; and (7) study or measurement of pollutant transport and fate, including diffusion models.

PERFORMANCE AUDITS: Procedures used to determine quantitatively the accuracy of the total measurement system or component parts thereof.

QUALITY ASSURANCE: The total integrated program for assuring the reliability of monitoring and measurement data. A system for integrating the quality planning, quality assessment, and quality improvement efforts to meet user requirements.

QUALITY ASSURANCE PROGRAM PLAN: An orderly assemblage of management policies, objectives, principles, and general procedures by which an agency or laboratory outlines how it intends to produce data of known and accepted quality.

QUALITY ASSURANCE PROJECT PLAN:
An orderly assembly of detailed and specific procedures which delineates how data of known and accepted quality is produced for a specific project. (A given agency or laboratory would have only one quality assurance program plan, but would have a quality assurance project plan for each of its projects.)

QUALITY CONTROL: The routine application of procedures for obtaining prescribed standards of performance in the monitoring and measurement process.

---

or maintain a list of approved laboratories. A number of steps that are part of many laboratory quality-control programs are outlined by the American Public Health Association.[54] These include written plans for:

- Organization of laboratory personnel
  - assignment of responsibilities
  - training programs
  - coordination between individuals and groups
- Conducting analytical procedures
  - policy for selection of methods
  - use of established analytical procedures
  - documentation of sample collection, presentation, handling, and chain of custody
  - data review and approval
  - record keeping

- Maintenance of equipment
  - regular maintenance checks
  - regular calibration
  - quality control statistics
  - performance, tests including duplicate analyses of samples, and reference and control sample analyses
- Participation in interlaboratory tests

In addition to state agency QA/QC programs the U.S. Environmental Protection Agency (EPA) policy requires participation by all EPA regional offices, program offices, EPA laboratories, and states in a centrally-managed quality assurance program. This requirement applies to all environmental monitoring and measurement efforts mandated or supported by EPA through regulations, grants, contracts, or other formalized means not currently covered by regulation.

Guidelines for Preparing QA Plans

Sixteen items should be included in a QA plan, and they can be grouped together as follows:[52]

A. Format
   (1) title page
   (2) table of contents
B. Project Overview (What is the purpose of the project?)
   (3) project description
   (4) project organization and responsibility
C. Data Quality Objectives (What will be required?)
   (5) QA objectives for measurement data in terms of precision, accuracy, completeness, representativeness, and comparability
D. Measurement Activities (How will it be done?)
   (6) sampling procedures
   (7) sample custody
   (8) calibration procedures and frequency
   (9) analytical procedures
   (10) data reduction, validation, and reporting
   (11) internal quality control checks, and frequency
   (13) preventive maintenance

E. Quality Assurance (Can the results be trusted?)
  - (12) performance and systems audits and frequency
  - (14) specific routine procedures to be used to assess data precision, accuracy, and completeness of specific measurement parameters involved
  - (15) corrective action
  - (16) quality assurance reports to management

This grouping could be useful in at least a couple of ways. For someone writing a QA plan, particularly for the first time, it might clarify the way in which the sixteen items relate to each other and to the plan as a whole. Additionally, there are occasions, particularly in small or short-term projects, when something less than a complete QA plan would be appropriate. In these cases, a smaller document organized around the major headings listed above might be in order.

The initial step for any sampling or analytical work should be to strictly define the program goals. Once the goals have been defined, a program must be designed that will meet these program goals. QC and QA measures will be the mechanisms used to monitor the program and to ensure that all data generated are suitable for their intended use. A knowledgeable person who is not directly involved in the sampling or analysis must be assigned the responsibility of ensuring that the QC/QA measures are properly employed. All program details should be put in writing and assignments made to appropriate personnel. If an appropriate program is designed, tasks are assigned to knowledgeable personnel, and sufficient QC/QA steps are employed, the program should meet and possibly surpass its goals in most cases; at worst the failure to meet the program goals will be detected and the usefulness of any data will be quantified.[53] The following subsections summarize the essential components of a QA plan.

Project Description

The plan should provide a brief general description of the project, including the experimental design. It must have sufficient detail to allow those individuals responsible for review and approval of the QA plan to perform their task. Where appropriate, include the following: (1) flow diagrams, tables and charts; (2) dates anticipated for start and completion; and (3) intended end use of acquired data.

Project Organization and Responsibility

The plan should include a table or chart showing the project organization and line authority. The key individuals should be listed, including the Quality Assurance Officer (QAO), who are responsible for ensuring the collection of valid measurement data and the routine assessment of measurement systems for precision and accuracy.

QA Objectives for Measurement Data in Terms of Precision, Accuracy, Completeness, Representativeness, and Comparability

For each major measurement parameter, including all pollutant measurement systems, list the QA objectives for precision, accuracy, and completeness. These QA objectives should be summarized in a table; Table 9.32 has an example of the format.

All measurements should be made so that results are representative of the media (soil, ground water, biota, etc.) and conditions being measured. Unless otherwise specified, all data must be calculated and reported in units consistent with other organizations reporting similar data to allow comparability of data bases among organizations. Data quality objectives for accuracy and precision established for each measurement parameter should be based on prior knowledge of the measurement system employed and method validation studies using replicates, spikes, standards, calibrations, recovery studies, etc., and the requirements of the specific project or program.

Sampling Procedures

For each major measurement parameter(s), including all pollutant measurement systems, a description should be provided of the sampling procedures to be used. Where applicable, include the following:

1. description of techniques or guidelines used to select sampling sites
2. specific sampling procedures to be used (by reference in the case of standard procedures and by actual description of the entire procedure in the case of nonstandard procedures)
3. charts, flow diagrams or tables delineating sampling program operations
4. a description of containers, procedures, reagents, etc., used for sample collection, preservation, transport, and storage

Table 9.32. Example of Format to Summarize Precision, Accuracy, and Completeness Objectives.

Measurement Parameter (Method)	Reference	Experimental Conditions	Precision, Std. Dev.	Accuracy	Completeness
NO_2 (Chemiluminescent)	EPA 650/4-75-011 February 1975	Atmospheric samples spiked with NO_2 as needed	10%	5%	90%
SO_2 (24 hr.) (Pararosaniline)	EPA 650/4-74-027 December 1973	Synthetic atmosphere	20%	15%	90%
.	.	.	.	.	.
.	.	.	.	.	.
.	.	.	.	.	.

5. special conditions for the preparation of sampling equipment and containers to avoid sample contamination (e.g., containers for organics should be solvent-rinsed; containers for trace metals should be acid-rinsed)
6. sample preservation methods and holding times
7. time considerations for shipping samples promptly to the laboratory
8. sample custody or chain-of-custody procedures (to be described later)
9. forms, notebooks, and procedures to be used to record sample history, sampling conditions, and analyses to be performed

Sample Custody

As a minimum, the following sample custody procedures should be addressed in QA plans:

1. Field sampling operations:
   - (a) documentation of procedures for preparation of reagents or supplies which become an integral part of the sample (e.g., filters, and absorbing reagents)
   - (b) procedures and forms for recording the exact location and specific considerations associated with sample acquisition
   - (c) documentation of specific sample preservation method
   - (d) pre-prepared sample labels containing all information necessary for effective sample tracking.
   - (e) standardized field tracking reporting forms to establish sample custody in the field prior to shipment
2. Laboratory operations:
   - (a) identification of responsible party to act as sample custodian at the laboratory facility authorized to sign for incoming field samples, obtain documents of shipment (e.g., bill of lading number or mail receipt), and verify the data entered onto the sample custody records
   - (b) provision for a laboratory sample custody log consisting of serially numbered standard lab-tracking report sheets
   - (c) specification of laboratory sample custody procedures for sample handling, storage, and dispersement for analysis

Calibration Procedures and Frequency

The QA plan should include calibration procedures and information as follows:

1. for each major measurement parameter including all pollutant measurement systems, reference the applicable Standard Operating Procedure (SOP) or provide a written description of the calibration procedure(s) to be used
2. list the frequency planned for recalibration
3. list the calibration standards to be used and the source(s), including traceability procedures

A large number of laboratory and field operations can be standardized and written as Standard Operating Procedures (SOP). When such procedures are applicable and available, they may be incorporated into the QA plan by reference. QA plans should provide for the review of all activities which could directly or indirectly influence data quality and the determination of those operations which must be covered by SOPs. Examples are:

1. general network design
2. specific sampling site selection
3. sampling and analytical methodology
4. probes, collective devices, storage containers, and sample additives or preservatives
5. special precautions, such as heat, light, reactivity, combustibility, and holding times
6. federal reference, equivalent or alternative test procedures
7. instrumentation selection and use
8. calibration and standardization
9. preventive and remedial maintenance
10. replicate sampling
11. blind and spiked samples
12. colocated samplers
13. QC procedures such as intralaboratory and intrafield activities, and interlaboratory and interfield activities
14. documentation
15. sample custody
16. transportation
17. safety

18. data handling procedures
19. service contracts
20. measurement of precision, accuracy, completeness, representativeness, and comparability
21. document control

Analytical Procedures

For each measurement parameter, including all pollutant measurement systems, reference the applicable Standard Operating Procedure (SOP) or provide a written description of the analytical procedure(s) to be used. Officially approved EPA procedures should be used when available.

Data Reduction, Validation and Reporting

For each major measurement parameter, including all pollutant measurement systems, the QA plan should include brief descriptions of the following:

1. the data reduction scheme planned on collected data, including all equations used to calculate the concentration or value of the measured parameter and reporting units
2. the principal criteria that will be used to validate data integrity during collection and reporting of data
3. the methods used to identify and treat outliers
4. the data flow or reporting scheme from collection of raw data through storage of validated concentrations (a flow chart will usually be needed)
5. key individuals who will handle the data in reporting scheme (if this has already been described under project organization and responsibilities, it need not be repeated here)

Internal Quality Control Checks

The QA plan should describe and/or reference all specific internal quality control ("internal" refers to both laboratory and field activities) methods to be followed. Examples of items to be considered include:

1. replicates

2. spiked samples
3. split samples
4. control charts
5. blanks
6. internal standards
7. zero and span gases
8. quality control samples
9. surrogate samples
10. calibration standards and devices
11. reagent checks

Preventive Maintenance

The following types of preventive maintenance items should be considered and addressed in the QA plan:

1. a schedule of important preventive maintenance tasks that must be carried out to minimize downtime of the measurement systems
2. a list of any critical spare parts that should be on hand to minimize downtime

Performance and System Audits

Each QA plan should describe the internal and external performance and systems audits which will be required to monitor the capability and performance of the total measurement system(s). The systems audit consists of evaluation of all components of the measurement systems to determine proper selection and use of these components. This audit includes a careful evaluation of both field and laboratory quality control procedures. Systems audits are normally performed prior to or shortly after systems are operational; however, such audits should be performed on a regularly scheduled basis during the lifetime of the project or continuing operation.

After systems are operational and generating data, performance audits are conducted periodically to determine the accuracy of the total measurement system(s) or component parts thereof. The QA plan should include a schedule for conducting performance audits for each measurement parameter, including a performance audit for all measurement systems. As part of the performance audit process, laboratories may be required to participate in analysis of performance evaluation samples related to specific projects. Project plans should also indicate, where applicable, scheduled participation in all other inter-laboratory performance evaluation studies.

Specific Routine Procedures Used to Assess Data Precision/Accuracy and Completeness

For each major measurement parameter, including all pollutant measurement systems, the QA plan should describe the routine procedures used to assess the precision, accuracy, and completeness of the measurement data. These procedures should include the equations to calculate precision, accuracy and completeness, and the methods used to gather data for the precision and accuracy calculations. Examples of these procedures include:

1. Central tendency and dispersion
   (a) arithmetic mean
   (b) range
   (c) standard deviation
   (d) relative standard deviation
   (e) pooled standard deviation
   (f) geometric mean
2. Measures of variability
   (a) accuracy
   (b) bias
   (c) precision; within laboratory and between laboratories
3. Significance test
   (a) u-test
   (b) t-test
   (c) F-test
   (d) Chi-square test
4. Confidence limits
5. Testing for outliers

Corrective Action

Corrective action procedures should be described in the QA plan. These procedures should include the following elements:

1. the predetermined limits for data acceptability beyond which corrective action is required

2. procedures for corrective action
3. for each measurement system, identify the responsible individual for initiating the corrective action and also the individual responsible for approving the corrective action, if necessary

Corrective actions may also be initiated as a result of other QA activities, including: performance audits; systems audits; laboratory/inter-field comparison studies; and QA program audits. A formal corrective action program is more difficult to define for these QA activities in advance and may be defined as the need arises.

Quality Assurance Reports to Management

QA plans should provide a mechanism for periodic reporting to management on the performance of measurement systems and data quality. The individual(s) responsible for preparing the periodic reports should be identified. As a minimum, these reports should include:

1. periodic assessment of measurement data accuracy, precision and completeness
2. results of performance audits
3. results of system audits
4. significant QA problems and recommended solutions

STATISTICAL ANALYSES OF GROUND WATER QUALITY DATA

Statistical analyses can serve three general purposes when applied to ground water quality monitoring data. First, appropriate statistics can serve to summarize large amounts of data. Summarizing data in this fashion allows for easier presentation and interpretation. The second purpose is to aid in interpreting what the data mean and/or allow comparisons of different sets of data. The third purpose of statistical procedures is to aid in determining sampling requirements. For example, Nelson and Ward[55] show how a statistical determination can be made of the number of samples required in both time and space to produce a statistically significant data set.

The remainder of this discussion will be heavily weighted towards analyses that aid in identifying data indicative of ground water contamination. The majority of work in the area of ground water statistical procedures has resulted from the requirements outlined for monitoring at RCRA facilities. The major objective of RCRA monitoring procedures is to detect ground water contamination (and its extent) if it exists. Some of the procedures outlined are also applicable to other types of monitoring networks.

Basic Statistics

The mean ($\bar{X}$) of a set of data and its variance (S^2) are basic statistics that give insight as to the properties of a whole population. The definitions are:

Mean ($\bar{X}$)—The arithmetic average of "N" observations.

$$X = \frac{X_1 + X_2 + \ . \ . \ . \ X_N}{N} = \frac{1}{N} \sum_{i=1}^{N} X_i$$

where:

N = total number of observations

X_i = individual observations

Variance (S^2) and Standard Deviation (S)—The variance is the sum of the squared deviations (differences) of the individual observations from the mean.

$$S^2 = \frac{1}{(N-1)} \sum_{i=1}^{N} (X_i - \bar{X})^2$$

The standard deviation is simply the positive square root of the variance.

The mean ($\bar{X}$) of the set of data is the best estimate which can be generated for the overall mean (μ) of the whole population of data. The mean gives an indication of the central tendency of the population. The variance (S^2) of the set of data provides the best estimate of the true variance (σ^2) of the whole population. The variance (and standard deviation) give insight as to the variability of the data within a population.

It should be noted that many of the statistical procedures outlined below require the use of the population mean (μ) or standard deviation (σ). Since it is impossible to gather every possible data point within a population, it is necessary to estimate the above parameters with $\bar{X}$ and S calculated for a given set of data. By utilizing the best estimate rather than the true value of a parameter, the accuracy of a given statistical procedure (or the confidence in a procedure) is diminished. In statistical terms, the degrees of freedom are reduced. Degrees of freedom will often be found in the statistical tables related to some of the procedures outlined below.

The above basic statistics can be applied to large amounts of ground water quality data in order to summarize the data. For example, one might have monthly samples for nitrates in a series of 6 wells. Rather than tabulating and comparing 72 (6 × 12) different data points, the data could be presented as yearly averages (means) and variances. Thus the data to be interpreted is reduced from 72 entities down to 12; 6 means and 6 variances.

In order to interpret the meaning of the above mentioned data, a number of tests or procedures have been developed for comparing sets of data. These procedures, outlined below, are all applied to the statistics (mean, variance, etc.) of a data set, rather than the individual data points with the set.

Student's t-Test

The Student's t-test is a statistical procedure for determining if the "mean" values of two sets of data vary significantly. An application of this procedure would be to compare the "mean" chemical concentrations of samples from wells upgradient and downgradient from a potential source of contamination. Statistical differences in their means might be indicative of contamination.

The t statistic is defined as

$$t = \frac{\bar{X}_2 - \bar{X}_1}{\sqrt{\frac{N_1 + N_2}{N_1 N_2} \quad \frac{(N_1 - 1)^{S_1^2} + (N_2 - 1)^{S_2^2}}{N_1 + N_2 - 2}}}$$

where:

$\bar{X}_1, \bar{X}_2$ = mean values of two different data sets
N_1, N_2 = number of data points in the two data sets
S_1^2, S_2^2 = variances for the populations from which the two data sets were obtained

Subject to certain restrictions, the t statistic will have a standard unit normal distribution and can be analyzed using normal distribution tables. These tables are found in most statistic textbooks.

Cochran's Approximation to the Behrens-Fisher t-Test

The RCRA ground water monitoring documents state that federal regulations require owner/operators to use a t-test to determine whether individual wells contain concentrations statistically greater than background.[56] One of the suggested variations of the t-test is the Cochran's Approximation to the Behrens-Fisher (CABF) t-test. The CABF t-test is used for comparing two groups of unpaired data that have different variances.

The Behrens-Fisher t statistic (t*) is calculated as follows:

$$t^* = \frac{\bar{X}_m - \bar{X}_b}{\sqrt{\dfrac{S_m{}^2}{N_m} + \dfrac{S_b{}^2}{N_b}}}$$

where:

N, $\bar{X}$ and S^2 = the number of data points (including replicates), mean and variance of a given data set

b = subscript referring to background data

m = subscript referring to monitoring well data

The value of t^* is compared with a critical value (t_c) calculated for a given level of significance by the following:

$$t_c = \frac{W_b\ t_b + W_m\ t_m}{W_b + W_m}$$

where:

$W_b = S_b{}^2/N_b$

$W_m = S_m{}^2/N_m$

t_b = tabulated critical value with $(N_b - 1)$ degrees of freedom

t_m = tabulated critical value with $(N_m - 1)$ degrees of freedom

Averaged Replicate t-Test

Another methodology suggested in the RCRA monitoring documents is the averaged replicate (AR) t-test. The main reason for using the AR t-test over the CABF t-test is that the AR t-test removes the heavy influence that split samples or replicate variability exerts on the overall estimate of variability in background data. Additionally, it is thought that the AR t-test may help alleviate statistical contributions to the problem of incorrect indication of contamination.[56]

The procedure for using the AR t-test is similar to that for the CABF t-test. The main difference in the two methodologies is that values for replicate measurements of a single sample are averaged. This averaged value is then used in the calculation of means and variances for a given data set.

The AR t statistic is calculated as:

$$t^* = \frac{\bar{X}_m - \bar{X}_b}{S_b{}^2\sqrt{1 - 1/m_b}}$$

where:

m_b = number of background data points with replicates averaged

The value of t^* is compared directly with a critical value (t_c). Values of t_c are read directly from a table corresponding to $(m_b - 1)$ degrees of freedom. Applications of both the AR and CABF t-tests, along with needed tables, can be found in the RCRA requirements.[56]

F-Distribution Test

Similar to the Student's t-test, the F-distribution test is a method of determining whether there exists a statistical difference between two sets of data. However, the F-distribution tests for differences in sample "variances" (S^2) rather than sample "means" ($\bar{X}$). In actual applications, the F-distribution test should be applied prior to the Student t-test in that equal variances is one of the assumptions for Student's t-test. The F-distribution test is described in most elementary statistics textbooks.

Non-Parametric Analyses

There are several statistical techniques available for analyzing data that are not dependent on the assumption that the data follow a given distribution. These distribution-free methods are sometimes referred to as non-parametric statistical tests. A popular non-parametric procedure is the Wilcoxon-Mann-Whitney (WMW) test. The WMW test is a procedure for comparing two sets of data to determine if one set tends to be smaller than a second set.[57] The WMW test is also described in many introductory level statistics textbooks.

Regression and Correlation

All of the preceding tests were concerned with comparing different sets of data for the same variable. Correlation and regression are two statistical methods available for comparing relationships between two variables within each sample (bivariate statistics). Correlation analysis is a method used to measure the degree of association between two variables. Regression analysis, on the other hand, provides equations for estimating individual values of one variable from given values of another variable. In practical applications, correlation analysis could be used to determine correlations between the occurrence of contamination and the concentration of various water quality constituents. Regression analysis might find its greatest application in estimating missing data. The correlation method requires assumptions not generally valid in hydrologic problems; therefore, use of regression analysis may be more appropriate.[58] Both regression and correlation are described in many statistics textbooks.

CONCERNS WITH STATISTICAL ANALYSES

There is a number of concerns associated with applying statistical methods to ground water quality data. The first and foremost concern should be the accuracy of the data to be analyzed. In other words, the statistical methods are only as accurate as the input data. Sgambat and Stedinger[57] developed Table 9.33 showing the sequence of steps in a ground water monitoring program. They also note that an increasing cumulative error results as the program progresses from step to step. Sgambat and Stedinger[57] also point out that quality control activities and integration of monitoring objectives, network design, and data analysis can keep the error small enough to avoid erroneous conclusions.

Southern Company Services, Inc.[58] (1985), in their ground water manual, identify large variability as another problem inherent to ground water quality data. Sources of this variability include:

1. A monitoring well is located close to a concentrated or diffuse contamination source. Source fluctuations over time will result in significant variations in contaminant concentration in the sample.
2. Monitoring wells are screened near the top of the saturated zone. This situation is sensitive to short-term or seasonal fluctuations in precipitation which flush contaminated soil water held in the unsaturated zone into the aquifer in sudden slugs.
3. Monitoring wells have long screened sections in two or more aquifers of differing qualities with resulting mixing and fluctuations in concentration.

Table 9.33. Steps Involved in Planning and Implementing a Ground Water Monitoring System.[57]

1. Establish Monitoring Objectives
2. Develop and Implement Monitoring Strategies
   a. design network
   b. construct wells
   c. collect samples
   d. analyze samples in laboratory
   e. massage data
   f. interpret data
   g. utilize information

The same study noted that variability could be reduced by frequent and long-term sampling. This would allow extreme values to be averaged out and would allow trend lines associated with seasonal or waste disposal fluctuations to be more easily recognized.

The RCRA regulations call for the placement of at least one monitoring well upgradient from a contaminant source in order to determine background water quality concentrations. However, the U.S. Environmental Protection Agency[56] has noted several reasons that may cause concentrations in supposedly upgradient (background) wells to increase statistically. These include:

1. Ground water flow direction was determined incorrectly and hazardous waste constituents have migrated into the upgradient wells.
2. Ground water flow direction was determined correctly, but hazardous waste constituents are moving in a direction that is opposite of the ground water flow.
3. Upgradient wells were located in a mound caused by the facility.
4. A source of contamination unrelated to the facility was detected.
5. An inconsistent methodology (e.g., well construction materials, sampling and analysis techniques) has been used which resulted in concentration differences that are unrelated to any change in the concentration of constituents in the ground water.
6. The t-test falsely indicated a difference between the background data and upgradient data when actually there was no difference.

Probably more important than noting the physical processes that can disrupt statistical analyses is to note the limitations of the statistical procedures themselves. The underlying assumptions of some of the statistical tests can serve to reduce their applicability and/or invalidate their conclusions. For example, Southern Company Services, Inc.[58] notes that the Student t-test is based on the assumption that the variance of the two populations from which the two sets of data are obtained must be equal. This assumption can and should be tested by the F-distribution prior to employing the Student's t-test.

Sgambat and Stedinger[57] and Hockman and Cibrik[59] both note that the Student's t-test is based on the assumption that the input data approximates a normal distribution. Both studies go on to cite several references which indicate that ground water quality data is not normally distributed. Sgambat and Stedinger[57] further point out that ground water constituents are non-negative, often highly skewed and not amenable to any suitable transformation. Hockman and Cibrik[59] note that a sample with a coefficient of variance less than one may or may not be normally distributed. Federal regulations allow an assumption of normally distributed data if the coefficient of variance is less than one.

Hockman and Cibrik[59] also point out that the use of replicates violates a basic assumption for use of a t-test—that being independent observations. They also point out that including replicates in the number of samples artificially (and incorrectly) inflates the number of degrees of freedom. Hockman and Cibrik[59] also take issue with the RCRA regulations which require background data be obtained for one (the first) year only. Citing seasonal and year-to-year fluctuations, the authors recommend a point-in-time-comparison technique that has been approved by the U.S. Environmental Protection Agency.

Finally, Rovers and McBean[60] point out that an essential first step in analysis of ground water quality data is to remove statistical bias. They note that unlike the sample mean, the variance of a sample is a biased estimator of the population variance and should be adjusted. In the same study, the authors recommend non-parametric analysis or data transformation when the assumptions of the t-test are violated.

SELECTED REFERENCES

1. Everett, L. G., et al. "Constraints and Categories of Vadose Zone Monitoring Devices," *Ground Water Monitoring Review,* 1984.
2. Morrison, R. D. "Ground Water Monitoring Technology," 1983, Timco Mfg., Inc., Prairie du Sac, Wisconsin.
3. Everett, L. G., Wilson, L. G., and McMillion, L. G. "Vadose Zone Monitoring Concepts for Hazardous Waste Sites," *Ground Water,* Vol. 20, No. 3, May–June, 1982, pp. 312–324.
4. Everett, L. G., et al. "Vadose Zone Monitoring Concepts at Landfills, Impoundments, and Land Treatment Disposal Areas," *Proceedings of the National Conference on Management of Uncontrolled Hazardous Waste Sites,* 1982, Hazardous Materials Control Research Institute, Silver Spring, Maryland, pp. 100–106.
5. Noel, M. R., Benson, R. C., and Glaccum, R. A. "The Use of Contemporary Geophysical Techniques to Aid Design of Cost-Effective Monitoring Well Networks," *The Second National Symposium on Aquifer Restoration and Ground Water Monitoring,* 1982, National Water Well Association, Worthington, Ohio, pp. 163–168.
6. Walther, E. G., et al. "Study of Subsurface Contamination with Geophysical Monitoring Methods at Henerson, Nevada," *Proceedings of the National Conference on Management of Uncontrolled Hazardous Waste Sites,* 1983, Hazardous Materials Control Research Institute, Silver Spring, Maryland, pp. 28–36.
7. DiNitto, R. G. "Evaluation of Various Geotechnical and Geophysical Techniques for Site Characterization Studies Relative to Planned Remedial Action Measures," *Proceedings of the National Conference on Management of Uncontrolled Hazardous Waste Sites,* 1983, Hazardous Materials Control Research Institute, Silver Spring, Maryland, pp. 130–134.

8. Benson, R., Glaccum, R., and Beam, P. "Minimizing Cost and Risk in Hazardous Waste Site Investigations Using Geophysics," *Proceedings of the National Conference on Management of Uncontrolled Hazardous Waste Sites,* 1981, Hazardous Materials Control Research Institute, Silver Spring, Maryland, pp. 84–88.
9. Gilkeson, R. H. and Cartwright, K. "The Application of Surface Geophysical Methods in Monitoring Network Design," *The Second National Symposium on Aquifer Restoration and Ground Water Monitoring,* 1982, National Water Well Association, Worthington, Ohio, pp. 169–185.
10. Feld, R. H., et al., "Geophysical Investigations of Abandoned Waste Sites and Contaminated Industrial Areas in West Germany," *Proceedings of the National Conference on Management of Uncontrolled Hazardous Waste Sites,* 1983, Hazardous Materials Control Research Institute, Silver Spring, Maryland, pp. 68–70.
11. Kopsick, D. A. and Stander, T. W. "Refinement of the Shallow Seismic Reflection Technique in Determining Subsurface Alluvial Stratigraphy," *The Third National Symposium on Aquifer Restoration and Ground Water Monitoring,* 1983, National Water Well Association, Worthington, Ohio, pp. 301–306.
12. Swallow, J. A. and Gschwend, P. M. "Volatilization of Organic Compounds from Unconfined Aquifers," *The Third National Symposium on Aquifer Restoration and Ground Water Monitoring,* 1983, National Water Well Association, Worthington, Ohio, pp. 327–333.
13. Harman, H. D., Jr. and Hitchcock, J. "Cost Effective Preliminary Leachate Monitoring at an Uncontrolled Hazardous Waste Site," *Proceedings of the National Conference on Management of Uncontrolled Hazardous Waste Sites,* 1982, Hazardous Materials Control Research Institute, Silver Spring, Maryland, pp. 97–99.
14. Koerner, R. M., et al., "Use of NDT Methods to Detect Buried Containers in Saturated Silt Clay Soil," *Proceedings of the National Conference on Management of Uncontrolled Hazardous Waste Sites,* 1982, Hazardous Materials Control Research Institute, Silver Spring, Maryland, pp. 12–16.
15. Lord, A. E., Jr., Koerner, R. M., and Brugger, J. E. "Use of Electromagnetic Wave Methods to Locate Subsurface Anomalies," *Proceedings of the National Conference on Management of Uncontrolled Hazardous Waste Sites,* 1980, Hazardous Materials Control Research Institute, Silver Spring, Maryland, pp. 119–124.
16. Peters, W. R., Schultz, D. W., and Duff, B. M. "Electrical Resistivity Techniques for Locating Liner Leaks," *Proceedings of the National Conference on Management of Uncontrolled Hazardous Waste Sites,* 1982, Hazardous Materials Control Research Institute, Silver Spring, Maryland, pp. 31–35.
17. Gilkeson, R. H. and Cartwright, K. "The Application of Surface Electrical and Shallow Geothermic Methods in Monitoring Network Design," *Ground Water Monitoring Review,* 1983.
18. McNeill, J. D. "Electromagnetic Resistivity Mapping of Contaminant Plumes," *Proceedings of the National Conference on Management of Uncontrolled Hazardous Waste Sites,* 1982, Hazardous Materials Control Research Institute, Silver Spring, Maryland, pp. 1–6.

19. Hitchcock, A. S. and Harman, H. D. "Application of Geophysical Techniques as a Site Screening Procedure at Hazardous Waste Sites," *The Third National Symposium on Aquifer Restoration and Ground Water Monitoring,* 1983, National Water Well Association, Worthington, Ohio, pp. 307–311.
20. Noel, M. R., Benson, R. C., and Beam, P. M. "Advances in Mapping Organic Contamination: Alternative Solutions to a Complex Problem," *Proceedings of the National Conference on Management of Uncontrolled Hazardous Waste Sites,* 1983, Hazardous Materials Control Research Institute, Silver Spring, Maryland, pp. 71–75.
21. Mehran, M., Nimmons, M. J., and Sirota, E. B. "Delineation of Underground Hydrocarbon Leaks by Organic Vapor Detection," *Proceedings of the National Conference on Management of Uncontrolled Hazardous Waste Sites,* 1983, Hazardous Materials Control Research Institute, Silver Spring, Maryland, pp. 94–97.
22. Turpin, R. D. "ERT's Air Monitoring Guides for Uncontrolled Hazardous Waste Sites," *Proceedings of the National Conference on Management of Uncontrolled Hazardous Waste Sites,* 1983, Hazardous Materials Control Research Institute, Silver Spring, Maryland, pp. 82–84.
23. Greenhouse, J. P., and Slaine, D. D. "Case Studies of Geophysical Contaminant Monitoring at Several Waste Disposal Sites in Southwestern Ontario," *The Second National Symposium on Aquifer Restoration and Ground Water Monitoring,* 1982, National Water Well Association, Worthington, Ohio, pp. 299–315.
24. Leis, W. M., and Sheedy, K. A. "Detailed Location of Inactive Disposal Sites," *Proceedings of the National Conference on Management of Uncontrolled Hazardous Waste Sites,* 1980, Hazardous Materials Control Research Institute, Silver Spring, Maryland, pp. 116–118.
25. Horton, K. A., et al., "The Complimentary Nature of Geophysical Techniques for Mapping Chemical Waste Disposal Sites: Impulse Radar and Resistivity," *Proceedings of the National Conference on Management of Uncontrolled Hazardous Waste Sites,* 1981, Hazardous Materials Control Research Institute, Silver Spring, Maryland, pp. 158–164.
26. Adams, W. M., Wheatcraft, S. W., and Hess, J. W. "Downhole Sensing Equipment for Hazardous Waste Site Investigations," *Proceedings of the National Conference on Management of Uncontrolled Hazardous Waste Sites,* 1983, Hazardous Materials Control Research Institute, Silver Spring, Maryland, pp. 108–113.
27. Campbell, M. D. and Lehr, J. H., 1973, *Water Well Technology,* McGraw-Hill Book Company, Inc., New York, New York.
28. Voytek, J. "Application of Downhole Geophysical Methods in Ground Water Monitoring," *The Second National Symposium on Aquifer Restoration and Ground Water Monitoring,* 1983, National Water Well Association, Worthington, Ohio, pp. 276–278.
29. Huber, W. F. "The Use of Downhole Television in Monitoring Applications," *The Second National Symposium on Aquifer Restoration and Ground Water Monitoring,* 1982, National Water Well Association, Worthington, Ohio, pp. 285–286.

30. Black, W. H., Davis, J. J., and Patton, F. D. "The Piezometric Permeability Profiler: A Tool for Hydrogeological Exploration," *The Second National Symposium on Aquifer Restoration and Ground Water Monitoring,* 1982, National Water Well Association, Worthington, Ohio, pp. 269–275.
31. Scalf, et al., "Manual of Ground-Water Sampling Procedures," 1981, Robert S. Kerr Environmental Research Laboratory, U.S. Environmental Protection Agency, Ada, Oklahoma.
32. Gillham, et al., "Groundwater Monitoring and Sample Bias," June, 1983, Department of Earth Sciences, University of Waterloo, Toronto, Ontario, Canada.
33. Miller, G. D. "Uptake and Release of Lead, Chromium and Trace Level Volatile Organics Exposed to Synthetic Well Casings," *The Second National Symposium on Aquifer Restoration and Ground Water Monitoring,* 1982, National Water Well Association, Worthington, Ohio, pp. 236–245.
34. National Water Well Association and Plastic Pipe Institute, 1980, Worthington, Ohio.
35. Nielsen, D. M. and Yeates, G. L. "A Comparison of Sampling Mechanisms Available for Small-Diameter Ground Water Monitoring Wells," *Proceedings of the Fifth National Symposium and Exposition on Aquifer Restoration and Ground Water Monitoring,* 1985, National Water Well Association, Worthington, Ohio, pp. 237–270.
36. Nacht, S. J. "Ground Water Monitoring System Considerations," *Ground Water Monitoring Review,* 1983.
37. Giddings, T. "Bore Volume Purging to Improve Monitoring Well Performance: An Often Mandated Myth," *The Third National Symposium on Aquifer Restoration and Ground Water Monitoring,* 1983, National Water Well Association, Worthington, Ohio, pp. 253–256.
38. Unwin, J. P. and Huis, D. "A Laboratory Investigation of the Purging Behavior of Small-Diameter Monitoring Wells," *The Third National Symposium on Aquifer Restoration and Ground Water Monitoring,* 1983, National Water Well Association, Worthington, Ohio, pp. 257–262.
39. Keely, J. F. "Chemical Time-Series Sampling: A Necessary Technique," *The Second National Symposium on Aquifer Restoration and Ground Water Monitoring,* 1982, National Water Well Association, Worthington, Ohio, pp. 133–148.
40. Keely, J. F. and Wolf, F. "Field Applications of Chemical Time-Series Sampling," *The Third National Symposium on Aquifer Restoration and Ground Water Monitoring,* 1983, National Water Well Association, Worthington, Ohio, pp. 373–381.
41. McTigue, W. H., and Kunzel, R. G. "A Technique for Renovating Clogged Monitoring Wells," *The Third National Symposium on Aquifer Restoration and Ground Water Monitoring,* 1983, National Water Well Association, Worthington, Ohio, pp. 247–249.
42. Hatayama, H. K. "Special Sampling Techniques Used for Investigating Uncontrolled Hazardous Waste Sites in California," *Proceedings of the National Conference on Management of Uncontrolled Hazardous Waste Sites,* 1981, Hazardous Materials Control Research Institute, Silver Spring, Maryland, pp. 149–153.

43. Johnson, M. G. "The Stratified Sample Thief—A Device for Sampling Unknown Fluids," *Proceedings of the National Conference on Management of Uncontrolled Hazardous Waste Sites,* 1981, Hazardous Materials Control Research Institute, Silver Spring, Maryland, pp. 154–157.
44. Rehtlane, E. A. and Patton, F. D. "Multiple-Ported Piezometers vs. Standpipe Piezometers: An Economic Comparison," *The Second National Symposium on Aquifer Restoration and Ground Water Monitoring,* 1982, National Water Well Association, Worthington, Ohio, pp. 287–295.
45. Jehn, J. L. "Design and Pump Testing of Individual Aquifers with Dual Completed Wells," *The Third National Sympoisum on Aquifer Restoration and Ground Water Monitoring,* 1983, National Water Well Association, Worthington, Ohio, pp. 428–433.
46. Kerfoot, W. B. "Comparison of 2-D and 3-D Ground Water Probes in Fully Penetrating Monitoring Wells," *The Second National Symposium on Aquifer Restoration and Ground Water Monitoring,* 1982, National Water Well Association, Worthington, Ohio, pp. 264–268.
47. Keith, S. J., Wilson, L. G., and Esposito, D. M. "Sources of Temporal-Spatial Variability in Ground Water Quality Data and Methods of Control: Case Study of the Contaro, Arizona, Monitoring Program," *The Second National Symposium on Aquifer Restoration and Ground Water Monitoring,* 1982, National Water Well Association, Worthington, Ohio, pp. 217–228.
48. Barcelona, M. J. "Chemical Problems in Ground-Water Monitoring," *The Third National Symposium on Aquifer Restoration and Ground Water Monitoring,* 1983, National Water Well Association, Worthington, Ohio, pp. 263–271.
49. Keith, S. J., Frank, M. T., and Mossman, G. "Dealing with the Problem of Obtaining Accurate Ground-Water Quality Analytical Results," *The Third National Symposium on Aquifer Restoration and Ground Water Monitoring,* 1983, National Water Well Association, Worthington, Ohio, pp. 272–283.
50. Ahern, J., and Grabowski, J. J. "Inaccurate Water Quality Data in Ground-Water Contamination Studies," *The Third National Symposium on Aquifer Restoration and Ground Water Monitoring,* 1983, National Water Well Association, Worthington, Ohio, pp. 284–291.
51. Dunbar, et al. "Ground Water Quality Anomalies Encountered During Well Construction, Sampling, and Analysis in the Environs of a Hazardous Waste Management Facility," *Proceedings of the Fifth National Symposium and Exposition on Aquifer Restoration and Ground Water Monitoring,* 1985, National Water Well Association, Worthington, Ohio, pp. 132–141.
52. Keith, S. J., and Wilson, L. G. "Stacking the Deck In Ground Water Quality Data," *Proceedings of the Arizona Section, American Water Resources Association Ground Water Quality Symposium,"* 1982, pp. 27–34.
53. U.S. Environmental Protection Agency, "Interim Guidelines and Specifications for Preparing Quality Assurance Project Plans," QAMS-005/80, 1980, Office of Monitoring Systems and Quality Assurance, Office of Research and Development, Washington, DC.

54. American Public Health Association, American Water Works Association, and Water Pollution Control Federation, 1981, *Standard Methods for the Examination of Water and Wastewater,* 15th ed., Washington, DC.
55. Nelson, J. D. and Ward, R. C. "Statistical Considerations and Sampling Techniques for Groundwater Quality Monitoring," *Ground Water,* Vol. 19, No. 6, 1981, pp. 617–625.
56. U.S. Environmental Protection Agency, "RCRA Ground-Water Monitoring Technical Enforcement Guidance Document (Draft)," 1985, Washington, DC.
57. Sgambat, J. P. and Stedinger, J. R. "Confidence in Groundwater Monitoring," *Ground Water Monitoring Review,* Vol. 1, No. 1, 1981, pp. 62–69.
58. Southern Company Services, Inc., "Groundwater Manual for the Electric Utility Industry," EPRI CS-3901, 1985, Electric Power Research Institute, Palo Alto, California.
59. Hockman, B., and Cibrik, J. "Point-In-Time Comparison: An Alternative to the Statistical Requirements of RCRA Accepted by EPA," *Proceedings of the Fifth National Symposium and Exposition on Aquifer Restoration and Ground Water Monitoring,* 1985, National Water Well Association, Worthington, Ohio, pp. 49–59.
60. Rovers, E. A. and McBean, E. A. "Significance Testing for Impact Evaluation," *Ground Water Monitoring Review,* Vol. 1, No. 2, 1981, pp. 39–43.

CHAPTER 10

Ground Water Pollution Control

As recently as the late 1970s the four most common perceptions of ground water cleanup or aquifer restoration measures were: (1) they were costly, (2) they were time consuming, (3) they were not always effective, and (4) pertinent information concerning them was not available. Although these perceptions have not been totally reversed, the state of the art is progressing significantly. The need for information concerning ground water cleanup has been responded to by: (1) an increasing number of symposia, specialty conferences, and technical short courses; (2) development of several information clearinghouses; (3) increasing amounts of information and trade literature concerning ground water contamination expertise and/or equipment; and (4) the referenced literature focusing more on aquifer restoration measures. The end result is that current aquifer restoration measures are now being designed to be rapid, economical, technically sound, and environmentally effective.

DESCRIPTION OF TECHNOLOGIES

Table 10.1 is a listing of some of the more common measures employed for preventing and/or cleaning up ground water contamination. Several key points need to be made concerning Table 10.1. First, although meant to be comprehensive, Table 10.1 is not complete. No single table could list all of the technologies or variations of technologies that have been (or could be) used in ground water contamination cases. Second, it should be noted that many of the technologies are not new and have been practiced for many years in other disciplines. Especially notable are the physical control measures that have been adapted from the construction and geotechnical engineering industries. Third, most of the technologies listed in Table 10.1 will, out

Table 10.1. Prevention and Remedial Measures.

Group/Technologies	Description
Source Control Strategies I. Volume reduction measures A. Recycling B. Resource recovery C. Centrifugation D. Filtration E. Sand drying beds II. Physical/chemical alteration measures A. Chemical fixation B. Detoxification C. Degradation D. Encapsulation E. Waste segregation F. Co-disposal G. Leachate recirculation	Source control strategies represent attempts to minimize or prevent ground water pollution before a potential polluting activity is initiated. The objectives of source control strategies are to reduce the volume of waste to be handled, or reduce the threat a certain waste possesses by altering its physical or chemical makeup.
Well Systems I. Well point systems II. Deep well systems III. Pressure ridge systems IV. Combined systems V. Immiscible (hydrocarbon) contaminant recovery systems	Well systems for ground water pollution control are based on manipulation of the subsurface hydraulic gradient through injection, and/or withdrawal of water. Well systems are designed to control the movement of the water phase directly and the subsurface contaminants indirectly. This approach is referred to as plume management. Well systems are also used for recovery of immiscible contaminants, usually hydrocarbons, that float on the water table.
Interceptor Systems I. Collector drains A. Leachate collection systems B. Interceptor drains C. Relief drains II. Interceptor trenches A. Actively pumped systems B. Gravity flow, skimmer pump systems	Interceptor systems involve excavation of a trench below the water table, and possibly the placement of a pipe in the trench. The trench can be left open (interceptor trench) or backfill can be placed on a pipe in the trench (collector drain). Interceptor trenches can be either active (pumped) or passive (gravity flow). These systems function similarly to an infinite line of extraction wells by effecting a continuous zone of depression running the length of the trench.
Surface Water Control, Capping and Liners I. Natural attenuation (no liner, no cap) II. Engineered liner III. Engineered cover IV. Engineered cover and liner	These three technologies are used in conjunction with each other, each serving a unique ground water pollution prevention purpose. Surface water control measures reduce potential infiltration by minimizing the amount of surface water flowing onto a site. Capping is designed to minimize the

Table 10.1, continued

Group/Technologies	Description
	infiltration of any surface water or direct precipitation that does come onto a site. Impermeable liners provide ground water protection by inhibiting downward flow of low-quality leachate and/or attenuating pollutants by adsorption processes.
Impermeable Barriers I. Steel sheet-piles II. Grout curtains or cutoffs III. Slurry walls	Barriers are measures designed to influence the subsurface hydraulic gradient by placing a low permeability material into the subsurface. Barriers typically are constructed with driven sheet piles, injected grouts, or dug slurry walls. Sheet-piles provide immediate impermeability, while grouts and slurries both are emulsions that require a hardening period to achieve impermeability.
In-situ Treatment I. In-situ chemical II. In-situ biological	The in-situ treatment methods involve adding materials to the subsurface so as to cause or increase the rate of a reaction that will render a contaminant immobile or remove the contaminant. The in-situ chemical technologies attempt to immobilize contaminants through some chemical reaction, while the in-situ biological techniques are designed to provide an environment suitable for microorganisms to utilize the contaminant as a food source.
Innovative Technologies I. Block displacement II. Envirowall III. Fly ash stabilization	Three of the innovative technologies that have yet to find wide applicability are the block displacement method, the Envirowall concept, and fly ash stabilization. The block displacement method is similar to grouting except the emulsion is injected laterally with the effect of raising or isolating a whole block of contaminated soil. The Envirowall concept involves vertical placement of a synthetic liner material within a slurry wall. Fly ash stabilization involves injection of fly ash into an existing impoundment to solidify the contained liquid or slurry.
Wastewater Treatment I. Air/steam stripping II. Carbon adsorption III. Biological treatment IV. Chemical precipitation	Wastewater treatment technologies are utilized at the surface to treat contaminated ground water. The technologies most widely applied to organic contaminants are air stripping, carbon adsorption and biological treatment. Chemical precipitation is used for inorganics and metals removal.

of necessity, have to be combined with other technologies. For example, a truly sound aquifer restoration strategy would utilize removal wells to complement a subsurface barrier. The final point to note about Table 10.1 is that several of the technologies are aimed at preventing or abating ground water contamination. This is a subtle hint as to where future efforts should be directed, i.e., prevention over cleanup. As will be shown, cleanup of contaminated ground water is expensive, but avoidable through sound planning practices.

PHYSICAL CONTROL MEASURES

The physical control measures described in Table 10.1 are again tabulated in Table 10.2 along with their relative advantages and disadvantages. The physical control measures are technologies whose primary aim is either pollution prevention or plume management by containment or removal.

Surface water control, capping, and liners are all technologies that are designed to directly or indirectly minimize low-quality percolation (leachate) from being discharged to underlying fresh ground water formations. Although surface water control and capping can be applied to an existing site or source of contamination, the feasibility of retrofitting a site with a liner, without totally removing the source material, is limited. It is emphasized that future state-of-the-art waste disposal sites should be designed to include all three of these technologies as well as leachate collection systems as an additional safeguard.

Capping of existing sites has become more common in recent years. Dowiak, et al.[1] outlined the capping process used at a lime-neutralized sludge disposal site. Emrich and Beck[2] itemize the components of a surface cap and revegetation program for a sanitary landfill, and Mutch, Daigler, and Clarke[3] discuss similar information for the capping of a 25-year-old industrial waste landfill. Sanning[4] reviews the construction of two different surface caps and also notes that from cost-effectiveness and ease of application standpoints, surface capping shows marked advantages over other remedial actions. Fung[5] provides a comprehensive overview of the materials and construction procedures used for surface capping.

In related works, Anderson, Brown, and Green,[6] and Brown and Anderson[7] both discuss the possible effects of organic components of hazardous waste leachate on the integrity of clay liners. Cooper and Schultz[8] discuss the feasibility of retrofitting existing liquid impoundments with synthetic membrane liners.

The three barrier technologies are all designed as containment measures in that their primary purpose would be to arrest the movement of a contaminant

Table 10.2. Advantages and Disadvantages of Physical Aquifer Restoration Technologies.

Technology	Advantages	Disadvantages
Source control strategies	1. Reduces the threat to the ground water environment.	1. Increased capital and maintenance costs.
	2. Accelerates the time for "stabilization" of waste disposal facilities.	2. Monitoring and skilled operator requirements.
	3. Offers opportunities for economic recovery.	
Well systems	1. Efficient and effective means of assuring ground water pollution control.	1. Operation and maintenance costs are high.
	2. Can be installed readily.	2. Require monitoring program after installation.
	3. Previously installed monitoring wells can sometimes be employed as part of well system.	3. Withdrawal systems necessarily remove clean (excess) water along with polluted water.
	4. Can sometimes include recharge of aquifer as part of the strategy.	4. Some systems may require the use of sophisticated mathematical models to evaluate their effectiveness.
	5. High design flexibility.	5. Withdrawal systems will usually require surface treatment prior to discharge.
	6. Construction costs can be lower than artificial barriers.	6. Application to fine soils is limited.
Interceptor systems	1. Not only easy but also inexpensive to install.	1. When dissolved constituents are involved, it may be necessary to monitor ground water downgradient of the recovery line.
	2. Useful for intercepting landfill side seepage and runoff.	2. Open systems require safety precautions to prevent fires or explosions.
	3. Useful for collecting leachate in poorly permeable soils.	3. Interceptor trenches are less efficient than well-point systems.

Table 10.2, continued

Technology	Advantages	Disadvantages
	4. Large wetted perimeter allows for high rates of flow.	4. Operation and maintenance costs are high.
	5. Possible to monitor and recover pollutants.	5. Not useful for deep disposal sites.
	6. Produces much less fluid to be handled than well-point systems.	
Collector drains	1. Operation costs are relatively cheap since flow to underdrains is by gravity.	1. Not well suited to poorly permeable soils.
	2. Provides a means of collecting leachate without the use of impervious liners.	2. In most instances, it is not feasible to situate underdrains beneath an existing site.
	3. Considerable flexibility is available for design of underdrains; spacing can be altered to some extent by adjusting depth or modifying envelope material.	3. System requires continuous and careful monitoring to assure adequate leachate collection.
	4. Systems are fairly reliable, providing continuous monitoring is possible.	
	5. Construction methods are simple.	
Surface water control, capping and liners		
a. Natural attenuation (no liner, no cap)	1. No leachate collection, transport and treatment costs.	1. Requires unusually favorable hydrogeologic setting.
	2. Reduced construction costs.	2. Regulatory acceptance difficult to obtain.
		3. Long-term liabilities.

b. Engineered liner	1. Lessens hydrogeologic impact. 2. Allows waste to stabilize quickly.	1. "Clay-bowl" effect. 2. Increased construction costs. 3. Chance for surface discharge.
c. Engineered cover	1. Lessens hydrogeologic impact after closure. 2. Reduces construction costs relative to liners.	1. Increases closure costs. 2. No leachate control during site operations. 3. Long-term monitoring and land surface care.
d. Engineered cover and liner	1. Lessens environmental impacts. 2. Minimizes post closure leachate collection, transport and treatment costs. 3. Politically and/or socially acceptable.	1. High cost for engineering and construction. 2. Need high-quality clay or synthetic material. 3. Lengthened time for waste stabilization.
Sheet-piles	1. Construction is not difficult; no excavation is necessary. 2. Contractors, equipment, and materials are available throughout the United States. 3. Construction can be economical. 4. No maintenance required after construction. 5. Steel can be coated for protection from corrosion to extend its service life.	1. The steel sheet-piling initially is not watertight. 2. Driving piles through ground containing boulders is difficult. 3. Certain chemicals may attack the steel.
Grouting	1. When designed on basis of thorough preliminary investigation, grouts can be very successful.	1. Grouting is limited to granular types of soils having a pore size large enough to accept grout fluids under pressure, yet small enough to prevent significant pollutant migration before implementation of grout program.

Table 10.2, continued

Technology	Advantages	Disadvantages
	2. Grouts have been used for over 100 years in construction and soil stabilization projects.	2. Grouting in a highly layered soil profile may result in incomplete formation of a grout envelope.
	3. Many kinds of grout to suit a wide range of soil types are available.	3. Presence of high water table and rapidly flowing ground water limits groutability through: a. extensive transport of contaminants. b. rapid dilution of grouts.
		4. Some grouting techniques are proprietary.
		5. Procedure requires careful planning and pretesting. Methods of ensuring that all voids in the wall have been effectively grouted are not readily available.
		6. Grouts may not withstand attack from specific pollutants.
Slurry walls	1. Construction methods are simple.	1. High cost of shipping bentonite from the west.
	2. Adjacent areas are not affected by ground water drawdown.	2. Some construction procedures are patented and require a license.
	3. Bentonite (mineral) will not deteriorate with age.	3. In rocky ground, overexcavation is necessary because of boulders.
	4. Leachate-resistant bentonites are available.	4. Bentonite deteriorates when exposed to high ionic strength leachates.

	5. Low maintenance requirements.	5. Adequate key to impermeable formation is critical.
	6. Eliminate risks due to strikes, pump breakdowns, or power failures.	6. Methods for assessing in-place integrity not available.
	7. Eliminate headers and other above ground obstructions.	

plume long enough for some removal measure to be employed. It should be noted that most regulatory agencies will no longer accept barriers as complete containment measures and will require that they be accompanied by some type of ground water removal system. Barriers are most effective as temporary containment measures aimed at minimizing the growth of existing problems.

The information on grout and grouting technology is dominated by studies concerning the compatibility of various grouts with different chemicals. Lord[9] examined the effects of chemicals on silicate-based grouts; Malone, Francigues, and Boa[10] examine grout compatibility with hazardous wastes; and Spooner, et al.[11] examine grout compatibility with hazardous waste and also suggests appropriate laboratory tests and required design factors. Huibregtse and Kastman[12] discuss an application of grouting through the development of their response vehicle for chemical spills.

The two most famous examples of slurry walls used for pollution containment are: (1) the Rocky Mountain Arsenal near Denver, Colorado,[13–14] and (2) the first Superfund slurry wall near Nashua, New Hampshire.[15] Monitoring data for the New Hampshire site is not yet available, but data from the Rocky Mountain Arsenal indicate some leakage either through or under the slurry wall. The effectiveness of impermeable barriers in general is dependent on establishing a truly impermeable formation, maintaining impermeability over time, and adequately keying the barrier into an underlying impermeable layer such as bedrock. Knox[16] discusses the ineffectiveness of barriers if they do not meet these criteria. Anderson and Jones[17] examine the effects leachate can have on the pore volume and permeability of clay barriers. The design, construction, and applications for slurry walls is discussed in D'Appolonia;[18] Ryan;[19] Millet and Perez;[20] and Spooner, Wetzel and Grube.[21] An in-depth discussion of all facets of slurry walls can be found in Xanthakos.[22]

Interceptor systems can be designed with pollution prevention or abatement, or containment removal as their goal. Leachate collection systems are designed to prevent low-quality percolation from reaching fresh ground water formations, while relief drains can be installed to lower water tables that have risen up into contaminated or waste disposal areas. Trenches can be used to intercept and remove contaminated ground water.

The use of interceptor trenches has found wide application for hydrocarbon recovery operations as discussed by the American Petroleum Institute.[23–24] Molsather and Barr[25] discuss the installation of a leachate collection system to catch toe seep at an existing landfill. Giddings[26] included a leachate interceptor drain as part of a ground water dam system to prevent shallow landfill leachate from reaching a nearby river. Jhaveri and Mazzacca[27] utilized an open trench as a means of recharging and flushing a shallow formation with its own treated effluent. Underwood and Thornton[28] give a good description of using a French drain system for contaminant recovery in a highly impermeable formation.

Despite the innovative technologies now being promoted, the most popular ground water cleanup measure remains removal and treatment. Removal is most often accomplished through removal wells or injection/extraction systems. Injection of fresh water or treated effluent is used more often for flushing formations than for plume containment. The use of wells for hydrocarbon recovery has seen increased popularity due to the economic incentive of recycling the product. The mechanics of ground water flow to wells are well known and readily applied. Well systems are almost universally applicable, with the possible exceptions of highly impermeable formations, extremely shallow problems, or contaminants restricted to the unsaturated (vadose) zone.

A listing of previous applications of well systems would be lengthy. One of the earliest applications is outlined by Gregg[29] who describes how pumping of an underlying saline formation helped to reduce upward intrusion into an overlying (overpumped) fresh water aquifer. Sheahan[30] describes a system where municipal wastewater is injected and extracted in coastal wells to prevent saltwater intrusion. Osiensky and Williams[31] discuss the merits of the ground water "pump back" system they designed for collecting seepage from a uranium tailings disposal pond. Bradley[32] outlines the case of a TCE-contaminated drinking water well becoming the contaminant extraction well. In this study it is pointed out that shutting off the pump of the contaminated drinking water well would have only caused the contamination to migrate to other parts of the aquifer. Blake and Lewis[33] give a concise overview of the three different types of well systems that can be used for hydrocarbon recovery.

IN-SITU TECHNOLOGIES

Treatment of polluted ground water by in-situ techniques is still relatively new; however, such treatment is receiving increased research attention, and practical applications of these technologies are increasing in number. The primary objective of the in-situ techniques is to achieve treatment of the contaminant(s) within the subsurface, either chemically or by biological processes.

In-Situ Chemical Treatment

In-situ chemical treatment of ground water can be considered only in cases where contaminants are known and the levels and extent of contaminated water in the aquifer are defined. In order to utilize chemical treatment, hydraulic control over the subsurface must be established and the limits of the contaminant plume delineated. A treatment agent is then introduced to the affected subsurface, which reacts with the contaminant to render it immobile. The type of reaction is dependent on the contaminant and could involve oxidation, reduction, precipitation, or polymerization. It is important to note that the contaminant is not removed, just immobilized.

An early application of in-situ chemical treatment was a process developed in Finland called the Vyredox Method. Basically, the method involves periodic oxygen injection into the saturated subsurface in order to precipitate iron and retain manganese.[34] Actually, the manganese is utilized by bacteria, so this method is really a hybrid chemical-biological scheme. Williams[35] describes how a catalyst and activator solution were percolated into the ground at the site of a spill of 4200 gallons of acrylate monomer. It is estimated that 85 to 90% of the liquid monomer contaminant was converted to a solidified polymer. Harsh[36] details the events that led to the cleanup of an acrylonitrile spill. Oxidation of acrylonitrile was accomplished by first raising the pH above 10 through lime addition and then spraying chlorine over the area. Matthess[37] outlines a case in which soluble arsenic was oxidized and caused to precipitate and/or coprecipitate with manganese or iron. Srivastava and Haji-Djafari[38] completed an overview study detailing the feasibility of in-situ detoxification of hazardous wastes.

In-Situ Biological Treatment

In-situ biological treatment is based on the same principle as traditional biological wastewater treatment operations, i.e., certain microorganisms, in the presence of adequate nutrients and under proper environmental conditions, will utilize organic constituents as their food source or substrate. The by-products of this biochemical reaction are simpler organics and ultimately carbon dioxide and water. Hence, the in-situ biological techniques simply attempt to turn the organically affected subsurface environment into a bioreactor.

There are two classes of biodegradation processes for ground water and soil applications. The first class is called "Enhanced Biodegradation." This process involves "enhancing" the subsurface environment through nutrient and/or oxygen addition and allowing the native or indigenous microorganisms to degrade the organic constituents. The second process classification is "Acclimated Biodegradation." In this process, microorganisms are identified and targeted to the specific constituent and environment to which they will be introduced and cultured. These acclimated microorganisms are then introduced into the subsurface to provide a biological "seed" for the treatment process.

Both of these in-situ processes are usually supplemented by several technologies, including: extraction of ground water, surface treatment of ground water, and reinjection or surface discharge of treated ground water. A schematic depiction of typical in-situ treatment operations is shown in Figure 10.1.

Previous applications of in-situ technologies are dominated by cases involving relatively small hydrocarbon leaks or spills into highly permeable soils with shallow water tables. Raymond, Jamison, and Hudson[39] state that permeability is not a limiting factor to in-situ degradation and that microorganism movement has been documented even in very fine sands. Conversely, Brubaker and O'Neill[40] say that the hydrogeological characteristics of a site represent the critical factors in that they affect the ability to move nutrients through the contaminated area. McKee, Laverty, and Hertel[41] note that degradation takes place more rapidly in the zone of aeration above the water table.

Although applications of in-situ enhanced biodegradation are dominated by examples involving petroleum hydrocarbons,[42] there is increased interest in other organic contaminants commonly found in ground water and soils. Brubaker and O'Neill[40] report a rule of thumb for assessing biodegradability as: "organics similar to those found in nature are degradable; highly chlorinated organics are difficult to degrade." Erlich, et al.[43] reported that naphthalene is not biodegraded in anaerobic environments, while both Lee and Ward[44] and Ogawa, Junk, and Svec[45] report naphthalene as biodegradable in aerobic environments. Roberts, et al.[46] found naphthalene to be biodegraded, after an initial time, in an aquifer recharged with reclaimed wastewater. Erlich, et al.[43] and Rees and King[48] document evidence of biodegradation of phenolic compounds under anaerobic conditions. Conversely, Raymond, et al.[49] reported phenol to be degraded only after aerobic conditions were achieved. Wilson and Wilson[50] have shown trichloroethylene (TCE) to degrade aerobically to carbon dioxide in unsaturated soil columns exposed to natural gas.

Several researchers have targeted microorganisms to certain contaminants. Edgehill and Finn[51] isolated a strain of Arthrobacter (ATCCB379) that could use pentachlorophenol as its sole carbon source. Walton and Dobbs[52] reported that O.H. Materials has used mutant bacteria to treat phenol and its derivatives.

The success of either type of in-situ biodegradation process depends on developing and maintaining a population of microorganisms capable of using the various contaminants. Zajik and Daugulis[53] suggest that a mixed micro-

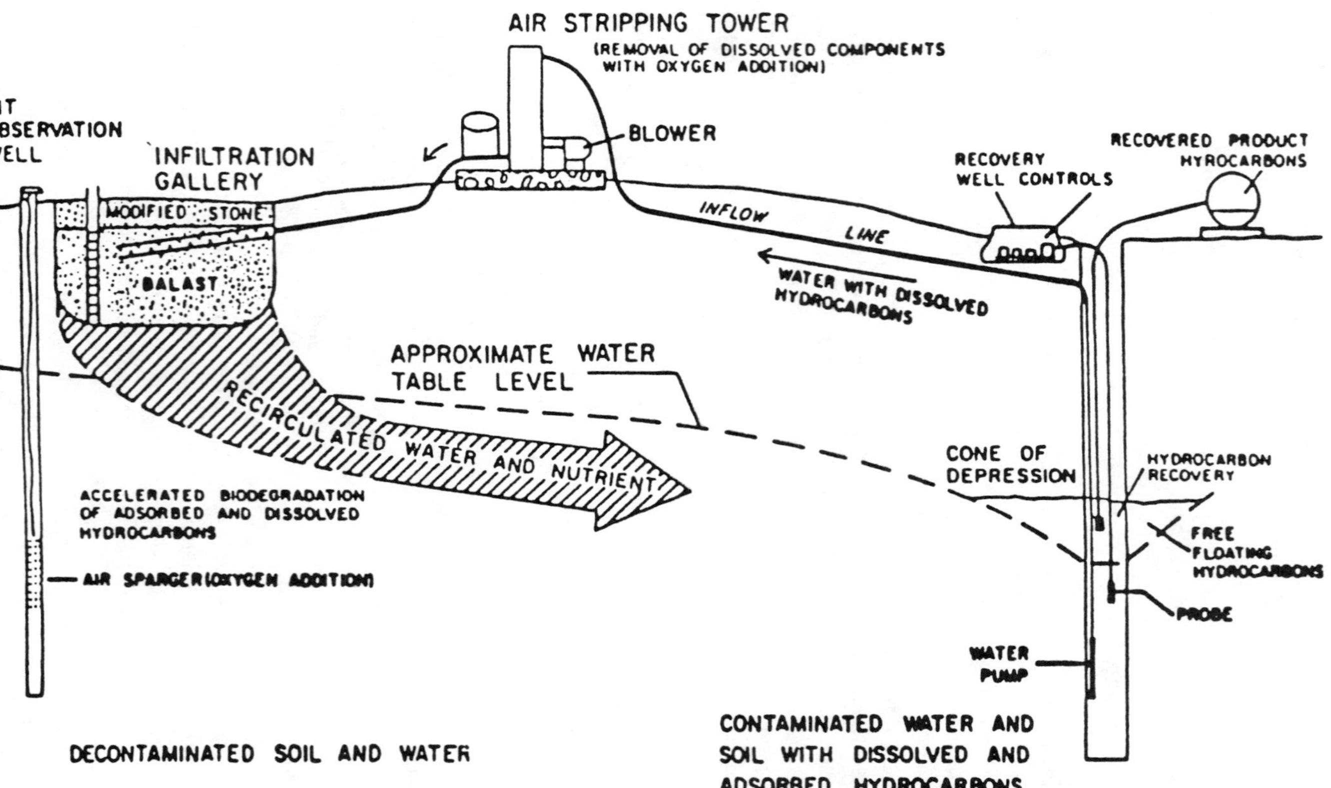

Figure 10.1. Typical operations for in-situ biodegradation process.[47]

bial population is necessary to degrade all the contaminants that might be present at a single site. Raymond, et al.[49] reinforces this point by noting that cooxidation seems to play the major role in hydrocarbon degradation because the microorganisms are not able to utilize any of the compounds as their sole carbon sources.

As depicted in Figure 10.1, in-situ technologies usually involve some type of injection/extraction system. The injection/extraction system is designed to develop hydraulic control over the movement of the contaminants or contaminated ground water. Additionally, an equally important objective of the injection/extraction system is to transport nutrients and oxygen to the subsurface microorganisms.

The time and cost of an in-situ treatment program are significantly affected by the amount of nutrients needed and the rate at which they can be injected.[40] As early as 1974, Raymond[54] suggested that nutrient addition be used to enhance subsurface biodegradation. The suggested nutrient solution contains nitrogen, phosphorus, and inorganic salts at concentrations up to 0.02% by weight. In related work, Brown, et al.[55] indicated that the nitrogen level is critical for biodegradation of oil sludges by landfarming, and they also suggest that carbon to nitrogen ratios be between 10:1 and 150:1. Oil sludge biodegradability was also examined in soils and maintenance of carbon to nitrogen and carbon to phosphorus levels at 60:1 and 800:1, respectively, were recommended.[56] Exact nutrient requirements for ground water applications have to be determined in the laboratory and have shown considerable variation between aquifers.[57]

Not only do the necessary nutrients vary from aquifer to aquifer; the form of the nutrients is also aquifer specific. Nutrients can be supplied by either of two modes; however, Raymond, et al.[49] report batch feed to be more economical than continuous feed. The total amount of nutrients can be estimated from a laboratory analysis and the extent of the subsurface contamination. At one site, the purchase of 87 tons of food grade quality chemicals was reported.[58]

The vast majority of successful in-situ biological treatment projects have relied on aerobic processes. Regardless of the type of in-situ process used, the efficiency and rate of removal will be dependent on oxygen addition. Dissolved oxygen may be the limiting factor for in-situ degradation.[49] Early measures for supplying oxygen to ground water systems involved simply injecting compressed air through carbonundum diffusers or DuPont™ perforated tubing. Oxygen addition rates were not specified, although it has been reported that these systems can deliver air at rates of from 1 to 10 standard cubic feet per minute.[57]

Because operation of mechanical air supply systems can be expensive and prone to malfunction, much interest has been expressed in chemical means of supplying dissolved oxygen. Texas Research Institute, Inc.[59] has examined the feasibility of injecting hydrogen peroxide to increase the dissolved oxygen content of contaminated ground waters. This research has shown that hydrogen peroxide in levels as low as 0.25% can be toxic to microorganisms and that mature organisms can utilize the peroxide better than young organisms. This suggests that the addition of hydrogen peroxide be phased in gradually. Nagel, et al.[60] report that contaminated ground water can be withdrawn and treated with ozone, then allowed to infiltrate back into the aquifer. This increases the dissolved oxygen level and the biodegradability of the ground water. Ozone dosages were about 1 gram of ozone per gram of dissolved organic carbon. Norris and Brown[61] found success in using specially stabilized hydrogen peroxide solutions. These solutions provided oxygen at 25 times the rate of air spargers and resulted in improved cost effectiveness.

The current state of the art for in-situ treatment of ground water and soils includes extraction of ground water and surface treatment followed by either nutrient addition and reinjection or surface discharge. Any conventional wastewater treatment process capable of treating organics could conceivably be used to treat the extracted ground water; however, some processes possess distinct advantages. The biological processes (activated sludge, biological disk, trickling filters) will provide an ample supply of acclimated organisms. Air stripping and ozonation provide for oxygenation of the water prior to reinjection. As indicated previously, oxygen is often the limiting factor for in-situ biodegradation processes.

A series of steps required in the design of an enhanced biodegradation process has been outlined by Suntech.[62] The first step is to investigate the hydrogeology and extent of contamination. Next, a laboratory study is undertaken to determine if the native microflora can degrade the contaminants and what nutrients will be needed. After the microbial requirements have been established, the injection/extraction system should be designed. Design of the injection/extraction system is critical, and it is required that the flow go through the zone of contamination. Recirculation of extracted water is suggested in that it eliminates the problem of disposal and allows for recirculation of unused nutrients. The advantages, disadvantages, and design factors for enhanced biodegradation are listed in Table 10.3.

The design steps of an acclimated biodegradation system are similar to those for enhanced biodegradation except the laboratory studies are more involved. It is necessary that the laboratory studies identify the specific microorganisms needed to degrade all of the contaminants of concern. Identification of these microorganisms is accomplished by taking a sample of contaminated soil or ground water and isolating the various cultures thought to be present. Once the necessary microorganisms have been identified, it is possible to develop a biological "seed" or population of organisms to be in-

Table 10.3. Advantages, Disadvantages, and Design Factors for Enhanced Biodegradation.

Advantages	Disadvantages	Factors Affecting Performance
(1) Useful for removal of hydrocarbon and other organics at low levels which would be difficult by other means.	(1) Not applicable to heavy metals and certain organics.	(1) Dissolved oxygen level
(2) Environmentally sound because no waste products are generated and no ecological changes occur because it utilizes indigenous microorganisms.	(2) Bacteria can plug soil and reduce recirculation.	(2) pH
(3) Fast, safe, and generally economical.	(3) Nutrients could adversely affect nearby surface waters.	(3) Temperature
(4) Treatment moves with plume.	(4) Residues may cause taste and odor problems.	(4) Oxidation-reduction potential
(5) Good for short-term treatment of contaminated ground water.	(5) Could be expensive if long-term injection of nutrients and equipment maintenance are needed.	(5) Available nutrients
	(6) For high concentrations of contaminants biodegradation is slower than physical recovery methods.	(6) Salinity
	(7) Long-term effects unknown.	(7) Contaminant concentrations
	(8) Regulatory acceptance difficult to obtain.	(8) Number and type of organisms
	(9) May not reduce parts per billion concentrations without complementary technology.	

jected into the subsurface. The biological seed can be developed through either of two processes: enrichment culturing or genetic manipulation. Enrichment culturing is a series of biochemical culturing steps designed to isolate contaminant-specific microorganisms. If an ideal microorganism does not exist or cannot be isolated, genetic manipulation may be used to artificially develop one. Radiation has been used in such an application.[63] Having identified and developed the biological seed, the design process proceeds as outlined previously for enhanced biodegradation. The advantages, disadvantages, and design factors for acclimated biodegradation are listed in Table 10.4.

GROUND WATER TREATMENT

By far the most common aquifer restoration measure is removal of the contaminated ground water followed by surface treatment and reinjection or discharge. Although a wide variety of technologies exists, the treatment options used for ground water contamination are usually limited to air stripping, carbon adsorption, or biological treatment for organics removal; and chemical precipitation for inorganics removal.

Air Stripping

Air stripping is a mass transfer process in which a constituent in solution in water is transferred to solution in a gas. The "strippability" of a given substance is a function of many factors, including the concentration of the contaminant and its Henry's law constant. In general, those compounds with high Henry's law constants are generally more easily "stripped" from water. Additionally, stripping efficiencies usually increase with temperature, thus making high temperature air or steam active stripping agents.

Air stripping has been successfully used for removing organic solvents from the drinking well water of a small town in New Jersey.[63] Lamarre, McGarry, and Stover[65] conducted pilot plant studies on the ground water extracted from a Superfund site in New Hampshire and found steam stripping to be most effective for organics removal. Allen and Parmele[66] present information on the feasibility of using air stripping to remove dissolved gasoline components from ground water.

Carbon Adsorption

Carbon adsorption occurs when an organic molecule is brought to an activated carbon surface and held there by physical and/or chemical forces.

Table 10.4. Advantages, Disadvantages, and Design Factors for Acclimated Biodegradation.

Advantages	Disadvantages	Factors Affecting Performance
(1) Microorganisms can be targeted to contaminants.	(1) Preliminary design is more complex, time consuming, and expensive than enhanced biodegradation.	(1) Dissolved oxygen level
(2) Degradation process proceeds faster.	(2) Injection of "seed" can cause plugging.	(2) pH
(3) "Seed" can be injected along with nutrients.	(3) Presence of toxicants can destroy seed.	(3) Temperature
(4) Possible to develop microorganism to degrade contaminant "immune" to indigenous population.	(4) Competition with native microbial population can destroy seed.	(4) Contaminant concentration
	(5) Regulatory acceptance difficult to obtain.	(5) Sufficent moisture
	(6) May not handle parts per billion concentrations without complementary technologies.	

Factors affecting the "adsorption" of a substance include:

(1) Adsorption increases with decreasing solubility.
(2) The pH of the water can affect adsorption—organic acids adsorb better under acidic conditions, whereas amino compounds favor alkaline conditions.
(3) Aromatic and halogenated compounds adsorb better than aliphatic compounds.
(4) Adsorption capacity decreases with increasing temperature although the rate of adsorption may increase.
(5) The character of the adsorbent surface has a major effect on adsorption capacity and rate—the raw materials and the process used to activate the carbon determine its capacity.

Adsorption isotherms are graphical depictions of the adsorption equilibrium of different compounds. From an isotherm test, it can be determined whether or not a particular contaminant can be removed effectively. The test will also show the approximate capacity of the carbon and provide a rough estimate of the carbon dosage required. Isotherm tests also provide a convenient way of studying the effects of pH and temperature on adsorption.[67]

Brunotts, et al.[68] used activated carbon for priority pollutant removal at one site, and Chaffee and Weimar[69] present two case histories where granular activated carbon was used to remove organics from a vapor phase and a liquid phase. Dobbs and Cohen[70] outline experimental methods for determining isotherms. Fochman[71] examined both biological treatment and carbon adsorption and found carbon adsorption not able to remove very large polynuclear aromatic compounds. Kaufmann[72] presents information on the cost effectiveness of carbon adsorption, and Parmele and Allen[73] describe a nondestructive technique for regenerating carbon.

Biological Treatment

Biological treatment involves providing an environment in which acclimated microorganisms will use organics in water as their food source. This process has been developed over many years and applied broadly to industrial and municipal wastewater treatment. The efficiency of a given biological treatment system is a function of many factors and can only be determined through laboratory treatment studies. Once rate constants have been obtained, design of a full-scale system can be completed for given flow rates and given influent and required effluent concentrations. Biological treatability studies for contaminated ground water are discussed by Kincannon and Stover.[67]

Chemical Precipitation

Chemical addition for removal of inorganic compounds is based on adjusting the pH of a solution to precipitate out the inorganics. The three common types of chemical addition systems are the carbonate systems, the hydroxide systems, and the sulfide systems. Sulfide systems remove more compounds than the others but have the disadvantages of difficulty in handling the chemicals, and metals can resolubilize out of sulfide sludges. The carbonate system uses soda ash and pH adjustment between 8.2 and 8.5. The carbonate system is difficult to control. The hydroxide system is the most widely used for inorganics and metals removal. This system uses either lime or sodium hydroxide to adjust the pH upward. The hydroxide system is easy to control, utilizes easily handled chemicals, and produces a low volume of sludge. However, the sludge is often hard to dewater. Kincannon and Stover[67] present data on metals removal from contaminated ground water using lime addition.

INNOVATIVE TECHNOLOGIES

Although the majority of aquifer restoration measures simply represent new applications of borrowed technologies, a few novel ideas have been developed. The Envirowall concept involves the placement of a synthetic membrane liner material vertically within a slurry wall to form a highly impermeable barrier to flow. Arlotta[74] and Cavalli, Arlotta, and Druback[75] both discuss the application of this technology at a test site. Although highly touted in both references, the Envirowall is plagued by the same problems as slurry walls plus the additional difficulty of fabricating seams. The block displacement method involves injecting a slurry laterally into the subsurface to actually displace a block of contaminated soil and effectively isolate it. Brunsing and Cleary[76] and Brunsing and Grube[77] both discuss a full-scale demonstration of the technology. A third innovative measure is the use of fly ash to consolidate and stabilize waste liquids or slurries. The fly ash is pneumatically injected into the waste and forms a solid mass with the intent of immobilizing the hazardous constituents. The literature on applications of this technology is sparse.

CLASSIFICATION OF METHODOLOGIES

Table 10.5 lists combinations of measures for achieving aquifer restoration organized by acute or chronic pollution problems. Acute pollution of an

Table 10.5. Methodologies for Aquifer Cleanup.

Pollution Problem	Goal	Methodologies
Acute	Abatement	1. In-situ chemical fixation
		2. Excavation of contaminated soil with subsequent backfilling with "clean" soil
	Restoration	1. Removal[a] wells, treatment of contaminated ground water, and recharge
		2. Removal wells, treatment of contaminated ground water, and discharge to surface drainage
		3. Removal wells and discharge to surface drainage
Chronic	Abatement	1. In-situ chemical fixation
		2. Excavation of contaminated soil with subsequent backfilling with "clean" soil
		3. Interceptor trenches to collect polluted water as it moves laterally away from site
		4. Surface capping with impermeable material to inhibit infiltration of leachate-producing precipitation
		5. Subsurface barriers of impermeable materials to restrict hydraulic flow from sources
		6. Modify pumping patterns at existing wells
		7. Inject fresh water in a series of wells placed around source or contaminant plume to develop pressure ridge to restrict movement of pollutants
Chronic	Restoration	1. Removal wells, treatment of contaminated ground water, and recharge
		2. Removal wells, treatment of contaminated ground water, and discharge to surface drainage
		3. Removal wells and discharge to surface drainage
		4. In-situ chemical treatment
		5. In-situ biological treatment

[a]Could also be referred to as interceptor wells.

aquifer may result from inadvertent spills of chemicals or releases of undesirable materials and chemicals, usually as a result of an accident or unplanned activity. Such events often require an emergency response. Chronic aquifer pollution comes from numerous point and area sources and involves conventional pollutants such as nitrates and bacteria, or more toxic compounds such as gasoline, metals, and synthetic organic chemicals.

Methodologies for aquifer cleanup can also be characterized in terms of the goals of abatement and restoration. Abatement refers to the application of methodologies which prevent or minimize pollutant movement into ground water, or prevent contaminant plume migration into usable aquifer horizons. Restoration refers to restoring the water quality to background (precontamination) levels, usually by removing both the source(s) of pollution and renovating the polluted portion of the aquifer. If the pollution source(s) has already been dissipated by time, restoration may involve only renovation of the polluted ground water.

FUTURE OF TECHNOLOGIES

Tables 10.6 and 10.7 are projections of the future use of the various technologies as developed by the Office of Technology Assessment (OTA) of the United States Congress.[78] Rather than analyzing the individual projections, it is more interesting and informative to observe the overall trends in the tables. Table 10.6 shows two important trends. First, the projected future use of the containment technologies parallels the pattern of today, i.e., extensive use of wells and slurry walls for containment, and utilization of surface water control and capping as preventative or leachate minimization steps. Second, the technologies projected for extensive future use are those that are known to be effective and are well proven. The future of the innovative measures may be limited.

Table 10.7 shows a trend similar to Table 10.6, that of future use paralleling current use of the technologies. The second trend to note from Table 10.7 is that projected future use of the technologies is inversely proportional to their complexity. Once again, the technologies to be used are those that are simple but effective and have been widely used for years in related industries.

Table 10.6. Future Use of Containment Technologies.

Technique	Applicability	Effectiveness	Confidence	Capital cost	Cap/O&M	Projected level of use
Barriers:						
Slurry wall	2	1	2	2	1	Extensive
Grout curtain	2–3	1	2	2–3	1	Limited
Vibrating beam	2	1	2–3	2–3	1	Moderate
Sheet-pile	3	1–2	2	2–3	1	Nil-Limited
Block displacement	3	1	4	3	1	Nil
Hydraulic controls (wells)	2	1, 3	1	1	3	Extensive
Subsurface drains	2	1	2	1	2	Moderate
Runon/runoff controls	1	3	1	1	2	Extensive
Surface seals and caps	1	2, 3	2	1	1	Extensive
Solidification, etc.	2	1, 3	3–4	2	1	Moderate-Limited

KEY:

Applicability:
1—Very broadly applicable; little or no site dependency
2—Broadly applicable; some sites unfavorable
3—Limited to sites of specific characteristics

Effectiveness:
1—Can produce "leak-tight" containment
2—Can reduce migration—some leakage likely
3—Used as supporting technique in conjunction with other elements

Confidence:
1—Well proven—long-term effectiveness—high
2—Well proven—long-term effectiveness—unknown
3—Limited experience; used in other applications
4—Developmental; little data

Capital cost for function provided:
1—Low
2—Normal
3—High

Capital to operation and maintenance (O&M) cost ratio:
1—Capital higher than O&M
2—Capital about same as O&M
3—Capital lower than O&M

Table 10.7. Future Use of Treatment Technologies.

Technique	Applicability	Effectiveness	Confidence	Capital cost	Cap/O&M	Secondary disposal	Projected level use of
Biological treatment:	Or, 1–2	2	1	1	1–2	3	Moderate
Chemical:							
Neutralization/precipitation	In, 1	1	1	1	2	4	Moderate-Extensive
Wet air oxidation	Or, 2	2	2	3	1–2	1	Limited
Chlorination	In, 3	1	2	2	2	1	Limited
Ozonation	Or, 3	2	3	3	2–3	3	Nil
Reduction (Cr)	In, 3	1	2	2	2	3	Limited
Physical:							
Carbon adsorption	Or, In, 1	1	1	2	2–3	2–3	Moderate-Extensive
Sedimentation/filtration	Or, In, 1	1	1	1	2–3	4	Moderate-Extensive
Stripping	Or, 2	1	1	1	2	4	Moderate
Flotation	Or, 2	2	1	1	1	4	Limited
Ion exchange	In, 3	1–3	3	3	3	4	Nil
Reverse osmosis	Or, In, 3	1–2	3	3	3	4	Nil
Gas stream controls:							
Thermal oxidation	Or, 1	1	1	3	3	1	Limited-Moderate
Carbon adsorption	Or, 1	1	1	3	2–3	2–3	Limited-Moderate
Incineration							
Onsite	Or, 1	1	2	3	1	3[a]	Limited
Offsite	Or, 1	1	1	3	NA	3[a]	Moderate
In-situ biodegradation	Or, 3	2	3	2	3	1	Limited

NOTES:
[a]Must dispose solid residues
[b]Depends on reactive material used

KEY:
Applicability:
Class:
Or—Organic compounds
In—Inorganic compounds
Range:
1—Broadly applicable to compounds in indicated class
2—Moderated applicable: depends on waste composition concentration
3—Limited to special situations

Effectiveness:
1—Highest levels available
2—Output may need further treatment; may have pockets untreated (in-situ)

Confidence:
1—Well proven—easily transferable to site cleanup
2—Well proven—but not in clean-up settings
3—Limited experience
4—Developmental; little data

Capital cost for function provided:
1—Low
2—Normal
3—High

Capital to operations and maintenance (O&M) cost basis:
1—Capital higher than O&M
2—Capital about the same
3—Capital lower than O&M

Secondary treatment or disposal:
1—None
2—Minor
3—Major, but does not require hazardous waste techniques.
4—Basically a separation process; must be used with subsequent hazardous waste treatment or used with secure disposal step

REFERENCES

1. Dowiak, M. J., et al., "Selection, Installation, and Post-Closure Monitoring of a Low Permeability Cover Over a Hazardous Waste Disposal Facility," *Proceedings of the National Conference on Management of Uncontrolled Hazardous Waste Sites,* 1982, Hazardous Materials Control Research Institute, Silver Spring, Maryland, pp. 187–190.
2. Emrich, G. H. and Beck, W. W., "Top-Sealing to Minimize Leachate Generation: Case Study of the Windham, Connecticut Landfill," *Proceedings of the National Conference on Management of Uncontrolled Hazardous Waste Sites,* 1980, Hazardous Materials Control Research Institute, Silver Spring, Maryland, pp. 135–140.
3. Mutch, R. D., Jr., Daigler, J., and Clarke, J. H., "Clean-up of Shope's Landfill, Girard, PA," *Proceedings of the National Conference on Management of Uncontrolled Hazardous Waste Sites,* 1983, Hazardous Materials Control Research Institute, Silver Spring, Maryland, pp. 296–300.
4. Sanning, D. E., "Surface Sealing to Minimize Leachate Generation at Uncontrolled Hazardous Waste Sites," *Proceedings of the National Conference on Management of Uncontrolled Hazardous Waste Sites,* 1981, Hazardous Materials Control Research Institute, Silver Spring, Maryland, pp. 201–205.
5. Fung, R., Editor, *Protective Barriers for Containment of Toxic Materials,* 1980, Noyes Data Corporation, Park Ridge, New Jersey.
6. Anderson, D. C., Brown, K. W., and Green, J., "Organic Leachate Effects on the Permeability of Clay Liners," *Proceedings of the National Conference on Management of Uncontrolled Hazardous Waste Sites,* 1981, Hazardous Materials Control Research Institute, Silver Spring, Maryland, pp. 223–229.
7. Brown, K. W. and Anderson, D., "Effect of Organic Chemicals on Clay Liner Permeability: A Review of the Literature," *Proceedings of the Sixth Annual Research Symposium on Disposal of Hazardous Waste,* 1980, U.S. Environmental Protection Agency, Cincinnati, Ohio, pp. 123–134.
8. Cooper, J. W. and Schultz, D. W., "Development and Demonstration of Systems to Retrofit Existing Liquid Surface Impoundment Facilities with Synthetic Membrane," *Proceedings of the National Conference on Management of Uncontrolled Hazardous Waste Sites,* 1982, Hazardous Materials Control Research Institute, Silver Spring, Maryland, pp. 244–248.
9. Lord, A. E., "The Hydraulic Conductivity of Silicate Grouted Sands with Various Chemicals," *Proceedings of the National Conference on Management of Uncontrolled Hazardous Waste Sites,* 1983, Hazardous Materials Control Research Institute, Silver Spring, Maryland, pp. 175–178.
10. Malone, P. G., Francingues, N. R., and Boa, J. A., "The Use of Grout Chemistry and Technology in the Containment of Hazardous Wastes," *Proceedings of the National Conference on Management of Uncontrolled Hazardous Waste Sites,* 1982, Hazardous Materials Control Research Institute, Silver Spring, Maryland, pp. 220–223.

11. Spooner, P. A., et al., "Compatibility of Grouts with Hazardous Waste," EPA-600/52-84-015, March 1984, U.S. Environmental Protection Agency, Cincinnati, Ohio.
12. Huibregtse, K. R. and Kastman, K. H. "Development of a System to Protect Groundwater Threatened by Hazardous Spills on Land," EPA-600/2-81-085, May 1981, National Technical Information Service, Springfield, Virginia.
13. Hager, D. G. and Loven, C. G., "Operating Experiences in the Containment and Purification of Groundwater at the Rocky Mountain Arsenal," *Proceedings of the National Conference on Management of Uncontrolled Hazardous Waste Sites,* 1982, Hazardous Materials Control Research Institute, Silver Spring, Maryland, pp. 259–261.
14. Pendrell, D. J. and Zeltinger, J. M., "Contaminated Ground-Water Containment/Treatment System: Northwest Boundary, Rocky Mountain Arsenal, Colorado," *Proceedings of the Third National Symposium on Aquifer Restoration and Ground Water Monitoring,* 1983, National Water Well Association, Worthington, Ohio, pp. 453–461.
15. Ayres, J. A., Barverik, M. J., and Lager, D. C., "The First EPA Superfund Cutoff Wall: Design and Specifications," *Proceedings of the Third National Symposium on Aquifer Restoration and Ground Water Monitoring,* 1983, National Water Well Association, Worthington, Ohio, pp. 13–22.
16. Knox, R. C., "Effectiveness of Impermeable Barriers for Retardation of Pollutant Migration," *Proceedings of the National Conference on Management of Uncontrolled Hazardous Waste Sites,* 1983, Hazardous Materials Control Research Institute, Silver Spring, Maryland, pp. 179–184.
17. Anderson, D. C. and Jones, S. G., "Clay Barrier-Leachate Interaction," *Proceedings of the National Conference on Management of Uncontrolled Hazardous Waste Sites,* 1983, Hazardous Materials Control Research Institute, Silver Spring, Maryland, pp. 154–160.
18. D'Appolonia, D. J., "Slurry Trench Cutoff Walls for Hazardous Waste Isolation," Undated, Engineered Construction International, Inc., Pittsburgh, Pennsylvania.
19. Ryan, C. R. "Slurry Cut-Off Walls: Methods and Applications," March 1980, Geo-Con, Inc., Pittsburgh, Pennsylvania.
20. Millet, R. A. and Perez, J. Y., "Current USA Practice: Slurry Wall Specifications," *Journal of the Geotechnical Engineering Division—ASCE,* Vol. 107, No. GT8, August 1981, pp. 1041–1056.
21. Spooner, P. A., Wetzel, R. S., and Grube, W. E., "Pollution of Migration Cutoff Using Slurry Trench Construction," *Management of Uncontrolled Hazardous Waste Sites,* 1982, Hazardous Materials Control Research Institute, Silver Spring, Maryland, pp. 191–197.
22. Xanthakos, P., *Slurry Walls,* 1979, McGraw-Hill Book Company, Inc., New York, New York.
23. American Petroleum Institute, "Underground Spill Cleanup Manual," API Publication 1628, June 1980, Washington, DC.

24. American Petroleum Institute, "The Migration of Petroleum Products in Soil and Groundwater—Principles and Countermeasures," Publication No. 4149, December 1982, Washington, DC.
25. Molsather, L. R. and Barr, K. D., "Retrofit Leachate Collection System for an Existing Landfill," *Proceedings of the Second National Symposium on Aquifer Restoration and Ground Water Monitoring,* 1982, National Water Well Association, Worthington, Ohio, pp. 316–322.
26. Giddings, M. T., "The Lycoming County, Pennsylvania Sanitary Landfill: State-of-the-Art in Ground Water Pollution," *Ground Water,* Vol. 15, No. 1, January–February 1977, pp. 5–14.
27. Jhaveri, V. and Mazzacca, A. J., "Bio-reclamation of Ground and Groundwater Case History," *Proceedings of the National Conference on Management of Uncontrolled Hazardous Waste Sites,* 1983, Hazardous Materials Control Research Institute, Silver Spring, Maryland, pp. 242–247.
28. Underwood, E. R. and Thornton, J. C., "Contaminated Ground-Water Recovery System Analysis, Design and Construction at a Waste Management Facility Located in a Gulf Coastal Plain," *Proceedings of the Third National Symposium on Aquifer Restoration and Ground Water Monitoring,* 1983, National Water Well Association, Worthington, Ohio, pp. 142–147.
29. Gregg, D. O., "Protective Pumping to Reduce Aquifer Pollution, Glynn County, Georgia," *Ground Water,* Vol. 9, No. 5, September–October 1971, pp. 21–29.
30. Sheahan, N. T., "Injection/Extraction Well System—A Unique Seawater Intrusion Barrier," *Ground Water,* Vol. 15, No. 1, January–February 1977, pp. 32–49.
31. Osiensky, J. L. and Williams, R. E., "Ground Water Pump-Back System for a Uranium Tailings Disposal Site," *Proceedings of the Second National Symposium on Aquifer Restoration and Ground Water Monitoring,* 1982, National Water Well Association, Worthington, Ohio, pp. 30–37.
32. Bradley, E., "Trichloroethylene in the Ground-Water Supply of Pease Air Force Base, Portsmouth, New Hampshire," Water Resources Investigations Open-File Report 80-557, 1980, U.S. Geological Survey, Boston, Massachusetts.
33. Blake, S. B. and Lewis, R. W., "Underground Oil Recovery," *Proceedings of the Second National Symposium on Aquifer Restoration and Ground Water Monitoring,* 1982, National Water Well Association, Worthington, Ohio.
34. Hallberg, R. O. and Martinelli, R., "Vyredox—In-Situ Purification of Ground Water," *Ground Water,* Vol. 14, 1976, pp. 88–93.
35. Williams, E. B., "Contamination Containment by In-Situ Polymerization," *Aquifer Restoration and Ground Water Rehabilitation,* Proceedings of the Second National Symposium on Aquifer Restoration and Ground Water Monitoring, May 26–28, 1982.
36. Harsh, K. M., "In Situ Neutralization of an Acrylonitrile Spill," *Proceedings of the 1978 National Conference on Control of Hazardous Material Spills,* 1978, Hazardous Materials Control Research Institute, Silver Spring, Maryland, pp. 187–189.

37. Matthess, G., "In Situ Treatment of Arsenic Contaminated Groundwater," *Studies in Environmental Science: Quality of Groundwater,* Vol. 17, 1981, Elsevier Scientific Publishing Company, Amsterdam, The Netherlands, pp. 291–296.
38. Srivastava, V. K. and Haji-Djafari, S., "In Situ Detoxification of Hazardous Waste," *Proceedings of the National Conference on Management of Uncontrolled Hazardous Waste Sites,* 1983, Hazardous Materials Control Research Institute, Silver Spring, Maryland, pp. 231–236.
39. Raymond, R. L., Jamison, V. W., and Hudson, J. O., "Final Report on Beneficial Stimulation of Bacterial Activity in Ground Waters Containing Petroleum Products," 1975, Committee on Environmental Affairs, American Petroleum Institute, Washington, DC.
40. Brubaker, G. R. and O'Neill, E., "Remediation Strategies Using Enhanced Bioreclamation," *Proceedings of the Fifth National Symposium and Exposition on Aquifer Restoration and Ground Water Monitoring,* 1985, Columbus, Ohio, National Water Well Association, pp. 505–509.
41. McKee, J. E., Laverty, F. B., and Hertel, R. M., "Gasoline in Ground Water," *Journal of Water Pollution Control Federation,* Vol. 44, 1972, pp. 293–302.
42. Canter, L. W. and Knox, R. C., *Ground Water Pollution Control,* 1985, Lewis Publishers, Chelsea, Michigan.
43. Erlich, G. G., et al., "Degradation of Phenolic Contaminants in Ground Water by Anaerobic Bacteria, St. Louis Park, Minnesota," *Ground Water,* Vol. 20, 1982, pp. 702–710.
44. Lee, M. D. and Ward, C. H., "Microbial Degradation of Selected Aromatics at a Hazardous Waste Site," *Developments in Industrial Microbiology,* Vol. 25, 1984.
45. Ogawa, I., Junk, G. A., and Svec, H. J., "Degradation of Aromatic Compounds in Groundwater and Methods of Sample Presentation," *Talanta,* Vol. 28, 1981, pp. 725–729.
46. Roberts, P. V., et al., "Organic Contaminant Behavior During Groundwater Recharge," *Journal of Water Pollution Control Federation,* Vol. 52, 1980, pp. 161–172.
47. Yaniga, P. M., Matson, C., and Demko, D. J., "Restoration of Water Quality in a Multiaquifer System Via In-Situ Biodegradation of the Organic Contaminants," *Proceedings of the Fifth National Symposium and Exposition on Aquifer Restoration and Ground Water Monitoring,* 1985, National Water Well Association, Worthington, Ohio, pp. 510–526.
48. Rees, J. F. and King, J. W., "The Dynamics of Anaerobic Phenol Biodegradation in Lower Greensand," *Journal of Chemical Technology and Biotechnology,* Vol. 31, 1980, pp. 306–310.
49. Raymond, R. L., et al., "Final Report-Field Application of Subsurface Biodegradation of Gasoline in a Sand Formation," American Petroleum Institute, 1978, Project No. 307-377.
50. Wilson, J. T. and Wilson, B., "Biotransformation of Trichloroethylene in Soil," *Applied and Environmental Microbiology,* Vol. 49, No. 1, January 1985, pp. 242–243.

51. Edgehill, R. U. and Finn, R. K., "Microbial Treatment of Soil to Remove Pentachlorophenol," *Applied and Environmental Microbiology,* Vol. 45, 1983, pp. 1122–1125.
52. Walton, G. C. and Dobbs, D., "Biodegradation of Hazardous Materials in Spill Situations," *Proceedings, 1980 National Conference on Control of Hazardous Material Spills,* Louisville, Kentucky, pp. 23–29.
53. Zajik, J. E. and Daugulis, A. J., "Selective Enrichment Processes in Resolving Hydrocarbon Pollution Problems," In A. W. Bourquin, D. G. Ahearn and S. P. Meyers, Eds., *Proceedings, Impact of the Use of Microorganisms on the Aquatic Environment,* EPA-660/3-75-001, 1975, U.S. Environmental Protection Agency, Corvallis, Oregon, pp. 169–182.
54. Raymond, R. L., "Reclamation of Hydrocarbon Contaminated Ground Waters," U.S. Patent Office 3,846,290, Patented November 5, 1974.
55. Brown, K. W., et al., "Factors Influencing the Biodegradation of API Separator Sludges Applied to Soil," In D. W. Shultz, Ed., *Proceedings, Land Disposal: Hazardous Waste,* Philadelphia, Pennsylvania, March 16–18, 1981, EPA 600/9-8-002b, U.S. Environmental Protection Agency, Cincinnati, Ohio, pp. 198–199.
56. Dibble, J. T. and Bartha, R., "Effect of Environmental Parameters on the Biodegradation of Oil Sludge," *Applied and Environmental Microbiology,* Vol. 37, 1979, pp. 729–739.
57. Ward, C. H. and Lee, M. D., "In-Situ Technologies," *Ground Water Pollution Control,* by Canter, L. W. and Knox, R. C., 1985, Lewis Publishers, Chelsea, Michigan, pp. 127–158.
58. Raymond, R. L., Jamison, V. W., and Hudson, J. O., "Beneficial Stimulation of Bacterial Activity in Ground Waters Containing Petroleum Products," *AICHE Symposium Series,* Vol. 73, 1976, pp. 390–404.
59. Texas Research Institute, Inc., "Feasibility Studies on the Use of Hydrogen Peroxide to Enhance Microbial Degradation of Gasoline," American Petroleum Institute, 1982, Washington, DC.
60. Nagel, G., et al., "Sanitation of Ground Water by Infiltration of Ozone Treated Water," *GWF-Wasser/Abwasser,* Vol. 123, 1982, pp. 399–407.
61. Norris, R. D., and Brown, R. A., "In-Situ Bioreclamation—A Complete On-Site Solution," FMC Corporation, Aquifer Remediation Systems, Princeton, New Jersey, (No date).
62. Suntech, Inc., "Environmental Bioreclamation Brochure," 1977.
63. Zitrides, T. G., "Mutant Bacteria for the Disposal of Hazardous Organic Wastewaters," Presented at Pesticide Disposal Research and Development Symposium Reston, Virginia, September 6–7, 1978.
64. Althoff, W. F., Cleary, R. W., and Roux, P. H., "Aquifer Decontamination for Volatile Organics: A Case History," *Ground Water,* Vol. 19, No. 5, September–October 1981, pp. 445–504.
65. Lamarre, B. L., McGarry, F. J., and Stover, E. L., "Design, Operation, and Results of a Pilot Plant for Removal of Contaminants from Ground Water," *Third National Symposium and Exposition on Aquifer Restoration and Ground Water Monitoring,* 1983, National Water Well Association, Worthington, Ohio.

66. Allen, R. D. and Parmele, C. S., "Treatment Technology for Removal of Dissolved Gasoline Components from Ground Water," *Proceedings of the Third National Symposium on Aquifer Restoration and Ground Water Monitoring,* 1983, National Water Well Association, Worthington, Ohio, pp. 51–59.
67. Kincannon, D. F. and Stover, E. L., "Treatment of Ground Water," *Ground Water Pollution Control,* 1985, Lewis Publishers, Chelsea, Michigan.
68. Brunotts, V. A., et al., "Cost Effective Treatment of Priority Pollutant Compounds with Granular Activated Carbon," *Proceedings of the National Conference on Management of Uncontrolled Hazardous Waste Sites,* 1983, Hazardous Materials Control Research Institute, Silver Spring, Maryland, pp. 209–216.
69. Chaffee, W. T. and Weimar, R. A., "Remedial Programs for Ground-Water Supplies Contaminated by Gasoline," *Proceedings of the Third National Symposium on Aquifer Restoration and Ground Water Monitoring,* 1983, National Water Well Association, Worthington, Ohio, pp. 39–46.
70. Dobbs, R. A. and Cohen, J. M., "Carbon Adsorption Isotherms for Toxic Organics," EPA-600/8-80-023, April 1980, U.S. Environmental Protection Agency, Cincinnati, Ohio.
71. Fochman, E. G., "Biodegradation and Carbon Adsorption of Carcinogenic and Hazardous Organic Compounds," EPA-600/2-81-032, March 1981, U.S. Environmental Protection Agency, Cincinnati, Ohio.
72. Kaufmann, H. G., "Granular Carbon Treatment of Contaminated Supplies," *Proceedings of the Second National Symposium on Aquifer Restoration and Ground Water Monitoring,* 1982, National Water Well Association, Worthington, Ohio.
73. Parmele, C. S. and Allen, R. D., "Activated Carbon Adsorption with Nondestructive Regeneration—An Economical Aquifer Restoration Technology, *Proceedings of the Second National Symposium on Aquifer Restoration and Ground Water Monitoring,* 1982, National Water Well Association, Worthington, Ohio, pp. 99–104.
74. Arlotta, S. V., Jr., "The 'Envirowall Cut-Off' Vertical Barrier," *Proceedings of the National Conference on Management of Uncontrolled Hazardous Waste Sites,* 1983, Hazardous Materials Control Research Institute, Silver Spring, Maryland, pp. 191–194.
75. Cavalli, N. J., Arlotta, S. V., and Druback, G., "Subsurface Pollution Containment with the 'Envirowall' Vertical Cutoff Barrier," *Proceedings of the Third National Symposium on Aquifer Restoration and Ground Water Monitoring,* 1983, National Water Well Association, Worthington, Ohio, pp. 23–27.
76. Brunsing, T. P. and Cleary, M., "Isolation of Contaminated Ground Water by Slurry-Induced Ground Displacement," *Proceedings of the Third National Symposium on Aquifer Restoration and Ground Water Monitoring,* 1983, National Water Well Association, Worthington, Ohio, pp. 28–38.
77. Brunsing, T. P. and Grube, W. E., Jr., "A Block Displacement Technique to Isolate Uncontrolled Hazardous Waste Sites," *Proceedings of the National Conference on Management of Uncontrolled Hazardous Waste Sites,* 1982, Hazardous Materials Control Research Institute, Silver Spring, Maryland, pp. 249–253.

78. Office of Technology Assessment, "Superfund Strategy," March 1985, U.S. Congress, Washington, DC.

CHAPTER 11

Ground Water Quality Management

Ground water quality management, particularly as related to the control of multiple man-made sources of pollution and the current emphasis on organic constituents, is a relatively new area of environmental concern. Proper management of the quality of ground water cannot be separated from management concerns related to the quantitative aspects of the resource and its usage in society for a variety of purposes. Effective ground water quality management requires the coordinated effort of federal, state, and local levels of government in the development and implementation of multifaceted programs. This chapter addresses ground water quality management from a brief historical perspective followed by a review and comparison of existing federal, state, and local initiatives and relationships. The basic requirements for building an institutional capacity for managing ground water quality are reviewed along with specific elements for inclusion in a comprehensive management program. Finally, a case study on the development of a protection and management plan for an aquifer system is summarized.

HISTORICAL PERSPECTIVE

Responsibility for the protection and management of ground water quality has not been clearly defined from an historical perspective largely because ground water boundaries themselves often transcend political boundaries, and because the development of ground water laws has basically been a patchwork affair, arising from existing surface water regulations. Possibly the major contributors to this patchwork protection have been the historical lack of knowledge of ground water processes and the increasing use of ground water over a relatively short period of time.

During the nineteenth century, courts were reluctant to extend to ground water users the legal protection granted users of surface waters because of the inherent difficulties in identifying the physical boundaries of subsurface waters. Instead, the courts divided underground waters into two categories: percolating waters and subterranean streams. Courts presumed that underground waters were percolating waters. The existence of a subterranean stream, they said, could be established only by surface observation. Once an underground stream was discovered, its use was governed by the riparian doctrine applicable to surface streams. Early common law designated that a landowner had the right to the absolute use of the percolating waters beneath the property without owing an obligation to a neighboring landowner.[1] As ground water usage increased and serious contamination of this previously pristine resource was discovered, the states began to address various aspects of ground water quantity and quality issues with statutes.

Growing evidence now reveals that ground water contamination is a serious problem nationwide, and federal attention has increased over the past few years. A comprehensive national ground water protection program does not exist, but several federal laws address ground water contamination problems. Many activities are currently underway at all levels of government to respond to this growing awareness of the importance of regulating ground water usage and managing ground water quality.

INSTITUTIONAL PROGRAMS—FEDERAL LEVEL

There are at least 16 federal statutes related either directly or indirectly to ground water quality management. Table 11.1 lists these statutes in accordance with pertinent management activities.[2] The majority of the statutes have been adopted or amended since 1970. The management activities listed in Table 11.1 are broadly divided into investigations/detection, correction, prevention, standards, and other. The relationship between the 33 sources of ground water contamination listed in Table 4.1 and the 16 federal statutes is shown on Table 11.2.[2]

One of the primary points relative to the 16 statutes is that none of them were developed to specifically address ground water quality protection and cleanup. These ground water quality management concerns are typically an outgrowth of pollution source control or other environmental concerns addressed in the statutes. The five most important statutes in terms of ground water quality management include the Resource Conservation and Recovery Act (RCRA), the Safe Drinking Water Act (SDWA), the Superfund program (CERCLA), the Surface Mining Control and Reclamation Act (SMCRA), and the Clean Water Act (CWA). Most states have accepted primacy over the programs authorized by these acts. For example, the majority have received interim authority to administer hazardous waste programs under

Table 11.1. Summary of Federal Statutes Related to the Protection of Ground Water Quality.[2]

	Investigations/detection				Correction		Prevention				
	Inventories of sources[a]	Ambient groundwater monitoring	Groundwater monitoring related to sources[a]	Water supply monitoring	Federally funded remedial actions	Regulatory requirements for sources[a]	Regulate chemical production	Standards for new/existing sources[a]	Aquifer protection	Standards	Other[b]
Atomic Energy Act			X		X	X		X		X	
Clean Water Act	X	X	X		X			X		X	X
Coastal Zone Management Act											X
Comprehensive Environmental Response, Compensation and Liability Act	X		X		X						
Federal Insecticide, Fungicide, and Rodenticide Act			X				X	X			
Federal Land Policy and Management Act (and associated mining laws)			X					X			
Hazardous Liquid Pipeline Safety Act	X							X			
Hazardous Materials Transportation Act	X							X			
National Environmental Policy Act											X
Reclamation Act					X						
Resource Conservation and Recovery Act	X		X			X		X			
Safe Drinking Water Act	X		X					X	X	X	
Surface Mining Control and Reclamation Act			X		X			X			
Toxic Substances Control Act			X				X	X			
Uranium Mill Tailings Radiation Control Act			X		X	X		X			
Water Research and Development Act											X

[a]Programs and activities under this heading relate directly to specific sources of ground water contamination. Table 11.2 summarizes the sources addressed by the statutes.
[b]This category includes items such as research and development and grants to the States to develop ground water-related programs.

Table 11.2. Relationship Between Sources of Contamination and Federal Statutes.[2]

	Federal statutes															
Sources	**AEA**	**CWA**	**CZMA**	**CERCLA**	**FIFRA**	**FLPMA**	**HLPSA**	**HMTA**	**NEPA[b]**	**RA**	**RCRA**	**SDWA**	**SMCRA**	**TSCA**	**UMTRCA**	**WRDA[c]**
Category I																
Subsurface percolation ...		E										A				
Injection wells (waste)				F								A				
Injection wells (non-waste) .												A				
Land application		D		F							A					
Category II																
Landfills				F							A, B			A		
Open dumps (including illegal dumping)				F							B					
Residential (or local) disposal					C											
Surface impoundments ...				F		A					A		A			
Waste tailings						A									A	
Waste piles				F		A					A		A			
Materials stockpiles					C											
Graveyards																
Animal burial																
Aboveground storage tanks		A		F							A			A		
Underground storage tanks		A		F							A			A		
Containers				F	C						A			A		
Open burning/detonation sites				F	C						A					
Radioactive disposal sites .	A															A, F
Category III																
Pipelines				F			A									
Materials transport/transfer operations				F				A								

Category IV							
Irrigation practices	C, E						
Pesticide applications	C, E		A				
Fertilizer applications	C, E						
Animal feeding operations	C, E						
Deicing salts applications							
Urban runoff	C, E						
Percolation of atmospheric pollutants							
Mining and mine drainage	C, E			A			A, F
Category V							
Production wells				A		A	
Other wells (non-waste)				A			
Construction excavation	C, E						
Category VI							
Groundwater-surface water interactions	C, E						
Natural leaching					F		
Salt-water intrusion/brackish water upconing	C, E	E					

[a]Key: A = Requires compliance with specified Federal requirements (some programs in this group may be implemented by States if they meet certain Federal criteria).
B = Authorizes funding of optional State programs that address specific sources.
C = Establishes Best Management Practices (BMPs) or recommended procedures for certain sources.
D = Establishes Federal criteria that must be met in order to receive funds for specific projects related to a source of contamination.
E = Establishes a grant program to States (funds may be used at the State or local level to address contaminants or sources).
F = Funds Federal cleanup of contaminated groundwater and associated sources.

[b]NEPA does not apply to any particular source. The environment impacts of projects involving the use of Federal funds may be subject to Federal agency review.

[c]WRDA does not apply to any particular source. The act provides research funds to States. Projects may focus on particular sources.

RCRA, and over half have received funds to administer the Underground Injection Control (UIC) Program authorized by the SDWA.[3,4]

The 16 statutes listed in Table 11.2 do not address all known sources of contamination, nor are sources treated in a uniform manner by the programs authorized by federal legislation.[2] All but four sources are covered by at least one statute; eighteen are covered by more than one. Most of the statutes, however, limit coverage by defining only subsets of facilities and/or activities of a given source type that are subject to their respective requirements. These definitions are based on various criteria, such as the presence of certain contaminants. Moreover, the statutory definition of sources is sometimes narrowed further by the regulations issued by the federal agencies responsible for implementing the statutes. Major sources not currently being adequately addressed by federal statutes include:[2]

(1) Surface impoundments used to contain nonhazardous wastes (e.g., wastes used in agriculture)

(2) Waste piles and materials stockpiles used to store nonhazardous wastes (except pesticides)

(3) Noncoal mining activities on private lands

(4) Pipelines not regulated by the Hazardous Liquid Pipeline Safety Act

An additional point relative to the 16 federal statutes is that they are not all administered by the same agency. For example, while the U.S. Environmental Protection Agency is the primary agency for many of the 16 statutes, other administering agencies include the U.S. Department of the Interior and the U.S. Nuclear Regulatory Commission. While not specifically involved in statute administration, still additional federal agencies and programs have activities related to ground water quality management. Table 11.3 contains a listing of ground water-related activities of 14 federal agencies or programs.[2] This listing is probably not complete. The important point is that in addition to the U.S. Environmental Protection Agency, there are also ground water responsibilities and interests held by the U.S. Geological Survey, the U.S. Department of Agriculture, the U.S. Department of Defense, and many other resource management agencies.

Two illustrations of recent federal initiatives related to ground water quality management will be presented. The first deals with a national strategy for ground water protection, and the second relates to protection measures and requirements for underground storage tanks as a source of contamination.

Table 11.3. Ground Water-Related Activities of Federal Agencies.[2]

Department of Agriculture—Agriculture Research Service: ARS is conducting a limited number of research projects related to groundwater recharge and the impacts of agricultural activities on groundwater quality.

Department of Agriculture—Forest Service: The Forest Service is conducting environmental research projects on the fate and transport of pesticides (under the National Agricultural Pesticide Impact Assessment Program).

Department of Commerce—National Bureau of Standards: NBS is responsible for projects regarding the development of quality assurance standards that are used by other Federal agencies (e.g., EPA and DOE) to monitor the analytical performance of laboratories.

Department of Defense: The Army, Navy, and Air Force are participating in a program to identify and evaluate hazardous waste disposal sites on military installations and to undertake remedial actions at certain sites to control the migration of wastes (Installation Restoration Program).

The Army Toxic and Hazardous Materials Agency (USATHAMA), Air Force Occupational Environmental Health Laboratory, Air Force Engineering and Service Center, and Navy Energy and Environmental Support Activity provide technical support for the Installation Restoration Program and conduct research related to these efforts.

The Army Medical Bioengineering Research and Development Laboratory develops water quality criteria for certain munitions compounds.

The Army Corps of Engineers is working with EPA (under an interagency agreement) on design and construction of remedial action projects for CERCLA-designated sites. Research projects are also being conducted to support these activities.

Environmental Protection Agency—Office of Research and Development: EPA's Environmental Photographic Interpretation Center in Warrenton, VA, is responsible for acquiring and interpreting overhead imagery to support programs of EPA as well as other Federal agencies. Activities include conducting inventories of abandoned wells, mines, and hazardous waste sites, identifying failures in septic tank systems, and supporting emergency (e.g., oil spills) response activities.

EPA's Environmental Monitoring Systems Laboratory in Las Vegas, NV, the Robert S. Kerr Environmental Research Laboratory in Ada, OK, and the Environmental Research Laboratory in Athens, GA are conducting studies related to prediction (e.g., studies of those characteristics of aquifers that influence contaminant behavior) and monitoring (e.g., protocols for designing groundwater sampling programs). Other research activities related to source control, health effects, and treatment technologies are also being conducted at other EPA facilities.[a]

Department of Energy: Programs have been established for identifying and decommissioning nuclear materials storage and processing facilities that have become contaminated. Hydrogeologic investigations are being conducted at some of these sites. These programs include the Formally Utilized Sites Remedial Action Program and the Surplus Facilities Management Program.

Department of Housing and Urban Development: Environmental assessments are conducted related to housing projects; groundwater impacts are considered.

Department of the Interior—Bureau of Land Management: BLM is conducting inventories of hazardous waste sites on public lands.

Table 11.3, continued

Department of the Interior—National Park Service: Groundwater monitoring studies are conducted at various national parks to develop baseline data and to determine the extent and impacts of groundwater contamination from sources such as septic tanks and agricultural activities.

Department of the Interior—U.S. Geological Survey: The Water Resources Division of USGS is responsible for collection and analysis of hydrogeologic information (including groundwater data), maintaining computerized data bases, conducting research, and coordinating Federal activities with respect to the use and acquisition of water data.

Department of the Interior—Fish and Wildlife Service: FWS is conducting inventories of hazardous waste sites for all FWS lands and facilities.

Department of the Interior—Bureau of Indian Affairs: BIA is planning to conduct inventories of hazardous waste sites on or near Indian reservations.

National Science Foundation: The Division of Civil and Environmental Engineering, Directorate for Engineering (the Hydraulics, Hydrology, and Water Resources Program, and the Environmental and Water Quality Engineering Program) supports research projects on topics such as subsurface transport and wastewater treatment. Policy-related research is conducted by the Division of Research and Analysis, Directorate for Scientific, Technological and International Affairs.

Nuclear Regulatory Commission: Research projects are conducted related to the fate and transport of radioactive substances in support of regulatory activities.

[a]EPA also supports several other types of activities related to groundwater. For example, EPA established a consortium called the National Center for Ground Water Research in September 1979. The consortium consists of the University of Oklahoma, Oklahoma State University, and Rice University; and the Ground Water Research Branch of the Kerr Laboratory serves as the center's immediate technical liaison. The primary objective of the center is to identify long-term problems and needs related to groundwater quality protection (e.g., transport and fate of contaminants and subsurface characterization). EPA also provides funding to the Ground Water Clearinghouse at the Holcomb Research Institute. The clearinghouse contains an extensive file of groundwater models and assists the States in model selection and application.

Ground Water Protection Strategy

A major initiative at the federal level toward managing the nation's ground water resources is represented by the Ground Water Protection Strategy issued by the U.S. Environmental Protection Agency in 1984.[5] The four major components included in this Strategy are:

(1) The provision of support by EPA to the states for program development and institution building relative to ground water quality management.
(2) The assessment of problems that may exist from currently unaddressed sources of contamination, in particular, leaking underground storage tanks, surface impoundments, and landfills.
(3) The issuance of guidelines for EPA decisions for consistency in ground water protection and cleanup programs.
(4) The development of an internal organizational structure within EPA for ground water management at both the headquarters and regional levels. This institutional structure should also strengthen EPA's cooperation with federal and state agencies.

The primary thrust of the EPA Strategy is to encourage states to assume the lead role in ground water quality management, with EPA providing background information and technical guidance as well as funding for program development. One of the critical aspects of the EPA Strategy is the concept of a ground water classification system, with the following classes being identified:[5]

Class I: Special ground waters—ground waters that are highly vulnerable to contamination because of the hydrological characteristics of the areas under which they occur and that are also characterized by either of the following two factors: irreplaceable, in that no reasonable alternative source of drinking water is available to substantial populations, or ecologically vital, in that the aquifer provides the base flow for a particularly sensitive ecological system that, if polluted, would destroy unique habitat.

Class II: Current and potential sources of drinking water and waters having other beneficial uses—these represent ground waters that are currently used or potentially available for drinking water or other beneficial use.

Class III: Ground waters not considered potential sources of supply and of limited beneficial use—these are ground waters that are heavily saline, with total dissolved solids levels over 10,000 mg/l, or otherwise contaminated beyond levels that would allow cleanup using methods reasonably employed in public water system treatment.

The basic concept of this classification system is that greater protection would be provided for Class I ground waters, with lesser degrees of protection associated with Class III. Many states are in the process of considering the development of aquifer classification systems, and several states have already adopted systems similar in concept to that contained in the EPA Ground Water Protection Strategy. Additional information on state classification systems will be presented later in this chapter.

Underground Storage Tanks

One of the primary sources of ground water contamination has been identified as leaking underground tanks storing many categories of liquids such as gasoline, fuels, processed chemicals, hazardous and toxic chemicals, and dilute waste.[6] The number of underground storage tanks in the United States is upwards of 3 million, and a conservative estimate is that 10% or more of these tanks may be leaking. Underground storage tanks may leak due to several reasons. Corrosion, both external and internal, is considered the most common cause of leaks. Structural failure, primarily from improper tank installation, can also cause leaks. In addition, tank contents that are incompatible with the tank's liner and/or construction materials may also induce leakage.[7] Poor operating practices have also been noted as a cause of tank leakage.[8]

The federal initiative for regulating underground storage tanks is contained in the Hazardous and Solid Waste Amendments to the Resource Conservation and Recovery Act (RCRA) of 1984 (P.L. 98-616). The Amendments added a new Subtitle I entitled, "Regulation of Underground Storage Tanks." Part of these Amendments require the U.S. Environmental Protection Agency to develop and establish a national regulatory program for the control of new and existing underground storage tanks.[9] The scope of this program is broad and applies to tanks and combinations of tanks with 10% or more of their volume underground, including the volume of underground piping, that are used to store petroleum products or other liquid materials defined as hazardous substances under Section 101(14) of the Comprehensive Environmental Response, Compensation and Liability Act (CERCLA, commonly known as Superfund). The following tanks are excluded from the Interim Prohibition:

(1) Farm or residential tanks of 1,100 gallons or less capacity used for storing motor fuel for noncommercial purposes

(2) Tanks used for storing heating oil for consumptive use on the premises where stored

(3) Septic tanks

(4) Flow-through process tanks

(5) Tanks above the floor level but still underground

Existing tanks are to be addressed by an inventory and monitoring program. For example, among Subtitle I's provisions is the requirement that underground tank owners notify designated state or local agencies of their tanks' existence. Under the law, EPA must also develop regulations for underground tanks addressing leak detection, corrective action, closure, recordkeeping, and reporting, and new tank performance standards.[10] In addition, any new tank installed after May 7, 1985, must be designed, installed, and operated so that:

(1) It will prevent releases due to corrosion or structural failure for the operational life of the tank.

(2) It is cathodically protected against corrosion, constructed of noncorrosive material, steel clad with a noncorrosive material, or designed in a manner to prevent the release or threatened release of any stored substance.

(3) The material used in the construction or lining of the tank is compatible with the substance to be stored.

In its interpretive rule, EPA has noted that the term "operational life" of a tank is "the time during which the tank stores regulated substances." The Interim Prohibition also has a limited exemption stating that a new tank does not have to be protected from corrosion if it is installed in a certain soil environment as follows:

> if soil tests conducted in accordance with ASTM Standard G57-78, or another standard approved by the administrator, show that soil resistivity in an installation location is 12,000 ohm/cm or more (unless a more stringent standard is prescribed by the Administrator by rule), a storage tank without corrosion protection may be installed in that location.

Approximately half of the 50 states have initiated regulatory programs for underground storage tanks, and additional states are anticipated to have programs in place in the near future. In addition, local governmental agencies have also developed regulatory programs.[11]

INSTITUTIONAL PROGRAMS—STATE LEVEL

State ground water quality management activities can be represented on a continuum from limited to comprehensive approaches. Limited activities include controlling a specific source of contamination (e.g., oil wells or landfills) or controlling a localized area (e.g., a recharge zone or an area with multiple contamination threats). Comprehensive approaches address ground water from a statewide perspective, ranging from an integrated quality-quantity surface-ground water perspective, to combining locational and source controls with specific organizational and management techniques. Where a state government falls on the continuum depends on the importance of ground water to the area, the level of knowledge about the resource, and the seriousness and pervasiveness of perceived threats.[12]

As of 1985, twelve states had enacted specific legislation dealing with ground water protection.[3] These twelve states are identified in Table 11.4. States generally have taken one of three approaches to ground water management: nondegradation, limited degradation, or differential protection approaches. Most states (38) have adopted only one of these approaches (Table 11.4). Only twelve states have adopted more than one approach, normally a mixture of nondegradation and limited degradation standards or limited de-

Table 11.4. State Ground Water Policy: Current Status and Characteristics.[3]

					Characteristics of Policies	
State	Specific State Statutes for Ground Water	Existing Policy for Protecting Ground Water Quality	Policy Under Development	Non-degradation	Limited Degradation	Differential Protection
Alabama			X			
Alaska						
Arizona	X	X			X	X
Arkansas			X			
California		X	X	X		
Colorado			X	X		X
Connecticut		X				X
Delaware		X	X			
Florida	X	X			X	X
Georgia	X	X		X		
Hawaii		X	X			
Idaho		X		X	X	X
Illinois			X			X
Indiana			X			
Iowa			X	X		
Kansas	X	X				
Kentucky			X			
Louisiana						
Maine	X	X	X	X		
Maryland					X	X
Massachusetts		X				X
Michigan		X	X	X		
Minnesota		X	X	X	X	
Mississippi			X			
Missouri			X	X		
Montana		X		X		

Nebraska			X			
Nevada		X		X	X	
New Hampshire		X		X		
New Jersey	X	X	X	X	X	
New Mexico	X	X			X	
North Carolina	X	X			X	
North Dakota		X	X			X
Ohio			X			
Oklahoma	X	X				
Oregon		X			X	
Pennsylvania			X		X	
Rhode Island			X	X		
South Carolina			X	X	X	
South Dakota			X			
Tennessee			X			
Texas						
Utah		X	X			
Vermont		X			X	X
Virginia	X					
Washington		X	X	X	X	X
West Virginia			X			
Wisconsin	X	X			X	
Wyoming	X	X			X	X
Total	12	27	27	16	16	12

gradation and differential protection.[3] A nondegradation policy protects the quality of ground water at existing levels. This policy is often accompanied by waivers for specific activities for which nondegradation would not be achievable. Limited degradation is designed to preserve ground water quality above a specified amount, and differential protection focuses on the present and potential uses of ground water. To accomplish differential protection, classifications based on various criteria may be utilized. They may be based on types of uses, degree of treatment, salinity-quality levels, vulnerability to contamination, environmental importance of aquifers, or availability of alternative water supplies. Twenty-two states have adopted or proposed a classification system, including eight states with specific ground water legislation.[3]

Several surveys of state ground water programs and laws have been conducted in recent years. In 1980, the U.S. Water Resources Council[13] surveyed all 50 states regarding their ground water planning and management activities. The Council found that while most states had some type of activity (e.g., permitting) underway, in general, state management of ground water was limited. Wickersham[14] conducted a survey of ground water laws in 24 states and found the following control measures to be common to many states: licensing of water well drillers, requirements for well permits before drilling, establishment of ground water management districts, use of ground water standards, development of water usage conservation programs, registration of water wells, and filing of annual pumpage reports. Ehrhardt[15] documented notable trends toward stricter permitting systems, conjunctive use management, development of quality standards, authorization of local governments to use zoning powers for ground water protection, and increased use of classification systems.

Pertinent issues related to state ground water protection programs include classification systems, ground water quality standards, wellhead protection and special aquifers, conjunctive use, and interstate ground water management. These issues will be discussed along with a summary of the development of a comprehensive ground water quality management program in Michigan.

Ground Water Classification

Ground water classification refers to a comprehensive system with which to classify waters for differential ground water protection strategies.[16] This concept is a basic component of the U.S. Environmental Protection Agency Ground Water Protection Strategy.[5] A classification system can be applied to ground water in several aquifer units or portions thereof and water basins.

This basic policy allows states to apply the appropriate levels of protection to ground water based on present and future uses; for example, high-water quality aquifers would receive the highest level of protection, whereas other less critical uses of ground water resources would receive lower levels of priority for protection. A classification system can:[16]

(1) Provide a focus for limited state resources on protection and restoration of valuable and sensitive aquifers
(2) Provide a basis for coordinated management of ground water resources and the activities that potentially have an impact on the resource at all levels of government
(3) Provide guidance for implementation of regulatory and enforcement programs

States can use a variety of protection mechanisms to implement classification systems.[17] Many states employ numerical quality standards as well as narrative descriptions for each class. States may require pollution source controls, such as permits, effluent limitations, mixing zones, waste facility design standards, and best management practices to keep ground water quality at the level prescribed in the classification system. Some states employ land use restrictions, such as zoning guidelines, siting regulations, and public acquisition of land to maintain designated levels.

Table 11.5 contains summary information on ten states and one territory which have adopted classification systems.[17] These systems are based on maintaining water quality standards which will ensure the protection of present and future uses of ground water resources. The highest designations pertain to ground water used as drinking water, with the quality standards in the Safe Drinking Water Act used as the basis for classification designations. The classification system used in Connecticut will be described in some detail.

In September 1980, Connecticut adopted a ground water classification system in pursuit of its statutory mandate to protect the public health and welfare, promote economic development, and preserve and enhance water quality for present and future use for public water supply, agriculture, industrial, and other legitimate uses.[17] Based upon hydrogeological, land use, and ground water use information, along with information on actual and potential sources of ground water contamination, the following four classes of ground water were designated:

(1) Class GAA—Existing or proposed drinking water use without treatment.
(2) Class GA— May be suitable for public or private drinking water use without treatment.
(3) Class GB— May not be suitable for public or private use as drinking water without treatment since the ground water is known or presumed to be degraded. High-density housing, waste disposal, or industrial sites form the basis for designating the ground water as GB.

Table 11.5. State and Territory Ground Water Classification Systems.[17]

State	Number of Classes	Criteria for Classification
Connecticut	4	• Suitable for public and private drinking water supplies without treatment; • May be suitable for public or private drinking water supplies without treatment; • May have to be treated to be potable; • May be suitable for waste disposal practices.
Florida	4	• Single-source aquifers for potable water use with total dissolved solids content (TDS) of less than 3,000 mg/l; • Potable water use with TDS content less than 10,000 mg/l; • Non-potable water use from unconfined aquifers with TDS content of 10,000 mg/l or greater; • Non-potable water use with TDS content of or greater than 10,000 mg/l from confined aquifers.
Guam	3	• Drinking water quality; • Saline; • (a) Less than 10,000 gpd, (b) greater than 10,000 gpd.
Maryland	3	• TDS less than 500 mg/l; • TDS between 500 and 6,000 mg/l; • TDS greater than 6,000 mg/l.
Massachusetts	3	• Drinking water quality; saline; below drinking water quality.
Montana	4	• Suitable for public and private water use with little or no treatment; • Marginally suitable for public and private use; • Suitable for industrial and commercial uses; • May be suitable for some industrial and other uses.
New Mexico	2	• TDS less than 10,000 mg/l; • TDS more than 10,000 mg/l.
New York	3	• Fresh water for potable use; • Saline waters containing between 250 mgCl/l and 1,000 mgCl; • Saline waters with chloride in excess of 1,000 mg/l.
North Carolina	5	• Fresh waters for drinking water use; • Brackish waters at greater than 20 feet below surface for recharge use;

Table 11.5, continued

State	Number of Classes	Criteria for Classification
		• Fresh waters at less than 20 feet below surface for recharge use; • Brackish waters less than 20 feet below surface; • Not suitable for drinking or food processing.
Vermont	2	• Aquifers that supply or in the future could supply community water supplies; • All other ground waters.
Wyoming	7	• Domestic; • Agricultural; • Livestock; • Aquatic life; • Industry; • Hydrocarbon and mineral deposits; • Unsuitable for any use.

(4) Class GC— May be suitable for certain waste disposal practices because land use or hydrogeologic conditions render this ground water more suitable for receiving permitted discharges than for development as a public or private water supply. Often, these areas have suffered waste disposal practices that have permanently made the ground water unsuitable for drinking without treatment. These are areas of stratified drift with greater than ten feet of unsaturated thickness that have the capability to provide optimum renovation for wastewater discharges.

The basic policy in Connecticut is to restore or maintain the quality of the ground water to a quality consistent with its use for drinking without treatment. In keeping with this policy, all ground water shall be restored to the extent possible to a quality consistent with Class GA. However, restoration of ground water to Class GA shall not be sought when: (1) the ground water is in a zone of influence of permitted discharge, (2) the ground water is designated as Class GB unless there is demonstrated need to restore ground water to a Class GA designation or where it can be demonstrated that restoration to Class GA can be reasonably achieved, or (3) the ground water is designated Class GC.[17] Connecticut implements its classification system by setting forth guidelines on land uses and discharges that are compatible with each class. The discharges allowed for each class are:[17]

(1) Class GAA—Restricted to wastewaters of human or animal origin and other minor cooling and clean water discharges.

(2) Class GA— Restricted to wastewaters of predominately human, animal or natural origin which pose minimal threat to drinking water supplies.

(3) Class GB— Wastewater discharges allowed for GAA and GA ground water and certain treated industrial wastewaters where the soils will be used as an integral part of the treatment system. (The intent of doing so is to allow the soil to be part of the treatment system for easily biodegradable organics and also function as a filtration process for inert solids.) Such discharges shall not cause degradation of ground water that could preclude its future use for drinking without treatment.

(4) Class GC— All discharges allowed for GAA, GA, or GB and industrial wastewater discharges that do not result in surface water quality degradation below established classification goals. The intent is to allow the soil to be part of the treatment process.

Ground Water Quality Standards

Standards for ambient ground water quality are intended to establish upper limits of concentration of designated pollutants in ground water consistent with the use of those waters for beneficial uses.[16] The task of developing standards is complex. State and local governments do not generally have the ability or resources to develop scientifically based standards. For this reason, most states have looked to the U.S. Environmental Protection Agency to perform this function or to provide technical information that the states can use in setting ambient standards for these pollutants. In turn, the states can use ambient standards as one basis for regulatory and enforcement action to limit discharges of those pollutants from point and nonpoint sources into ground water.

Under the federal Safe Drinking Water Act (SDWA) of 1974, the U.S. Environmental Protection Agency is supposed to prepare and promulgate regulations or standards for constituents of health concern in public water supplies. The SDWA calls on EPA to set two different kinds of standards for water used for human consumption: recommended maximum contaminant levels (RMCLs) and maximum contaminant levels (MCLs). The RMCLs represent maximum concentrations of pollutants based solely on health concerns. Under the SDWA, EPA may not enforce these limits; they are primarily informational and represent long-term goals. By contrast, the MCLs are enforceable. If a public water supply exceeds a MCL for a pollutant, the purveyor is required to take action to reduce concentrations of that pollutant below the MCL.[16]

Many states have used the RMCLs and MCLs as a basis for setting ambient ground water quality standards. However, EPA has adopted RMCLs and MCLs for only a limited number of constituents, including very few organic chemical compounds. The states find themselves having to deal with an increasing number of compounds for which standards have not been developed. Where MCLs and RMCLs are not available, states must develop their own standards based on their own analysis or based on suggested no adverse response levels (SNARLs), which are guidelines suggested by EPA for a limited number of organic chemicals.[16]

A comprehensive review of state numerical and narrative ground water quality standards has been prepared.[18] Examples of standards from two states (Wisconsin and New York) will be highlighted. Wisconsin uses a two-tiered approach for substances already detected in ground water or substances that have a reasonable probability of entering ground water. For each pertinent substance, an enforcement standard and a preventive action limit (PAL) has been or will be established. Enforcement standards define when violations of ground water quality standards have occurred and apply to all state-regulated activities that have an impact on ground water quality. When a substance is detected in ground water in concentrations equal to or greater than its enforcement standard, the source is subject to immediate enforcement action. The preventive action limit (PAL) represents a ground water quality standard that is a lower concentration of the substance than the enforcement standard. PALs are intended to function as warning levels and as standards in facility design. These limits have been established by the state for each of the substances with enforcement standards. For the substances of public health concern (volatile organics and heavy metals, for example), PALs are set at 20% of the enforcement standard except where the substance is reported to be carcinogenic, mutagenic, or teratogenic. For these substances, the PAL is set at 10% of the enforcement standard. PALs for the public welfare-related substances (chloride, turbidity, and TDS, for example) are set at 50% of their enforcement standards.[16]

An issue of concern relative to ground water quality standards is associated with where they are to be applied. In Wisconsin, PALs apply everywhere ground water is monitored. Enforcement standards are applicable at any point of ground water use, at any point beyond the property boundary of a regulated facility, and at any point outside the subsurface attenuation zone provided by what are known as design management zones (DMZs) for regulated activities. DMZs are subsurface zones extending horizontally a specified distance from the particular source and vertically downward from the water table through the entire saturated zone.[16]

In 1978, the state of New York established a three-tiered ground water classification system as follows:[17]

(1) Class GA— The best usage of Class GA water is as a source of potable water supply. Class GA waters are fresh ground waters found in the saturated zone of unconsolidated deposits and consolidated rock or bedrock.

(2) Class GSA—The best usage of Class GSA waters is as a source of potable mineral waters, for conversion to fresh potable waters, or as raw material for the manufacture of sodium chloride or its derivatives or similar products. Such waters are saline waters found in the saturated zone.

(3) Class GSB—The best usage of Class GSB waters is as a receiving water for disposal of wastes. Such waters are those saline waters found in the saturated zone which have a chloride concentration in excess of 1,000 mg/l or a TDS concentration in excess of 2,000 mg/l.

Ground water quality management in New York is based on using the classification system in conjunction with numerical and narrative quality standards, effluent limitations, and monitoring requirements. Numerical quality standards for Class GA have been identified for 83 pollutants, including metals, chloride, foaming agents, nitrates, pH, numerous pesticides, and some organic solvents. Table 11.6 summarizes the pertinent numerical and narrative standards.[19] New York also has effluent standards for discharges to Class GA waters. These standards apply to any discharge from a point source or outlet that enters the saturated or unsaturated zone. Effluent samples are collected at a point where the effluent emerges from a treatment works, disposal system, outlet, or point source prior to being discharged into the ground.[17] The ground water monitoring policy emphasizes a combination of effluent monitoring and monitoring of drinking water sources.

As noted earlier, nondegradation and limited degradation policies have been used by several states in their approach to ground water quality management. However, Hoffman[20] indicated that nondegradation and zero pollution of ground water are goal statements which in practice cannot be achieved. Therefore, some measurable parameters must be set so that activities necessary for the existence of society can be evaluated as to compliance with a reasonable standard. Hoffman suggests a model framework for ground water protection such that: (1) no activity will cause a statistically significant deviation from the main geochemistry of the ground water existing beneath disposal facilities, (2) a compliance point be defined as the lesser of 1,200 feet from the source or the project property line, and (3) an intervention point be established at one-half of the above distance. The intervention point is the place where measurements against the standards are to be taken and corrective action employed if necessary. For hazardous wastes, or large activities which involve a site of more than 100 acres, a computer model of ground water flow and contamination should be required.[20]

Wellhead Protection and Special Aquifers

The amendments to the Safe Drinking Water Act which were passed in May 1986, included two provisions relating to ground water concerns. These provisions involve the development of a program to require all states to establish a wellhead protection program and also to develop programs to provide protection to special aquifers such as those uniquely used for water sup-

Table 11.6. Ground Water Quality Standards for New York Ground Water Class GA.[19]

Items	Specifications
1. Sewage, industrial waste or other wastes, taste or odor producing substances, toxic pollutants, thermal discharges, radioactive substances or other deleterious matter.	1. None which may impair the quality of the ground waters to render them unsafe or unsuitable for a potable water supply or which may cause or contribute to a condition in contravention of standards for other classified waters of the State.
2. The concentration of the following substances or chemicals:	2. Shall not be greater than the limit specified, except where exceeded due to natural conditions:
(1) Arsenic (As)	(1) 0.025 mg/l
(2) Barium (Ba)	(2) 1.0 mg/l
(3) Cadmium (Cd)	(3) 0.01 mg/l
(4) Chloride (Cl)	(4) 250 mg/l
(5) Chromium (Cr) (Hexavalent)	(5) 0.05 mg/l
(6) Copper (Cu)	(6) 1.0 mg/l
(7) Cyanide (CN)	(7) 0.2 mg/l
(8) Fluoride (F)	(8) 1.5 mg/l
(9) Foaming Agents	(9) 0.5 mg/l
(10) Iron (Fe)[2]	(10) 0.3 mg/l
(11) Lead (Pb)	(11) 0.025 mg/l
(12) Manganese (Mn)	(12) 0.3 mg/l
(13) Mercury (Hg)	(13) 0.002 mg/l
(14) Nitrate (as N)	(14) 10.0 mg/l
(15) Phenols	(15) 0.001 mg/l
(16) Selenium (Se)	(16) 0.02 mg/l
(17) Silver (Ag)	(17) 0.05 mg/l
(18) Sulfate (SO_4)	(18) 250 mg/l
(19) Zinc (Zn)	(19) 5 mg/l
(20) pH Range	(20) 6.5–8.5
(21) Aldrin. or 1,2,3,4,10,10 hexachloro-1,4,4a,5,8,8a-haxahydro-*endo*-1, *4*-exc-5.8 dimethanonaphthalene	(21) not detectable
(22) Chlordane, or 1,2,4,5,6,7,8,8-octachloro-2,3,3a,4,7,7a-hexahydro-4,7-methanoindene	(22) 0.1 μ/l
(23) DDT, or 2,2-bis-(p-chlorophenyl)-1, 1,1-trichloroethane and metabolites	(23) not detectable[3]
(24) Dieldrin, or 6,7-epoxy aldrin	(24) not detectable[3]
(25) Endrin, or 1,2,3,4,10,10-hexachloro-6,7-epoxy-1,4,4a,5,6,7,8,8a-octahydroendo-1,4-*endo*-5,8-dimethanonaphthalene	(25) not detectable[3]
(26) Heptachlor, or 1,4,5,6,7,8,8-heptachloro-3a,4,7,7a-tetrahydro-4,7-methanoindene and metabolites	(26) not detectable[3]

Table 11.6, continued

Items	Specifications
(27) Lindane and other Hexachloro-cyclohexanes or mixed isomers of 1,2,3,4,5,6-hexachlorocyclohexane	(27) not detectable[3]
(28) Methoxychlor, or 2,2-*bis*-(*p*-methoxyphenyl)-1,1,1-trichloro-ethane	(28) 35.0 μ/l
(29) Toxaphene (a mixture of at least 175 chlorinated camphene derivatives)	(29) not detectable[3]
(30) 2,4-Dichlorophenoxyacetic acid (2,4-D)	(30) 4.4 μ/l
(31) 2,4,5-Trichlorophenoxypropionic acid (2,4,5-TP) (Silvex)	(31) 0.26 μ/l
(32) Vinyl chloride (chloroethene)	(32) 5.0 μ/l
(33) Benzene	(33) not detectable[3]
(34) Benzo(a) pyrene	(34) not detectable[3]
(35) Kepone or decachlorooctahydro-1, 3,4-metheno-2H-cyclobuta (cd) pentalen-2-one (chlordeone)	(35) not detectable[3]
(36) Polychlorinated biphenyls (PCB) (Aroclor)	(36) 0.1 μ/l
(37) Ethylene thiourea (ETU)	(37) not detectable[3]
(38) Chloroform	(38) 100 μ/l
(39) Carbon tetrachloride (tetrachloro-methane)	(39) 5 μ/l
(40) Pentachloronitrobenzene (PCNB)	(40) not detectable[3]
(41) Trichloroethylene	(41) 10 μ/l
(42) Diphenylhydrazine	(42) not detectable[3]
(43) bis (2-chloroethyl) ether	(43) 1.0 μ/l
(44) 2,4,5-Trichlorophenoxyacetic acid (2,4,5-T)	(44) 35 μ/l
(45) 2,3,7,8-Tetrachlorodibenzo-p-dioxin (TCDD)	(45) 3.5×10^{-5} μ/l
(46) 2-Methyl-4-chlorophenoxyacetic acid (MCPA)	(46) 0.44 μ/l
(47) Amiben, or 3-amino-2,5-dichloro-benzoic acid (chloramben)	(47) 87.5 μ/l
(48) Dicamba, or 2-methoxy-3,6-dichlorobenzoic acid	(48) 0.44 μ/l
(49) Alachlor, or 2-chloro-2′,6′-diethyl-*N*-(meth oxymethyl)-acetanilide (Lasso)	(49) 35.0 μ/l
(50) Butachlor, or 2-chloro-2′,6′-diethyl-N-(butoxymethyl)-acetanilide (Machete)	
(51) Propachlor, or 2-chlor-N-isopropyl-N-acetanilide (Ramrod)	(51) 35.0 μ/l
(52) Propanil, or 3′,4′-dichloropropion-anilide	(52) 7.0 μ/l

Table 11.6, continued

Items	Specifications
(53) Aldicarb, [2-methyl-2-(methylthio) propionaldehyde *0*-(methyl carbamoyl) oxime] and methomyl [1-methylthioacetaldhyde *0*-(methyl-carbamoyl) oxime]	(53) 0.35 μ/l
(54) Bromacil, or 5-broma-3 secbutyl-6-methyluracil	(54) 4.4 μ/l
(55) Paraquat, or 1,1′beta-dimethyl-4,4′-dipyridylium	(55) 2.98 μ/l
(56) Trifluralin, or α,α,α-trifluoro-2.6-dinitro-N-dipropyl-p-toluidine (Treflan)	(56) 35.0 μ/l
(57) Nitralin, or 4-(methylsulfonyl)-2.6-dinitro-N,N-dipropylaniline (Planavin)	(57) 35.0 μ/l
(58) Benefin, or N-butyl-N-ethyl-α,α,α-trifluoro-2.6-dinitro-p-toluidine (Balan)	(58) 35.0 μ/l
(59) Azinphosmethyl, or *0,0*-dimethyl-*S*-4-oxo-1,2,3-benzotriazin-3(4H)-yl methylphosphorodithioate (Guthion)	(59) 4.4 μ/l
(60) Diazinon, or *0,0*-diethyl *0*-(2-isopropyl-4-methyl-6-pyrimidinyl)-phosphorothioate	(60) 0.7 μ/l
(61) Phorate (also for Disulfoton), or *0,0*-diethyl-*S*-[(ethylthio)methyl]-phosphorodithioate (Thimet R), and disulfoton, or *0,0*-diethyl-*S*-[(2-ethylthio)ethyl]phosphorodithioate (Di-System R)	(61) not detectable[3]
(62) Carbaryl, or 1-naphthyl-N-methyl-carbamate	(62) 28.7 μ/l
(63) Ziram or zinc salts of dimethyl-dithiocarbamic acid	(63) 4.18 μ/l
(64) Ferbam, or iron salts of dimethyl-dithiocarbamic acid	(64) 4.18 μ/l
(65) Captan or *N*-trichloromethylthio-4-cyclohexene-1,*2*-dicarboximide	(65) 17.5 μ/l
(66) Folpet, or *N*-trichloromethylthio-phthalimide	(66) 56.0 μ/l
(67) Hexachlorobenzene (HCB)	(67) 0.35 μ/l
(68) Paradichlorobenzene (PDB) (also orthodichlorobenzene)	(68) 4.7 μ/l
(69) Parathion (and Methyl parathion), or (*0,0*-diethyl-*O*-*p*-nitrophenyl-phosphorthioate, and methyl parathion or *0,0*-dimethyl-*O*-*p*-nitrophenylphosphorothioate	(69) 1.5 μ/l

Table 11.6, continued

Items	Specifications
(70) Malathion, or *S*-1,2-*bis* (ethoxy-carbonyl) ethyl-*0,0*-dimethylphos-phorodithioate	(70) 7.0 μ/l
(71) Maneb, or manganese salt of ethylene-bis-dithiocarbamic acid	(71) 1.75 μ/l
(72) Zineb or zinc salt of ethylene-bis-dithiocarbamic acid	(72) 1.75 μ/l
(73) Dithane, or zincate of manganese ethylene-bis-dithiocarbemate	(73) 1.75 μ/l
(74) Thiram, or tetramethylthiuramdi-sulfide	(74) 1.75 μ/l
(75) Atrazine, or 2-chloro-4-ethylamino-6-isopropylamino-S-triazine	(75) 7.5 μ/l
(76) Propazine, or 2-chloro-4,6-diiso-propyl-amino-S-triazine	(76) 16.0 μ/l
(77) Simazine, or 2-chloro-4,6-diethyl-amino-S-triazine	(77) 75.25 μ/l
(78) Di-n-butylphthalate	(78) 770 μ/l
(79) Di (2-ethylhexyl) phthalate (DEHP)	(79) 4.2 mg/l
(80) Hexachlorophene, or 2,2′-methyl-ene-bis (3,4,6-trichlorophenol)	(80) 7 μ/l
(81) Methyl methacrylate	(81) 0.7 mg/l
(82) Pentachlorophenol (PCP)	(82) 21 μ/l
(83) Styrene	(83) 931 μ/l

1. Foaming agents determined as methylene blue active substances (MBAS) or other tests as specified by the Commissioner.
2. Combined concentration or iron and manganese shall not exceed 0.5 mg/l.
3. "Not detectable" means by tests or analytical determinations referenced elsewhere.

ply and/or having ecologically important characteristics. In effect, both of these provisions are related to the development of land usage programs.

Some states and certainly many local governmental entities have already developed wellhead protection programs. One example of a local program is in Dade County, Florida.[21] The Wellfield Protection Program in Dade County includes five discrete elements: water management, water and waste treatment, public awareness and involvement, environmental regulation and enforcement, and land use policy. Water management refers particularly to the construction of works to physically manage ground and surface water levels. Water and waste treatment includes all aspects of treatment of both water and waste, including the possibility of cleaning up contaminated ground water at the wellhead. Public awareness and involvement refers to the efforts made to accomplish ground water protection through some level of education. Environmental regulatory programs involve regulating and gaining direct control of polluting activities. Typically. these programs involve design standards, operating procedures, management practices, monitoring, recordkeeping, inspections, and enforcement actions where standards are not being met. Land use control includes master planning, zoning, and restrictive covenants which can be used to limit the types of activities that would occur which may ultimately cause ground water pollution within defined protection zones. Land use controls which have or are being incorporated into the Dade County program include:[21]

(1) Restriction of all activities using or storing hazardous materials or generating hazardous waste within the cones of influence of well fields
(2) Restriction of nonresidential land uses to certain low-risk activities within cones of influence
(3) Identification of all well field protection areas on development master plans
(4) Restriction of residential density development in cones of influence
(5) Restriction of lake excavation in proximity to landfills
(6) Restriction of access to limestone mining sites
(7) Cleanup of known existing areas of contamination in well field areas

A critical issue in any wellhead protection program involves the technical and policy rationale for establishing the cones of influence of the wells in a well field. For example, Biscayne aquifer is the primary ground water system used in Dade County, Florida. It is a highly prolific water table aquifer in porous limestone. Within the area of influence around the large wells tapping the Biscayne aquifer, ground water flow velocities range from 1 to 3 feet per day in the outer boundaries of the area and increase to 100 feet per day in the steep gradient area near the wells. Therefore, it was determined that more stringent controls were necessary within the inner zones where contaminants would move rapidly toward the wells.[16] A multitiered protection system has been developed, ranging from total protection within 100 feet of the well to lesser levels of protection between 100 feet and the 10- to 30-day travel time distance and the 30- to 210-day travel time distance. The inner zones are based on the die-off of bacteria in soil and the ground water environment. A second consideration was dilution provided by recharge. The 210-day travel time interval is the longest period with no rainfall on record. Thus, protection of the 210-day zone was viewed as protecting all the ground water that could reach a well undiluted by recharge. The 210-day travel time radius is up to 2 miles for a large wellfield.

Conjunctive Use

O'Hare[22] identified conjunctive use as an important state or local legislative issue. In many parts of the western states in particular, ground water and surface water supplies are physically interrelated; where this occurs, ground water withdrawals can affect steam flow and vice versa, both in terms of quantity and quality. Where appropriation applies to both surface and ground water, any surface-ground water conflicts are often resolved on the basis of priority. Where different allocation rules apply, the adoption of a subflow doctrine can determine how conflicts are resolved. Various overlying rights theories apply in each state, as does prior appropriation. Where surface and ground waters constitute a common source of supply, the rights to use should be correlated. Whether the basis for interrelating surface and ground water rights should be prior appropriation or correlative rights depends on the nature of the hydrologic system.[23]

Interstate Ground Water Management

Ground water systems do not necessarily coincide with political boundaries, thus both quantitative and qualitative ground water management issues can arise between states sharing a common aquifer system. Regarding interstate water law issues, Heath[24] states that the principal legal avenues for resolving interstate water disputes involve either legislation, litigation, or interstate cooperation. Examples of one option in each of these categories that may be available to state governments are: (1) unilateral state legislation designed to restrict outsiders from access to a state's water resources, (2) original actions against a neighboring state in the U.S. Supreme Court, and (3) interstate compacts.[24] Interstate compacts are likely the preferred enforceable alternative for both quantitative and qualitative issues.

Ground Water Quality Management—The Michigan Example

The Ground Water Management Strategy for Michigan represents one of the most comprehensive programs in the United States.[22] As early as 1929, the Water Resources Commission Act (Act 245) created a Water Resources Commission to protect and conserve the water resources of the state and to prohibit the pollution of any waters of the state and the Great Lakes.[25] In 1949, an amendment to Act 245, which stated that the commission shall have control of the pollution of surface or underground waters of the state, initiated the Michigan Department of Natural Resources' activities toward ground water protection. In 1968, the first water quality standards were adopted for the state. In response to the Federal Clean Water Act of 1972, Act 245 was amended again to give the commission authority to issue permits to regulate all those who discharge or store polluting substances and to set permit restrictions to comply with the federal law. All of this legislation, in addition to a series of serious ground water contamination problems in the 1970s, led the Michigan government to develop a Ground Water Management Strategy, a major portion of which is the Water Resources Commission's Groundwater Quality Rules.[25]

Some of the issues covered in the development of the rules included: (1) ground waters to be protected, (2) the basis of protection, (3) the approach to controlling discharges, and (4) the party responsible for determining ground water quality.[25] Major issues addressed by a proposal for considering degradation and nondegradation of usable aquifers included: (1) the approach to be taken in dealing with hazardous and nonhazardous materials in a discharge, (2) methods by which to determine degradation and nondegradation and set effluent limits for ground water discharge permits, and (3) points at which compliance with discharge permits are to be determined.[25] In developing these rules, an important consideration was the inclusion of public participation.

Another of the 23 tasks included in the Ground Water Management Strategy was a predetermination of to what extent the ground water resources have been defined by describing the types, location, and quantity of ground water information in tabular form.[26] In reference to incident response, two primary management tools have been described: (1) general response procedures—guidance as to how and when state actions are initiated in response to incidents of ground water pollution concern, and (2) incident tracking procedures—an Information Management System to facilitate the recording, tracking, and distribution of information on the status of each contamination incident.[27]

Listings of sites where ground water is known to be contaminated or suspected of being contaminated has also been compiled. The purposes of this listing were to document the overall scope of the problem, identify the numbers and locations of the sites, make the information readily available to the public, and serve as a basis for the allocation of resources and for development of problem investigation and resolution strategies.[28] The Hydrologic Study Handbook was published to provide assistance to the people of the state in the planning and conducting of hydrogeologic studies and ground water quality monitoring.[29] Case studies have also been conducted to help state agencies, local agencies, and citizens deal more effectively with future ground water problems that may develop, and to provide recommendations based on the results.[30]

INSTITUTIONAL PROGRAMS—LOCAL LEVEL

Ground water quality management strategies at the local level are typically implemented for the protection of municipal water supplies.[22] In addition, local strategies may be used to coordinate activities which affect a common aquifer in order to ensure that water supplies are maintained at a quality level acceptable to the communities dependent on the aquifer. Local strategies may also be used by industries at sites exhibiting pollution or pollution potential. The three different kinds of local strategies are those developed for municipalities, for areas that share a common aquifer, and for pollution-generating sites. They may arise as a result of contamination or as preventative measures.

The development of ground water quality management strategies at the local level requires site-specific considerations. Their success depends on a thorough knowledge of the individual needs of the areas and adaptation of the strategies to those specific needs. Some generalizations may be made, however, regarding basic requirements and certain situations. Any plan should, at least, outline a strategy for protection of the drinking water source and provide for necessary backup information for the various municipal agencies responsible for its implementation.[31] The responsible agencies should all be identified, and should be involved in the initial stages of the plan, as well as its implementation.

An example of a general process which may be applied to individual localities seeking ground water protection is the Aquifer Protection Planning process developed for northeastern communities. In 1981, the Massachusetts Legislature reported that 31 communities had lost at least one water supply source to chemical contamination. As a result, many communities in Massachusetts began adopting individual regulations and by-laws to protect their public water supplies.[32] Consequently, an Aquifer Protection Planning (APP) process has been identified to facilitate development of local ground water protection strategies in the northeast. The APP outlines a strategy for the protection of a community's ground water supply and provides necessary technical support for its adoption.[32] Effective implementation of the plan requires coordination and cooperation among several municipal agencies, and these agencies must have early involvement in the development of the plan.[32]

Figure 11.1 illustrates the steps involved in the preparation of an APP. Many of these steps can proceed concurrently. For example, surveys of existing and planned development and hydrogeological studies to define recharge areas can be performed as existing water quality data are gathered and reviewed. The identification of potential contaminant sources and delineation of the well field recharge areas aid in choosing sampling locations. An awareness of potential contaminant sources can also assist in selecting which parameters should be evaluated in the samples.[32]

As the sampling program is underway, alternative protection measures can be identified, and existing town regulations and management practices can

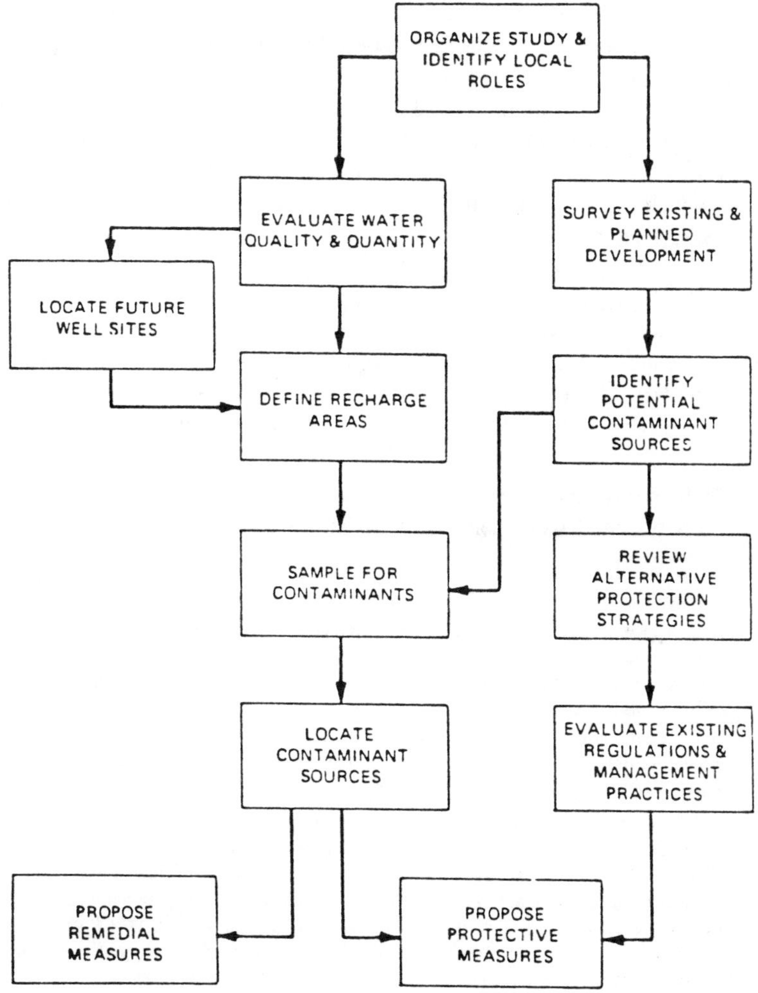

Figure 11.1. Aquifer protection planning process.[32]

be evaluated for their effectiveness in protecting ground water. Town boards and committees become acquainted with ground water concerns and can begin drafting amendments to regulations and identifying costs and personnel needs for changes in management or enforcement practices. Capital improvement needs may be identified, and some remedial measures may be undertaken before the study is complete.[32]

Zimmerman[33] has described local government responses to four cases of pollution of drinking water supplies. These four cases include pesticide contamination of ground water from aldicarb on Long Island, New York, and DBCP in California; migration of contaminants into ground water from abandoned hazardous waste sites at Love Canal, New York; and contamination of ground water from industrial waste spills and discharges in North Miami Beach, Florida. The three strategies used by local governments in these cases included:[33]

(1) The use of short-term emergency oriented strategies, such as abandoning water supplies, providing treatment of well water for consumption, and providing emergency water supplies
(2) The conducting of negotiations with the private sector to eliminate sources of contamination in the long term and to provide resources for cleanup
(3) The use of long-term land use and environmental planning for the protection of ground water quality.

Generalizations about what factors were needed to make the three strategies work well were delineated by Zimmerman[33] as follows:

(1) If standards or criteria exist that specify a given threshold or an allowable concentration of a contaminant in ground water, then short-term strategies to abandon a well or provide treatment are easier to implement. The existence of such thresholds can shift action away from a debate about standards and toward remedial and preventive strategies. In the DBCP case, some towns in California refused to close wells without a standard. In the aldicarb case, the county health department was able to act more rapidly in the absence of a standard; however, a considerable amount of time was spent arguing over the standard as a basis for providing granular activated carbon (GAC) filters to homeowners.
(2) If the nature of a ground water contamination problem changes while solutions are being explored, then the rapid implementation of a solution can be impeded. Changes in the nature of the problem can occur for the following reasons: new chemicals can appear that have greater risks than the contaminant originally detected (for example, the discovery of epichlorhydrin in association with DBCP and the conversion of aldicarb into nitrosoaldicarb), new risks from the original chemical can be discovered (for example, DBCP was originally thought to produce sterility—in the early 1970s, it was implicated in cancer as well), and the location and degree of contamination can change.
(3) If solutions to ground water contamination problems are monitoring intensive, then costs of such monitoring may act as a barrier to the feasibility of such a solution. In the aldicarb case, the provision of filters to homeowners with water supplies exceeding the 7 ppb threshold for aldicarb required an extensive and costly monitoring network to establish the threshold for each homeowner.
(4) If a solution to a ground water problem is to be feasible, then resources from state and federal agencies for both monitoring and cleanup or treatment services associated with the solution are usually a necessary prerequisite to implementation by local government.
(5) If solutions to ground water contamination problems involve negotiation with the

owners of sources of contamination, then active state and federal participation is often necessary to accomplish such negotiations. This is especially true where large corporations are involved.

COMPARISON OF INSTITUTIONAL RESPONSIBILITIES

To be effective, ground water quality management requires coordinated action by all levels of government.[34] Multiple agencies and programs within federal, state, and local levels of government may be involved in one to several elements of ground water quality management. There is no consensus on exact roles for each governmental level; however, Table 11.7 contains one possible scenario for the various roles of governmental units relative to information, planning, standards and enforcement, and the public process.[16] Considerable developmental work remains on appropriate institutional responsibilities for ground water quality management.

INSTITUTIONAL CAPACITY FOR MANAGING GROUND WATER QUALITY

Six requirements have been identified as necessary requisites for an effective ground water quality management program.[12] These requirements, which are applicable for an appropriate institutional capacity at any level of government, include:

(1) The political will to take action
(2) A legal/regulatory framework that defines the ground water program and establishes its statutory authority
(3) An organizational structure to implement the program and coordinate the activities of the various agencies and levels of government
(4) Information requirements to help define the status of the ground water resource, identify current problems, and to monitor the impacts of management strategies
(5) Financing needs and sources in order to fund the ground water program
(6) The personnel and expertise needed to implement all phases of the program

Political Will

Political will is the existence of a political climate which supports the development of a public sector response to a given problem. Four factors or conditions appear to be necessary for the development of political will: awareness, executive pressure, bureaucratic or internal pressure, and exter-

Table 11.7. Potential Roles of Government Units in the Protection of Ground Water.[16]

Governmental Level	Information	Planning	Standards and Enforcement	Public Process
Federal				
USGS	National water quality data, surface/ground.	Conducts local studies or portions of studies jointly with states or local agencies.	Provides technical data or resource conditions to EPA and states.	Provides information from data storage and studies.
EPA	Assesses national ground water conditions and provides technical advice to states.	Provides grants conditioned on establishing self-sufficient continuing planning. Conducts basic national health research.	Establishes national standards and provides technical and scientific basis for state standards. Supports states in enforcement of standards. Initiates national source control.	Is national focus of scientific and public dialogue on ground water protection strategy.
State[a]	Maintains central record of local quality and quantity of ground water, chemical usage, and disposition and present and projected land use.	Conducts continuing statewide planning and provides assistance and grants to regional or local agencies preparing ground water management plans.	Sets state or regional standards, including prohibitions when required national standards have not been set, but based on national scientific assessment. Primary enforcement agency for state and federal standards and source control regulations.	Provides information, education, and continuing forum for public dialogue on goals, standards, control strategies, and investments.

Regional[b]	Provides same information as state for a specific ground water basin serving several local agencies.	Prepares and implements specific basin management plans.	Accepts state delegation for enforcement as appropriate, including standard setting source controls, class systems, and land use plans to reflect actual environmental conditions.	Is primary focus of public process, including citizens, committees, working groups, and information gathering and dissemination.
Local	Provides basic information to individual users.	Prepares and implements local ground water plans.	Enforces appropriate standards. Assists state or regional entities in specific local situations.	Is primary contact with individual citizens.

[a]Each state's structure varies; ground water protection responsibilities may also vary; clear central state coordination is needed.
[b]Interstate Basin Agency: coordinates roles in several states where ground and related surface water problems exist.

nal pressure. Awareness means that sufficient recognition exists that problems require a public sector response beyond what is currently being done. This could include recognition that problems are widespread, serious, long-term, or irreversible. Executive pressure means that pressure for change usually must be supported by the executive officer of a state or local government. Internal pressure means that support for change also must be within the existing institutions of government. Typically, this occurs at the bureaucratic level; that is, within the agency or agencies which will have responsibility to manage a new policy. A fourth condition for political will is outside pressure—political pressure external to the governmental level. While outside pressure is seldom sufficient for change, it often serves as a catalyst. Perhaps the most significant source of outside pressure in state environmental policy is federal programs which require states to develop implementation procedures, monitor, and enforce federal regulations, or make funds available for new projects. In addition, media coverage, consensus in the research community, and organized pressure from interest groups can all be important in pushing a governmental level to adopt a new policy.

Table 11.8 identifies several examples of states which have developed one or more components of strong political will.[12] Perhaps the best example is Arizona, which combined strong political leadership with emerging economic interests, the use of an innovative task force to develop compromises, and external pressures from the federal government to pass one of the most comprehensive ground water laws in the country.

Table 11.8. Examples of Positive Political Will for Ground Water Quality Management.[12]

Condition	State
Problem severity over long term; sustained public awareness	Florida
Pressures from new economic interests combined with political leadership	Arizona
Strong executive leadership	Arizona
Use of institutional mechanisms (e.g., task forces) to develop consensus	Arizona
Centralization of administrative institutions	Alabama Georgia Wisconsin
Development of symbolic policy statements regarding the significance of the resource or the need for an active public sector role	Maryland Vermont Wisconsin

Legal/Regulatory Framework

The legal/regulatory framework for ground water quality management should clearly delineate the organizations responsible for administering the system, the distribution of responsibilities within that system, its relationship to existing programs and agencies, and enforcement responsibilities and procedures.[12]

State ground water legal/regulatory systems derive from three sources: common law, specific state statutes, and federal environmental law. State ground water regulation has its legal basis in common law. Traditionally, disputes over the quality or allocation of ground water have been settled by state courts with reference to common law concepts such as reasonable use, correlative rights, prior appropriation, and nuisance. However, while the states have developed a detailed body of law concerning ground water allocation, the courts have typically used common law tort doctrines to deal with ground water pollution issues.[34]

Three basic tort doctrines have been used to deal with ground water pollution problems: nuisance, negligence, and strict liability for abnormally dangerous activities. Each provides some relief from ground water pollution by instructing the defendant to abate the pollution or allowing for the recovery of monetary damages commensurate with the damage done. Nuisance is perhaps the most common basis of liability for ground water pollution. Nuisances can either be public—a criminal type activity defined as an "unreasonable interference in the right common to the general public"—or private—a "substantial and unreasonable interference with the use and enjoyment of . . . (an individual's) . . . own land."[34]

Negligence involves the absence of "due care" in the creation of an "unreasonable risk of harm to others."[34] Traditionally, the courts have turned to policymaking bodies or customary practice for guidance to determine what constitutes negligence. In some ground water cases, the "foreseeability" of the harm done—that is, the extent to which an individual knew or should have known that his or her activities would lead to the pollution of the ground water—has played a major role in determining the liability. Strict liability arises when the courts consider an individual's activities to be "abnormally dangerous." In such cases, the court makes the polluter strictly liable for the harm done, regardless of "due care" or "reasonably nuisance" concerns.[34]

Organizational Development and Coordination

To exercise the authority provided by the legal/regulatory framework, responsible governmental levels must develop an administrative structure to implement the ground water protection program and mechanisms to coordinate within and among the levels of government. This structure and coordination includes specification of agency responsibilities; planning, management, and evaluation components; mechanisms for coordinating activities and responsibilities across federal, regional, state, and local entities; and coordination of activities within departments and agencies. Programs for ground water pollution control range from regulation of a single pollution source to comprehensive programs that address all the major aspects of the problem. The administrative arrangements to implement these programs also vary greatly and include very centralized structures with primary responsibility in one or a few governmental agencies, fragmented administrative structures involving many agencies (with and without coordinating mechanisms), and frameworks which involve state and local entities with major responsibilities at the city, county, or substate regional level.

Georgia has developed one of the most integrated administrative systems in the country.[12] All surface and ground water management authority resides in a single agency, the Environmental Protection Division (EPD) of the Department of Environmental Regulation. The EPD has water use management authority and has been assigned water quality functions previously performed by the Department of Mines, Department of Health, and the State Water Quality Board. The EPD has a wide range of ground water protection responsibilities, including setting ground water policy and developing rules and standards, federally delegated programs, data collection, aquifer mapping, soil conservation, and mining. The EPD shares responsibility with the Department of Agriculture for activities under the federal pesticides enforcement programs and with the Transportation Department which provides protection from road deicing.[3,35]

Texas is an example of a state with a large number of agencies involved in ground water quality management. The Texas Department of Water Resources has primary responsibility for protecting and planning the use of the state's ground water, including implementing of the water quality chapter of the Texas Water Code. In addition, nine other state agencies (primarily the Department of Health and the Railroad Commission) and nine Underground Water Conservation Districts also manage, control, and protect the ground water. The Conversation Districts, created and funded at the local level, have broad powers, but most of the state's ground water resources are not within the jurisdiction of these districts.[36]

Although somewhat less fragmented than many states, several Wisconsin state agencies historically were involved in nonregulatory activities related to ground water, such as education, laboratory analysis, data management, and research, with no mechanism for coordinating their activities. In 1983, Wisconsin Act 410 established the Ground Water Coordinating Council to coordinate the nonregulatory activities of state agencies and the exchange of information relating to ground water. The Council is an independent body and is attached to the Department of Natural Resources (the major ground water quality agency) for administrative purposes.[37]

Florida has an administrative system characterized by a strong state role in conjunction with substate districts. For the most part, responsibility for directing, managing, and protecting ground water resources has been given to the Department of Environmental Regulation (DER). Historically, the DER has had the major responsibility for water quality, and the five regional water management districts have dealt with water use. Over the past few years, the regional districts have been taking more responsibility for ground water quality concerns. Florida also has actively pursued coordination mechanisms between the state and regional districts. Both formal and informal agreements and operating procedures exist among the Divisions of the DER, the district offices of DER, and the regional water management districts. Further, the regional districts develop cooperative arrangements with the local governments and regional planning councils.[3,12]

Information Needs and Uses

The availability of accurate and sufficient information is a basic need in developing an institutional capacity for ground water management. Information needed for ground water management is of two types: technical data, such as resource characterization and pollution source assessment, and that which concerns the functioning of the legal/regulatory system. Table 11.9 provides a listing of information needs for ground water planning activities involving resource usage.[38] To provide ground water usage-related information on a regional scale, the U.S. Geological Survey has initiated a series of hydrologic investigations—The Regional Aquifer System Analysis (RASA) Program. This program represents a systematic effort to study 29 regional aquifers which cover much of the country and represent most of the national water supply. For interconnected aquifers, the study objective will be to define the regional hydrology and establish a framework—geologic, hydraulic, and geochemical—for detailed local studies. For systems composed of independent aquifers, the study objective will be to establish common principles governing the occurrence, movement, and quality of ground water in the individual units. Every project will use computer models to analyze existing hydrologic regimes and provide predictive capabilities to estimate effects of future development.

Technical information needs relative to ground water quality management are identified in Tables 11.10 and 11.11.[16] Table 11.10 addresses relevant

Table 11.9. Technical Information Useful for Ground Water Planning.[37]

Category	Specific Type
Physical Framework	Hydrogeologic maps showing extent and boundaries of all aquifers and nonwaterbearing rocks. Topographic map showing surface-water bodies and land forms. Water table, bedrock-configuration, and saturated-thickness maps. Transmissivity maps showing aquifers and boundaries.[a] Map showing variations in storage coefficient.[a] Relation of saturated thickness to transmissivity. Hydraulic connection of streams to aquifers.
Hydrologic Stresses	Areas of discharge, recharge basins, recharge wells, natural areas. Surface-water diversions. Ground water pumpage (distribution in time and space). Precipitation. Areal distribution of water quality in aquifer (ambient ground water quality). Streamflow quality (distribution in time and space). Geochemical and hydraulic relations of rocks, natural water, and artificially introduced water or waste liquids.
Model Calibration	Water level change maps and hydrographs. Streamflow, including gain and loss measurements. History of pumping rates and distribution of pumpage.
Prediction and Optimization Analysis	Economic information on water supply and demand. Legal and administrative rules. Environmental factors. Other social considerations.

[a]Mapping this information involves considerable field work. Usually, only single values for these parameters are available.

hydrogeological, ground water usage, and pollution source information needs. Table 11.11 illustrates detailed chemical information needs for addressing pollutant transport and fate in the subsurface environment.

Ground water monitoring is a major activity required for collection of technical information. The monitoring program can document impacts from pollution sources, indicate adverse trends in ground water levels or quality, help assess the effectiveness of control measures, and can verify or validate impact prediction techniques.[12] While an abundance of monitoring data exists for some areas, much of it still needs to be identified, aggregated, assessed for completeness and appropriateness, and interpreted. At least 30 states recently indicated a need for additional monitoring information, including increased funding to obtain timely and sufficient data.[2]

Table 11.10. Information Base Components for Ground Water Quality Management Decisions.[16]

Hydrogeology

- Soil and unsaturated zone characteristics
- Aquifer characteristics
  - Depths involved
  - Flow patterns
  - Recharge characteristics
  - Transmissive and storage properties
  - Ambient water quality
  - Interaction with surface water
  - Boundary conditions
  - Mineralogy, including organic content

Water Extraction and Use Patterns

- Locations
- Amounts
- Purpose (domestic, industrial, agricultural)
- Trends

Potential Contamination Sources and Characteristics

- Point sources
  - Industrial and mining waste discharges
  - Commercial waste discharges
  - Hazardous material and waste storage
  - Domestic waste discharges
- Nonpoint sources
  - Agricultural
  - Septic tanks
  - Land applications of waste
  - Urban runoff
  - Transportation spills (may also be considered a point source)
  - Pipelines (energy and waste water) (may also be considered a point source)

Population Patterns

- Demographic
- Economic trends
- Land use patterns

Since ground water monitoring can be expensive, carefully planned monitoring networks need to be established. One example of a prioritized network is in Idaho, with 565 wells included in a statewide program.[39] Frequencies of sampling at the different sites are assigned at quarterly, semiannual, annual, and 5-year intervals. The network is designed to: (1) enable water managers to keep abreast of the general quality of the state's ground

Table 11.11. Information on Properties of Chemicals of Interest to Ground Water Quality Management Programs.[16]

Physical Properties
Water solubility
Octanol/water partition coefficients
Vapor pressure (Henry's law constant)
Sorptive characteristics
Density
Viscosity
Chemical Properties
Structure
Isomeric forms, homologs
Transformation potential and end products
Other commingled compounds and carriers
Biological Properties
Biodegradability
Aerobic
Anaerobic
Metabolic products
Toxicologic Properties (The Chemical and Its Transformation Products)
Acute effects
Chronic effects

water, and (2) serve as a warning system for undesirable changes in ground water quality. A "hydrologic unit priority index" was used to rank 84 hydrologic units of the state for monitoring according to pollution potential. Emphasis for selection of monitoring sites was placed on the 15 highest ranked units. The potential for pollution is greatest in areas of privately owned agricultural land. Other areas of pollution potential are residential development, mining and related processes, and hazardous waste disposal sites.

Other types of technical information are needed for ground water quality management that are not obtained from monitoring programs, including assessments of the health effects of various contaminants that are found in ground waters. Usually, this type of information is obtained from epidemiological or toxicological research studies conducted by federal agencies. Other technical information needed includes assessments of pollutant sources, pollutant transport and fate under various conditions that occur in the subsurface, and technical options or management strategies for pollution source control and cleanup.[12]

Information needed on the legal/regulatory framework includes an understanding of the extent of the various federal, state, and local laws and regulations pertinent to ground water management. For example, information on federal policies to protect interstate aquifers is needed by several states, including Arkansas and Oklahoma. Periodic assessments are needed to identify regulatory constraints on ground water pollution control and also to identify the effects of regulations on the economy of an area. Regulatory gaps need to be identified where pollution sources are not or cannot be adequately controlled. This includes gaps in institutional systems.[12]

Financing Needs and Sources

Ground water quality management programs may include a variety of activities, such as mapping of aquifers, monitoring for ground water quality, identification of pollution sources, enforcement of regulatory programs, cleanup of contaminated aquifers, conducting information transfer programs, and coordination with local, state, and federal institutions sharing ground water protection responsibilities. Sufficient funding is necessary to conduct these activities in a comprehensive and responsible manner. Federal, state, and local ground water quality management programs all have financial requirements. For example, the major financial needs for state programs were recently explored by the Office of Technology Assessment.[2] The financial needs and the 39 states identifying them are shown in Table 11.12.

Multiple sources of potential funding are available for meeting the financial requirements of ground water quality management programs.[12] Examples include budgetary allocations from general revenue funds; special taxes, fees, or charges levied at various government levels; special single purpose allocations for ground water cleanup or protection programs; and federal grants and assistance programs. Direct budgetary allocations from general revenue funds represent the primary source of funding for most ground water programs. Due to the longer time cycle, careful planning in terms of identifying budgetary needs and expenditures are particularly necessary for those state and local governments that operate on biannual cycles. In addition, many governmental entities are faced with increasing budgetary demands in situations with reduced sources of revenue due to economic downturns. Therefore, it may be necessary to reallocate personnel assignments and funds from existing environmental programs to meet the increasing requirements of ground water programs. Special taxes, fees, or charges can be integrated with permitting programs for potential ground water pollution sources or activities such as hazardous waste sites, underground storage tanks, and pesticide applications. Some states, for example Wisconsin, have established a state fund for remedial action programs.

Table 11.12. Examples of State-Identified Financial Needs.[2]

Financial Need	States Identifying Need[a]
General improvement of capabilities and detection (monitoring) of ground water problems	Alaska, Arkansas, California, Colorado, Connecticut, Florida, Hawaii, Idaho, Illinois, Indiana, Iowa, Kansas, Maine, Maryland, Massachusetts, Michigan, Minnesota, Montana, Nebraska, Nevada, New Hampshire, New Jersey, New Mexico, New York, North Carolina, North Dakota, Ohio, Oklahoma, Oregon, Pennsylvania, Rhode Island, South Carolina, South Dakota, Utah, Vermont, Virginia, West Virginia, Wisconsin, and Wyoming
Enforcement	Alabama, Alaska, Arkansas, Missouri, New Hampshire, Ohio, Oklahoma, and Pennsylvania
Cleanup or correction of problems	Alabama, Maine, Michigan, South Carolina, South Dakota, Tennessee, and Utah
Problem identification and geophysical evaluations	Colorado, Idaho, Indiana, Kentucky, Maryland, and Washington
Laboratory equipment and facilities	Arizona, Indiana, New Mexico, New York, and Oklahoma
Other:	
Research and development	Arizona, Connecticut, and Nebraska
Special studies	Oklahoma, Tennessee (siting of nonhazardous waste facilities)
Prevention	Wisconsin
Ground water usage	Indiana
Education	Oklahoma
Program coordination	Ohio

[a]The information reflects the views of the state personnel involved in ground water quality programs who responded to the survey, and the fact that only a few states identified a particular financial needs does not necessarily imply that the need does not exist in other states.

Federal grant programs exist for state implementation of federal programs to protect ground water. Statutory authorities basic to these grant programs include Sections 106, 205(g), 205(j), and 208 of the Clean Water Act; the UIC Program of the SDWA; and Section 23(a)(1) of the Federal Insecticide, Fungicide, and Rodenticide Act (FIFRA). Finally, it should be noted that federal funds can be used to supplement state funding for several elements within state-originated ground water protection programs. An illustration for Connecticut for eight program activities is as follows:[3]

(1) Ground water strategy—state funds and funds authorized from Sections 106 and 205(j) of the Clean Water Act (CWA)
(2) Ground water monitoring—state funds and funds authorized from Sections 106 and 205(j) of CWA, Section 3011 of RCRA, and the UIC program of the SDWA
(3) Ground water resource assessment/aquifer study/mapping—state funds and U.S. Geological Survey matching funds
(4) Agricultural contamination control—state funds, funds authorized from Section 106 of CWA and from SDWA, and other sources related to pesticide control
(5) Permits and control of discharges to ground water—state funds and funds authorized from Section 106 of CWA
(6) Septic tank management program—state funds and funds authorized from Section 106 of CWA
(7) Bulk storage and underground storage tank program—state funds and funds authorized from Section 106 of CWA
(8) Contamination response program—state funds and funds authorized from Section 106 of CWA, Section 3011 of RCRA and the UIC program of SDWA

One of the key issues associated with financing ground water quality management programs is the need to balance expanding requirements with available revenues over time. A new program would be expected to have certain startup costs which would be in excess of their continuing costs. Examples of startup costs include the development of a ground water protection strategy, drilling of monitoring wells in an initial network, and the planning and implementation of an information storage and retrieval system. Initial staffing of a program also represents a startup cost. As a ground water protection program expands its areas of responsibilities and activities, it is necessary to expand the available financial resources. In order to maintain institutional capacity, careful and systematic planning is necessary in balancing needs and resources in an expanding program, particularly when the rate of expansion is high.[12]

Personnel and Expertise Needs

The implementation of a ground water quality management program requires a variety of scientific-technical and administrative-legal personnel. Scientific-technical positions may include field inspectors, hydrogeologists, soil scientists, engineers, microbiologists, chemists, computer programmers, mathematicians, and others. The education level for job entry into these positions may range from a high school graduate (for a field inspector) to a Ph.D. chemist conducting toxicological research. The years of experience required for entry into these positions also varies.[12] Examples of various staffing needs are shown in Table 11.13.

Table 11.13. Sample Elements and Staffing Needs of a Ground Water Program.[12]

Program Element	Staffing Category: Scientific-Technical[a]	Staffing Category: Legal/ Administrative[b]
Detection		
Respond to complaints	X	X
Verify contamination	X	
Who is at fault, who will pay	X	X
What to do about the contamination		X
Correction		
Technological options (containment, withdrawal/treatment, etc.)	X	
Management options (purchase alternative supplies, limit aquifer use, etc.)		X
Prevention		X
Permits for discharge		
Best management, facility siting	X	X
Assessment of pollution sources	X	
Aquifer classification	X	X
Standards (well construction, water quality)	X	X
Research		
Health effects/toxicology	X	
Risk assessment	X	X
Inexpensive treatment techniques	X	

[a]This category includes field inspectors, hydrogeologists, soil scientists, engineers, microbiologists, chemists, computer programmers, mathematicians, and consultants.
[b]This category includes lawyers, managers, administrators, economists, planners, and clerks.

Having an adequate number of people to implement ground water quality management programs appears to be an increasing problem because ground water problems are growing, yet resources for hiring and training typically are restricted. In a national survey of 50 states compiled by the Office of Technology Assessment,[2] the main ground water agencies in 13 states cited the need for an increase in staff size as shown in Table 11.14. In addition, 24 states indicated that the expertise of their staffs was insufficient to adequately deal with ground water pollution problems.

ELEMENTS IN A GROUND WATER QUALITY MANAGEMENT PROGRAM

The elements of a comprehensive ground water quality management pro-

Table 11.14. State Responses to Staffing/Training Needs.[2]

States	Insufficient Staff Size	Insufficient Staff Expertise
Alabama		X
Arizona	X	X
Arkansas		X
Colorado		X
Connecticut		X
Delaware		X
Florida	X	
Idaho	X	X
Illinois		X
Indiana		X
Kentucky		X
Maine	X	
Michigan		X
Minnesota	X	
Missouri	X	X
Montana		X
Nebraska	X	
New Mexico	X	
Nevada		X
North Carolina	X	
North Dakota		X
Ohio	X	X
Oklahoma		X
Oregon	X	
Rhode Island	X	X
South Dakota		X
Texas		X
Virginia		X
Washington		X
West Virginia		X
Wyoming	X	X
Total	13	24

gram can be considered in two categories: (1) foundation elements and (2) action elements associated with an ongoing program.[40]

Foundation Elements

There are four elements that can be identified as basic to the successful implementation of a ground water quality management program. They can be viewed as providing a foundation to the entire program, while at the same time being subject to ongoing implementation and change as appropriate. The elements include the development of ground water standards, identification and usage of an appropriate database and information management system, performance of research and development and special studies, and the initiation and continuation of an educational program relating to ground water quality management.

Historical ground water quality standards have primarily addressed inorganic constituents; however, with the current emphasis on synthetic organic and metal constituents, numerical standards for these components are being promulgated. Standards are necessary for development and use of aquifer classification systems, pollution source evaluations, interpretation of ground water quality monitoring data, and the issuance of permits.

As ground water monitoring programs multiply at federal, state, and local government levels, there is an expanding need for the development of compatible databases and information management systems to facilitate data and information use and dissemination. Considerable time can be required for identifying existing data and information bases and planning for the incorporation of new data and information. Whether to utilize a single database and information management system for the state or local government or to develop interchanges among existing systems to access data and information must be decided. Significant expenditures for both hardware and software can be associated with this foundation element.

As technical knowledge related to ground water pollution and the transport and fate of specific constituents from both man-made and natural sources has increased, technical and policy-related research and development needs have expanded. Research is needed on rational scientific and policy bases for numerical limits used as ground water standards. In addition, basic information is needed on pollution source evaluation techniques and the effectiveness of remedial actions. Therefore, one foundation element must be associated with both the analysis and incorporation of research findings beyond the geographical scope of the program, as well as the conduction of special studies and research within the specific geographical area.

Another foundation element for a ground water protection program should be associated with information transfer and education. Education and training needs can be viewed from both the perspective of staffs involved in program planning and implementation, as well as the perspective of the general public. One of the frequently identified national issues related to ground water protection is the need for technically trained individuals. Technical staffs, as well as policy and legal staffs, are necessary in the planning and implementation of ground water quality management programs. There is also a major need for increasing public awareness relative to the importance of ground water as a resource and appropriate measures for ground water protection and remediation. Finally, individuals employed by local, state, and federal government agencies, as well as elected officials and their staffs, have needs for training relative to resource awareness and general policy implications.

Action Elements

Commonly occurring action elements in an ongoing ground water quality management program include: (1) planning and conducting ground water monitoring programs, including the provision of appropriate quality control and quality assurance related to monitoring well location, design, and construction, and sample collection and analysis; (2) ground water resource assessment relative to ground water uses, aquifer mapping, and pollution source prioritization; (3) source control and ground water prevention programs, including specific program elements directed toward sources such as septic tank systems, underground storage tanks, and agricultural usage of chemicals; (4) permits and inspection and enforcement programs; (5) contamination response and remedial action programs for both acute and chronic pollution events; and (6) pollution incident tracking programs directed toward the identification of ground water pollution occurrences.

Traditional ground water monitoring has been focused on inorganic constituents and water level measurements. There is an increasing awareness of the need for collection of more extensive information on both ground water quantity and quality, with particular emphasis being given to quality constituents such as synthetic organics, metals, bacteria, and viruses. As ground water monitoring programs are expanding in both geographical coverage and extent of quality constituents addressed, there is the concomitant need for planning of monitoring networks, including the placement of monitoring wells and usage of appropriate standards of practice relative to monitoring well design, construction, and completion, and ground water sampling and subsequent chemical or bacteriological analyses. Coordination of the existing monitoring programs of multiple governmental agencies at local, state, and federal levels is needed.

A major element in a ground water quality management program is associated with the assessment of the pattern of ground water resource usage. Considerable requirements will probably be associated with mapping the physical and geographical characteristics of both major and minor aquifers. While extensive information has already been developed on aquifer characterization, in many instances it is not as specific as is needed for ground water protection, nor does it address the spatial variability of individual physical or chemical characteristics. Aquifer mapping is often perceived as an initial and basic activity in the development of a ground water quality management program.

With expanding documentation of ground water pollution resulting from man-made sources, it is important to prioritize ground water pollution source categories and individual sources. Prioritization schemes may include consideration of the type of pollutant, its transport and fate characteristics, and its potential for limiting ground water usage. Several methodologies have been or are being developed to provide a systematic basis for source prioritization.[41]

Development of source control measures, or preventive measures, directed toward minimizing future ground water pollution from new sources is important in ground water quality protection. In addition, ongoing programs of source control and permitting continue to require considerable time and effort. Three examples of pollutant source categories that have been or are being considered for subjection to prevention programs include septic tank systems, underground storage tanks, and the use of agricultural chemicals. Permits for septic tanks based on tank density and local soil characteristics represent an ongoing program activity throughout the United States. Underground storage tanks represent a source category of recent national, state, and local attention. A relatively new issue is the ground water pollution potential of agricultural chemicals, including commercial fertilizers and a variety of pesticides and herbicides. This source category is being recognized since it is a widespread nonpoint source of ground water pollution; in addition, it is of potential major significance in states that are highly dependent upon large-scale agricultural operations in their economic base.

Ground water protection also involves permitting new potential sources and inspection and enforcement of permit requirements on both new and existing sources. This program area, as related to the above mentioned pollution sources and others such as hazardous waste sites, liquid impoundments, and sanitary landfills, will become an increasingly important element in ground water quality management programs.

Ground water quality management programs should also address responses to both acute and chronic ground water pollution problems. Acute problems are reflected by accidental chemical spills, and immediate and appropriate remedial action measures are necessary to minimize ground water pollution. Chronic ground water pollution has occurred over time as a result of leachates from hazardous waste sites and landfills, as well as leakage from underground storage tanks and liquid impoundments. Remedial action efforts for chronic ground water pollution are expanding, with the best known efforts associated with the cleanup of hazardous waste sites through the federal Superfund program. A comprehensive state or local ground water protection strategy should include a contamination response program. Implementation of a pollution incident tracking program is also important. It is anticipated that there will be more incidents of ground water pollution as a result of both inadvertent waste disposal practices and the gathering of heretofore unknown information relative to ground water quantity and quality. The contamination response program must be based in part upon a pollution incident tracking program. Ground water pollution incident tracking can range from a passive program in which individuals respond on their own initiative to provide information to a central receiving point, to an active program in which appropriate individuals and organizations are periodically contacted relative to their knowledge associated with ground water pollution occurrences.

CASE STUDY OF DEVELOPMENT OF AN AQUIFER PROTECTION AND MANAGEMENT PLAN

The Garber-Wellington aquifer system in central Oklahoma will be used as an example of the development of a ground water quality management plan. The Garber-Wellington (G-W) Aquifer is the major ground water resource in central Oklahoma. The aquifer contains over 60×10^9 m^3 (50 mil acre-ft) of freshwater, with approximately two-thirds of this potentially available for development. About 30% of the total water usage in central Oklahoma is from ground water sources, with the G-W Aquifer being the major one. The total water usage from the G-W Aquifer is currently less than 308×10^6 m^3 (250,000 acre-ft) per year; however, usage is anticipated to increase. The G-W Aquifer consists of lenticular beds of sandstone alternating with shale. Sandstone comprises about 50% of the aquifer. The thickness of the freshwater zone ranges from 45 to 260 m (150 to 850 ft).

Water from the G-W Aquifer is generally suitable for drinking, but individual samples may be high in sulfates, chlorides, total dissolved solids, or other mineral constituents. Excessive amounts of sodium may limit water usage for irrigation. Numerous potential ground water pollution sources, including sanitary landfills, septic tank systems, and oil and gas well operations, exist in the central Oklahoma area. Intrusion of saline water into the freshwater zone is also a potential threat to water quality if excessive drawdown occurs. The distance from the water table down to the saltwater-freshwater interface varies locally between 30 and 300 m (100 and 1000 ft).

A preliminary study of the G-W aquifer was conducted in 1979.[42] The preliminary study elements consisted of the procurement of published information on the G-W Aquifer; contacts with federal, state, and local government agencies having responsibilities or interests in the G-W Aquifer; solicitation of information on unpublished data relative to the aquifer; and literature searches of computer-based information storage and retrieval systems. Based on the information procured from these studies and input from an advisory group meeting, work tasks for a comprehensive study of the aquifer were identified.

One of the first activities of the preliminary study was to procure published information and data on the G-W Aquifer and to review it in terms of applicability and completeness. Examples of the type of subject matter investigated include: geologic features and hydraulic characteristics of the aquifer; soil features in the G-W area; quantity of available ground water; quality of ground water; existing or potential pollution sources; surface area and volume of existing and planned surface water reservoirs in central Oklahoma; current and projected land use and water use patterns in central Oklahoma; climatological records; and applicable Oklahoma laws and regulations related to ground water resources.

Federal, state, and local agencies having jurisdiction or interests in the G-W Aquifer were visited in conjunction with the preliminary study to: (1) describe the preliminary study being conducted, (2) learn about the interests and responsibilities of each agency relative to the aquifer, (3) identify and procure any published reports or information on the aquifer available through the agency, (4) identify future plans of the agency regarding continuing studies related to development and use of the aquifer, and (5) locate unpublished information that was contained in agency files and that could be procured for analysis in the comprehensive study. Contacts were also made with some relevant companies and individuals for these same purposes.

Following the assemblage and review of published and unpublished information and the contacts with government agencies, several data gaps and work tasks were identified. Although considerable information existed on the aquifer, there had never been a coordinated and comprehensive study for the specific purpose of gathering information that could be used to develop a management plan for the aquifer. The major work tasks identified are shown in Table 11.15.[43] To solicit feedback and additional input from the interested government agencies, a briefing on the preliminary findings and the list of work tasks for the comprehensive study was held. The format of the briefing encouraged a free interchange of information, and suggestions for refinement of several of the tasks were made.

A major component of the preliminary study was a comprehensive, computer-based survey of published literature on ground water quality management. Systematic searches were conducted and refined to gain information on both the G-W Aquifer and other pertinent materials in ground water quality management. The basic approach was to begin with a series of key words and to search a bibliographic database for references with those same key-

Table 11.15. Work Tasks for Development of Protection and Management Plan for Garber-Wellington Aquifer System.[43]

Task 1:	Physical Characteristics of the Aquifer • Assemble and synthesize existing data. • Develop and implement data storage and retrieval system. • Construct aquifer definition wells. • Collect and analyze data. • Maintain interagency communication.
Task 2:	Background Water Quality • Assemble and synthesize existing data. • Establish a ground water monitoring system. • Construct water quality monitoring wells. • Collect and analyze data. • Identify and prioritize ground water pollution sources. • Maintain interagency communication.
Task 3:	Present and Future Water Uses • Assemble and synthesize existing data. • Develop and implement a municipal records system. • Inventory existing wells. • Modify existing wells. • Maintain interagency communication.
Task 4:	Modeling • Develop criteria for model selection. • Select model. • Establish model, enter data and determine critical parameters. • Calibrate model.
Task 5:	Public Participation Program • Identify objectives. • Identify "public." • Select appropriate techniques. • Conduct program.
Task 6:	Management Plan • Technical considerations. • Institutional factors. • Environmental factors. • Interagency coordination.
Task 7:	Reports • Progress reports. • Protocol report. • Technical report.

word descriptors. Combinations of words were also used to provide the greatest searching effectiveness. Printouts of the abstracts of references containing the selected key words as identifiers were then procured. Information from the reference sources was used in defining the data gaps and in developing and refining the list of tasks in Table 11.15.

A comprehensive study of the G-W Aquifer was conducted by the Association of Central Oklahoma Governments.[43] The major accomplishments of this study included:

(1) Development of a uniform municipal recordkeeping and data collection system. This system provides a uniform means to collect data on municipal water wells and place this data into a computerized data system. It provides a cost savings to municipalities through close monitoring of well fields, charting well efficiencies, and elimination of major breakdowns through early detection of problems.

(2) Development of a suggested oil and gas ordinance for consideration and use by cities and towns located on the aquifer. This ordinance was developed with the cooperation of the state oil and gas regulatory agency as an additional step toward protecting the aquifer from contamination which would result from oil and gas drilling, development and production, or subsurface injection of liquids.

(3) Identification of a water quality problem involving high levels of chromium, selenium, arsenic, and uranium. Gamma ray logging tools and other test procedures are being used to attempt to delineate areas of uranium mineralization. Various methods for mitigating these problems are being explored. Also, multidepth monitoring wells have been installed to give data on water quality at various depths in the aquifer.

(4) Examination of the pollution potential from various man-made sources of contamination. The sources include oil field activities, urban runoff, pits, ponds, and lagoons, and septic tank systems.

One of the purposes of the comprehensive study of the G-W Aquifer was the generation of a protocol for the development and implementation of a plan for protecting, developing, and managing aquifers. This generic protocol, shown in Figure 11.2, provides a basis for ground water quality management planning for any aquifer system.

SELECTED REFERENCES

1. Ballew, W. W., III, "Groundwater Laws: Opportunities for Management and Protection," *Journal of the American Water Works Association,* Vol. 75, No. 6, June 1983, pp. 280–287.
2. Office of Technology Assessment, "Protecting the National's Groundwater from Contamination," OTA-0-233, October 1984, U.S. Congress, Washington, DC.
3. U.S. Environmental Protection Agency, "Overview of State Ground Water Strategies," Volume 1, 1985, Office of Ground Water Protection, Washington, DC.
4. Henderson, T. R., Trauberman, J., and Gallagher, T., "Groundwater Strategies for State Action," 1984, Environmental Law Institute, Washington, DC.
5. U.S. Environmental Protection Agency, "A Ground Water Protection Strategy for the Environmental Protection Agency," August 1984, Office of Ground Water Protection, Washington, DC.

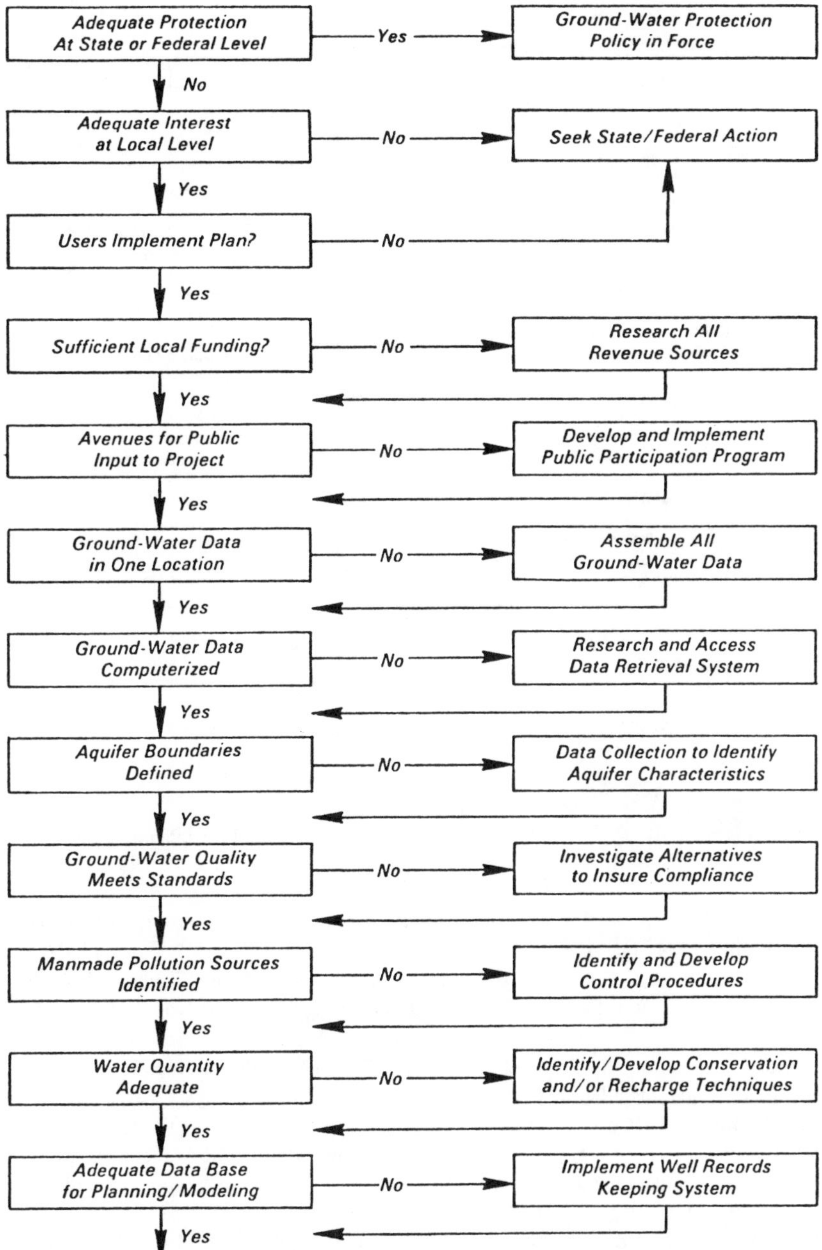

Figure 11.2. Protocol flowchart for establishment of a ground water protection and management plan for an aquifer system.[43]

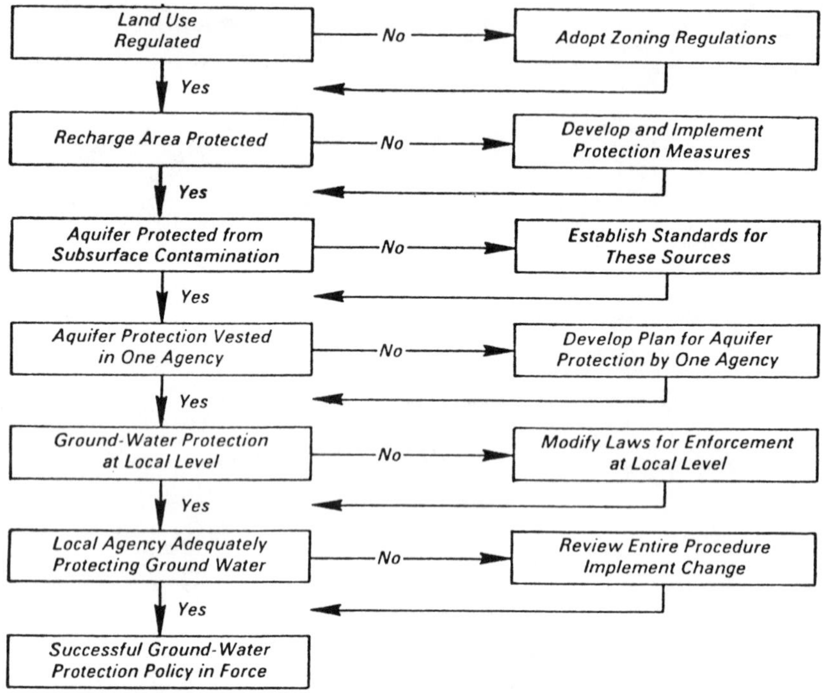

Figure 11.2. Continued

6. Cheremisinoff, P. N., Casana, J. G., and Ouellette, R. P., "Underground Storage Tank Control," *Pollution Engineering,* Vol. 18, No. 2, Feb. 1986, pp. 22–29.
7. Plehn, S. W., "An Introduction to LUST," *Environmental Forum,* Vol. 3, No. 3, July 1984, pp. 5–9.
8. Woods, P. H., and Webster, D. E., "Underground Storage Tanks: Problems, Technology, and Trends," *Pollution Engineering,* Vol. 16, No. 7, July 1984, pp. 30–40.
9. U.S. Environmental Protection Agency, "The Interim Prohibition: Guidance for Design and Installation of New Underground Storage Tanks," August, 1985, Washington, DC.
10. Miller, L. A. and Taylor, R. S., "The Enemy Below: EPA Plans Action in Leaking Underground Storage Tanks," *Environmental Law Reporter,* Vol. 15, No. 5, May 1985, p. 10135.
11. Zakheim, B. and Ehrhardt, R. F., "Assessment of State and Local Regulatory Programs for Underground Storage Tanks," paper presented at 21st Annual American Water Resources Association Conference and Symposium, August 1985, Tucson, Arizona.

12. James, T., et al., "Elements of Institutional Capacity for Ground Water Pollution Control," *Proceedings of a National Symposium on Institutional Capacity for Ground Water Pollution Control,* Sept. 1985, University of Oklahoma, Norman, Oklahoma, pp. 1–53.
13. U.S. Water Resources Council, "State of the States: Water Resources Planning and Management, Ground Water Supplement," 1981, U.S. Government Printing Office, Washington, DC.
14. Wickersham, G., "A Preliminary Survey of State Ground Water Laws," *Ground Water,* Vol. 19, No. 3, May/June 1981, pp. 321–327.
15. Erhardt, F., et al., "Trends in U.S. Groundwater Law, Policy, and Administration," Edison Electric Institute, 1984, Washington, DC.
16. National Research Council, "Ground Water Quality Protection: State and Local Strategies," 1986, National Academy Press, Washington, DC.
17. U.S. Environmental Protection Agency, "Selected State and Territory Ground Water Classification Systems," May 1985, Office of Ground Water Protection, Washington, DC.
18. American Petroleum Institute, "Guide to Ground Water Standards of the United States," Pub. No. 4366, July 1983, Washington, DC.
19. Pye, V. I., Patrick, R., and Quarles, J., *Ground Water Contamination in the United States,* 1983, University of Pennsylvania Press, Philadelphia, Pennsylvania, pp. 217–226.
20. Hoffman, J. I., "A Model Provision for Non-degradation of Groundwater: What is a Detrimental Effect and Where is it Measured?" *Proceedings of Third Annual Conference on Applied Research and Practice on Municipal and Industrial Waste,* Sept. 1980, University of Wisconsin-Extension, Madison, Wisconsin, pp. 168–173.
21. Yoder, D., "Protection of Wellfields and Recharge Areas in Dade County, Florida," *Proceedings of a National Symposium on Local Government Options for Ground Water Pollution Control,* June 1986, University of Oklahoma, Norman, Oklahoma, pp. 183–198.
22. O'Hare, M. P., "State Ground Water Protection Strategies," May 1986, Environmental and Ground Water Institute, University of Oklahoma, Norman, Oklahoma.
23. Aiken, J. D., "Western Ground Water Rights: An Overview," Department of Agricultural Economics Staff Paper No. 7, 1980, University of Nebraska, Lincoln, Nebraska.
24. Heath, Jr., M. S., "Interstate Water Law Issues: Unilateral State Legislation Original Suits in the Supreme Court Interstate Compacts," OWRT-A-121-NC(B)(2), Sept. 1983, Office of Water Research and Technology, U.S. Department of the Interior, Washington, DC.
25. Michigan Department of Natural Resources, "Groundwater Management Strategy for Michigan: Development of Administrative Rules for Groundwater Quality," MI/DNR/GW-82/03, Jan. 1982, Lansing, Michigan.
26. Michigan Department of Natural Resources, "Groundwater Management Strategy for Michigan. Task 4: Resource Definition," MI/DNR/GW-81-01, Nov. 1981, Lansing, Michigan.

27. Michigan Department of Natural Resources, "Groundwater Management Strategy for Michigan: Draft Response and Incident Tracking Procedures," MI/DNR/GW-82/05, July 1982, Lansing, Michigan.
28. Michigan Department of Natural Resources, "Assessment of Groundwater Contamination: Inventory of Sites, Groundwater Management Strategy for Michigan," MI/DNR/GW-83-01, July 1982, Lansing, Michigan.
29. Michigan Department of Natural Resources, "Groundwater Management Strategy for Michigan: Hydrogeologic Study Handbook," MI/DNR/GW-82/04, Mar. 1982, Lansing, Michigan.
30. Michigan Department of Natural Resources, "Groundwater Management Strategy for Michigan: Economic and Social Impacts of Groundwater Contamination; A Case Study in East Bay Township, Grand Traverse County, Michigan," MI/DNR/GW-82/07, June 1982, Lansing, Michigan.
31. Kilner, S. M., "Groundwater Plan Sidesteps Contamination Woes," *Water Engineering Management,* Vol. 131, No. 3, Mar. 1984, pp. 27–29.
32. Kilner, S. M., Rizzo, Jr., W. J., and Shawcross, J. F., "Aquifer Protection Planning in the North Eastern United States," *Journal of the Institution of Water Engineers and Scientists,* Vol. 38, No. 3, June 1984, pp. 247–258.
33. Zimmerman, R., "Performance of Management Strategies for Ground Water Protection: Some Illustrative Cases Involving Local Government Action," *Proceedings of a National Symposium on Local Government Options for Ground Water Pollution Control,* 1986, University of Oklahoma, Norman, Oklahoma, pp. 250–263.
34. Tripp, J. T. and Jaffe, A. B., "Preventing Ground Water Pollution: Towards a Coordinated Strategy to Protect Critical Recharge Zones," *The Harvard Environmental Law Review,* Vol. 3, 1979, pp. 1–47.
35. Kundell, J. E., and Breman, V. A., "Regional and Statewide Water Management Alternatives," Institute of Government, 1982, University of Georgia, Athens, Georgia.
36. U.S. General Accounting Office, "Federal and State Efforts to Protect Ground Water," GAO/RCED-84-80, Feb. 1984, Washington, DC.
37. Patronsky, M., and Bogar-Rieck, A., "The New Law Relating to Groundwater Management: 1983 Wisconsin Act 410," Information Memorandum 84-11, Wisconsin Legislative Council, 1984, Madison, Wisconsin.
38. U.S. Water Resources Council, "Essentials of Groundwater Hydrology Pertinent to Water Resources Planning," Bulletin 16, 1980, Washington, DC.
39. Whitehead, R. L. and Parliman, D. J., "A Proposed Ground Water Quality Monitoring Network for Idaho," Open File Report 79-1477, Oct. 1979, U.S. Geological Survey, Boise, Idaho.
40. Canter, L. W., "Functions and Activities of Groundwater Protection: Implications for Institutional Coordination," *The Environmental Professional,* Vol. 8, 1986, pp. 219–224.
41. Canter, L. W., "Methods for Assessment of Ground Water Pollution Potential," In: *Ground Water Quality,* Ward, C. H., Giger, W., and McCarty, P. L., Editors, 1985, John Wiley and Sons, Inc., New York, New York, pp. 270–306.

42. Canter, L. W., "Groundwater Quality Management," *Journal American Water Works Association,* Oct. 1982, pp. 521–527.
43. Association of Central Oklahoma Governments, "Protocol for Establishment of a Ground Water Management and Protection Plan," EPA-600/S2-84-053, April 1984, Robert S. Kerr Environmental Research Laboratory, U.S. Environmental Protection Agency, Ada, Oklahoma.

Index